THE SOLVAY CONFERENCES ON PHYSICS

Aspects of the Development of Physics since 1911

The Solvay Conferences on Physics

ASPECTS OF THE DEVELOPMENT OF PHYSICS SINCE 1911

by

Jagdish Mehra

With a Foreword by Werner Heisenberg

D. Reidel Publishing Company

Dordrecht-Holland / Boston-U.S.A.

1975

Library of Congress Cataloging in Publication Data

Mehra, Jagdish.
    The Solvay Conferences on Physics.

    Includes bibliographical references and index.
    1. Solvay Conference on Physics.   2.   Solvay, Ernest, 1838–1922.
    3.   Physics—History.
QC1.S792M43   530′.09   75-28332
ISBN 90–277–0635–2

Published by D. Reidel Publishing Company,
P.O. Box 17, Dordrecht, Holland

Sold and distributed in the U.S.A., Canada and Mexico
by D. Reidel Publishing Company, Inc.
Lincoln Building, 160 Old Derby Street, Hingham,
Mass. 02043, U.S.A.

Printed in The Netherlands by D. Reidel, Dordrecht

Foreword

Jagdish Mehra's historical account of the Solvay Conferences from 1911 to 1973 demonstrates not only the great influence which these conferences have had on the development of modern physics, but it also shows clearly how far-sighted and well-planned were the intentions of Ernest Solvay when he took the initiative for organizing a new type of international conferences. In contrast to the conventional meetings in which reports are given on the successful solution of scientific problems, the Solvay Conferences were conceived to help directly in solving specific problems of unusual difficulty. The importance of the quantum structure of Nature had become well understood already by 1911, but at that time there was no hope for an answer to the extremely difficult new questions posed by the atomic phenomena. The new conferences should therefore be devoted primarily to thorough discussions of such problems between a small number of the most competent physicists, and Ernest Solvay was guided by the hope that the discussions would eventually lead to a real and substantial progress.

The earliest Solvay Conferences which I attended were those of 1927, 1930 and 1933, and they served this purpose extremely well. In 1926 the mathematical formalism of quantum- and wave-mechanics approached its final shape, but the interpretation was still controversial. Schrödinger hoped that his matter waves could be considered as waves in three-dimensional space and time, and that the discontinuous feature of quantum 'jumps' could be avoided thereby. Born, in his theory of collisions, had given the statistical interpretation of the waves in a many-dimensional configuration space. The members of the Copenhagen group, primarily Bohr, Kramers, Pauli and I, after a thorough analysis of the uncertainty principle and the concept of complementarity, had come to the conviction that the paradoxa of quantum theory could be finally resolved within their philosophy, and that the new interpretation would answer all the hard questions for a well-defined experimental situation. However, there were many problems in which this final answer had not yet been given explicitly.

Therefore the discussions at the 1927 Solvay Conference, from the very beginning, centred around the paradoxa of quantum theory. The Compton effect emphasized the apparent wave-particle duality; the flexibility of the mathematical formalism demonstrated that the two pictures, waves and particles, may be compatible, if the limited range of their applicability is taken properly into account. Einstein criticized this very limitation because it seemed to undermine the ideal of an objective description of Nature, which had been considered to lie at the basis of physics. Besides it introduced a statistical element into the foundations of physics, which Einstein would not admit. Einstein therefore suggested special experimental arrangements for which, in his opinion, the uncertainty relations could be evaded. But the analysis carried out by Bohr and others during the Conference revealed errors in Einstein's arguments. In this situation, by means of intensive discussions, the Conference contributed directly to the clarification of the quantum-theoretical paradoxa.

The next meeting (1930) dealt largely with the applications of quantum mechanics to problems of general interest in magnetism, such as the magnetic behaviour of solid bodies or the fine structure of spectral lines. The discussions made visible the wide field that had been opened up by the final understanding of quantum theory and they spread the knowledge of the new methods and their use in many parts of physics. The discussions on the paradoxa of quantum theory were taken up again between Bohr and Einstein. When Einstein discussed an experiment in which the energy of a photon was measured by its gravity, Bohr was able to demonstrate that the influence of the gravitational field on the frequency of light, as described in Einstein's theory of general relativity, was indeed just sufficient to guarantee the uncertainty relations; the inner consistency of quantum mechanics could not be better demonstrated. The whole weight of the Conference lay on the discussions, not on the reports, and the results justified the hopes of Ernest Solvay that this style would immediately foster progress in physics.

Three years later, in the meeting of 1933, the interest had changed from quantum theory to the structure of the atomic nucleus. The discovery of the neutron by Chadwick, and of the positron by Dirac, Anderson and Blackett, had raised entirely new theoretical problems. If the nucleus consists of protons and neutrons, does it also contain electrons? Or are the electrons created in β-decay out of energy, as the positron-electron pairs are created by γ-quanta? Pauli enunciated his hypothesis of the neutrino. Again, theoretical research was actually carried forward at the Conference by means of discussions between those who had the best insight into the difficult new problems. There can be no doubt that in those years the Solvay Conferences played an essential role in the history of physics.

I have taken up these reminiscences in this foreword in order to emphasize that the historical influence of the Solvay Conferences on the development of physics was connected with the special style introduced by their founder: a small group of the most competent specialists from various countries discussing the unsolved problems of their field and thereby finding a basis for their solution.

During the period following the Second World War the situation in physics had changed. Progress was mainly due to the new experimental results, e.g. the observations concerning the spectrum and the interaction of particles. After the radical changes brought about by the discovery (in 1932) of anti-particles and anti-matter had been interpreted and understood, no fundamental difficulty had appeared which would foreshadow new radical changes in the foundation of physics. The main obstacle to further progress seemed to be the high degree of complexity in the spectrum of particles, and with this obstacle the methods of the Solvay Conferences were perhaps less efficient than with the fundamental problems of the early 1920s. Nevertheless the

Solvay Meetings have stood as an example of how much well-planned and well-organized conferences can contribute to the progress of science, and this book provides a testimony of that progress.

Munich, 19 November 1974

Acknowledgements

On several occasions during the 1970 Solvay Conference (on the symmetry properties of nuclei) I found myself answering various questions about the Solvay Conferences on Physics: their origin, the fundamental problems which had come up for discussion and highlights of the encounters between famous physicists at the earlier Conferences, etc. It was possible to answer some of these questions because in previous visits to Brussels I had received the opportunity of studying many documents* relating to the Conferences. Moreover, I had had the benefit of conversations with Professors Niels Bohr, P. A. M. Dirac, Werner Heisenberg, Wolfgang Pauli, and Léon Rosenfeld†, all of whom attended the 1927 and 1930 Conferences (on quantum mechanics and magnetism, respectively) and had witnessed the Einstein–Bohr discussions. Professors Bohr, Heisenberg, Pauli and Rosenfeld continued to take an active interest in the later Conferences, and I had been able to gather much interesting and useful information from them.

Hitherto the only published accounts of the early Solvay Conferences, other than references to them in letters and memoirs, were: (1) Maurice de Broglie, *Les Premiers Congrès de Physique Solvay* (Editions Albin Michel, Paris 1951), which gave short biographical sketches of the participants in the first *Conseil Solvay* and a very brief report on the first three Conferences, and (2) N. Bohr, 'The Solvay Meetings and the Development of Quantum Physics' (presented at the 1961 Solvay Conference), in which Bohr discussed the fundamental problems which came into focus at different times. During the 1970 Conference Professor Ilya Prigogine, Director of the Instituts Internationaux de Physique et de Chimie (Solvay), expressed to me the wish that the Solvay Institutes would welcome a book giving a survey of the scientific content of the reports and discussions at the Solvay Conferences, and he thought that it might also be of interest to physicists and historians of science. Mr Jacques Solvay, President, the Administrative Council of the Solvay Institutes and great-grandson of Ernest Solvay, concurred in the project.

I take great pleasure in thanking Professor Ilya Prigogine for inviting me to write this work as a scientific member of the Solvay Institutes during the academic year

---

* Extracts from a number of these documents are quoted in Jean Pelseneer's (unpublished) *Historique des Instituts Internationaux de Physique et de Chimie (Solvay)*.
† Léon Rosenfeld was not an invited participant in the 1927 and 1930 Conferences. He went to the former in order to take up contact with Max Born about the possibility of working with him in Göttingen. He worked with Born during 1927–29, then went to Pauli in Zurich. At the 1930 Conference, Rosenfeld recalled, 'I did not attend it, but being in Brussels at the time, I hovered around, especially at the Club of the Fondation Universitaire, where the participants gathered after the sessions.'

1973–74, and for his warm hospitality during my stay in Brussels. I also wish to thank Professor André Jaumotte, President, Université libre de Bruxelles, for election to an Invited Professorship and hospitality of the University. Mr and Madame Jacques Solvay took a personal interest in my work and encouraged me throughout, for which I am grateful.

JAGDISH MEHRA

Contents

Foreword, by Werner Heisenberg   V

Acknowledgements   IX

List of Plates   XII

Introduction   XIII

1   Ernest Solvay and the Origin of Solvay Conferences on Physics   1
2   Radiation Theory and the Quanta   13
3   The Structure of Matter   75
4   Atoms and Electrons   95
5   The Electrical Conductivity of Metals   115
6   Electrons and Photons   133
7   Magnetism   183
8   The Structure and Properties of Atomic Nuclei   207
9   Towards the Spectrum of Elementary Particles and the Hierarchy of Interactions   227
10   The Elementary Particles   239
11   Quantum Field Theory   269
12   Fundamental Problems in Elementary Particle Physics   299
13   Symmetry Properties of Nuclei   323
14   Solid State Physics   339
15   Astrophysics, Gravitation, and the Structure of the Universe   381

Index of Names   405

List of Plates

Ernest Solvay   XXXIV

First Solvay Conference (1911)   12
Second Solvay Conference (1913)   74
Third Solvay Conference (1921)   94
Fourth Solvay Conference (1924)   114
Fifth Solvay Conference (1927)   132
Sixth Solvay Conference (1930)   182
Seventh Solvay Conference (1933)   208
Eighth Solvay Conference (1948)   238
Ninth Solvay Conference (1951)   338
Tenth Solvay Conference (1954)   356
Eleventh Solvay Conference (1958)   382
Twelfth Solvay Conference (1961)   268
Thirteenth Solvay Conference (1964)   388
Fourteenth Solvay Conference (1967)   298
Fifteenth Solvay Conference (1970)   322
Sixteenth Solvay Conference (1973)   396

Signatures of Some Famous Physicists, from the Fifth Solvay Conference (1927)   180

Some Drawings and Doodles, by H. J. Bhabha   266

Introduction

In a meeting with the distinguished German physico-chemist Walther Nernst in spring 1910, Ernest Solvay learnt about the difficulties of reconciling the consequences of the Maxwell–Boltzmann molecular-kinetic theory with the quantum conceptions of Planck and Einstein in the domains of specific heats and the theory of radiation. To discuss and hopefully to solve these difficulties, Ernest Solvay convened an international *conseil de physique* in October 1911, to which he invited some of the most prominent physicists of the day – including Lorentz, Planck, Einstein, Sommerfeld, Wien, Kamerlingh Onnes, Rutherford, Marie Curie, Jean Perrin, Langevin, Marcel Brillouin and Poincaré.

The formal reports at the *conseil* covered a large range of problems and opinions, from James Jeans's attempt to explain all the apparent failures of the classical theory without invoking the quantum ideas to Einstein's arguments in favour of the absolute inevitability of the quantum structure of radiation.

H. A. Lorentz served as chairman at all the sessions of the *conseil*. The discussions after each report were free, intense and often very pointed. For instance, Poincaré dismissed Jeans's attempts with the remark: 'That is not the role of physical theories. One must not introduce as many arbitrary constants as there are phenomena to be explained. The goal of physical theory is to establish a connection between diverse experimental facts, and above all to predict.' Einstein complained that Planck's way of using the probability was 'somewhat shocking', because it deprived the Boltzmann relation of any physical content. Lorentz needed all his vast scientific knowledge, mastery of languages, and incomparable tactfulness, to keep the discussions focused on the problems at hand and yet allowing each participant's views to come through.

The *conseil* succeeded in sharpening the issues. Paul Langevin acknowledged the power of quantum theory to discover totally unexpected relationships among phenomena as apparently distinct as optical absorption frequencies and specific heats. Marcel Brillouin concluded that, 'From now on we will have to introduce into our physical and chemical ideas a discontinuity, something that changes in jumps, of which we had no notion at all a few years ago.' Planck's fear, expressed to Nernst before the *conseil*, that hardly anyone would feel the 'urgent necessity' of a reform leading to a solution of the 'unresolved problems', could now be dismissed. Ernest Solvay's foresight and initiative, Walther Nernst's organizing ability, and Hendrik Lorentz's scientific authority had played the decisive role in the success of the *conseil*.

Einstein's attitude towards the *conseil* seems to have been ambiguous. To Nernst he had written: 'I shall gladly prepare the report assigned to me. The whole enterprise appeals to me, and I hardly doubt that you are the spirit behind it.' But in letters to his friend Michele Besso, he complained that he was 'plagued with my tittle-tattle' (his report on specific heats) for the 'witches' sabbath in Brussels'.[1] A few weeks after the *conseil*, he wrote to Besso again: 'I have not made any further progress in electron theory. In Brussels, too, one lamented at the failure of the theory without finding a remedy. This Congress had an aspect similar to the wailing at the ruins of Jerusalem. Nothing positive came out of it. My treatment of fluctuations aroused great interest, but elicited no serious objection. I did not benefit much, as I did not hear anything which was not known to me already.'[2]* However, in his letter to Ernest Solvay (from Prague, 22 November 1911), Einstein wrote: 'I thank you sincerely for the extremely beautiful week which you provided us in Brussels, and not the least for your hospitality. The Solvay Congress shall always remain one of the most beautiful memories of my life.'

The sentiment which Einstein expressed in his letter to Solvay reflected the feelings of the others as well, all of whom sent enthusiastic messages of appreciation for the opportunity they had had of participating in the 'Conseil Solvay'.

---

* Immediately on his return to Prague from the first Solvay Conference, Einstein wrote to his friend Heinrich Zangger, Director of the Institute for Forensic Medicine of Zurich University: 'I returned home yesterday from Brussels, where I spent much time together with Jean Perrin, Paul Langevin and Madame Curie, and I am just † delighted with these people. The latter even promised me to visit us with her daughters. The thriller † spread around in the newspapers is nonsense. That Langevin wants to get a divorce has been known for some time. If he loves Madame Curie and she him, then they don't need to elope as they have enough opportunity of meeting each other in Paris. I did not at all get the impression that there was anything special between them; rather [they] enjoyed being together in a harmless way. I also don't believe that Madame Curie is either domineering or has some other such affliction. She is a straightforward, honest person whose duties and burdens are just too much for her. She has a sparkling intelligence, but in spite of her passionateness she is not attractive enough to become dangerous for anyone... H. A. Lorentz presided [at the Conference] with incomparable tact and incredible virtuosity. He speaks all three languages [French, German, English] equally well and possesses unique scientific acumen. I was able to convince Planck to a large extent about my views, now that he has resisted them for years. He is a totally honest man who does not worry about himself.'

Einstein continued his comments on the *conseil* in another letter to Zangger, dated Prague, 16 November 1911: 'It was most interesting in Brussels. Other than the French – Curie, Langevin, Perrin and Poincaré –, and the Germans – Nernst, Rubens, Warburg and Sommerfeld –, Rutherford and Jeans were present. Of course, H. A. Lorentz and Kamerlingh Onnes as well. Lorentz is a miracle of intelligence and subtle tact – a living work of art. In my opinion he was the most intelligent of all the theoreticians present. Poincaré was altogether simply negative about the relativity theory and, in spite of all [his] acumen, showed little understanding for the situation. Planck is untractable about certain preconceived ideas which are, without any doubt, wrong..., but nobody really knows. The whole thing [the *conseil*] would have been a delight for the diabolical Jesuit fathers.'

Note: Einstein's letters to Heinrich Zangger, quoted in Carl Seelig (Ed.), *Helle Zeit – Dunkele Zeit, in Memorium Albert Einstein*, Europa Verlag, Zurich 1956, have been translated into English and used by permission of the publisher.

† The reference is to the 'breath of scandal' concerning the Curie–Langevin 'affair' which first made its way into *Le Journal* of Paris on 4 November 1911 while Madame Curie and Paul Langevin were still in Brussels. The newspaper implied that the two had eloped from Paris. *L'Indépendance belge* took up the story the next day in Brussels, and printed a vehement declaration of Madame Curie and Jean Perrin's clarification of the circumstances and his expression of complete support for Curie and Langevin. For details of the Curie–Langevin relationship, see *A travers deux siècles: Souvenirs et rencontres* by Camille Marbo (daughter of the French mathematician Emile Borel), Editions Bernard Grasset, Paris 1968, and Robert Reid, *Marie Curie*, Collins, London 1974.

Encouraged by the success of this conference Ernest Solvay, with the counsel and help of H. A. Lorentz, established a foundation on 1 May 1912, initially for a period of thirty years, to be called the *Institut International de Physique*, with the goal 'to encourage the researches which would extend and deepen the knowledge of natural phenomena'. The new foundation was intended to concentrate on 'the progress of physics and physical chemistry', including the problems pertaining to them in other branches of the natural sciences.[3] The inclusion of physical chemistry, as one of the fields to be encouraged by the Institute, was intended as a tribute to the role which Walther Nernst had played in the organization and success of the first Solvay Conference.

The activities of the Institute were to be directed by two committees: (a) An Administrative Council, consisting of three Belgian members: (1) Ernest Solvay or a person designated by him, (2) a member designated by the King of the Belgians, and (3) a member designated by the Université libre de Bruxelles. In the absence of Mr Solvay, the other two members would invite one of his descendants to become a member; (b) An International Scientific Committee, consisting of nine regular members, to which an 'extraordinary' member could be added. This Committee was given the responsibility for directing the scientific activities, and at the time of the establishment of the Institute it consisted of: H. A. Lorentz, President, Madame Marie Curie, Marcel Brillouin, R. B. Goldschmidt, H. Kamerlingh Onnes, M. Knudsen, W. Nernst, E. Rutherford, and E. Warburg.

Article 10 of the Statutes[3] required that, 'At times determined by the Scientific Committee a *Conseil de Physique*, analogous to the one convened by Mr Solvay in October 1911, will gather in Brussels, having for its goal the examination of significant problems of physics or physical chemistry.'

In 1913, after a series of exchanges with Wilhelm Ostwald and William Ramsay, Ernest Solvay established another foundation, the *Institut International de Chimie*, and the responsibility for activities related to chemistry was passed on to it. The two foundations were ultimately united into *Les Instituts Internationaux de Physique et de Chimie*, each of them having its own Scientific Committee.

The first Solvay Conference on Physics had set the style for a new type of scientific meetings, in which a select group of the most well informed experts in a given field would meet to discuss the problems at its frontiers, and would seek to define the steps for their solution. But for the interruptions caused by the two World Wars, these international conferences on physics have taken place almost regularly since 1911 in Brussels. They have been unique occasions for physicists to discuss the fundamental problems which were at the centre of interest at different periods, and have stimulated the development of physical science in many ways.

*   *   *

H. A. Lorentz, in his capacity as President of the Scientific Committee, directed the work of the first five conferences. These conferences dealt with radiation theory and the quanta (1911), the structure of matter (1913), the atoms and electrons (1921), the electrical conductivity of metals (1924), and the electrons and photons (1927).

At the second Solvay Conference (1913), the centre of interest was Laue's discovery in 1912 of the diffraction of X-rays in crystals. The structure of matter was discussed on the basis of J. J. Thomson's atomic model, to which Madame Curie took serious objection in view of the laws of radioactive transformations. The reference to Rutherford's nuclear model of the atom was made by Rutherford himself.

At the third Solvay Conference Rutherford gave a detailed account of his atomic model and of the many phenomena which, in the meantime, had been successfully interpreted on its basis. During the discussions Rutherford mentioned his conception of the neutron. 'It has occurred to me', he declared, 'that the hydrogen of the nebulae might consist of particles which one might call "neutrons", which would be formed by a positive nucleus with an electron at a very short distance. These neutrons would hardly exercise any force in penetrating into matter .... The electron is much closer to the nucleus in the neutron than in hydrogen.'

Also during the discussions at the third Solvay Conference, Madame Curie contended that while the electrostatic forces could explain the speeds of β-particles, they were not compatible with the stability of the nucleus. The latter would require non-Coulomb forces acting within a very short distance of the nucleus.

The new quantum mechanics, complete with the Copenhagen interpretation, was presented at the fifth Solvay Conference (1927). After Compton's lecture, Madame Curie made the prophetic remark that the Compton effect might have important applications in biology, and that the high voltage technique employed in the production of high frequency X-rays would find important uses for 'therapeutic' purposes.

Louis de Broglie gave an exposition of his theory of the 'double solution' in the form of the 'pilot wave', but 'It received hardly any attention. Physicists such as Planck, Lorentz and Langevin, accustomed to the old methods, were hoping for an interpretation of wave mechanics in accordance with classical concepts but they made no pronouncement upon its nature. Schrödinger remained faithful to a pure wave interpretation. Only Einstein encouraged me somewhat on the path I wished to tread. But I was faced with redoubtable adversaries: Niels Bohr and Max Born, scientists of world renown. And there was also the group of young researchers who formed the Copenhagen School, amongst them, in particular, Pauli, Heisenberg and Dirac, who were already authors of remarkable works. They interpreted the duality of corpuscles and waves by the theory of complementarity recently proposed by Bohr and, no longer attributing to the arbitrarily normalised wave of Schrödinger any more than the role of representing the probability of certain observations being obtained, they concluded by abandoning any clear picture of a wave or a particle. I was very distressed. I found Bohr's complementarity quite obscure and I did not like abandoning physical images which had guided me for many years. However, the probabilistic interpretation of "quantum mechanics", developed by numerous young keen researchers possessing great facility in mathematical calculations, rapidly took the form of elegant and rigorous mathematical formalism.'[4]

The new orthodoxy had already taken shape. For instance, in the general discussion following Bohr's report on 'The quantum postulate', Dirac commented at length on the essential differences between the classical and quantum descriptions of physical processes. Quantum theory, he said, describes a state by a time-dependent wave function ψ, which can be expanded at a given time t_1 in a series containing wave functions ψ_n with coefficients c_n. The wave functions ψ_n are such that they do not interfere at an instant $t > t_1$. Now Nature makes a choice sometime later and decides in favour of the state ψ_n with the probability $|c_n|^2$. This choice cannot be renounced and determines the future evolution of the state. Heisenberg opposed this point of view by asserting that there was no sense in talking about Nature making a choice, and that it is our observation that gives us the reduction to the eigenfunction. What Dirac called a 'choice of Nature', Heisenberg preferred to call 'observation', showing his predilection for the language he and Bohr had developed together.

During the general discussion, Lorentz did his best to give the floor to one speaker at a time. However, many of the participants felt strongly about the interpretation of the new theory and wished to express their views. The discussion became very animated, with several participants speaking at the same time, each in his own language. Ehrenfest went to the blackboard, which successive speakers had used, and wrote on it: 'The Lord did there confound the language of all the Earth.'

The Einstein–Bohr dialogue on deterministic description versus statistical causality also started at the fifth Solvay Conference. (See Chapter 6, Appendix.) During one of the lectures, Paul Ehrenfest passed on a note to Einstein, saying 'Don't laugh! There is a special section in purgatory for professors of quantum theory, where they will be obliged to listen to lectures on classical physics ten hours every day.' To which Einstein replied, 'I laugh only at their naïveté. Who knows who would have the laugh in a few years?'

Upon Lorentz's death in February 1928, Paul Langevin became President of the Scientific Committee and the two following conferences, on magnetism (1930) and on the structure and properties of atomic nuclei (1933), were held under his leadership.

During the sixth Solvay Conference (1930) Pauli remarked on the impossibility of measuring the magnetic moment of a free electron, and Bohr emphasized that 'the direct inobservability of the intrinsic magnetic moment of the electron does not imply that the concept of spin has lost its significance as a means of explaining the fine structure of spectral lines and the polarization of electron waves. Only the manner in which the concept of spin appears in the formalism of quantum mechanics is such that it does not lend itself to an independent interpretation based on classical notions.'

The Einstein–Bohr discussions, on the question whether the quantum mechanical description of physical reality is 'complete', were continued at the sixth Solvay Conference. These discussions took place at the Club of the Fondation Universitaire. Léon Rosenfeld has recalled: 'It was the occasion when Einstein thought to have found a counter-example of the uncertainty principle with his well-known box from which a photon is emitted at a certain time, and a weighing of the box before and after the emission determines the energy of the emitted photon. It was quite a shock for Bohr to be faced with this problem; he did not see the solution at once. During the whole evening he was extremely unhappy, going from one to the other and trying to persuade them that it couldn't be true, that it would be the end of physics if Einstein were right; but he couldn't produce any refutation. I shall never forget the vision of the two antagonists leaving the Club: Einstein, a tall, majestic figure, walking quietly, with a somewhat ironical smile, and Bohr trotting near him, very excited, ineffectually pleading that if Einstein's device would work, it would mean the end of physics. The next morning came Bohr's triumph and the salvation of physics; Bohr had found the answer: the displacement of the box in the gravitational field used for the weighing would disturb the frequency of the clock governing the photon emission just to the amount needed to satisfy the uncertainty relation between energy and time .... Who knows, this may change. It is one of the great open questions whether gravitation does or does not play any part in the interactions at the subnuclear level.'[5]

Rosenfeld also remembered the occasion during the sixth Solvay Conference when, 'I met Louis de Broglie* in the street near the Club, and we got into conversation. I told him that Heisenberg and Pauli had started building up quantum electrodynamics and were encountering great difficulties. With an expression of great surprise he asked: "What difficulties?" '[5]

Ernest Solvay had died on 26 May 1922. His monument, erected near the Université libre de Bruxelles on what is now called Avenue Franklin Roosevelt, was unveiled on 16 October 1932. On this occasion a national homage, under the patronage of King Albert of the Belgians, was rendered to Solvay, and Paul Langevin gave one of the main eulogies.[6]

Armand Solvay, Ernest's eldest son, carried on the tradition of his father; he represented the family on the Administrative Council of the Institute at the 1924 and 1927 Conferences. Armand died in February 1930 and his son, Ernest-John Solvay, took over his functions at the 1930 Conference. Eduard Herzen, professor at the Ecole des Hautes Etudes de Bruxelles, represented the Solvay family at the 1933 Conference.

At the seventh Solvay Conference (1933), Rutherford expressed his pleasure at the development of the 'modern alchemy', the science of nuclear and particle physics. Rutherford reported the results of his recent experiments, with Oliphant, on the bombardment of lithium with protons and deuterons, yielding evidence about the existence of hitherto unknown isotopes of hydrogen and helium with atomic mass 3.

Chadwick reported on the discovery of the neutron. Heisenberg had immediately grasped this discovery as the new foundation of nuclear structure, with neutrons and protons as the proper nuclear constituents. He had developed a theory of nuclear structure on its basis, and gave a report on it at the seventh Solvay Conference.

The Joliot-Curies discussed the assumption concerning the complex structure of the proton, it being formed of a neutron and a positron. They would soon (1934) go on to discover artificial radioactivity.

Blackett told the story of the discovery of the positron by C. D. Anderson and

---

* L. de Broglie was not an invited participant at the sixth Solvay Conference (1930), but he was visiting Brussels at the time.

himself in cosmic ray researches, and its interpretation in terms of Dirac's relativistic quantum theory.

Dirac, in his report on the theory of the positron, announced a logarithmically divergent charge renormalization. But he also pointed out a fascinating *finite* correction to electrodynamics, to the polarization of the vacuum. This was the beginning of a new development which was to bear its most valuable fruits fifteen years later in renormalization theory.

Niels Bohr had suggested earlier that the laws of energy and momentum conservation might possibly be violated in β-decay. At the 1933 Conference, Pauli expressed his belief that the conservation principles of energy, momentum, angular momentum, and statistics were valid in *all* elementary processes, and that his conception of the neutrino, which he had proposed in June 1931 at Pasadena, would help in upholding them in β-decay.

Assuming that energy and momentum conservation could be applied, and that only one neutrino was emitted in β-decay, Francis Perrin made a relativistic calculation and reported that the intrinsic mass of the neutrino had to be zero. Fermi (1934) would use the conception of the neutrino in his theory of β-decay, but twenty-two years later an important conservation law (that of parity) would be found to break down in weak decays.

*   *   *

The Solvay Conferences had been held every three years since 1921, and plans were made to hold the eighth Solvay Conference in 1936. The Scientific Committee, consisting of Langevin, Bohr, Cabrera, Debye, De Donder, Richardson, and Verschaffelt met on 29–31 October 1935 at the Club of the Fondation Universitaire in Brussels to decide upon a programme for the meeting. They selected 'Cosmic Rays and Nuclear Physics' as its theme, to be discussed from 26 to 31 October 1936, and left the rest in the hands of Paul Langevin. A long period of illness during 1935–36 prevented Langevin from taking the initiative for holding the eighth Solvay Conference in 1936, and it was postponed to 1937. Rapid scientific developments in nuclear physics and Langevin's continuing bad health made a further delay necessary.

The Scientific Committee met again on 22–23 October 1938 to devise a programme and to prepare a list of speakers and invited participants. It was decided that the conference would deal with the problems of elementary particles and their mutual interactions, and would be held from 22 to 29 October 1939.

The invited speakers and their subjects were: W. Pauli and W. Heisenberg (general properties of elementary particles); G. Wentzel (interactions between protons, neutrons and electrons); W. Heitler (the heavy electron, i.e. meson, from a theoretical point of view); P. Blackett (the meson from an experimental point of view); J. Solomon (the beta-spectrum and the theory of the neutrino); P. Auger (the showers from the experimental viewpoint); L. de Broglie (the theory of the photon); C. F. von Weizsäcker (astronomical indications concerning the properties of particles); and F. Bloch (the magnetic moments of protons and neutrons).

The other invited participants included: C. D. Anderson, W. Bothe, A. H. Compton, P. A. M. Dirac, E. Fermi, G. Gamow, H. Geiger, Irène Joliot-Curie, F. Joliot, H. A. Kramers, Lise Meitner, F. Perrin, A. Proca, F. Rasetti, B. Rossi, M. A. Tuve, E. J. Williams, and H. Yukawa. The Scientific Secretaries were to be: E. Stahel, H. J. Bhabha, M. Cosyns, J. Géhéniau, C. Møller, L. Rosenfeld, and J. Solomon.

World War II started on 3 September 1939. In a letter addressed on the same day to Charles Lefébure, Secretary of the Administrative Council, Paul Langevin wondered whether the Conference could still be held; he thought that a final decision should be taken by 15 September, and hoped that it would be favourable. On 5 September 1939 the Administrative Council declared that the circumstances then prevailing had made it necessary to postpone the eighth Solvay Conference indefinitely.

Early in August 1939 Pauli and Heisenberg were still collaborating by mail on their joint report for the Solvay Conference. But on 29 August 1939 Pauli informed Heisenberg that the meeting which he had organized in Zurich (to be held in September) had been cancelled on account of the difficulties of transportation and communication. Pauli said: 'It's a great pity; I would have so much liked once again to talk about physics in detail with you and the others.' For the next six years at least, normal contacts between scientific colleagues were suspended and, in many cases, personal relationships were permanently altered.

*   *   *

The first Solvay Conference (1911), the 'witches' sabbath in Brussels' as he had described it to Besso, was not Albert Einstein's first contact with Belgium. During his years at the Patent Office in Bern, he visited his 'favourite uncle' Cäsar Koch in Antwerp several times in his vacations. Nor was it the first occasion of some importance which he had attended – Geneva and Salzburg had preceded Brussels. It was certainly the first to which he had gone as a young celebrity on whom many eyes were focused.

In anticipation of his first academic appointment at Zurich University, starting October 1909, Einstein had resigned on 6 July from the Federal Department of Justice and Police where he was formally employed at the Patent Office. From 7 to 9 July 1909 he attended the celebrations commemorating the 350th anniversary of the founding of the University of Geneva* by Calvin. Over two hundred delegates from universities and learned societies had been invited as guests of honour. It rained so much during the celebrations that the journalist, covering the official procession as it wended its way on 8 July to St Peter's Cathedral, reported: 'Le défilé était un peu trop silencieux et même funèbre.'[7] The convocation to confer honorary degrees took place in Victoria Hall under the aegis of the Federal President Adolphe Deucher. In a veritable orgy of academic dissipation, and a most un-Calvinistic display of ostentation, the five Geneva faculties bestowed over one hundred and ten honorary doctorates. Among those who received them were Ernest Solvay, Marie Curie, Wilhelm Ostwald (who a few months later was awarded the Nobel Prize in chemistry for his work on catalysis),

---

* The *Academia Genevensis*, founded by John Calvin in 1559, had two sections: (1) *Schola Privata*, i.e. 'collège' or secondary school, and (2) *Schola Publica*, also called 'l'académie', for higher education.

Albert Einstein, and Ernst Zahn, the restaurant keeper of Göschnen (in Canton Uri), whose coarse prose was considered in those days as the authentic representation of the voice of the Swiss mountain folk.

Einstein had received his doctorate *honoris causa* thanks to the enthusiasm of Charles Eugène Guye, Professor of Experimental Physics (1900–1930) at the University of Geneva. Guye had taught (1894–1900) electrical engineering and done research on alternating currents, polyphase generators and hysteresis phenomena at the E.T.H. in Zurich, where Einstein had attended his course.[8] With the advent of Lorentz's theory of the electron and Max Abraham's rival theory, Guye became interested in using his accurate instrumental techniques to test for evidences of the Fitzgerald–Lorentz contraction hypothesis and transformation equations. He thought that a crucial experiment ought to be possible to decide between Lorentz's conception of a deformable electron, with a shape dependent on its velocity, and Abraham's notion of permanently spherical electrons. The opportunity seemed all the more inviting when Einstein's theory of special relativity, which had a close connection with Lorentz's work, began to stir controversy after 1905 – W. Kaufmann's experiments appeared to support Abraham and contradict the predictions of Lorentz and Einstein.

Guye became an admirer of Einstein's work. For over a decade, beginning in 1907, he carried on a series of experiments with charged particles moving through electromagnetic fields. In collaboration with M. Ratnowsky and Charles Lavanchy, Guye developed very precise techniques for measuring particle deflections in carefully controlled electric and magnetic fields. In 1916 and 1921 Guye published the accounts of his experimental techniques and pronounced results in favour of the Lorentz transformations and Einstein's theory. He had, however, demonstrated his confidence in the latter by persuading the University of Geneva to honour Einstein already in 1909. Charles Guye served as a member of the Scientific Committee of the Solvay Institute for Physics from 1925 to 1934.

Two months after his visit to Geneva, Einstein attended the 81st Congress of German Natural Scientists and Physicians in Salzburg (19–25 September 1909). Already in summer 1908 Rudolf Ladenburg had gone from Berlin to Bern to invite Einstein to address the Congress. Einstein delivered his lecture, 'On the development of our views concerning the nature and structure of radiation', on 21 September 1909. In Salzburg he met Planck, Wien, Rubens, Sommerfeld ('a splendid chap'), Max Born, and Ludwig Hopf who became his assistant in Zurich and Prague. As he told a colleague on this occasion, he had 'never met a real physicist' before he was thirty. He would take up his own first academic appointment in the following month.

In January 1911 H. A. Lorentz invited Einstein and his wife, Mileva, to visit Leyden. They stayed with the Lorentz family in February for a few days, and also stopped by in Antwerp to see Uncle Cäsar. Shortly afterwards Einstein was formally invited to attend the first Solvay Conference in Brussels.

In Brussels, Einstein met Lorentz, Planck and Nernst on equal terms for the first time, and made an enormous impression on his colleagues. Frederick Lindemann, who had accompanied Walther Nernst as one of the scientific secretaries to the Conference, wrote home to his father the day after the meeting. Lindemann, the future Viscount Cherwell and scientific adviser to Winston Churchill, described Ernest Solvay as 'a very nice man, unfortunately though with rather liberal views.' He continued, 'I got on very well with all the people here, even with Madame Curie who is quite a good sort when one knows her. I got on very well with Einstein who made the most impression on me except perhaps Lorentz .... He says he knows very little mathematics, but he seems to have had a great success with them.'[9] And, almost half a century later, Lindemann wrote: 'I well remember my co-secretary, M. de Broglie, saying that of all those present Einstein and Poincaré moved in a class by themselves.'[10]

Others present at the Conference also had the opportunity of sizing up Einstein.* Among those in Brussels who were most deeply impressed with his ability were Max Planck and Walther Nernst, twin pillars of the Prussian scientific establishment. A year and a half after the first Solvay Conference, Planck and Nernst, together with Rubens and Warburg who had also attended the Conference, recommended to the Prussian Academy of Sciences to elect Einstein (he was thirty-four) to full membership and award him a research professorship. What had motivated them, in addition to the theory of relativity, was Einstein's work on the quantum theory of matter, especially on specific heats, about which he had spoken at the Conference. In the minds of these people, in particular Planck's, who drafted the recommendation, Einstein's ideas on the structure of radiation were in doubt, for they remarked, 'That

---

* After the first Solvay Conference Einstein's friends, Heinrich Zangger and Marcel Grossmann, initiated the moves to bring Einstein from the German University of Prague to the E.T.H. (the Swiss Federal Institute of Technology) in Zurich. In order to further these plans Pierre Weiss, Professor of Physics at the E.T.H., had the brilliant idea of asking the opinions of Marie Curie and Henri Poincaré, both of whom had just recently become personally acquainted with Einstein in Brussels.

In her reply, dated Paris, 17 November 1911, Madame Curie wrote: 'I have greatly admired the papers published by Mr Einstein on questions dealing with modern theoretical physics. Moreover, I believe that the mathematical physicists all agree that these works are of the highest order. In Brussels, where I attended a scientific conference in which Mr Einstein took part, I was able to appreciate the clarity of his mind, the vastness of his documentation and the profundity of his knowledge. If one considers that Mr Einstein is still very young, one has every right to justify the greatest expectations from him, and to see in him one of the leading theoreticians of the future....'

And Henri Poincaré wrote: 'Mr Einstein is one of the most original thinkers I have ever met. In spite of his youth, he has already achieved a very honourable place among the leading savants of his age. What one has to admire in him above all is the facility with which he adapts himself to new concepts and knows how to draw from them every possible conclusion. He has not remained attached to classical principles, and when faced with a problem of physics he is prompt in envisaging all its possibilities. A problem which enters his mind unfolds itself into the anticipation of new phenomena which may one day be verified by experiment. I do not mean to say that all these anticipations will withstand the test of experiment on the day such a test would become possible. Since he seeks in all directions one must, on the contrary, expect most of the trails which he pursues to be blind alleys. But one must hope at the same time that one of the directions he has indicated may be the right one, and that is enough. This is indeed how one should proceed. The role of mathematical physics is to ask the right questions, and experiment alone can resolve them. The future will show more and more the worth of Mr Einstein, and the university intelligent enough to attract this young master is certain to reap great honour.'

Note: These letters of Marie Curie and Henri Poincaré are quoted in original French in Carl Seelig, *Albert Einstein, Eine Dokumentarische Biographie*, Europa Verlag, Zurich 1954, pp. 162–163, and have been re-translated into English and reproduced here by permission of the publisher.

he may sometimes have missed the target in his speculations, as for example, in his hypothesis of light quanta, cannot really be held too much against him, for it is not possible to introduce really new ideas, even in the most exact sciences, without sometimes taking a risk.'[11]

In spring 1913 Einstein was invited to attend the second Solvay Conference, to be held in October. On 29 April 1913 he informed the Scientific Committee that he would be glad to participate in the Conference, but requested that he should not be asked to give a report as he was 'already very overloaded beyond the call of professional duties'.[12] Einstein's talk on 'the status of the problem of specific heats' at the first meeting would remain as the only report which he presented at the Solvay Conferences.

Plans were made during summer 1920 for the third Solvay Conference to be held in April 1921. Following World War I the feeling against the participation of German scientists in international meetings was still strong. 'An exception was made,' Tassel wrote, 'in the case of Einstein, of ill-defined nationality, Swiss I believe, who was roundly abused in Berlin during the war because of his pacifist sentiments which have never varied for a moment.'[13] Rutherford put it differently: 'The only German invited is Einstein who is considered for this purpose to be international.'[14] Einstein accepted 'with great pleasure'.[15] Lorentz informed Rutherford that Einstein would speak on 'The electron and magnetism; gyroscopic effects'.[16] However, on 22 February 1921 Einstein wrote to Lorentz, sending his regrets that he would not be present at the Conference. 'The Zionists have planned to establish a university in Jerusalem,' he wrote, and they had requested Einstein to speak for them in the United States in March and April so that through his personal cooperation 'the rich American Jews will be persuaded to pay up.'[17] Nevertheless, Einstein wished the Conference every success. At the Conference, W. J. de Haas gave a report on 'the angular momentum in a magnetic body', based on the notes supplied by Einstein.

Niels Bohr, who had been invited to attend the 1921 Conference, was not able to do so on account of illness. His report on the 'application of the theory of quanta to atomic problems' was presented by Paul Ehrenfest.

Two years later the 1924 Conference was being planned, and again only Einstein was to be invited from Germany. On 16 August 1923 he wrote to Lorentz from Lautrach in southern Germany: 'This letter is hard for me to write but I have to write it. I am here together with Sommerfeld. He is of the opinion that it is not right for me to take part in the Solvay Congress because my German colleagues are excluded. In my opinion it is not right to bring politics into scientific matters, nor should individuals be held responsible for the government of the country to which they happen to belong. If I took part in the Congress I would by implication become an accomplice to an action which I consider most strongly to be distressingly unjust.' But he still looked forward to a future international cooperation, and continued: 'I should be grateful if you would see to it that I do not even receive an invitation to the Congress. I want to be spared the necessity of declining – an act which might hinder the gradual reestablishment of friendly collaboration between physicists of various countries.'

International scientific relations began to return to normal after Germany joined the League of Nations in 1926. 'Now,' Lorentz made a private note, 'I am able to write to Einstein.'[18] The fifth Solvay Conference was being planned and Einstein's name had been proposed as a member of the Scientific Committee to replace Kamerlingh Onnes who had died on 21 February 1926. However, it was thought proper that the approval of King Albert of the Belgians should be sought. On 2 April 1926, Lorentz was given an audience at which the King specifically approved Einstein's nomination to the Scientific Committee. The proposal to invite Planck and the other German physicists was also approved.

Lorentz subsequently reported that, 'His Majesty expressed the opinion that, seven years after the war, the feelings which they aroused should be gradually damped down, that a better understanding between peoples was absolutely necessary for the future, and that science could help to bring this about. He also felt it necessary to stress that in view of all that the Germans had done for physics, it would be very difficult to pass them over.'[18a] In any case, the 1927 Conference was planned to discuss quantum mechanics, and it could hardly have proceeded without Heisenberg, Born, Planck, and Einstein. Surprisingly, Sommerfeld was not invited.

The fifth Solvay Conference witnessed the start of the discussions between Einstein and Bohr, which became the nucleus of the ever-growing literature on the interpretation of the quantum principle. Einstein, to the end, did not like Heisenberg's uncertainty and Bohr's complementarity principles.

In spring 1929 Einstein made one of his regular trips from Berlin to Leyden and, as usual, he called on his Uncle Cäsar in Antwerp. Here he received an invitation to visit Queen Elizabeth of the Belgians at the Château de Laeken on Monday, 20 May 1929. King Albert, who had a genuine interest in science, was absent only because he was visiting Switzerland. The Queen, formerly Princess Elizabeth of Bavaria, was unconventional and artistic. On 20 May Einstein, with his violin, 'spent the first of many musical afternoons' at the Palace, Her Majesty 'playing second fiddle'.[19] 'There followed tea under the chestnuts and a walk in the grounds, followed by dinner at 7:30.'[20] A few days later Einstein received prints of the photographs which the Queen had taken. In a covering note she expressed the King's regrets that he had been away, and added: 'It was unforgettable for me when you came down from your peak of knowledge and gave me a tiny glimpse into your ingenious theory.'[20]

An unusual friendship thus began between Queen Elizabeth and Albert Einstein and, during the next four years until he left Europe for ever for the U.S.A., he was always invited to Laeken when he visited Belgium. A royal chauffeur would be sent to meet Einstein at the train, who at times would miss the man with the violin case. Einstein would then go to a small café nearby, request the use of the telephone, call the Palace and ask to speak to the Queen to announce his arrival. At the Palace there would be music, trios and quartets, with Einstein, the Queen, her lady-in-waiting and another Palace guest keeping it going for several hours. On one occasion he wrote to his (second) wife Elsa: 'Then they all went away, and I stayed behind alone for dinner with the King, vegetarian style, no servants. Spinach with hard-boiled eggs and potatoes, period.'[21]

Einstein would write a post-card or send a note to the Queen from wherever he was, and she would always reply. Before going to Pasadena in the winter of 1932–33, Einstein called on the King and Queen at Laeken. In a letter of thanks to Her Majesty, he wrote: 'It was a great happiness for me to explain to you something of the mysteries in front of which the physicists stand silent.'[22] From the Rayben Farm Hope Ranch in California, which Queen Elizabeth had visited during a state visit to the United States in 1919, Einstein sent a post-card to her with the lines: 'A tree stands in the cloister garden / Which was planted by your hand. / It sends a little twig as greeting / Because there it must forever stand. / It sends a friendly greeting with Yours, A. Einstein.'[23]

In December 1932 Kurt von Schleicher had become Chancellor of Germany, and for some weeks he desperately tried to form a stable government. He failed, the third man to do so in as many years. Then on 30 January 1933 President Hindenburg turned over the government to the leader of the National Socialists. Einstein's reaction in far-away Pasadena was immediate and unqualified: he could not return to Germany under the circumstances. On 10 March, the eve of his departure from Pasadena, Einstein made his decision public in an interview with Evelyn Seeley of the *New York World-Telegram*: 'As long as I have any choice in the matter,' he said, 'I shall live only in a country where civil liberty, tolerance, and equality of all citizens before the law prevail. Civil liberty implies freedom to express one's political convictions, in speech and in writing; tolerance implies respect for the convictions of others whatever they may be. These convictions do not exist in Germany at the present time.'[24]

Einstein and his wife crossed the Atlantic on the *Belgenland*, which docked in Antwerp on 28 March 1933. Einstein was immediately driven to Brussels where, in the German embassy, he formally surrendered the rights of full German citizenship that he had taken after the First World War. He retained his Swiss nationality, and handed in his German passport. As he descended the steps of the embassy, Albert Einstein, 'the Swabian from Ulm, left German territory for the last time.'[25]

On the same day, 28 March, he wrote to Berlin formally announcing his resignation from the Prussian Academy of Sciences, on the grounds that he was no longer able to serve the Prussian state. An important personal reason for his immediate resignation from the Academy was his concern for the difficult position into which his old friends Planck and Nernst would be thrust; if he were expelled they would find it dangerous to protest, yet disloyal not to. These were the men who had persuaded the Academy in the first place to elect him to full membership in 1913, thereby conferring on him 'the greatest boon', and freeing him 'from the distractions and cares of a professional life and so making it possible for me to devote myself entirely to scientific studies.'[26] This connection was now severed finally.

Albert and Elsa Einstein took temporary refuge at 'Canteroy', a historic manor house outside Antwerp, which belonged to the family of Professor A. de Groodt of Ghent University. Then they made a base, for their last six months in Europe, at Le Coq-sur-mer, a small resort near Ostend on the Belgian coast. King Albert wrote to Einstein:

My dear Professor,

...

We are delighted that you have set foot on our soil. There are men who by their work and intellectual stature belong to mankind rather than to any one country, yet the country they choose as their asylum takes keen pride in that fact.

The Queen joins me in sending you best wishes for a pleasant stay in Belgium. Please accept my expression of high esteem.

Albert[27]

Europe had begun to prepare itself for a tumult, and the few months which Einstein spent at Le Coq were hectic. He had great concern about all that was happening and anxiety about what might happen. From Le Coq, he wrote to his friend Maurice Solovine: 'My great fear is that this hate and power epidemic will spread throughout the world. It comes from below to the surface like a flood, until the upper regions are isolated, terrified and demoralized, and then also submerged.'[28] Einstein had much to do, numerous visitors called on him, and he did much travelling. His science was still his main concern and on three separate evenings he gave lectures on general relativity and the unified theory to invited audiences at the Club of the Fondation Universitaire in Brussels.

There were, now and then, some moments of relaxation as well. He was invited to Laeken, where he and the Queen played the violin together.

In meetings with Abraham Flexner at Oxford and Caputh (Einstein's summer home) during summer 1932, Einstein had agreed to visit, on a regular basis, starting in the fall of 1933, the Institute for Advanced Study which Flexner was planning to establish in Princeton. After the events leading to his refuge in Le Coq-sur-mer, the arrangement in Princeton was going to be permanent.

Einstein left Le Coq on 9 September 1933 and took a Channel boat from Ostend to England. There he spent a month full of meetings with scientists and politicians. He boarded the *Westerland*, bound for New York, on the evening of 7 October 1933; his wife Elsa, Miss Helen Dukas, his secretary, and Dr. Walther Mayer, his assistant, had already embarked in Antwerp. As the *Westerland* sailed up the approaches to New York Harbour after an uneventful voyage, a tugboat, with two trustees of the Institute for Advanced Study in it, came aside, and Einstein and his party were helped aboard. The *Westerland* continued on its way and, long before it docked, Einstein and his 'family' had been transferred to a car and were being driven to Princeton.

The seventh Solvay Conference (22–29 October 1933) took place in Brussels two weeks after Einstein had left Europe for good. In a letter (Le Coq-sur-mer, 12 August 1933) to Charles Lefébure he had sent his regrets concerning his inability to attend the Conference in view of his impending departure for Princeton at the beginning of October. Einstein's absence was felt keenly at the Conference and Paul Langevin, on behalf of all the participants, sent Einstein a message of greetings. On 26 November, Einstein wrote to Lefébure to thank him for the group photograph of the Conference.

On Friday, 27 October 1933, King Albert of the Belgians received the participants in the Conference for a dinner* at Laeken. The royal couple missed Einstein's presence.

A few days after the dinner, King Albert wrote to Lefébure: 'The Queen and I were very happy to receive "ces Messieurs les Physiciens", and to see that the great Ernest Solvay's noble efforts continue to perpetuate themselves in the most effective service of science and the relations between the savants of all countries.'[29]

Einstein settled down in Princeton. 'Princeton is a wonderful little spot', he wrote, giving his first impressions to Queen Elizabeth of Belgium, 'a quaint and ceremonious village of puny demigods on stilts. Yet, by ignoring certain special conventions, I have been able to create for myself an atmosphere conducive to study and free from distraction. Here, the people who compose what is called "society" enjoy even less freedom than their counterparts in Europe. Yet they seem unaware of this restriction since their way of life tends to inhibit personality development from childhood.'[30] And, more than a year later, he wrote to her: '... as an elderly man, I have remained estranged from the society here ....'[31] As always, he would remain an outsider.

King Albert died in a mountaineering accident in 1934. Queen Elizabeth, now the Queen Mother, and Einstein continued to exhange letters for another twenty years. Early in 1939 he wrote to her: 'The work has proved fruitful this past year. I have hit upon a hopeful trail, which I follow painfully but steadfastly in company with a few youthful fellow workers. Whether it will lead to truth or fallacy – this I may be unable to establish with any certainty in the brief time left to me. But I am grateful to destiny for having made my life into an exciting experience so that life has appeared meaningful....'[32]

The eighth Solvay Conference was scheduled for 22–29 October 1939. Einstein was, of course, invited but he had informed the Solvay Institute already on 20 April that, 'Unfortunately, it would not be possible for me to come to Europe this year.' To the Belgian Queen Mother he wrote: 'The moral decline we are compelled to witness, and the suffering it engenders, are so oppressive that one cannot ignore them even for a moment. No matter how deeply one immerses oneself in work, a haunting feeling of inescapable tragedy persists.'[32]

The tragedy had begun, and the 1939 Conference was cancelled.

*   *   *

After World War II Sir Lawrence Bragg became President of the Scientific Committee of the Solvay Institute for Physics. He directed the Conferences on the elementary particles (1948), the solid state (1951), the electrons in metals (1954), the structure and evolution of the universe (1958), and quantum field theory (1961). Upon Bragg's resignation, J. Robert Oppenheimer became President of the Committee, and directed the Conference on the structure and evolution of galaxies (1964). After Oppenheimer died on 18 February 1967, C. Møller acted as President of the Scientific Committee of the fourteenth Solvay Conference on fundamental problems in elementary par-

---

* Dîner du 27 Octobre 1933, Laeken: Crême à la Sévigné; Saumon sauce Genèvoise, Pommes à l'Anglaise; Filet de boeuf à la Française, Chicorées de Bruxelles glacées; Poulardes rôties cresson, Salade coeurs de laitues, Compote de reinettes; Bombe Fédora; Corbeilles de fruits; Desserts.

ticle physics (1967). Edoardo Amaldi then took over the Presidency of the Scientific Committee, and directed the fifteenth and sixteenth Conferences on symmetry properties of nuclei (1970) and astrophysics and gravitation (1973).

Ernest-John Solvay continued to represent the Solvay family at all the Conferences until 1967. At the 1967 Conference his son, Jacques Solvay, also became a member and, later on President (1970), of the Administrative Council of the International (Solvay) Institutes of Physics and Chemistry.

Until 1958 the administrative functions of the Solvay Institutes were executed by the successive Secretaries of the Administrative Council (E. Tassel, J. E. Verschaffelt, C. Lefébure, F. H. van den Dungen). Ilya Prigogine became a member of the Council in 1958 and, in 1959, was appointed its Secretary. Under the new Statutes, Prigogine was appointed Director of the International (Solvay) Institutes of Physics and Chemistry in 1970.

*   *   *

The eighth Solvay Conference, and the first after World War II, took place in 1948 and discussed the problems of elementary particles. This field had grown immensely since the 1933 Conference, especially on account of the appearance of mesons as carriers of the nuclear interactions. At the 1948 Conference, the latest sensation 'was the production of artificial mesons in Berkeley, about which Serber was supposed to give exciting details. However the technical aids were not then as refined as they are now: there were no microphones, so that Serber could hardly be heard, and the projection apparatus was so bad that it was hardly possible to see the meson tracks that he showed on the screen. This somewhat subdued the expected excitement, and at the banquet concluding the Conference, poor Serber was lampooned in a "meson song",[33] an adaptation by Teller of another composition, sung to the music composed by O. R. Frisch for the occasion. Here is the relevant verse (See Chapter 10, Appendix).

> Some beautiful pictures are thrown on the screen;
> Though the tracks of the mesons can hardly be seen,
> Our desire for knowledge is most deeply stirred,
> When the statements of Serber can never be heard.

> What, not heard at all?
> No, not heard at all!
> Very dimly seen
> And not heard at all!

The conclusion of the song gave an adequate description of the degree of understanding that had been reached at the time:

> From mesons all manners of forces you get,
> The infinite part you simply forget,
> The divergence is large, the divergence is small,
> In the meson field quanta there is no sense at all.

> What, no sense at all?
> No, no sense at all!
> Or, if there is some sense,
> It's exceedingly small.

In 1929 Heisenberg and Pauli had laid a systematic foundation of quantum field theory. When Léon Rosenfeld wrote to Pauli (in 1929) whether he could come to Zurich to work under his guidance, Pauli sent a friendly reply, pointing out that the time was quite favourable since Heisenberg and he had just started work on the quantization of electrodynamics – 'ein Gebiet,' he said, 'das noch nicht ganz abgedroschen ist' (a domain that is not yet threshed out). When Rosenfeld arrived in Zurich, he immediately got the proof-sheets of their first paper to read. 'The first thing I did,' said Rosenfeld, 'was to correct a minor mistake in it. I was very proud of it, and wrote a little note which started with the words: "In ihrer grundlegenden Arbeit, haben Heisenberg und Pauli ..." (In their fundamental paper ...) When he saw this, Pauli laughed and remarked: "Es ist ein ziemlich morscher Grund, den wir da gelegt haben!" (It is a rather swampy foundation we have laid.)'[34]

The renormalization technique in quantum electrodynamics had just come into view at the time of the 1948 Conference. It developed very rapidly, and quantum field theory continued to follow its adventurous course until the 1961 Conference. The twelfth Solvay Conference in 1961 was also the occasion at which the fiftieth anniversary of the first *Conseil Solvay* was celebrated. Feynman took note of the distance one had travelled since then when he remarked: 'Fifty years ago at this Conference one of the problems most energetically discussed was the apparent quantum nature of the interaction of light and matter. It is a privilege to be able, after half a century, to give a report on the progress that has been made in its solution. No problem can be solved without it dragging in its wake new problems to be solved. But the incompleteness of our present view of quantum electrodynamics, although presenting us with the most interesting challenges, should not blind us to the enormous progress that has been made. With the exception of gravitation and radioactivity, all of the phenomena known to physicists and chemists in 1911 have their ultimate explanation in the laws of quantum electrodynamics.'[35]

At the 1961 Conference, Rosenfeld recalled, 'The heroes of the day were our friends Chew and Gell-Mann. We then witnessed the birth of the bootstrap idea and the canonization of Regge. There was some alarm when Regge poles were first mentioned, and I remember Professor Wigner's polite and ineffectual efforts to induce somebody to define these Regge poles. However, the discussion went on happily without such a definition. At one point, Chew and Gell-Mann seemed preoccupied with understanding what the old people could have expected from such queer concepts as Lagrangians. Gell-Mann was suggesting: "They may have thought this and this ...," and Chew replied: "No, probably they thought that and that...." It did not occur to them that Heisenberg was sitting there and that it might have been simpler to ask him. What I found most remarkable, however, was that Heisenberg himself did not volunteer any statement about this point.' In Rosenfeld's view, the 'swampy foundation' which Heisenberg and Pauli had laid over thirty years previously had not yet become terra firma, and 'we are still plodding in this mud, some of us trying to dig a little deeper in the hope of finding some firmer ground, others trying to extricate themselves by pulling their own bootstraps.'[36]

All the Solvay Conferences until 1948 had been devoted to the discussion of the development of quantum physics. After the 1961 Conference on quantum field theory, the fundamental problems in elementary particle physics again came up for discussion at the fourteenth Solvay Conference (1967). The symmetry properties of nuclei were discussed in the following Conference (1970).

After World War II the field of solid state physics had emerged into prominence and many fundamental problems occupied physicists working in it in different countries. Thus the 1951 Conference was devoted to the solid state, and that of 1954 to the electrons in metals.

Technological progress during the Second World War had also opened new horizons in the study of astronomy, especially since micrometer wavelength radiation became observable. The realm of radio stars and very distant radio galaxies was now ready for investigation. The eleventh Solvay Conference (1958) was devoted to the discussion of the structure and evolution of the universe. The thirteenth Conference (1964) dealt with the structure and evolution of galaxies, the recent discovery of quasars engendering appropriate enthusiasm. The discussion of black holes and pulsars at the sixteenth Solvay Conference (1973) showed how some of the most fundamental concepts of physics had become important for the understanding of problems in astrophysics and gravitation.

*   *   *

Numerous participants in the various Solvay Conferences have given accounts of the direct beneficial effects which they experienced on these occasions. Reports given by the most well-informed scholars and discussions among the authoritative experts of a field could not but stimulate all who came together. Take Henri Poincaré. As the direct result of his participation in the first Solvay Conference, Poincaré wrote two memoirs on quantum theory a few months before his death on 17 July 1912, taking the reports of Planck and Sommerfeld as his points of departure.[37] Extensive references to Solvay Conferences in the correspondence or recollections of various participants – such as Lorentz, Einstein, Rutherford, Madame Curie, Langevin, Joffé, Debye, Bohr, Compton, Gamow, Pauli, Heisenberg, de Broglie, Rosenfeld, and others – bear witness to the importance of these meetings as occasions for strong scientific interaction in an agreeable setting. For some they became the happiest memory of their lives.

The indirect effects of the Solvay Conferences are incalculable. Niels Bohr, for instance, heard a first-hand account of the proceedings of the first Solvay Conference from Rutherford when Bohr visited him in Manchester a few weeks after Rutherford's return from Brussels.[38] The nineteen-year-old Louis de Broglie, whose older brother, Maurice, was one of the scientific secretaries of the first Solvay Conference, read the discussions as the manuscript was being prepared for publication. 'With all the ardour of my youth,' he has written, 'I was swept away by my enthusiasm for the problems discussed and I resolved to devote all my efforts to understanding the true nature of the mysterious quanta that Max Planck had introduced ten years earlier.'[39]

It is quite possible that others may also have been inspired, or might be in the future, by the knowledge of what some of the greatest physicists of the twentieth century sought to understand at these Conferences. As Niels Bohr once said: 'The careful recording of the reports and of the subsequent discussions at each of these meetings will in the future be a most valuable source of information for students of the history of science wishing to gain an impression of the grappling with the new problems raised in the beginning of our century. Indeed, the gradual clarification of these problems through the combined effort of a whole generation of physicists was in the following decades not only so largely to augment our insight in the atomic constitution of matter, but even to lead to a new outlook as regards the comprehension of physical experience.'[40]

'A new outlook as regards the comprehension of physical experience,' – this is precisely what Ernest Solvay had been thinking when he convened his first *conseil* and then took the initiative to provide the forum and the means for the dialogue to go on.

REFERENCES

1. A. Einstein to M. Besso, 21 October 1911. See *Albert Einstein – Michele Besso, Correspondance 1903–1955* (Ed. P. Speziali), Hermann, Paris 1972.
2. A. Einstein to M. Besso, 26 December 1911, Ref. 1.
3. Institut International de Physique (Solvay), *Status*, 1 May 1912. Revised 1 July 1919; the earlier Article 10 now became Article 13.
4. L. de Broglie, 'Beginnings of Wave Mechanics', in *Wave Mechanics, the First Fifty Years* (Eds. W. C. Price, S. S. Chissick, and T. Ravensdale), Butterworths, London 1974, pp. 16–17.
5. L. Rosenfeld, 'Some Concluding Remarks and Reminiscences', in *Fundamental Problems in Elementary Particles Physics*, Proceedings of the fourteenth Solvay Conference on Physics, John Wiley–Interscience, New York 1968, p. 232.
6. *Hommage National à Ernest Solvay, Inauguration du Monument*, 16 October 1932, Brussels 1933, pp. 31–44.
7. C. Seelig, *Albert Einstein*, Staples Press, London 1956, p. 93.
8. Biographical Notice of Charles Eugène Guye in *Dictionary of Scientific Biography*, Volume V, Charles Scribner's Sons, New York 1972, pp. 597–598.
9. F. W. F. Smith, Earl of Birkenhead, *The Professor and the Prime Minister*, Houghton Mifflin, Boston 1962, p. 43.
10. Quoted in R. W. Clark, *Einstein, The Life and Times*, World Publishing Company, New York 1971; Avon Books, New York, 1972. See the latter edition, p. 185.
11. Ref. 7, p. 145.
12. A. Einstein to E. Tassel, 29 April 1913.
13. E. Tassel to M. Huisman, 23 March 1921.
14. E. Rutherford to B. B. Boltwood, 28 February 1921.
15. A. Einstein to H. A. Lorentz, 15 June 1920.
16. H. A. Lorentz to E. Rutherford, 16 December 1920.
17. A. Einstein to H. A. Lorentz, 22 February 1921.
18. H. A. Lorentz's private note on the back of a telegram (documents of the Solvay Institutes).
18a. H. A. Lorentz, Report to the Administrative Council, Solvay Institute, 3 April 1926.
19. Queen Elizabeth's personal notes in her agenda book: Royal Archives, Brussels. Quoted in Ref. 10, p. 511.
20. Ref. 19. Quoted in Ref. 10, p. 512.
21. A. Einstein's letter to his wife Elsa, quoted in O. Nathan and H. Norden (Eds.), *Einstein on Peace*, London 1963, p. 662.
22. A. Einstein to Queen Elizabeth, 19 September 1932, Royal Archives, Brussels; Ref. 10, p. 546.

23. A. Einstein to Queen Elizabeth, 19 February 1933, Royal Library, Brussels; Ref. 10, p. 555.
24. Interview with Evelyn Seeley, *New York World-Telegram*, 10 March 1933; Ref. 10, p. 557.
25. R. W. Clark, Ref. 10, p. 565.
26. A. Einstein, inaugural address at the Prussian Academy of Sciences, 2 July 1914, *Sitz. Ber. Kgl. Akad. Wiss.* **28**, 739–742 (1914).
27. King Albert to A. Einstein, 24 July 1933, Royal Archives, Brussels; quoted in Ref. 10, p. 595.
28. A. Einstein to M. Solovine, 28 April 1933, quoted in Ref. 7, p. 192.
29. King Albert to Charles Lefébure, 5 November 1933 (documents of the Solvay Institutes).
30. A. Einstein to Queen Elizabeth, 20 November 1933, Royal Archives, Brussels; Ref. 10, p. 643.
31. A. Einstein to Queen Elizabeth, Ref. 10, p. 643.
32. A. Einstein to Queen Elizabeth, 9 January 1939, Royal Library, Brussels; Ref. 10, p. 656.
33. L. Rosenfeld, Ref. 5, pp. 232–233.
34. L. Rosenfeld, Ref. 5, p. 231.
35. R. P. Feynman, 'The Present Status of Quantum Electrodynamics', in *La Théorie Quantique des Champs*, Douzième Conseil de Physique, 1961, Interscience, New York, and R. Stoops, Brussels, 1962, p. 61.
36. L. Rosenfeld, Ref. 5, p. 231.
37. H. Poincaré, 'Sur la théorie des quanta', *J. Phys. Théor. Appl. (Paris)* **2**, 5 (1912); 'The Quantum Theory', Chapter VI in *Mathematics and Science: Last Essays*, Dover Publications, New York, 1963 (originally published in *Dernières Pensées*, Ernest Flammarion, 1913).
38. N. Bohr, *Essays 1958–1962 on Atomic Physics and Human Knowledge*, Interscience, New York 1963, pp. 31, 83.
39. *Louis de Broglie, Physicien et Penseur*, Editions Albin Michel, Paris 1953, p. 458; see p. 425 for the comments of Maurice de Broglie. Also M. de Broglie, *Les Premiers Congrès de Physique Solvay*, Editions Albin Michel, Paris 1951.
40. N. Bohr, 'The Solvay Meetings and the Development of Quantum Physics', in *La Théorie Quantique des Champs*, Douzième Conseil de Physique, 1961, Interscience, New York, and R. Stoops, Brussels, 1962, p. 13.

Note: I would also like to cite M. J. Klein, 'Einstein, Specific Heats, and the Early Quantum Theory', *Science* **148**, 173–180 (1965). I take special pleasure in thanking my friend Ronald W. Clark, the author of *Einstein: The Life and Times*, for allowing me to use some information which he had painstakingly gathered.

Ernest Solvay
16 April 1838 – 26 May 1922

1

Ernest Solvay and the Origin of Solvay Conferences on Physics

1. SOLVAY, THE MAN AND ENTREPRENEUR

Ernest Solvay, industrial chemist and social reformer, was born at Rebecq-Rognon, Belgium, on 16 April 1838. His father, Alexandre Solvay, was the owner of a small salt refinery. Young Ernest attended school as a boarder at the Institut des Frères at Malonne. He read widely in chemistry and physics. Even at school he had converted his room into a chemical laboratory, and spent his vacations working on experiments.

The atmosphere at home was warm and inspiring, and his father imbued him with a remarkable sense of social responsibility. Father Alexandre lived by certain rules which he inculcated in his son; among these were the maxims: 'Paradise on Earth is a good book, a good conscience, and a home where peace reigns'; 'The most important intellectual quality is good sense'; 'Work is a debt which every citizen owes to society.' The family atmosphere in which Ernest Solvay grew up had a lifelong influence upon him.

His father had hoped that Ernest would pursue a brilliant academic career in engineering at the University at Liège, but an attack of pleurisy forced him to give up plans for University studies. His father then thought of a career in commerce for him. A maternal uncle, Florimond Semet, who was director of the gas company of Saint-Josse-ten-Noode, offered him a place in his small factory. Ernest Solvay, who was then twenty-one years old, immediately accepted the offer because it gave him the opportunity of satisfying his enthusiasm for chemical experimentation.

At Saint-Josse, Solvay worked on all aspects of gas manufacture, and very soon rose to the position of deputy director. He conducted experiments which led to the ammonia process for producing sodium carbonate. On 15 April 1861, the eve of his twenty-third birthday, Ernest Solvay took out his first patent on 'the industrial manufacture of the carbonate of soda by means of sea salt, ammonia and carbonic acid'. It was one of the most important advances in industrial chemistry in the nineteenth century.

With the aid of modest funds made available by his family, and the assistance of his brother Alfred, Ernest Solvay set up a testing laboratory at Schaarbeek and rapidly obtained results that were satisfactory enough so that he could offer his process for the manufacture of sodium carbonate to Belgian companies dealing with chemical products. From them, however, he encountered only indifference.

Solvay and Company was formed as a family enterprise in 1863. A factory was constructed at Couillet, which started functioning in January 1865. After five months of installation and hard work, it produced only 200 kg of sodium carbonate daily, as against the 12 tons that were expected. There were numerous initial difficulties until Solvay's ammonium process of manufacture was perfected, replacing the old Leblanc 'black ash process'. Thanks to Solvay's efforts, and the improvement of his technical arrangement (such as using vertical columns for the absorption of ammonia), the daily production rose to 1500 kg by June 1866, and to 3 tons in 1867. The viability of the ammonium process had thus been demonstrated, and by 1869 the size of the plant was doubled.

Ernest Solvay immediately sought to give an international character to his enterprise. In 1872 he established a factory in the locality of Dombasle in France, which became one of the most important producers of soda in the world. During the same year the German chemist Ludwig Mond, who had been using the Leblanc process in England and who saw the advantages of the Solvay process, approached Solvay. In 1873 an English factory, operating under licence from Solvay and Company, was established at Northwich. Thereafter, factories were established at Syracuse, U.S.A., at Bernburg in Germany, at Ebensee in Austria, at Beresniki in Russia, and soon they numbered more than twenty.

Ernest Solvay continued to improve the technical aspects of the manufacturing process and to guide his developing enterprise in different countries. In all his efforts he enjoyed the support of his family, especially of his brother Alfred. Solvay's efforts were crowned with extraordinary success. The Solvay process was adopted all over the world, making him immensely wealthy. For Ernest Solvay, the pursuit of wealth was not an end in itself; it was a means to help society. He employed the power of his wealth and prestige in the service of humanity by promoting social and scientific projects and institutions. He devoted much attention to educational and social problems and contributed to various philanthropies and research organizations.

Ernest Solvay regarded science, in its various forms, as the key that would open the door to a richer life for man. On the occasion of the unveiling of Solvay's monument near the Université libre de Bruxelles, the French physicist Paul Langevin said: 'He attributed to ignorance the origin of most of the social conflicts, if not international conflicts. He regarded the diffusion of popular instruction, the creation of an educational system based on the interest of all concerned, as the ushering in of justice.'

Ernest Solvay had himself remarked once: 'Among the new paths of science, I undertook to follow three directions; three problems which, in my view, form a certain unity. First, a general problem of physics – the constitution of matter in time and space; then a problem of physiology – the mechanism of life from its most humble manifestations up to the phenomena of thought processes; and finally, the third one, a problem complementary to the first two – the evolution of the individual and that of social groups.'

Solvay sought to realize this programme in his life. In 1894 he endowed the Institut de Physiologie and the Institut des Sciences Sociales (which in 1901 became the Institut de Sociologie) at the Université Libre de Bruxelles. He also endowed, in 1903, a School of Commerce, and a workman's educational centre. In 1912 Ernest Solvay founded the Instituts Internationaux de Physique et de Chimie, concerning which Sir

William Pope remarked: 'In order to find a parallel example of a wise and princely endowment of a new foundation, supported by the genuine sympathy and personal assistance of the donor, one would have to look back to the 15th century, to the epoch when the Medici family made of Florence the most learned city in the whole world, not by the force of wealth alone but by the active collaboration of the genius of its members.'

Ernest Solvay opposed the amassing of 'unearned' inherited wealth. He noted that there were three major socio-economic classes in society: one, the producers; second, those who have authority over the first; and a third, which struggles to survive by serving the other two. Solvay thought that the well-being of both the individual and humanity would be advanced if all people became producers. Countries with higher productivity would help less advanced countries in increasing their productivity. He argued that there would be no need for money if all people were producers. He proposed a system of social accounting, *comptabilisme*, which would replace money; this would be a system of bookkeeping in which an individual's possessions would be entered into an account book at a national office, and when goods changed hands the value would be debited or credited in the account books of the two parties.

Ernest Solvay's works include: *Science contre religion* (1879); *Du rôle de l'électricité dans les phénomènes de la vie animale* (1894); *Le comptabilisme social* (1897); *L'Energétique considérée comme principe d'orientation rationnelle pour la sociologie* (1904); *Questions d'énergétiques sociales* (1910); and *Gravito-Matérialitique* (1911).

Throughout his life Ernest Solvay maintained an absolute simplicity, and had no love of luxury or honours. He died in Brussels on 26 May 1922, leaving behind an honoured memory of his life and works. Alone the Solvay Conferences on Physics and Chemistry would have perpetuated his memory in the world of science.

BIBLIOGRAPHY

1. *Florilège des Sciences en Belgique*, Académie Royale de Belgique, 1968. Biographical notice on E. Solvay by L. D'Or, pp. 385–406.
2. Paul Langevin's tribute to E. Solvay in *Hommage National à Ernest Solvay, Inauguration du Monument, 16 Octobre 1932*, Brussels, 1933.

2. THE ORIGIN OF SOLVAY CONFERENCES

Ernest Solvay was particularly attracted to the study of the structure of matter in the context of a theory of gravitation, and by 1887 he had developed his own ideas on this subject. He submitted his work to be read by 'three illustrious savants' – Houzeau, Stas and Hirn – who praised it for its 'originality, generality of the conceptions, and intimate knowledge of the subject'.

From 1887 to 1910 Ernest Solvay continued to develop his ideas on 'gravitation and self-gravitation' and on 'gravitation and matter (*gravito-matérialitique*)' with the help of his collaborators E. Tassel, E. Herzen, and G. Hostelet. During this period he pursued numerous scientific interests, but the 'fundamental considerations and theoretical deductions' of his '*gravito-matérialitique*' specially appealed to his imagination. He discussed, albeit qualitatively, the relationship between inertia and gravitation, and sought to understand the basic unity underlying natural phenomena, such as gravitation and electromagnetism.

In the spring of 1910 Solvay encountered Walther Nernst at the house of his collaborator Robert Goldschmidt in Brussels. Nernst, the former disciple of Ludwig Boltzmann and assistant of Wilhelm Ostwald, was then a professor of chemistry in the University of Berlin, director of the University's Second Chemical Institute, and one of the most distinguished physical chemists of the day. In 1889 Nernst had devised the theory of electric potential and conduction in electrolytic solutions, and in 1906 had developed the Third Law of Thermodynamics – according to which, at the absolute zero of temperature, the entropy of every substance in perfect equilibrium is zero; thus pressure, volume, and surface tension all become independent of temperature. He had also designed an electric piano and invented an electric incandescent lamp, Nernst's lamp, which used little current and required no vacuum. He had sold the patent of the latter invention to Siemens for a huge sum of money, thus laying the foundation of his wealth. He had enjoyed considerable academic and political influence in Göttingen before going to Berlin in 1905. In Berlin he was at the very top of German science.

At Goldschmidt's house in Brussels, Solvay mentioned to Nernst about his ideas on gravitation and the structure of matter, and wondered whether they could be brought to the attention of the great physicists. The physicists Solvay had in mind were men like Planck, Lorentz, Poincaré and Einstein. Solvay was also interested in the crisis that had been developing in classical physics by the advent of relativity and quantum theories.

Great entrepreneur that he was, Nernst saw in Solvay's desire to submit his work* to the attention of Europe's leading physicists the possibility of holding an international conference on the current problems of kinetic theory of matter and the quantum theory of radiation. The idea struck an immediate responsive chord in Solvay's mind, and he charged Nernst to explore it further with Planck, Lorentz, Einstein, and the other prominent physicists. Nernst was quick to pursue the idea immediately on his return from Brussels to Berlin.

Nernst and Planck had already discussed, early in 1910, the possibility of holding a conference dealing with the reform of the classical theory of matter and radiation in the light of the new ideas, although Planck had wondered whether the time was ripe and whether one should not wait another few years. Nernst's meeting with Solvay in Brussels gave fresh encouragement to holding the conference.

Nernst wrote to Planck about the kind of conference he had envisaged and Solvay's encouragement of it. Nernst's ideas had already taken shape. To these, Planck replied in a letter dated Grünewald, 11 June 1910. He wrote:

---

* Ernest Solvay, *Sur l'Etablissement des Principes Fondamentaux de la Gravito-Matérialitique*, Brussels, 1911.

Dear Colleague:

To the marginal notes that I have already made, with your permission, on your manuscript [of the proposal] allow me to add a few more generalities.

Your idea corresponds, fully and completely, to the problem whose solution is envisaged, and I can only associate myself with it with full conviction. However I am not able to hide my great concern about its execution. As I have already mentioned in my marginal notes, such a conference will be more successful if you wait until more factual material is available.

For practical reasons, in my view, a more decisive argument would suggest the postponement for another year. It is that the convocation of the conference is based on the presumption, which, by the way, is also emphasized in your proposal, that the current situation of the theory which has been created by the laws of radiation, [the problem of] the specific heat, etc., is full of holes, and intolerable for every true theoretician, making it therefore necessary to take a joint action and to find a remedy together. Now, based on my experiences, I am of the opinion that hardly half of the participants you have in mind are conscious of a sufficiently active concern for the urgent necessity of a reform [of the theory] to justify their coming to the conference. As for the oldsters (Rayleigh, van der Waals, Schuster, Seeliger) I shall not discuss in detail whether they will be excited by the thing. But even among the younger people the urgency and the importance of these questions has hardly been recognized. Among all those mentioned by you I believe that, other than ourselves, only Einstein, Lorentz, W. Wien and Larmor will be seriously interested in the matter.

However, let one or better two years go by, and we shall see how the gap which begins to open in the theory shall develop, and how finally those who still stand at a distance will be forced to join in. I don't believe that such a process can really be accelerated, the matter must and will take its course; and if you would then take the initiative for [holding] such a conference, it would attract the attention of a hundred times as many eyes and, above all, it would realize what I consider doubtful at the moment.

I need hardly mention that whatever is done in this regard I shall take the greatest interest, and I promise my fullest collaboration in any project of this kind. Because I can say without exaggeration that for ten years nothing in physics, without a break, has absorbed, excited and attracted me as these quanta of action.

With cordial greetings, yours

Planck

Nernst decided not to share either Planck's caution or his conservatism, and construed his letter as indicating full support. He took up personal contact with Lorentz and Knudsen, securing the assurance of their participation in the proposed conference. And just six weeks after hearing from Planck he wrote to Ernest Solvay proposing details of the project and the draft of an invitation which Solvay should send out:

Berlin, 26 July 1910

Respected Mr Ernest Solvay:

In this letter I am taking the liberty of submitting to you a proposal, and I would like to request you right away that if for any reason you would not wish to occupy yourself with it to consign this letter and the enclosure to the wastepaper basket. But, perhaps, the project might interest you a little, and I am therefore sending it to you, honoured Mr Solvay, especially because I know how much you are interested in all the important general problems, and also because I would like to propose to you that for an international conference Belgium, in particular Brussels, seems to me to be better suited than Berlin, Paris or London. The details are contained in the enclosed draft of an invitation. My friend Goldschmidt, with whom I have discussed the project in detail, would also be glad to provide further information.

The 'conseil' can function most effectively only if those researchers take part in it who are closest to the subject and have a deep interest for the problems in question. In the following I am taking the liberty of proposing a tentative list:

Honorary President: Solvay

President: Lord Rayleigh (England)

Secretaries: Dr Goldschmidt and a younger person. Members: Einstein (Switzerland); Knudsen (Denmark); Hasenöhrl (Austria); Lorentz (Holland); Langevin, Perrin (France); Van der Waals (Holland); Larmor, Jeans, Schuster, J. J. Thomson, Rutherford (England); Nernst, Planck, W. Wien, Röntgen, Seeliger (Germany).

Thus far I have talked quite confidentially, without any mention of time and place, only with Lorentz, Knudsen and Planck. All of these gentlemen will take part willingly, and based on these initial contacts the holding of the conference is not subject to doubt.

It would be desirable to allocate a reasonable sum for travel expenses in order to assure the participation of all the invitees.

Allow me to presume that, apart from the lectures mentioned in the enclosed [programme], you would yourself open the conference with an address of welcome and also perhaps give a lecture.

I presume and hope that the conference, which will last about one week (for instance, right after Easter 1911), will represent a milestone in the history of science, especially when the lectures with corresponding discussions are later on published in a volume. It also appears that there has hardly been a more appropriate time than the present when such a conference could more favourably influence the development of physics and chemistry, because we stand [now] at a decisive turning point in the development of our latest theoretical conceptions.

I request you again to treat this proposal as 'à faire ou à laisser'. I remain,

Most respectfully yours,
Walther Nernst

Together with this letter, Nernst attached, marked 'confidential', the draft of an invitation [in the name of E. Solvay]. It read:

Invitation to an 'International scientific conference to elucidate certain current questions of the kinetic theory'.

It appears that we find ourselves at present in the midst of an all-encompassing re-formulation of the principles on which the erstwhile kinetic theory of matter has been based.

On the one hand, this theory leads to a logical formulation – which nobody contests – of a radiation formula whose validity is contradicted by all experiments; on the other, there follow from the same theory certain results on the specific heat (constancy of the specific heat of a gas with the variation of temperature, the validity of Dulong and Petit's law up to the lowest temperature), which are also completely refuted by many measurements.

As Planck and Einstein in particular have shown, these contradictions disappear if one places certain limits (doctrine of energy quanta) on the motion of electrons and atoms in the case of their oscillations around a position of rest. But this interpretation, in turn, is so far-removed from the equations of motion of material points employed until now, that its acceptance would incontestably lead to a far-reaching reformation of our erstwhile fundamental notions.

The undersigned – though a stranger to such special questions because of his other activities, but aroused with a sincere enthusiasm for all questions the study of which deepens our understanding of nature – believes that a personal exchange of views on these problems between the researchers who are more or less directly concerned with them, would lead, by considered critique, if not perhaps to a definitive resolution of these questions, [at least] to point the direction of their solution. A great step in this direction of the orderly continuous development of atomistics would have already been taken if one could clearly establish as to which of the interpretations of the erstwhile kinetic theory are in accord with observations and which others should undergo a complete transformation.

With this purpose in mind the undersigned invites you, among others, to participate in a scientific conference which, in a very restricted group, will bring together some of the most eminent specialists to Brussels on ....

On receiving acceptance [of this invitation], the following subjects will be distributed among the participants of the conference, on which [after every lecture] an extensive discussion will take place:

1. Derivation of Rayleigh's radiation formula.
2. To what extent does the kinetic theory of ideal gases agree with experiments?
3. The kinetic theory of specific heats due to Clausius, Maxwell and Boltzmann.
4. Planck's radiation formula.

5. Theory of energy quanta.
6. Specific heat and quantum theory.
7. Consequences of quantum theory for a series of physico-chemical and chemical problems.

It is intended that the reports and discussions [of the conference] will be gathered together in a volume and published.

(Signed) E. Solvay

Nernst had provided a list of eighteen distinguished participants for the conference, and proposed the name of Lord Rayleigh for the chairmanship.

Solvay indicated his approval of Nernst's proposals in a letter dated 5 August 1910, but wondered whether the date of the conference could not be postponed beyond Nernst's suggestion of 'immediately after Easter 1911'. In a further letter dated 21 September 1910, Solvay conveyed to Nernst his choice of October 1911 for the conference. Nernst's enterprise was beginning to take shape, and, in his reply, he wrote:

Berlin, 27 November 1910

Respected Mr Ernest Solvay:

I thank you most sincerely, with regrets for my great delay, for your kind letter of 21 September. I am writing only today because the detailed messages which you had sent me through our common friend Goldschmidt, because of unexpected delay, have reached me just a few days ago. I am very happy that you are favourably disposed towards the idea of the conference. The period during October 1911 selected by you would just suit perfectly. If the invitations are sent about six months earlier the various participants would have, on the one hand, sufficient time to prepare their reports; on the other, the interval of time [until the conference] would not be too long to give rise to unnecessarily protracted discussions about the conference among the scientists, something which is always better to avoid. For the same reason, it would be better to send the invitations marked 'confidential'.

Permit me to reiterate that you should not pay any special attention to my tentative proposal for the convocation [of the conference], but rather give it entirely the form you like. I would like to express my wish that, in the invitations, you should not mention me as the initiator of the idea of the conference. I would prefer that you should not refer to me at all – and, if at all necessary, then only in your address of welcome. Since Goldschmidt will be coming to Berlin soon, I suggest for your convenience that you might confide in him any other wishes you may have.

Very soon I expect to send you certain publications dealing with the theme of the conference. I shall also be very pleased to learn more about your own ideas which you kindly indicated to me in your letter; then by the time of the conference there will certainly be ample opportunity of studying them in detail.

Very sincerely yours,

W. Nernst

Ernest Solvay answered this letter immediately.

Brussels, 7 December 1910

Dear Mr Nernst:

I entirely agree with your good letter of 27 November.

If something would need changing in the letter of invitation to the conference, it would be quite insignificant, and I shall mention it to you in due course. We have enough time before us, and at the moment I am rather occupied. When I am [more free], I shall try to read the publications which you intend to send me, and I am sure they would interest me greatly.

I work, in an independent fashion, on a kind of research on the objective basis of physics. My method is deductive, and I am sure you know its dangers better than I, and you have expressed them correctly. Naturally I start from a phenomenon that I consider as having been defined rigorously. I

am advancing, but I have not yet arrived at a point where one could say that I could talk to you about my results – they are still tentative. I ask you to give me time until we have to send out the invitations.
I have hardly seen Goldschmidt in particular, but I am sure I shall see him before his next departure.

Cordially yours,

E. Solvay

Please remember me to Madame Nernst.

There was little else for Nernst to do but wait until Ernest Solvay would recover from his myriad preoccupations and give the go-ahead to send out the invitations for the conference. But he kept in touch with the intermediary, Robert Goldschmidt. During the following weeks Solvay discussed the details of the plans of the conference with Goldschmidt, and wrote to him in March:

12 March 1911

My dear Goldschmidt:

I have now thought about the conference since last Sunday, as I had promised you. I have decided – in accordance with the conditions that you had indicated to me, i.e. charging you with the responsibility of all the invitations, and all that would concern the meeting later on, while I shall have only to sign, approve, etc. – that we may proceed for next October as envisaged.

I would like to make an observation regarding the [intended] invitees. It seems to me that France is much less represented compared to England, Germany and Austria, and one could certainly add Marcel Brillouin among the French, and perhaps eliminate one Englishman and an Austrian, if not also a German. Have a look at it with Mr Nernst; I would not like to offend him, but I would still have to defend my selections.

I shall write to Mr Nernst a little later on concerning what I would myself like to say at the conference.

Wishing you rest and relaxation,

Very cordially yours,

E. Solvay

On 13 March 1911 Edouard Herzen, Ernest Solvay's scientific collaborator at the laboratory of Solvay and Co., wrote a memorandum to Solvay proposing the final text of the letter of invitation. Herzen suggested that in a personal covering letter Solvay should give a list of invited scientists, offering to cover their expenses for travel and stay (of about one week), which for each participant should come to about 1000 Belgian francs.* Compared to the earlier list of names, this memorandum now mentioned the name of Planck[†] as the President of the Conference, and divided the other names into two categories as follows:

Country	First Category	Second Category
England	Lord Rayleigh	Jeans
	Larmor	Schuster
	J. J. Thomson	
	Rutherford	

* The Société Générale de Banque S. A. has informed me that this sum in today's currency (1975) would be equivalent to 57000 Belgian francs.
† Ernest Solvay invited H. A. Lorentz (Leyden) to be the President of the Conference.

Germany	Nernst	Seeliger
	W. Wien	
	Röntgen	
France	Langevin	
	Brillouin	
	Perrin	
Austria	Einstein	Hasenöhrl
Holland	Van der Waals	
	Lorentz	
Denmark	Knudsen	

Herzen suggested that those in the second category could well be left out if necessary, that Goldschmidt should be the Secretary, and that Nernst would propose that Solvay should accept to be the Honorary President of the Conference.

To Nernst's original draft of the invitation, Herzen added in closing: 'The reports received before 1 September [1911] will be distributed among the participants as soon as their copies are made. These papers together with the discussions based on them will be published in a volume later on.'

And, as a postscript, the invitation would say: 'Not being a man of science I do not expect to participate in the detailed discussions of the specialized subjects indicated above. However, in the opening address I shall discuss certain points of view to which I have been led; in particular, the phenomena of gravity in relation to those of physics and chemistry. E.S.'

The invitations, signed by Ernest Solvay, and marked 'confidential', were sent to the prospective speakers on 9 June 1911, and on 15 June to the other participants. H. A. Lorentz was invited to assume the presidency of the conference.

The acceptances came immediately, showing great enthusiasm. As Solvay himself noted, 'It is a pleasure to read them.' There was not a single refusal. Nernst wrote to Solvay: 'You will be glad to learn that almost all the acceptances for the conference have welcomed the idea with enthusiasm. Rubens told me that he found your kind invitation just grand. Rutherford declares that the idea is excellent. Kamerlingh Onnes has especially asked me to inform you of his appreciation for the imaginative new direction which your beautiful idea of the conference would give rise to. And Einstein is just delighted with the whole thing.'

Some extracts from the letters received by Solvay give an impression of the appreciation his invitation had aroused. 'I applaud the pleasant initiative that you have taken,' wrote Henri Poincaré. Marcel Brillouin informed Solvay: 'The initiative which you have taken appears to me of the highest interest. It is in a meeting of limited [number of participants] such as the one you envisage that it is possible to exchange profitably one's views about new ideas, which are far from being clear. I therefore accept your generous invitation with pleasure.'

W. Wien wrote: 'You have found a completely new way of pushing scientific progress forward. Indeed the questions that will be treated at the conference are extremely difficult, and it is impossible, it seems to me, for a single person to find their solution. But the united efforts of the scientists of all nations, when they get together at such an excellent occasion for discussion, would provide the best chance of finding a way around the large number of molecular questions.'

J. D. van der Waals expressed his great interest in the conference, but wondered whether his health would permit him to attend. Just before the conference he decided that he was unable to make the trip.

Einstein, writing to Nernst from Prague on 20 June 1911, informed him that he accepted the invitation with pleasure. 'I shall gladly prepare the report assigned to me. The whole enterprise appeals to me, and I hardly doubt that you are the spirit behind it.'*

Lorentz addressed himself to Solvay from Leyden on 3 July 1911: 'Although I have already, in my reply to Professor Nernst, accepted the invitation to the scientific conference that will take place in Brussels at your initiative, I would like to thank you sincerely for the honour you have done me by calling upon me. Allow me to tell you also that I greatly appreciate your generous efforts for the progress of science, and your project has my complete sympathy. In fact, this meeting of a limited number of physicists, all of whom are seriously occupied with the important questions on the programme, would serve without doubt greatly to elucidate and to state the difficulties precisely, as well as to prepare the way for their solution. I am sure that this combined work, in which we shall be very glad to have you yourself take part, would be for us extremely useful, and it would stimulate us in new experimental and theoretical researches. As for myself, I shall be very happy if I should be able to contribute a little to the success of your plans.'

Jean Perrin was particularly enthusiastic. He wrote to Nernst: 'I regard it as a very great honour to be associated with the savants, whom I admire the most, in assisting toward the solution of a problem which might well be the most important of modern physics.'

Writing also to Nernst, Joseph Larmor noted: '... I have to confess that, do what I may, I cannot find time to keep up with recent progress in the subject in which you have contributed so much of the essential experimental data.'

In his response, date 13 July 1911, Lord Rayleigh wrote: '... I should much like to meet some of the gentlemen invited. But, being a very poor linguist, I am not effective on occasions of this sort.'

The care with which Nernst had gone about the planning of the conference is evident from a letter he wrote to Solvay on 26 August 1911. Other than a reception by the King, he hoped that there would not be too many others. He pointed out that the meetings would be very tiring for the participants, and a free evening of rest would be indispensable. He also thought that an intimate inaugural evening with all the participants together would allow them to get to know each other.

---

* Einstein was certainly amused with the idea of the Conference. He referred to it as the 'Witches' Sabbath [*Hexensabbat*] in Brussels' in a letter to his friend Michele Besso on 21 October 1911 and again two days later. [See *Albert Einstein–Michele Besso, Correspondance 1903–1955*, Ed. P. Speziali, Hermann, Paris 1972.]

On 27 September, Ernest Solvay wrote to King Albert, sending him the details of the planned scientific conference. The King replied on 29 September 1911, from the Château de Laeken:

Dear Mr Solvay:

I sincerely thank you for kindly sending me the dossier concerning the scientific conference which will take place soon in Brussels.

It is with great pleasure that I have taken note of it. This initiative seems to me to be remarkable, and would lead to fruitful results.

I shall be particularly happy to interest myself in these important scientific meetings, and to follow attentively this occasion devoted to the higher science .... With my warm wishes that your project may succeed to the complete satisfaction of the eminent scientists who would be coming together,

Yours affectionately,

Albert

The participants in the Conference, the first 'Conseil Solvay', stayed at the Hotel Metropole in the heart of Brussels. Maurice de Broglie arrived on 27 October; Martin Knudsen and Emil Warburg on 28 October. All the other participants arrived in the Belgian capital on 29 October 1911, and the first reception took place the same evening at the Metropole. The first Solvay Conference in physics took place in Brussels from 30 October to 3 November 1911.

The proceedings of the Conference were edited by P. Langevin and M. de Broglie under the title: 'La théorie du rayonnement et les quanta. Rapports et discussions de la réunion tenue à Bruxelles, du 30 octobre au 3 novembre 1911, sous les auspices de Monsieur E. Solvay.' It was published by Gauthier-Villars, Paris, 1912.

FIRST SOLVAY CONFERENCE 1911

GOLDSCHMIDT   PLANCK   RUBENS   LINDEMANN   HASENOHRL
NERNST   BRILLOUIN   SOMMERFELD   DE BROGLIE   HOSTELET
                  SOLVAY   HERZEN   JEANS   RUTHERFORD
                              LORENTZ   KNUDSEN   WIEN   EINSTEIN   LANGEVIN
                                             WARBURG   Madame CURIE   POINCARÉ   KAMERLINGH ONNES
                                       PERRIN

2

Radiation Theory and the Quanta*

1. INTRODUCTION

The first Solvay Conference on Physics took place in Brussels from 30 October to 3 November 1911. Its general theme was 'The Theory of Radiation and the Quanta'. The President of the Conference was H. A. Lorentz (Leyden, Holland), and the participants were:

Germany: W. Nernst (Berlin), M. Planck (Berlin), H. Rubens (Berlin), A. Sommerfeld (Munich), E. Warburg (Berlin-Charlottenburg), W. Wien (Würzburg).

England: J. H. Jeans (Cambridge), E. Rutherford (Manchester).

France: M. Brillouin, Madame Curie, P. Langevin, J. Perrin, H. Poincaré (all from Paris).

Austria: A. Einstein (Prague), F. Hasenöhrl (Vienna).

Holland: H. Kamerlingh Onnes (Leyden).

Denmark: Martin Knudsen (Copenhagen).

The Scientific Secretaries of the Conference were: R. Goldschmidt (Brussels), M. de Broglie (Paris), and F. A. Lindemann (Berlin).

As we have already noted, the theme of the first Solvay Conference in 1911, Radiation Theory and the Quanta, indicated the central question in physics in those days. The most important advances in physics in the 19th century were perhaps the development of Maxwell's electromagnetic theory, which offered a far-reaching explanation of radiative phenomena, and the statistical formulation and interpretation of thermodynamics which culminated in Boltzmann's relation between the entropy and probability of the state of a complex mechanical system. However, as Rayleigh's analysis of black-body radiation had shown, the physical and mathematical description of the spectral distribution of cavity radiation in thermal equilibrium presented unsuspected difficulties.

Planck's discovery of the universal quantum of action in 1900 was a turning point in the development. It revealed a feature of 'discreteness' in atomic processes, which was completely foreign to classical physics. Einstein not only emphasized the apparent paradoxes which had arisen in the detailed description of the interaction between matter and radiation, but provided support for Planck's ideas by his investigation of the specific heat of solids at low temperature. Einstein also introduced the idea of light

---

* *La Théorie du Rayonnement et les Quanta*, Rapports et Discussions de la Réunion tenue à Bruxelles, du 30 Octobre au 3 Novembre 1911, Publiés par MM. P. Langevin et M. de Broglie, Gauthier-Villars, Paris, 1912.

quanta or photons as carriers of energy and momentum in elementary radiative processes, and successfully used it in his explanation of the photoelectric effect.

The purpose of the first Solvay Conference was thus two-fold: first, there was the need to examine whether classical theories (molecular-kinetic theory and electro-dynamics) could, in some undiscovered ways, provide an explanation of the problem of black-body radiation and of the specific heat of polyatomic substances at low temperatures; secondly, to consider phenomena in which the theory of quanta could be successfully used.

The discussions at the Conference were initiated after the report of H. A. Lorentz. He developed the arguments based on classical ideas leading to the principle of the equipartition of energy between the various degrees of freedom of a physical system, including not only the motion of its constituent material particles but also the normal modes of vibration of the electromagnetic field associated with the electric charge of the particles. These arguments, which followed the lines of Rayleigh's analysis of thermal radiative equilibrium, led to the well-known paradoxical result that no temperature equilibrium was possible since the whole energy of the system would be gradually transferred to electromagnetic vibrations of steadily increasing frequencies.

James Jeans suggested that the only way to reconcile radiation theory with the principles of statistical mechanics was that under experimental conditions one did not have to deal with a true equilibrium but with a quasi-stationary state, in which the production of high frequency radiation would escape notice. The acuteness with which the difficulties in radiation theory were felt was brought home by a letter from Lord Rayleigh, which was read at the conference, in which he recommended that Jeans' suggestion should be given careful consideration. But on closer examination it became evident that Jeans' argument could not be upheld.

After the reports of Warburg and Rubens on the experimental evidence supporting Planck's law of temperature radiation, Planck himself gave an exposition of the arguments which had led him to the discovery of the quantum of action. Planck was deeply concerned with the problems of harmonizing this new feature with the conceptual framework of classical physics. He emphasized that the essential point was not the introduction of a new hypothesis of energy quanta, but rather a reformulation of the very concept of action. Planck expressed the conviction that the principle of least action, which had also been upheld in the theory of relativity, would serve to guide the further development of quantum theory.

Walther Nernst, in his report on the application of quantum theory to various problems of physics and chemistry, considered the properties of matter at very low temperatures. Nernst remarked that his theorem regarding the entropy at absolute zero, of which he had made important applications since 1906, now appeared as a special case of a more general law derived from the theory of quanta.

Kamerlingh Onnes reported on the discovery of superconductivity of certain metals at very low temperatures. This phenomenon presented a great puzzle, and would find its explanation only several decades later.

Nernst's idea of the quantized rotation of gas molecules, which would receive confirmation in the measurements of the fine structure of infra-red absorption lines, was commented upon from various sides at the conference.

Paul Langevin reported on the variation of the magnetic properties of matter with temperature. He made special reference to the idea of the magneton, which had been introduced by Pierre Weiss to explain the remarkable numerical relations between the strength of the elementary magnetic moments of atoms deduced from the analysis of his measurements. Langevin showed that the value of the magneton could be approximately derived on the assumption that the electrons in atoms were rotating with angular momenta corresponding to a Planck quantum of action.

Arnold Sommerfeld discussed the production of X-rays by high speed electrons as well as problems involving the ionization of atoms in the photoelectric effect and by electronic impact. Sommerfeld considered the existence of Planck's quantum of action as fundamental for any approach to questions of the constitution of atoms and molecules.

Martin Knudsen reported on kinetic theory and the experimental properties of perfect gases, while Jean Perrin gave a detailed report on the proofs of molecular reality and the counting of atoms by the use of statistical fluctuations.

The last report of the conference was given by Albert Einstein. He summarized many applications of the quantum concept and dealt in particular with the fundamental arguments used in his explanation of the anomalies of specific heats at low temperatures.

We shall now treat the individual reports presented at the first Solvay Conference.

2. APPLICATION OF THE ENERGY EQUIPARTITION THEOREM TO RADIATION

(H. A. Lorentz)

In the first scientific lecture at the conference, Lorentz declared right away that among physical phenomena there were hardly any more mysterious ones than those of thermal and light radiation. He discussed Kirchhoff's empirical laws of black-body radiation and the 'felicitous' application of the principles of thermodynamics by Boltzmann and Wien, which had led to experimentally verifiable general laws. 'But, in spite of all this', said Lorentz, 'the ideas which prevailed until the end of the last century could not explain why a piece of iron, for instance, [certainly absorbs but] does not emit radiation at ordinary temperatures.' With this he undertook to point out the difficulties which arise in the classical theory of radiation.

Lorentz gave a brief account of Rayleigh's radiation law[1], which had been established by assuming the validity of the equipartition theorem. But this law, which is quite satisfactory for long wavelengths (up to infra-red), is in complete disagreement with observations in the short wavelength region of the black-body spectrum. Lorentz asked if there was any means of avoiding the equipartition theorem in this particular case or, for that matter, generally.

In order to answer such questions Lorentz was led to analyze the assumptions made in the establishment of the equipartition theorem. He recognized two assumptions: first, the use of statistical physics, i.e. the existence of most probable values; second, the requirement of stationarity for the ensemble under consideration. He thought that the first assumption was certainly true even though it could not be completely proved; while the stationarity of the ensemble follows from Liouville's theorem, which in itself is a consequence of Hamilton's equations. Lorentz was thus led to examine whether the phenomena under consideration could be described by the Hamiltonian formalism.

The problem thus was to define the Hamiltonian for a system consisting of ether and matter in a cavity surrounded by perfectly reflecting walls. Lorentz thought that it was not at all necessary to give a mechanical explanation of the electromagnetic phenomena; the only requirement was that the equations which determine these phenomena be written in the Hamiltonian form. Maxwell's equation for the system, together with the boundary conditions (at the walls of the cavity) for the fields, lead to

$$\delta \int_{t_1}^{t_2} (\mathscr{L} - \mathscr{U})\, \mathrm{d}t = 0 \tag{1}$$

where $\mathscr{U}$ is the electrical energy and $\mathscr{L}$ the magnetic energy, a result which can be easily deduced from Maxwell's equations with the given boundary conditions.

In order to pass from this result to Hamilton's equations, Lorentz introduced a coordinate system giving the positions of the particles and the electric field. With q_1 and q_2, the coordinates of the particles with and without charge, respectively, and q_3, the coordinates of the field, he obtained,

$$\mathscr{L} = \mathscr{L}_0 + \frac{fgh}{64\pi^2 c^2} \sum \lambda^2 \dot{q}_3^2 + \sum_{ij} l_{ij} \dot{q}_{2i} \dot{q}_{3j}, \tag{2}$$

where $\mathscr{L}_0$ is a homogeneous quadratic function of $\dot{q}_1$ and $\dot{q}_2$; f, g, h, are the lengths of the parallelepiped [cavity] under consideration; λ, a wavelength given by

$$\lambda = \frac{2}{\sqrt{\left(\dfrac{u^2}{f^2} + \dfrac{v^2}{g^2} + \dfrac{w^2}{h^2}\right)}} \tag{3}$$

with u, v, w as integers; and l_{ij} are some functions of the $\dot{q}_{2i}$.

Lorentz assumed that the electrons are spheres of radius R, carrying a surface charge e, and that the system has an infinite number of degrees of freedom. In order to avoid the difficulties that would result from this, he introduced the constraint that all fields disappear for $\lambda < \lambda_0$, and that physical results would be obtained in the limit $\lambda_0 \rightarrow 0$. For this fictitious system he showed that the Hamiltonian equations are evidently valid. Moreover, since most of them are cyclic, the equipartition theorem is valid, and from this by an easy reasoning he could obtain Rayleigh's radiation formula.

The important point was that the formula thus obtained by Lorentz was independent of the nature of the particles and took into account all the vibrational and translational modes of the electrons. Also, implicitly, the influence of radiation on the motion of the electrons and the change they produce on the field (a kind of Doppler effect) had been considered.

Lorentz applied his result to the calculation of the mean square velocity of the electron, $\overline{v^2}$, and readily obtained

$$\tfrac{1}{2}m'\overline{v^2} = \tfrac{3}{2}kT, \tag{4}$$

where

$$m' = m - \frac{e^2}{6\pi^3 c^2 R^2} \int_{\lambda_0}^{\infty} \sin^2 \frac{2\pi R}{\lambda} \, d\lambda. \tag{5}$$

With the assumption that the lower limit, λ_0, is much larger than the diameter of the electron, one may put $\sin(2\pi R/\lambda) \simeq 2\pi R/\lambda$, and the last term becomes $2e^2/3\pi c^2 \lambda_0$. This term is very small compared to m, and Lorentz thus reproduced the well-known result,

$$\tfrac{1}{2}m\overline{v^2} = \tfrac{3}{2}kT. \tag{6}$$

This result indicated that, for λ_0 much larger than the diameter of the electron, it was possible to consider the electrons as molecules. However, for $\lambda_0 \sim R$, in the limit $\lambda_0 \to 0$, Lorentz showed that $m' = 0$, leading to an infinite value for $\overline{v^2}$.

Lorentz recalled that J. D. van der Waals Jr. had once pointed out to him that when electrons are considered as massless, their speeds are fully determined by the knowledge of the electromagnetic field. However, since the initial velocities are not given by Hamilton's equations, it would be impossible to build a canonical ensemble. The only way to avoid this difficulty was to introduce the restriction $\lambda < \lambda_0$. In this way the electron's mass would remain finite and the difficulty pointed out by Van der Waals Jr. would disappear.

This manner of dealing with the problem was not free from objections, but it was physically plausible because the wavelengths which occur experimentally are much smaller than λ_0. Lorentz showed that Jeans' ideas concerning an infinitely slow process, leading to thermal equilibrium after a very long time, were not realistic.

Lorentz pointed out that the distribution of radiation must depend on two constants, one giving the total intensity and the other determining the position of the wavelength maximum. He explained that Planck's constant, h, which is a universal constant, will certainly not arise from classical considerations.

Lorentz then discussed the problem of black-body radiation from another angle. The energy is never distributed absolutely uniformly; on the contrary, the intensity of the electromagnetic field will vary irregularly from one point to another. This irregularity arises from the interference of incoherent waves, and these interferences may have great importance if the radiation possesses a certain *structure*. Lorentz considered the following example: Let a small mass m be placed in a space occupied by black-body radiation. It will experience non-isotropic pressure and describe motions similar to the Brownian motion. As Einstein had shown, the amplitude of the motion is a function of the temperature, leading one to suppose that this would also be the case for m. Assuming, with Einstein, that the mass m can move only in the X-direction, having speeds v and v' at two instants separated by an interval τ (τ being large compared to the periods of vibration, and yet sufficiently small, so that the difference $v - v'$ is small), Lorentz obtained

$$\tfrac{1}{2}m\overline{v^2} = \frac{\overline{\chi^2}}{4\mathscr{A}\tau}, \tag{7}$$

where $\mathscr{A}$ is a constant and χ, the part of the force exerted on the mass due to the inequality of the field strength at different points.

By applying this result to Planck's linear resonator Einstein and Hopf had obtained a value for $\tfrac{1}{2}m\overline{v^2}$ considerably less than $\tfrac{1}{2}kT$. If one could be sure that this is what really happens in a resonator, one could conclude the existence of a *structure* for radiation, such as the one required by the hypothesis of energy elements [quanta].

For a single electron Lorentz obtained

$$\tfrac{1}{2}m\overline{v^2} = \frac{15}{64\pi}\frac{\displaystyle\int \lambda^4 F^2 \, d\lambda}{\displaystyle\int F \, d\lambda}, \tag{8}$$

where $F(\lambda, T)$ is the black-body distribution function. By taking F consistent with the equipartition theorem he obtained

$$\tfrac{1}{2}m\overline{v^2} = \tfrac{15}{8}kT. \tag{9}$$

If F is taken as the Planck distribution function, then

$$\tfrac{1}{2}m\overline{v^2} = 0.315 \, kT. \tag{10}$$

By considering the inequalities of the interferences, Lorentz concluded that the excitation energy transmitted to the electrons by black-body radiation cannot give the value $\tfrac{3}{2} \, kT$.

Lorentz also presented the results of other cases he had computed, but none of them could explain the black-body radiation. He pointed out that all mechanisms of explaining the structure of black-body radiation would lead to the same result as obtained from the equipartition law.

DISCUSSION

In the discussion following Lorentz' report, Brillouin wondered whether the partition of energy could be done by avoiding a certain number of frequencies which are unsuitable. According to Lorentz, such phenomena are also included in the Hamiltonian formalism.

The expression for the energy is always a sum of squares, and Brillouin asked whether it could be made subject to physical constraints. To which Poincaré answered: 'The old [classical] theories are, by definition, those that admit Hamilton's equations without restriction, and the arguments of Mr. Lorentz show that all of them lead to the same result.'

Langevin presented a long calculation, obtaining a result which partly contradicted Lorentz's conclusion, to which Lorentz was able to point out that this was because Langevin had not included the term representing the light pressure. Poincaré agreed with this and showed that indeed it was Langevin's formula which had contradictions.

Wien wondered whether it was possible to find a system of proper vibrations which does not satisfy Hamilton's equations, and yet does not have any discontinuous changes. Thus far, he said, only those systems of proper vibrations had been proposed as models for spectral lines which satisfied Hamilton's equations. Wien thought that, Liouville's theorem being more general than Hamilton's equations, it should be possible to consider more general systems than just those obeying the latter.

REFERENCE

1. Lord Rayleigh, 'Remarks upon the law of complete radiation', *Phil. Mag.* 5th series, **59**, 539 (1900).

3. LETTER FROM LORD RAYLEIGH AND ITS DISCUSSION

Lord Rayleigh, who, in June 1900, had been the first person to apply the classical molecular kinetic theory to the problem of black-body radiation, was not able to attend the Solvay Conference. He thought that he did not have more to contribute to the subject than what he had already written, except certain clarifications. This he did in a letter addressed to Nernst, which was read at the conference following Lorentz's report.

Rayleigh emphasized the difficulty that arises in employing generalized coordinates. The possibility of representing the state of a body by a finite number of appropriate coordinates (less than the total number of molecules), rests on the hypothesis that the body can be considered as rigid or incompressible – in any case, simplified. The justification, satisfactory in most cases, of this hypothesis is that a displacement with respect to the given normal state would correspond to a much larger change of the potential energy than can be produced under the action of the forces present. However, the law of equipartition requires that each degree of freedom must have its corresponding kinetic energy. If we first consider an almost rigid elastic body, the corresponding vibrations play their full role which does not diminish with the increment of rigidity. From this point of view, the simplification disappears, which is to say that the method of generalized coordinates cannot be applied, and the reasoning becomes contradictory.

One might regard this defect, Rayleigh said, as an argument in favour of Planck's views, that the laws of dynamics in their usual form cannot be applied to the particles of the body. Rayleigh did not like this solution of the difficulty. Of course, he did not mind that one should pursue the consequences of the theory of energy quanta, because 'this method has already led to interesting results, thanks to the ingenuity of those who have applied it. But I find it difficult to regard it [the theory of energy quanta] as an image of reality.'

Rayleigh drew attention to a gas of diatomic molecules. Under the action of collisions a diatomic molecule acquires, easily and rapidly, a rotational motion. Why does a vibration not arise along the line joining the two atoms? 'If I understand rightly,' Rayleigh said, 'Planck's response is that it is because of the rigidity of the connection between the two atoms; the amount of energy that could be acquired in each collision falls below the necessary minimum, and, therefore, nothing is absorbed, an argument which appears to be really paradoxical.'

On the other hand, Rayleigh pointed out, Boltzmann and Jeans would consider that it was only a matter of time, and that vibrations necessary for a complete statistical equilibrium could not be established before thousands of years. 'Jeans' calculations seem to show that there is nothing arbitrary in this view. I would like to know whether it is contradicted by precise experimental facts. As far as I know,' wrote Lord Rayleigh, 'the usual laboratory experiments have not demonstrated anything decisive in this matter.'

DISCUSSION

Nernst pointed out that one had never observed continually increasing values in the measurement of specific heats, in particular for gases which do not obey the equipartition law, and on which very precise experimental data were now available. Using the explosion method, one had found that the time necessary for the variation of temperature was only a few milliseconds. From thermodynamics it follows that the melting point and the vapour pressure would be considerably modified if the specific heat and, therefore, the energy content will change with time; but one has never noticed a difference in the melting point between natural minerals and synthetic compounds. One must suppose that the equilibrium condition required by the equipartition theorem 'would not arise for four hundred million years', while an equilibrium of energy is obtained immediately; it is hardly probable. In numerous organic compounds, all with extremely small atomic heats, one has never found a variation of the melting point that would suggest a further variation of the energy.

One could hold the opinion, said Nernst, that a part of the energy is absorbed in such a long time as to be of no interest to the experimentalist. As for the energy that plays an actual role, the classical theory of equipartition is not sufficient. One has to look for another theory, as for instance of *quanta*.

Madame Curie wondered whether one could suppose, in the context of the usual kinetic theory, that the molecules in a diatomic gas are absolutely rigid but that this rigidity disappears progressively in the more condensed state. Kamerlingh Onnes pointed out that such questions had been considered in Van der Waals' theory of the equation of state, but that in a majority of cases it would be impossible to deduce the incompressibility of molecules, because the sharpness of the effect of the incompressibility of individual molecules is lost because of the presence of other complex phenomena.

4. KINETIC THEORY OF SPECIFIC HEAT ACCORDING TO MAXWELL AND BOLTZMANN

(J. H. Jeans)

James Hopwood Jeans gave a report on the Maxwell–Boltzmann kinetic theory of specific heat. For a system of N identical particles of mass m, and average energy E, at temperature T, the specific heat at constant volume is given by

$$C = \frac{1}{Jm} \frac{dE}{dT}, \tag{1}$$

where J is the mechanical equivalent of heat. In the derivation of this formula, it is assumed that N and m remain constant with the change of temperature. Experimentally one finds that dE/dT is constant for a large domain of the variation of temperature. Putting $dE/dT=\frac{1}{2} Rs$, the above formula becomes,

$$E = \tfrac{1}{2}RTs, \qquad C = \tfrac{1}{2}\frac{R}{Jm}s, \tag{2}$$

where R is the gas constant and s takes the values 3, 5 or 6 according as one considers mono-, di-, or tri-atomic molecules (or solids). Dulong and Petit's law deals with the approximate constancy of s for solids, the value $s=6$ corresponding to the atomic heat 5.95.

Maxwell and Boltzmann had given an explanation of these regularities by deriving their theorem of the equipartition of energy. Briefly, their explanation is that the value of E given by the above formula corresponds to a contribution of $\frac{1}{2} RT$ for each effective square term in the energy of a particle, where s is an integer giving the number of these square terms. The equipartition theorem, on which this explanation is based, says that each effective square term makes a contribution of exactly $\frac{1}{2} RT$ to the average energy E.

As in many other theorems, the difficulties lie not so much in its demonstration, but in the conditions that must be imposed so that the proof should be valid. In the case of the equipartition theorem, it is the exact significance of the expression 'effective square term' that must be defined. Jeans reproduced the demonstrations of the equipartition theorem as given by Maxwell and Boltzmann, and gave a simple proof himself.

If a dynamical system is represented by the coordinates $\theta_1, \theta_2,..., \theta_n$, then for a large class of phenomena the total energy can be written as

$$W = \alpha_1\theta_1^2 + \alpha_2\theta_2^2 +\cdots+ \alpha_n\theta_n^2 +\cdots, \tag{3}$$

where α_i are constants. Or, since θ_i are defined up to a constant,

$$W = \theta_1^2 + \theta_2^2 +\cdots+ \theta_n^2 +\cdots. \tag{4}$$

Having introduced this notation, Jeans stated the conditions that must be satisfied to establish the equipartition theorem, these being: For an interval of time t_0, (1) if the system is not absolutely conservative, the energy lost during the interval t_0 must be small compared with the total energy; (2) each term $\theta_1^2, \theta_2^2,...$, of the energy must belong to one of two classes: (a) The time necessary for a change of θ^2 is *small* compared to t_0; (b) The time necessary is *long* compared to t_0.

According to Jeans the equipartition theorem can be applied only to systems for which such a separation in two opposite classes is possible. He pointed out that the restriction is very serious and showed that the applications of the theorem are therefore very limited. He then established the equipartition theorem for class (a) of the coordinates.

Jeans then discussed the relation between entropy and the equipartition theorem. He showed, following Boltzmann, that the entropy can be characterized by the different possible distributions of the coordinates in configuration space, the equipartition of energy taking place at the maximum of entropy.

After a discussion of the classes (a) and (b), and of 'effective' and 'non-effective' terms in the cases of gases and solids, Jeans proceeded to discuss the subject of 'ether and radiation'. The equipartition theorem assumes a somewhat different aspect when applied to the degrees of freedom of a continuous medium, because 'it is in this form that it must be considered when we attempt to analyze the energy of the ether'.

The energy of any continuous medium, subject to vibrations of small amplitude, may be expressed as a sum of squares, two square terms corresponding to each independent free vibration of the medium. Each vibration has a definite period, and corresponds, if the medium is homogeneous, to a definite wavelength in the medium, and the energy of any disturbance may be analyzed into the energy of a series of waves or oscillations of different wavelengths.

From physical dimensional considerations it can be shown that the number of free vibrations, of wavelengths between λ and $\lambda+d\lambda$, in any medium must be of the form

$$c\lambda^{-4}\,d\lambda \tag{5}$$

per unit volume, where c is a constant. For ether, the value of c is 8π, and for a gaseous medium, carrying sound waves, the value of c is 4π. Each vibration contributes two square terms to the energy, and in the state of maximum entropy, these two terms represent an amount of energy equal to RT on the average. Hence the total energy per unit volume of the medium, when in a state of thermodynamic equilibrium with matter at temperature T, must be

$$c \int RT\lambda^{-4}\,d\lambda. \tag{6}$$

When this integral is represented graphically it is at once obvious that the energy runs entirely into the vibrations of infinitesimal wavelength. Or, if vibrations of infinitesimal wavelength cannot exist owing to the medium having a definite structure, then the major part of the energy runs into vibrations of short wavelength. 'Thus the [equipartition] theorem becomes little more than a mathematical expression of the general tendency of energy in a continuous medium to become degraded into irregular disturbances.'

An illustration of the equipartition theorem and its meaning is provided by considering the energy of sound waves in a closed vessel containing air. Suppose that the vessel is absolutely impervious to energy, so that the air forms a conservative system. Suppose also that originally a system of sound waves is started in the vessel, and the vessel is then closed up and left to itself. In the language of the older physics, the waves will be gradually dissipated by viscosity until their energy becomes transformed into heat spread uniformly throughout the vessel. In the language of the molecular kinetic theory, 'we may say that the regularity of the mass-motion of the air becomes impaired by collisions between the molecules, and is finally degraded into an irregular heat motion.' The final state, in whatever way one regards it as being reached, is one in which the molecules move with random velocities, distributed according to Maxwell's law.

This random motion, just as any other motion, can be resolved by Fourier analysis into the motion of a series of regular trains of waves. On carrying out the calculations, it is found that the energy of random motion, resolved into the energy of trains of waves, has an energy per unit volume of amount

$$4\pi \int RT\lambda^{-4}\, d\lambda, \tag{7}$$

until one comes to wavelengths so small as to be comparable with molecular distances. This is exactly the partition of energy predicted by the equipartition theorem.

Similarly, from the equipartition theorem, one should anticipate a final state in which the energy per unit volume of the ether would be

$$8\pi \int RT\lambda^{-4}\, d\lambda. \tag{8}$$

This is the partition of energy which gives the full value $s=2$ to each vibration, no matter what its wavelength may be. And it is exactly the partition of energy observed experimentally for very large values of λ, but the formula is found to fail for smaller values of λ.

By means of a hydrodynamical analogy Jeans sought not so much to explain why the equipartition theorem fails as to develop an extension of this theorem not contemplated by the classical theory of Maxwell and Boltzmann. As a mechanism of the transfer of energy between matter and ether, Jeans recalled that the only hypothesis on the basis of which the theory had been tested thus far was that radiation and absorption were caused by 'the motion of free electrons through the interstices of solid matter'. For the partition of energy in the steady state in which emission and absorption balance is given by $\int E_\lambda\, d\lambda$, Jeans obtained the general formula

$$E_\lambda\, d\lambda = 8\pi RT\lambda^{-4} f(\chi/\lambda) d\lambda, \tag{9}$$

where χ_λ is the conductivity of the medium for disturbances of frequency $2\pi c/\lambda$, and is such that $\chi/2\pi c$ is comparable with the time of collision, c being the velocity of light.

Jeans concluded that Planck's formula, regarded as an expression of the observed facts of radiation, is also contained in the above general formula. However, in order that this formula should agree with Wien's law, the value of χ, and, therefore, the time of a collision, must be exactly proportional to $1/T$, 'a condition which cannot be easily reconciled with any reasonable view of the motion of free electrons'. If the numerical value of χ is determined by comparing the general formula with Planck's law, it is found that at normal temperatures the time of a collision would have to be of the order of 10^{-14} seconds, 'which is a time too large to reconcile with our knowledge of the dimensions of molecular structure'. Also, 'the value of χ and the time of collision would have to be exactly the same for all substances, a condition which is almost unthinkable'. Jeans mentioned that the experiments of Richardson and Brown had shown that the velocities of free electrons are distributed according to Maxwell's law, so that the values of χ must be different for different collisions, and, on integrating for all velocities, one obtains for $f(\chi/\lambda)$ a limiting form which vanishes not as $e^{-\chi/\lambda}$ but as $e^{-\sqrt{(\chi/\lambda)}}$, a result which is contrary to observation.

As a result of these considerations, Jeans concluded that any such extension of the Maxwell–Boltzmann theory cannot account for the phenomena of radiation on the assumption of the proposed mechanism of radiation, and that the theory could not be tested further until some other mechanism of radiation were suggested.

DISCUSSION

The comments of Lorentz, Rutherford, Langevin, Lindemann and Wien emphasized the difficulties present in Jeans' extension of the Maxwell–Boltzmann theory. Referring to Jeans' hydrodynamical analogy, with his use of pipes, tanks, and leakages, Poincaré remarked sharply that with all this 'Mr Jeans could account for any experimental result. However, that is not the role of physical theories. One must not introduce as many arbitrary constants as there are phenomena to be explained. The goal of physical theory is to establish a connection between diverse experimental facts, and above all to predict.'

5. THE LAW OF BLACK-BODY RADIATION AND THE HYPOTHESIS OF THE ELEMENTARY QUANTUM OF ACTION

(M. Planck)

The laws of classical mechanics, electrodynamics and electron theory, have been found to apply with remarkable accuracy to all phenomena in which they have been tested experimentally. The statistical methods and results of the kinetic theory of gases seem to apply even in the domain of atoms and electrons. But the application of laws of classical theory, even with the extension provided by the Lorentz–Einstein principle of relativity, encounters insurmountable difficulties for understanding the experimental laws of black-body radiation.

5.1. CLASSICAL THEORY AND RADIATION LAWS

If u_λ is the energy density of black-body radiation within the wavelengths λ and $\lambda + d\lambda$, then the Stefan–Boltzmann law for total radiation states that,

$$\int_0^\infty u_\lambda \, d\lambda = CT^4, \tag{1}$$

where C is a constant and T the absolute temperature. Wien's displacement law requires that,

$$u_\lambda \, d\lambda = \frac{1}{\lambda^5} f(\lambda, T) \, d\lambda, \tag{2}$$

and the Rayleigh–Jeans law gives,

$$u_\lambda \, d\lambda = \frac{CT}{\lambda^4} \, d\lambda. \tag{3}$$

However, the results of measurements [of F. Paschen, O. Lummer, E. Pringsheim,

H. Rubens and F. Kurlbaum]* are best represented by Planck's formula,

$$u_\lambda \, d\lambda = \frac{C_1}{\lambda^5} \frac{d\lambda}{e^{C_2/\lambda T} - 1}. \tag{4}$$

Planck thought that his formula (4), even if it were not exact, did possess sufficient validity as to be in agreement with experimental results. For large values of the product λT, the expressions (3) and (4) become identical; while for small values, they disagree completely with each other.

The question arises as to how to bring the theory into accord with facts, and Planck addressed himself to this question. He briefly discussed the different approaches that had been made in this direction. Thus Jeans' theory[1] had sought to remove the contradiction between the formulas (3) and (4) by not admitting that the measured quantity (4) corresponds to the normal black-body radiation. According to Jeans, the radiant energy escaping from a small opening in a cavity at uniform temperature does not correspond to the equilibrium radiation of a completely isolated cavity – because, in the latter, the density of radiation according to the formula (3) must increase indefinitely as the wavelength λ diminishes. According to this idea, the question does not arise of a given spectral distribution of a finite quantity of radiant energy in equilibrium, because the integral in (3) over all the wavelengths is infinite.

According to the theory of Jeans, one would have to assume that the observed phenomena correspond to a continuous transformation in which the radiant energy inside the cavity passes continuously from long toward the small wavelengths so slowly that the newly formed radiations always have the time to escape across the walls of the cavity, leading to the establishment of a kind of stationary state of the transformation whose character varies from one case to another. No available experimental evidence justifies such a conception – while many contradict this, among them the almost perfect opacity of the cavity walls [used in the measurements] for the radiation of very small wavelengths, as well as the complete independence of the observed radiation of the nature of the substance inside the cavity or the material of the cavity walls.

For these reasons, Jeans' hypothesis was not received favourably among the physicists. Almost all the investigations of the theory of radiation are based on the supposition – introduced by Kirchhoff and Boltzmann, and verified by Wien and Lummer – of a veritable state of equilibrium, in the thermodynamic sense, inside a cavity at uniform temperature. Planck accepted this point of view.

An especially important confirmation is supplied by the experimental verification of all the consequences obtained by the laws of thermodynamics and electrodynamics at equilibrium. Some of these consequences are particularly remarkable and fruitful: among them, other than Kirchhoff's law of the proportionality of the powers of emission and absorption, are the Stefan–Boltzmann law of total radiation, Equation (1), and Wien's displacement law, Equation (2). The latter is compatible with Equa-

---

* E. Warburg and H. Rubens gave brief reports on the experimental verification of Planck's blackbody radiation formula (4) at the first Solvay Conference, just prior to Planck's report.

tions (3) and (4), while the Stefan–Boltzmann law is in contradiction with Equation (3).

The application of the general principles of thermodynamics and electrodynamics does not lead beyond Wien's displacement law. The form of the function f in Equation (2) can only be determined by a detailed analysis of the [molecular] mechanism of the emission and absorption of radiation. For instance, Lorentz[2] had calculated, for a metallic conductor, the emissivity on the basis of the acceleration of the electrons, and the absorptivity by considering the electrical conductivity as being due to the motion of the same electrons. By dividing the quantities thus obtained, one by the other, Lorentz had obtained the emissive power of the black body.

In any case, one can almost certainly foresee that all methods analogous to the foregoing must lead to the Rayleigh–Jeans radiation law, particularly since the motions of and forces between the molecules and the electrons are calculated by means of classical dynamics and electrodynamics. In Planck's view, this would apply also to the radiation law given by J. J. Thomson[3], in which he had introduced the special hypothesis of a repulsion between the electrons and the molecules proportional to the inverse cube of the distance [the dipole interaction]. It applies also to Ritz' theory of retarded potentials[4] to the extent that this theory is compatible with classical dynamics.

From the foregoing it follows that in order to escape from the [Rayleigh–Jeans] radiation law (3) it is indispensable to make a fundamental modification of the classical theory. One can easily recognize that, above all, it is necessary to introduce a completely new notion of the dynamical significance of the temperature. In fact, according to (1), the energy of radiation, for all wavelengths, is proportional to the temperature, while according to (4) when the temperature [in the first order] becomes infinitely small the energy of radiation becomes infinitesimal.

The most general relation between energy and temperature, Planck concluded, can only be obtained by considerations of probability. Consider two physical systems capable of exchanging energy, whose state is determined by a large number of independent variables. They shall be in statistical equilibrium if further exchange of energy would not correspond to an increment of probability. If $W_1 = f(E_1)$ is the probability that the first system has an energy E_1, and $W_2 = \varphi(E_2)$, the probability that the second system has energy E_2, then the probability that the two systems have the energies E_1 and E_2 respectively is $W_1 W_2$; and the condition that this be a maximum is

$$d(W_1 W_2) = 0,$$

or

$$\frac{dW_1}{W_1} + \frac{dW_2}{W_2} = 0$$

under the condition

$$dE_1 + dE_2 = 0.$$

Thus, the general condition for statistical equilibrium is

$$\frac{1}{W_1} \frac{dW_1}{dE_1} = \frac{1}{W_2} \frac{dW_2}{dE_2}.$$

If this condition of statistical equilibrium is identified with the thermodynamic condition according to which the two systems have the same temperature, then the general definition of temperature is given by

$$\frac{1}{T} = k \frac{\mathrm{d} \log W}{\mathrm{d}E} \tag{5}$$

where the universal constant k depends only on the units of energy and temperature.

Thus, for Planck, the problem of the black-body radiation reduced itself to the calculation of the probability W for which the energy of radiation has a given value E.

5.2. DOMAINS OF PROBABILITY AND THE QUANTUM OF ACTION

The probability for a given value of a continuously variable quantity can be obtained if one can define the *independent elementary domains of equal probability*. The probability that a physical system described by a large number of variables possesses an energy E is thus represented by the number of ways of distributing [complexions] compatible with the energy E, of the independent variables of the system among the different elementary domains of equal probability.

In order to determine these domains, one employs, in classical mechanics, the theorem that two states of a system, which succeed each other according to the laws of motion, have equal probability. If the state of a system is defined by the independent coordinates, q, and their corresponding momenta, p, then Liouville's theorem states that the domain given by $\iint \mathrm{d}q \, \mathrm{d}p$ remains invariant with time provided the evolution of q and p obeys Hamilton's equations. Moreover, at a given instant, q and p may attain all possible values independently of each other. Consequently, the infinitesimally small elementary domain of probability is given by

$$\mathrm{d}q \, \mathrm{d}p. \tag{6a}$$

The new hypothesis must therefore be chosen in such a manner as to introduce certain limitations in the system of the variables q and p, whether one considers discontinuous variation for these quantities or whether one regards them to some extent related to each other. In any case, one is led to a diminution in the number of independent domains of equal probability, and obtains the extension of each of these domains. The hypothesis of the elementary quanta of action obtains this change in a precise manner by introducing, in place of infinitesimally small elementary domains, the domains of finite extension,

$$\int \int \mathrm{d}q \, \mathrm{d}p = h. \tag{6b}$$

The quantity h, the elementary quantum of action, is a universal constant of the dimension of energy multiplied by time. If, for the calculation of the probability, W, of an energy density, u_λ, one employs the finite value $\iint \mathrm{d}q \, \mathrm{d}p$ instead of the infinitesimally small value $\mathrm{d}q \, \mathrm{d}p$, one obtains – with the help of Equations (5) and (2) – the formula (4) [i.e. Planck's formula] in place of Equation (3). Thus the theoretical radiation law is made to agree with the results of measurements.

One has to content oneself with the principle that the elementary domain of probability has an extension h, and put aside all questions concerning the physical significance of this remarkable constant. J. Larmor[5] had accepted this phenomenological point of view, and P. Debye[6] had also taken this attitude. Planck thought that this was indeed how one should introduce the essential content of the hypothesis of quanta. However, it seemed to Planck that one would not know thereby how to stick to it without endangering the further development of the theory, and it was of the utmost importance to look for relations which might exist between the quantum of action and the other physical constants in order to determine and extend its significance.

5.3. CALCULATION OF PROBABILITY AND THE PHYSICAL NATURE OF THE CONSTANT h

Planck, therefore, tried to examine closely the physical nature of the constant h. First of all there arises a question of principle: Does this element of action possess a physical significance for the propagation of radiant energy in the vacuum, or – because of its nature – it plays a role only in the processes of emission and absorption of light? The further development of the theory would certainly follow completely different routes depending upon the answer to this question.

The former point of view had been adopted by A. Einstein[7] in his hypothesis of the light quanta, and J. Stark[8] had followed him. According to this hypothesis, the energy of a light ray of frequency v is not distributed continuously in space, but propagates in a straight line in quanta of magnitude hv in the same manner as the light corpuscles in the emission theory of Newton. As a confirmation of this hypothesis, one invokes the fact that the velocity of secondary cathode rays produced by Röntgen rays is independent of the intensity of the latter.

J. J. Thomson[9] had been led to an analogous conclusion from a study of photoelectric phenomena. He found that the small number of emitted electrons and the independence of their velocities from the intensity of the incident light could only be explained by admitting local accumulations of this energy, rather than a uniform distribution of energy, in the wavefronts of light.

It goes without saying that such hypotheses are irreconcilable with Maxwell's equations and all the electromagnetic theories of light that have been proposed. All of them suppose, in effect, that the smallest luminal perturbation propagates itself throughout all space, if not in all directions with the same intensity then at least with a continuous distribution over concentric spheres whose radius increases with the speed of light. Planck noted that, 'If one considers the complete experimental confirmation which Maxwell's electrodynamic theory obtained by means of the most delicate interference phenomena, and if one considers the extraordinary difficulties which its abandonment would entail for the entire theory of electric and magnetic phenomena, then one senses a certain repugnance in ruining its very fundamentals.' For this reason, Planck thought, 'We shall leave aside the hypothesis of light quanta, especially since it is still quite early in the development of this notion.'

Planck insisted that all phenomena which take place in vacuum are governed exactly by Maxwell's equations, which have no connection of any kind with the constant h. From this he was led to the following conclusions: Thermal radiation, enclosed in an evacuated cavity with perfectly reflecting walls, will indefinitely conserve its initial distribution of energy in the spectrum. One cannot admit that this distribution would slowly evolve toward that of black-body radiation. A fundamental difference manifests itself here between the theory of radiation and kinetic gas theory. For a gas enclosed in a cavity, an initial arbitrary distribution of velocities transforms itself with time into the most probable distribution determined by Maxwell's law. This difference arises because the molecules of the gas collide with each other, while thermal radiation merely traverses through the cavity. The results of the collisions can only be calculated by the methods of probability, while these methods cannot be applied to the radiation in vacuum because each ray always conserves its initial energy. This energy is given to it forever at the moment of emission and can only be modified by absorption or by a new emission. If an arbitrary distribution of the energy is conserved indefinitely in an absolute vacuum, then, on the other hand, the introduction [into the cavity] of the smallest quantity of a substance capable of absorbing or emitting suffices to modify progressively the composition of the radiation and to transform it into the indefinitely stable black-body radiation. From this point of view it is not possible to evaluate the probability for the energy of radiation without taking into account the phenomenon of emission itself, and one is therefore obliged to examine in detail the mechanism of the emission and absorption of radiant energy.

Since, according to Kirchhoff, the intensity of black-body radiation enclosed in a cavity is independent of the emitting and absorbing substance, one must conclude that all mechanisms compatible with the principles of thermodynamics and electrodynamics would provide a correct expression for the [spectral] composition of black-body radiation. The simplest radiating system is a linear oscillator of a given period. Its energy is of the form

$$E = \tfrac{1}{2}Kq^2 + \tfrac{1}{2}L\left(\frac{dq}{dt}\right)^2, \tag{7}$$

where q is the dielectric moment of the oscillator, and K and L are positive constants. The frequency of the oscillations is given by

$$v = \frac{1}{2\pi}\sqrt{\frac{K}{L}}. \tag{8}$$

This, together with the relation (6b), allows one to calculate the magnitude of the energy corresponding to the elementary domain of probability, i.e. the magnitude of the element of action h. Thus,

$$h = \int_{E}^{E+\varepsilon} dq\, dp,$$

and, since,

$$p = L\frac{dq}{dt},$$

one obtains,

$$E = \tfrac{1}{2}Kq^2 + \tfrac{1}{2}\frac{1}{L}p^2.$$

(9)

The double integral giving h represents the surface in the qp-plane between the ellipse $E=$const, and the ellipse $E+\varepsilon=$const. With this, one has

$$h = 2\pi\sqrt{\frac{L}{K}}\,\varepsilon = \frac{\varepsilon}{\nu}.$$

Hence, for an oscillator of a given proper frequency ν, there exist elements of energy,

$$\varepsilon = h\nu,$$

(10)

in the sense that the probability of a given value of the energy depends only on the number of energy elements which it contains.

The question now arises as to how to interpret the elements of energy physically or, in other words, which dynamical law should be taken as the basis for the vibrations of the oscillator in order to obtain the statistical law that has been enunciated. The simplest assumption is that the energy of the oscillator is always an integral multiple of the energy element $h\nu$. Thus it becomes relatively simple to calculate the probability such that a system consisting of a large number N of identical oscillators possesses a given energy E_N. If

$$P = \frac{E_N}{\varepsilon} = \frac{E_N}{h\nu}$$

(11)

represents the number of energy elements contained in the total energy E_N, then the probability W is measured by the number of complexions in which the oscillators may be distributed among the domains of energy corresponding to the integral multiples of ε. This number is equal to the number of ways of distributing P energy elements among N oscillators, if one takes into account only the number and not the individuality of the energy elements which appertain to each oscillator in each distribution considered. One obtains[10]

$$W = \frac{(N+P)!}{N!\,P!} = \frac{(N+P)^{N+P}}{N^N P^P},$$

(12)

and by using Equations (6) and (11),

$$E_N = \frac{Nh\nu}{e^{h\nu/kT} - 1}.$$

(13)

The calculation can be made in other ways which are not different in principle from the preceding one. One can represent each complexion of the system of oscillators by a point in the $2N$-dimensional phase space of Gibbs. The probability W is then represented by an extension of the surface in this space, determined by the condition

$$E = E_N.$$

(14)

The calculation is notably simplified if one considers a canonical distribution of the system of oscillators in phase space with an average energy E_N. This is because one can easily identify, without appreciable error, the number of systems which possess an average energy E_N with the total number of systems of the canonical distribution; and the modulus of this distribution is kT. Thus, for the energy of the system of oscillators considered, one obtains,

$$E_N = \frac{\int E e^{-E/kT}\, d\sigma}{\int e^{-E/kT}\, d\sigma}, \tag{15}$$

where $d\sigma = dq_1\, dp_1\, dq_2\, dp_2 \ldots dq_N\, dp_N$, and the integration is over the entire extension in the phase space of $2N$-dimensions; moreover, the energy E must be expressed as a function of the variables $q, p, \ldots$.

If we now introduce the hypothesis that E can only be an integral multiple of $\varepsilon = h\nu$, the integrations may be transformed into summations and we finally obtain[11],

$$
\begin{aligned}
E_N &= N\, \frac{0 + \varepsilon e^{-\varepsilon/kT} + 2\varepsilon e^{-2\varepsilon/kT} + \cdots}{1 + e^{-\varepsilon/kT} + e^{-2\varepsilon/kT} + \cdots} \\
&= \frac{N h\nu}{e^{h\nu/kT} - 1}.
\end{aligned}
\tag{16}
$$

A third way of calculating the probability W proceeds from the first one in a manner opposite to that of Gibbs. Thus, in the canonical distribution one introduces not only the complexions compatible with the given energy E_N, but also all the complexions relative to the energies comprised between $E = 0$ and $E = \infty$; while, following Boltzmann, the sought for probability, W, is determined by only one part of the complexions compatible with the energy E_N, i.e. one which corresponds to the most probable distribution of this energy among different oscillators. The latter definition leads to the same expression for W as the two preceding ones, since, compared to the most probable distribution, the other distributions comprising all ensembles correspond merely to a negligible number of complexions.

Let $N_0, N_1, N_2 \ldots$ be the number of oscillators of the system which, for an arbitrary distribution of the quantity of energy E_N, possess the energies, $0, \varepsilon, 2\varepsilon, \ldots$. Following Boltzmann, the probability of this distribution is

$$W = \frac{N!}{N_0!\, N_1!\, N_2!\, \ldots} = \frac{N^N}{N_0^{N_0} N_1^{N_1} N_2^{N_2} \ldots}. \tag{17}$$

The condition that W be a maximum, keeping in mind that $E_N = Ph\nu$, gives the following values for the most probable distribution:

$$N_0 = N^2\, \frac{1}{N + P}, \qquad N_1 = N^2\, \frac{P}{(N + P)^2},$$

$$N_2 = N^2\, \frac{P^2}{(N + P)^3}, \qquad \ldots, \tag{18}$$

and these values, introduced in the expression for W, again lead to the formula (12).

A fourth manner of obtaining the formula (12), less exact but with a more concrete physical significance, was proposed by Nernst.[12] He looked for a distribution of energy in a system of oscillators having a circular vibration, subject to collisions of molecules of an ideal gas and in statistical equilibrium with it. The hypothesis that the energy of an oscillator is necessarily an integral multiple of the quantum, ε, is then introduced. One assumes further that in the most probable stationary distribution the number of oscillators which possess the energy, $n\varepsilon$, is equal to the number of oscillators whose energy, for the same temperature, would comprise between $n\varepsilon$ and $(n+1)\varepsilon$ if the ordinary Maxwellian distribution were applicable. This then gives, for the numbers N_0, N_1, N_2, etc. of the oscillators which possess 0, 1, 2,... quanta of energy, the values

$$N_0 = N(1 - e^{-\varepsilon/kT}), \qquad N_1 = N(e^{-\varepsilon/kT} - e^{-2\varepsilon/kT}),$$
$$N_2 = N(e^{-2\varepsilon/kT} - e^{-3\varepsilon/kT}), \qquad \ldots, \quad \text{etc.} \tag{19}$$

The total energy of the system of oscillators,

$$E_N = N_0 \times 0 + N_1 \cdot \varepsilon + N_2 \cdot 2\varepsilon + \cdots,$$

then takes the form given by Equation (13).

As the concordant results of the different procedures of calculation show, the relations (12) and (13) necessarily follow from the hypothesis that the energy of an oscillator is an integral multiple of the element of energy ε.

In order to pass from formula (13) to a law of radiation capable of experimental verification, one must again know the relation between the average energy of the oscillator, $E_N/N = \bar{E}$, and the density, u_ν, corresponding to the frequency of radiation present in space. Maxwell's electrodynamics leads to the equation[13]

$$u_\nu \, d\nu = \frac{8\pi\nu^2}{V^3} \bar{E} \, d\nu. \tag{20}$$

In combination with Equation (13) one obtains for the law of black-body radiation,

$$u_\nu \, d\nu = u_\lambda \, d\lambda = \frac{8\pi h\nu^3}{V^3} \frac{d\nu}{e^{h\nu/kT} - 1}$$
$$= \frac{8\pi hV}{\lambda^5} \frac{d\lambda}{e^{hV/k\lambda T} - 1}, \tag{21}$$

which is exactly of the same form as the experimental law (4), in agreement with the measurements.

If, in order to calculate the two constants k and h, one employs the number used by Lummer and Pringsheim,

$$\lambda_{max}T = 0.294 \text{ cm deg}, \tag{22}$$

and Kurlbaum's number

$$S_{100} - S_0 = 7.31 \times 10^5 \text{ ergs/cm}^2 \text{ sec}, \tag{23}$$

where S_t represents the total energy radiated per second per cm^2 of a black-body at temperature t degrees, one obtains[14]

$$k = 1.346 \times 10^{-16} \text{ erg degree} \tag{24}$$

and

$$h = 6.548 \times 10^{-27} \text{ erg sec.} \tag{25}$$

The quantity k, since it results from an application of Equation (5) to the statistical equilibrium of the molecules of an ideal gas, is equal to the *gas constant*, related not to a gram molecule but really to a single isolated molecule. Thus there results a method of calculating the number of molecules which is much more precise than all others employed earlier. Since, however, these other methods had been perfected only just prior to Planck's calculations, the agreement of their results with Planck's provided a remarkable confirmation of the considerations developed by him.

Planck thought that in spite of its apparent success, the new theory of radiation could not in any way be considered satisfactory, since the hypotheses on the basis of which Equations (13) and (20), respectively, have been derived are mutually contradictory. In order to obtain the former, one assumes that the energy of the oscillator is an integral multiple of hv, while the reasoning which leads to the second supposes that this energy is continuously variable. It is not possible to decide in favour of one of these cases without noting, at least on a first consideration, that the equation based on the other alternative becomes illusory.

Planck pointed out that all models which had been proposed for representing the properties of an oscillator, capable of emitting and absorbing radiant energy in agreement with the theory of quanta, suffered from the same inner contradiction. A. E. Haas[15], for instance, considered as the oscillator the uniform sphere of J. J. Thomson, inside which an electron can oscillate around a centre. The maximum energy for this oscillation, which is attained when the amplitude becomes equal to the radius of the sphere, must be equal to the quantum of energy $\varepsilon = hv$. For larger amplitudes, periodic oscillation is impossible because the electron leaves the sphere completely; while for smaller amplitudes, the oscillation is periodic and its energy is continuously variable. The grounds on which the demonstration of Equation (13) is based do not admit these conditions, and this equation is no longer applicable. This is particularly evident if one considers the oscillations of Haas' system of oscillators in a stationary field of radiant energy, sufficiently weak, such that the average energy of an oscillator is small compared with the quantum ε. Each oscillator, in a stationary state, vibrates with a small amplitude like a dipole governed by the laws of Maxwell's electrodynamics, emitting and absorbing radiant energy in a continuous manner. None of these oscillations obtains the energy ε, the radius of the sphere becomes immaterial, and the elements of energy play no role. The existence of the fluctuations of free radiation caused by interference do not affect this conclusion, because these fluctuations are too feeble to explain the distribution (18) of energy among the oscillators corresponding to the correct laws of radiation. The same difficulties occur in the modification introduced by A. Schidlof[16] in Haas' oscillator.

5.4. Atomic Models and the Law of Radiation

In Planck's view, the models of Schidlof or Haas led necessarily to the Rayleigh–
Jeans law of radiation for the simple reason that these models rigorously admitted all
the laws of classical dynamics. In order that an oscillator should provide radiation
according to Equation (4), it is necessary to introduce in the law of its behaviour a
special physical hypothesis which, at a fundamental point, is in contradiction with
classical mechanics, either explicitly or tacitly.

The oscillator model proposed by M. Reinganum[17] was much closer to this condition. In it, an electron was assumed to be completely immobile until it had absorbed a
whole quantum of energy. This at least guaranteed that the energy of the oscillator was
always an integral multiple of ε.

However, in this case also, as in all cases where the energy of an isolated oscillator is
supposed to vary in a discontinuous manner, it is impossible to understand where the
energy absorbed by the oscillator comes from since (as it should often happen at low
temperatures) its energy increases suddenly from 0 to $h\nu$. The intensity of the thermal
radiation in space, below a certain wavelength, is too feeble at low temperatures to
supply the necessary energy. According to the laws of Maxwell's electrodynamics, the
time required for an oscillator placed in a radiation field to absorb the energy, $\varepsilon = h\nu$,
corresponding to its frequency, is

$$t = \frac{e^{h\nu/kT} - 1}{2\sigma\nu}, \tag{26}$$

where σ is the average decrease of the oscillations. This value of the time, t, increases so
rapidly as T decreases, that at relatively low temperatures one could hardly speak of a
sudden absorption of an element of energy.

This difficulty becomes still more serious if one considers an oscillator subject to the
action of non-stationary radiation. In this case, one has in effect no possibility of
knowing whether the oscillator would start to absorb at all, because one does not
know whether the duration of radiation would be sufficiently long enough to allow the
oscillator to extract a whole quantum.

The hypothesis that the absorbed energy does not come from absorbed radiation,
but from the electrons encountered, for instance, does not offer any help here. If,
indeed, the free radiation is not absorbed by virtue of the laws of the stationary state,
then it can also not be emitted, and the fundamental hypothesis of an exchange of
energy between the oscillator and free radiation cannot be preserved.

5.5. Continuous Absorption

Faced with these difficulties, it seemed inevitable to Planck to renounce the supposition
that the energy of the oscillator should be an integral multiple of the element of
energy $\varepsilon = h\nu$. He concluded that *the phenomenon of the absorption of free radiation is
essentially continuous*. From this point of view, one could preserve the fundamental
hypothesis of quanta by supposing that the emission of thermal radiation by an

oscillator of frequency v is discontinuous, and is produced in integral multiples of the elements of energy $\varepsilon = hv$.

Even though the energy of an oscillator is continuously variable, one can still define the domains of elementary probability by means of the finite quantum of action h. Let the energy of the oscillator be expressed in the form

$$E = n\varepsilon + \varrho \qquad (27)$$

in such a way that the oscillator possesses n whole quanta, ε, and the remaining $\varrho < \varepsilon$; the integral number n alone is subject to the law of chance while ϱ, which naturally assumes different values for different oscillators in a field of stationary radiation, increases continuously and uniformly as a function of time. The probability of the energy E does not depend on the quantity ϱ, which is continuously variable in a regular and familiar manner, but only on the integral number n which alone is subject to the law of chance. If $n = 0$ and $E < \varepsilon$, the oscillator does not emit at all and its energy increases constantly by absorption until it becomes equal to the first quantum, after which emission occurs sooner or later.

This hypothesis of the quanta of emission[18] also leads to formula (21) for the radiation, but the relation between the average energy of the oscillator and the temperature is no longer given by (13), but by the equation,

$$\bar{E} = \frac{E_N}{N} = \frac{hv}{2} \frac{e^{hv/kT} + 1}{e^{hv/kT} - 1}. \qquad (28)$$

At very low temperatures, $\bar{E}$ is equal to $hv/2$, i.e. almost all of the oscillators possess only the energy ϱ which cannot be lost, and its average value is

$$\frac{\varepsilon}{2} = \frac{hv}{2}.$$

Equation (20) is thus replaced by,

$$u_v \, dv = \frac{8\pi v^2}{V^3} \, dv \left(\bar{E} - \frac{\varepsilon}{2} \right). \qquad (29)$$

The consequence that the energy of an oscillator does not tend to zero as the temperature decreases indefinitely, but only remains less than ε, seems to offer a satisfactory solution of the difficulty mentioned earlier, which led J. J. Thomson, Einstein and Stark to the hypothesis of a discontinuous structure of free radiation.

When light waves or Röntgen rays fall on a metal and release electrons, it is not necessary that radiation should supply, according to the hypothesis of the quanta of emission, the total energy which comes into play. The radiation would only have to complete the energy, ϱ, of an oscillator up to a whole quantum, ε, in order that the emission of an electron becomes possible. The more feeble the intensity of the external radiation, the less will be the number of oscillators whose energy will thus be completed, and consequently the smaller will be the number of electrons emitted. On the other hand, one understands readily that the velocity of the electrons depends only on the frequency of light or the duration of Röntgen rays, if one assumes that the emission of electrons, just as that of radiation, is produced by quanta of energy whose magnitude depends only on the oscillators which emit and, consequently, on the nature of the radiation absorbed by these oscillators.

5.6. Quantum Oscillators and the Specific Heat

One obtains a new verification of the theory of quanta by taking the derivative, with respect to the temperature, of Equations (13) or (28) which express the energy of the oscillator as a function of the temperature. In both cases one obtains, for the specific heat of the oscillators,

$$\frac{dE_N}{dT} = Nk \frac{(hv/kT)^2 \, e^{hv/kT}}{(e^{hv/kT} - 1)^2} . \tag{30}$$

From this A. Einstein[19] had derived a formula giving the variation with temperature of the specific heat of solids, by identifying the latter with systems of oscillators capable of vibrating with the same frequency along the three directions of the axes, and his formula agreed [at least roughly] with the results of measurements. The deviations which persist may well be because the simple hypothesis admitted by Einstein does not correspond exactly to the facts. Since the laws of black-body radiation are completely independent of the type of oscillator employed, it is probable that simple oscillators, convenient for establishing the law of radiation, differ notably from more complex ones which exist in nature and which determine the specific heat of solids.

5.7. Theory of Radiation and Aperiodic Phenomena

Finally, Planck mentioned that a completely satisfactory theory of radiation would necessarily have to explain non-stationary phenomena, the discussion of which lay beyond the limits of his report. He made only some general remarks in this connection.

Above all, it is important to insist that the hypothesis of quanta is not, strictly speaking, a hypothesis of *energy* but could be called an hypothesis of *action*. The fundamental conception is that of an elementary domain of probability of extension h. The quantum of energy or radiation, hv, is derived from that, and has only a significance for periodic phenomena of a given frequency, v. There is no doubt that, to the extent that the hypothesis of quanta possesses a profound significance, the element of action h must also have a fundamental importance for non-periodic and non-stationary phenomena. Sommerfeld[20] has shown it directly for some particular cases. It might perhaps be possible, thanks to the introduction of this element of action h, to deduce the laws of black-body radiation for all wavelengths starting from non-periodic phenomena, as H. A. Lorentz had done for the large wavelengths starting from electron collisions.

The extension of the theory of quanta to phenomena of ordinary mechanics raises a question of fundamental importance. Do the quanta play no role in these phenomena because the acceleration is too small or because the theory of quanta cannot be applied? In other words, is the difference between the laws of ordinary mechanical and electrical phenomena and those of emission from the optical oscillator fundamental or merely qualitative?

Planck expressed his preference for the first alternative, i.e. the admission of an essential difference between the phenomena produced by the quantum of action and those which arise in a continuous manner in agreement with the laws of classical dynamics. Planck even felt tempted to see in this the separation or differentiation between physical and chemical phenomena. The atoms and molecules, and also perhaps the free electrons, move according to the laws of classical dynamics; however, the atoms or the electrons subject to a molecular interaction obey the laws of the theory of quanta. Physical forces such as gravitation, electrical and magnetic attractions and repulsions, and cohesion, act in a continuous manner; chemical forces, on the other hand, act through quanta. The physical laws are of the same kind as allow the masses in physics to interact with each other to any extent, while in chemistry they can act only in definite proportions and vary in a discontinuous manner.

Planck thought that a complete understanding of the physical significance of the element of action, h, will only be obtained by means of the principle of least action, which seems to govern all fundamental phenomena and whose importance has been affirmed even in the theory of relativity. The theory of quanta, in Planck's view, must be brought into harmony with the principle of least action. It would only be necessary to give the latter principle a more general form that would make it applicable to discontinuous phenomena.

Planck concluded that if one considers the general result of the efforts made thus far to interpret theoretically the laws of radiation, one cannot, in any way, regard it as being satisfactory. One should not be surprised by this if one recognizes that the solution of the problem cannot be obtained except by the introduction of a completely new hypothesis which is in direct contradiction with the conceptions accepted hitherto. Still, it is without doubt that one shall, by continued effort, be able to formulate a hypothesis free of contradictions that will preserve in a definitive manner a certain number of the ideas introduced thus far.

DISCUSSION

At the conclusion of Max Planck's report Albert Einstein expressed the opinion that the manner in which the probability W had been introduced by Planck was 'somewhat shocking', because in doing so the Boltzmann principle did not have a physical meaning any longer. Even if one succeeded in defining the probability in a way that the entropy derived from Boltzmann's principle would agree with the empirical definition, it seemed to Einstein that the manner in which Planck had introduced this principle would not allow one to conclude the correctness of the theory by basing it on its [i.e. Boltzmann's principle's] agreement with experimental thermodynamic properties. H. A. Lorentz agreed that it would be desirable to start with a good definition of probability.

Planck was aware of the problem but thought that in the state of knowledge of the day it was not possible to give such a definition. He summed up the problem as follows: What is the method of calculation which, by an application of Boltzmann's principle, leads to an entropy of radiation in agreement with experiment? Once the general solution has been obtained, Planck thought, one would also have a general physical definition of the probability.

F. Hasenöhrl emphasized the difference between the points of view of Boltzmann and Planck by noting that it arose because Boltzmann took the element of extension in phase as being infinitesimally small, while Planck gave it a finite value.

Henri Poincaré raised questions concerning the elements of action and stated the following difficulty. Since h is the area of a certain domain of the pq-plane, the decomposition of the plane can be done in many different ways according to the chosen shape of the area. What would be the influence of this choice on Planck's result? Planck maintained that the process was introduced to evaluate the probability of a given energy, hence the shape of the elements was determined by the lines of constant energy. Poincaré then wondered whether there exists a conservation of the action in the same way as there exists conservation of matter or of electricity.

Lorentz pointed out that historically the element of energy was introduced by Planck *before* the element of action, and wanted to know the relation between the statistical methods of Planck and Gibbs. In Planck's view the difference was that Gibbs had infinitesimal extensions in phase while he himself had introduced finite ones, leading to a limitation in the use of Hamilton's equations.

J. H. Jeans and P. Langevin raised the question concerning the existence of a quantum of ether. Planck thought that if it meant that a finite element of action plays a role in the propagation of electromagnetic disturbances in ether, the answer would be no.

The discussion of Planck's report then concentrated on the question as to how legitimate was the use of statistical methods for radiation and, in general, for ether. W. Wien pointed out that a ray of light emitted by a black body has a temperature and entropy given by the laws of probability.

Poincaré returned to his question concerning the decomposition of the pq-plane and gave the example of a three-dimensional harmonic oscillator for which different decompositions lead to a contradiction. He wondered how Planck's treatment of the harmonic oscillator and its quantum condition, according to which the equation $\iint dq\ dp = h$ determines the elementary region of a priori probability in the pq-plane, should be extended to systems with more than one degree of freedom. In his reply, Planck expressed the belief that the formulation of a quantum condition for systems with more than one degree of freedom, although it did not exist at that moment, would soon be possible.*

Lorentz pointed out that the assumption, $\varepsilon = h\nu$, made by Planck was the only possible one that was not in contradiction with Carnot's principle. He wondered whether in the new theory it was possible to see that the relation $\varepsilon = h\nu$ was the only one ensuring agreement with the second law of thermodynamics. Langevin thought that the resaon for this agreement had to be sought in Liouville's theorem, which is still valid in the decomposition of Planck, at least for a decomposition in which the definition of probability is still correct.

Lorentz emphasized the importance of Arthur Haas' model which had been rejected by Planck, to which Langevin pointed out that this model would lead to absorption of energy varying with the intensity of radiation.

W. Nernst wondered whether, in addition to Planck's resonator, it would not be important to introduce the assumption of a charged particle forced to remain at a fixed distance from a given point. Hasenöhrl responded that in such a model the period is not independent of the energy; if the elements of extension in phase space are equal, the elements of energy are different, and vice versa.

---

* Planck and Sommerfeld formulated the generalization to many degrees of freedom only *four years later*. [M. Planck, 'Die Quantenhypothese für Molekeln mit mehreren Freiheitsgraden', *Verh. Deut. Phys. Ges.* **17**, 407–418, 438–451 (1915); 'Die Physikalische Struktur des Phasenraumes', *Ann. Physik* **50**, 385–418 (1916); A. Sommerfeld, 'Zur Theorie der Balmerschen Serie', *Münchener Ber.* (1915), 425–458, 'Die Feinstruktur der Wasserstoff und wasserstoffsähnlichen Linien', *Münchener Ber.* (1915), 459–500, 'Zur Quantentheorie der Spektrallinien', *Ann. Physik* **51**, 1–94, 125–167 (1916).]

Planck considered dynamical systems of f degrees of freedom, the equations of motion of which admit regular intervals. He generalized his treatment of the harmonic oscillator by dividing the phase space by means of surfaces $F(p_k, q_k) = $ constant, defined by the integrals, into regions of volume h^f Planck postulated that the stationary states correspond to the f-dimensional intersections of these surfaces.

Paul Epstein showed later on [P. S. Epstein, 'Über die Struktur des Phasenraumes bedingt periodischer Systeme', *Berliner Ber.* (1918), 435–446] that Planck's conditions for characterizing the stationary states were equivalent to the quantum conditions of Sommerfeld. Kneser, who wrote a thesis on quantum theory under David Hilbert in 1921 [H. Kneser, 'Untersuchungen zur Quantentheorie', *Math. Ann.* **84**, 277–302 (1921)] also proved rigorously the equivalence of the quantum conditions of Planck and Sommerfeld for the characterization of stationary states.

Wien thought that the difficulty with Planck's first theory consisted in the great amount of time which would be necessary in order that an oscillator could receive an element of energy in the case of a weak radiation. Wien suggested the introduction of a coupled oscillator. Planck thought that even in the case of a coupled oscillator the absorption time would remain too large.

Poincaré pointed out that if only a fixed resonator and ether were considered, avoiding the Doppler effect, it would be impossible to reach an equilibrium without the presence of matter or of some medium the energy of which varied in an arbitrary manner. Poincaré then asked Planck to justify an assumption which he had introduced in the derivation of the absorption formula – specifically, why the energy between nhv and $(n+1)\,hv$ had to vary linearly with the time? Planck replied that if the oscillator would only absorb radiant energy, its energy would increase proportionally with the time, but if the exchange of energy were caused by collision with an atom or an electron nothing precise could be said about the mechanism of this exchange.

Madame M. Curie thought that, since in Planck's theory the emission of energy was instantaneous, Maxwell's equations would not hold even in vacuum. Lorentz responded that emission might take place during a large number of vibrations.

Madame Curie then asked how one could imagine a mechanism that would allow this emission to be interrupted. 'It is quite probable', she said, 'that such a mechanism will not exist at our [human] level, but rather be comparable to Maxwell's demon. It would allow one to obtain the deviations based on the laws of radiation as envisaged by statistics just as Maxwell's demon allows one to obtain the deviations following from the consequences of Carnot's principle.' Poincaré gave the example of the discharge of an Hertzian oscillator which starts instantaneously but lasts a certain time.

Planck recognized that if the emission from an oscillator takes place in quanta then Maxwell's equations will have to be modified within the oscillator and its immediate vicinity. Einstein also pointed out that if an oscillator emits in a different manner than the one corresponding to Planck's original theory, Maxwell's equations must be abandoned in the vicinity of the oscillator, because the application of Maxwell's equations to the quasi-static oscillating dipole field necessarily yields the emission of energy in the form of spherical waves.

Nernst drew attention to the fact that Planck's new hypothesis concerning the zero point energy will probably require that, at the absolute zero of temperature, the atoms will still have motions, and thereby solid bodies will still have vapour pressure. Planck, however, thought that the vapour pressure at absolute zero should be null.

Arnold Sommerfeld expressed his belief that the hypothesis of emission quanta, as well as the initial hypothesis of the quanta of energy, should be considered more as a form of explanation [a model] rather than as physical reality.

Lorentz was interested in finding out how Planck would seek to modify the fundamental equations for the electron. Planck did not have a precise answer, but he indicated the direction in which he was searching. For Planck, the equations of the electromagnetic field were exact outside the linear oscillator, but in between two emissions the vibrations were governed by the law,

$$m\frac{\mathrm{d}^2x}{\mathrm{d}t^2} + nx = efx,$$

where the symbols have their usual significance. However, Lorentz insisted on the fact that one should not attach too much importance to the units of h, as the quantity e^2/V, if e is measured in electrostatic units, has also the dimension of an action.

The discussion that ensued after Planck's report at the first Solvay Conference provided the beginning of a new way of thinking, and increasingly subtle arguments would be pursued in the succeeding decades. With the work of Einstein on light quanta, Niels Bohr's work on the theory of the hydrogen atom and the search for an atomic mechanics, leading finally to the creation of a unified quantum mechanics in 1926, the physical reality of quanta and the statistical laws for the description of atomic phenomena would become central questions in the epistemology of one of the greatest scientific revolutions in history.

REFERENCES

1. J. H. Jeans, *Phil. Mag.* **19**, 209 (1909).
2. H. A. Lorentz, *Proc. Akad. Wet. Amsterdam*, (1903), p. 666.
3. J. J. Thomson, *Phil. Mag.* **20**, 238 (1910).

4. W. Ritz, *Physik. Z.* **9**, 903 (1908).
5. J. Larmor, *Proc. Roy. Soc.* **A 83**, 82 (1909).
6. P. Debye, *Ann. Phys.* **33**, 1427 (1910).
7. A. Einstein, *Physik. Z.* **10**, 185, 817 (1909).
8. J. Stark, *Physik. Z.* **11**, 25 (1909).
9. J. J. Thomson, *Electricity and Ether*, Manchester University Lectures, No. 8, 16 (1908).
10. Max Planck, *Verh. Deut. Phys. Ges.* **2**, 237 (1900).
11. A. Einstein, *Ann. Physik*, **22**, 180 (1907).
12. W. Nernst, *Z. Elektrochemie*, **17**, 265 (1911).
13. M. Planck, *Sitzber. Preuss. Akad. Wiss.* (18 May 1899), p. 461, *Physik. Z.* **2**, 533 (1900–1901).
14. M. Planck, *Verh. Deut. Phys. Ges.* **2**, 239–241 (1900).
15. A. E. Haas, *Wien Sitzber. math. naturw. Klasse*, **119**, IIa (February 1910).
16. A. Schidlof, *Ann. Phys.* **35**, 90 (1911).
17. M. Reinganum, *Physik. Z.* **10**, 351 (1909).
18. M. Planck, *Verh. Deut. Phys. Ges.* **13** (1911); *Sitzber. Berliner Akad. Wiss.* (13 July 1911).
19. A. Einstein, *Ann. Phys.* **22**, 180 (1908).
20. A. Sommerfeld, *Sitzber. Bayr. Akad. Wiss.* (7 January 1911).

6. KINETIC THEORY AND THE EXPERIMENTAL PROPERTIES OF PERFECT GASES

(M. Knudsen)

Martin Knudsen gave an account of the status of kinetic theory and the experimental properties of perfect gases. A pure gas is composed of molecules in motion, all of the same mass m. The interaction between the molecules is appreciable only when the distance is smaller than a certain limit which, for perfect gases, is negligible compared to the path during which the molecules move freely.

Having stated this fundamental assumption, Knudsen recalled that according to the general laws of collision, and using Avogadro's law, there exists the relation,

$$p = \tfrac{1}{3} \varrho \overline{c^2},$$ (1)

where p is the pressure, ϱ the density ($\varrho = Nm$), and $\overline{c^2}$ the mean square speed of the particles.

In order to confirm this theory many different quantities can be computed and compared with experimental results. Knudsen discussed the derivation of the law $c_p/c_v = \tfrac{5}{3}$ for specific heats, a result which had been verified for monatomic gases but was a little too high for polyatomic gases. He also discussed certain results in which the nature of the particles was irrelevant.

The mass G of a gas flowing out, through a hole in a surface A, during a time τ at a stationary regime, is given by

$$G = A\tau \frac{1}{\sqrt{2\pi}} \sqrt{\varrho_0} \sqrt{\frac{273}{T}} (p' - p''),$$ (2)

where p' and p'' are the pressures on the two sides of the surface, and ϱ_0 is the density of the gas before expansion. Knudsen illustrated the validity of this law with experimental examples.

If the mutual collisions between molecules may be neglected, the theory shows that the condition of equilibrium between two containers filled with the same gas at absolute temperatures T' and T'', communicating through a hole, is given by

$$\frac{p'}{p''} = \sqrt{\frac{T'}{T''}}.$$ (3)

This relation had been verified for a tube in which the temperature varies gradually.

According to the theory, two plates of the same area but slightly different temperatures must repel each other with a force F given by

$$F = \frac{p}{2}\left(\sqrt{\frac{T'}{T''}} - 1\right) A,$$ (4)

and Knudsen reported that this law had been verified to be correct for all gases on which experiments had been done.

The mass G of a gas flowing through a cylindrical tube of length L, radius R, during time τ is given by

$$G = L\tau\sqrt{\varrho_0}\sqrt{\frac{273}{T}}\frac{R^3}{L}(p' - p''),$$ (5)

where p' and p'' are the pressures at the extremities of the tube. This law was also found to be in perfect agreement with the experiments.

In conditions where collisions between the molecules are sufficiently rare, the theory shows that the amount of heat transferred between two plates of surface area A is given by

$$Q = A\tau(T_1 - T_2)p\varepsilon,$$ (6)

where ε, called the molecular thermal conductivity, is given by

$$\varepsilon = \tfrac{1}{4}\sqrt{\frac{2}{273\pi\varrho_0\tau}}\frac{c_p + c_v}{c_p - c_v}.$$ (7)

Experiments showed that the quantity Q is very sensitive to the polish of the plates and the nature of the gas.

In the absence of collisions between the molecules, the resistance K, which acts opposite to a plate in motion with a small speed v within the gas, is theoretically given by

$$K = k\sqrt{\frac{273}{T}}Ap\sqrt{\varrho_1 v},$$ (8)

where k is a numerical coefficient depending upon the angle between the plate and the current. The experimental result did not allow one to conclude the verification of this law.

The experiments thus led to a quantitative verification of kinetic theory but the domain of its validity was found to be limited to small ranges of temperature and pressure, i.e. to those for which the mutual interaction of the molecules could be neglected. Knudsen was led to introduce these interactions, and thereby to discuss the problem of the calculation of the coefficients of viscosity, diffusion, and thermal conductivity by means of the kinetic theory. He concluded that there were so many facts in agreement with kinetic theory that it could not be rejected.

DISCUSSION

Following Knudsen's report, the discussion was initiated by Nernst. Nernst thought that the law of collisions (Maxwell's distribution) should be modified even for monatomic gases in view of the theory of quanta; because if it were not done, and the molecules of the gas were charged, there would arise an inadmissible radiation. An idea of this could be obtained from Knudsen's result according to which, when a molecule strikes the wall of a solid, all directions of reflection have equal probability. One would imagine that if two molecules of a gas collide, they rotate around each other in circles and then separate. Under such conditions, one should examine whether Maxwell's distribution law is obtained exactly or only in an approximate form.

Einstein replied to Nernst that Maxwell's distribution is based on theorems which take into account only the momentum and energy, and these would remain valid even if the mechanics had to be modified.

E. Warburg gave an account of some experiments and observations which are not in agreement with Stokes' law, and Rutherford recalled that Stokes' law led to values for the electronic charge that were relatively high.

Jean Perrin drew attention to the fact that Millikan's method [for measuring the electronic charge] was based on the assumption that the real speed of the droplets is obtained by multiplying by $(1 + A(\lambda/R))$ the speed which would be given by Stokes' law, the parameters A and λ being then fitted with the experiment. Knudsen reported on an experiment done in his laboratory showing that A was not a constant.

Further discussion bypassed Knudsen's report and dealt with questions raised by the determinations of the value of the elementary electronic charge.

7. THE PROOF OF MOLECULAR REALITY

(J. Perrin)

Jean Perrin began his report by pointing out that the molecular hypothesis had existed for over two thousand years. However, the ancient Greek philosophers, who assumed that matter consisted of indestructible atoms in ceaseless motion, had not made it clear as to how they had arrived at this hypothesis or how it could be demonstrated.*

Perrin stated certain properties of mixtures of different fluids that could give proof of the existence of molecules, or at least could be understood clearly on the basis of the molecular hypothesis. He noted that the discontinuities observed in the constitution

---

* The nineteenth century bore witness to the application of Dalton's atomic theory to chemistry, and the birth and development of kinetic theory in the hands of Herapath, Waterston, Krönig, Clausius, Maxwell and Boltzmann, saw the coming to fruition of the atomic idea as one of the cornerstones of a mechanistic description of natural phenomena. Yet numerous distinguished scientists and natural philosophers – among them Mach, Duhem, Helm, and Ostwald – while admitting the fruitfulness of the atomic hypothesis, did not regard it as possessing physical reality. The atomic conception would be established anew through the work of Boltzmann, J. J. Thomson, the Curies, Einstein, Smoluchowski, Rutherford and Bohr – and, of course, Jean Perrin.

of molecules required one to introduce the notion of chemically indivisible elements called atoms, and cited the ratio of the molecular masses of different gases varying as integers in support of this idea. He also discussed Avogadro's law and its consequences, pointing out the remarkable agreement between this law and experiments.

As one of the proofs of molecular reality furnished by the kinetic theory, Perrin cited the average value of the kinetic energy due to molecular translations, w, which can be written as

$$w = \tfrac{3}{2} \frac{R}{N} T,\qquad\qquad (1)$$

and whose deviation from the average value is given by the Maxwell distribution function.

Perrin reported on an experiment, due to Michelson, showing the enlargement of spectral rays due to kinetic motion of the molecules, in perfect agreement with the theory. Another proof of molecular reality was related to the viscosity. As established by Maxwell, the viscosity, ζ, density, δ, the mean velocity of the molecules of the gas, Ω, and the mean free path, L, are related by

$$\zeta = 0.31\ \delta\Omega L. \qquad\qquad (2)$$

Since the mean free path is proportional to δ^{-1}, the viscosity is independent of the pressure, a fact which had been established experimentally.

Identifying the molecules with spheres of diameter D, Clausius, in 1858, had obtained a formula (improved later on by Maxwell) relating the mean free path L and the diameter of the molecule,

$$\pi N D^2 = \frac{v}{L\sqrt{2}}, \qquad\qquad (3)$$

where v is the volume of the gram-molecule and N is the Avogadro number. It is necessary to have another relation in order to calculate N and D. This relation is provided by the equation of Van der Waals,

$$\left(p + \frac{a}{v^2}\right)(v - b) = RT, \qquad\qquad (4)$$

and one obtains

$$\tfrac{1}{6}\pi N D^3 = \frac{b}{4}, \qquad\qquad (5)$$

where b can be defined experimentally. For Argon, this calculation yields, $N = 6.2 \times 10^{23}$ and $D = 2.85 \times 10^{-8}$. However, errors of 5% in L can have large influence on the values of N and D, and conclusions from such experiments could be doubtful.

7.1. EMULSIONS

Jean Perrin discussed phenomena of a different kind which, by their nature, could furnish information about the agitation of molecules. He recalled the discovery of

Brownian motion and its character. Wiener (1863) thought that the agitation did not have its origin in the molecules nor in a cause external to the liquid, but must be attributed to some internal motions specific to the state of the fluid. Gouy (1888) had pointed out that the Carnot principle is only a statistical result, and in view of the Brownian motions it cannot be true at the microscopic scale. After Wiener and Gouy, Siedentopf, Exner (1901), and finally Einstein (1905) and Smoluchowski (1906) investigated Brownian motion.

Perrin, convinced that Brownian motions were important to the understanding of molecular motions, pursued his researches on emulsions. He extended the law of perfect gases to emulsions consisting of small but visible particles, an idea which was suggested to him by similar results obtained in the theory of osmotic pressure.

Perrin gave a classical derivation of the formula for the rarefaction of air with altitude. He obtained, as expected,

$$\log \frac{p_0}{p} = \frac{Mgh}{RT},$$ (6)

M being the mass of the gram-molecule. Taking into account the fact that the small spheres of the emulsion are in a liquid, he then obtained the formula

$$\log \frac{n_0}{n} = \frac{N}{RT} m \left(1 - \frac{\delta}{\Delta}\right) gh,$$ (7)

where n_0 and n are the numbers of 'grains' corresponding to the pressures p_0 and p, δ and Δ are the densities of the substance and the fluid, respectively, and h is the height of the observation.

Perrin described a long experiment in which he first made some emulsions, then determined the mass of the particles and their rarefaction with height. The distribution was found to be exponential, leading at the same time to a confirmation of the origin of Brownian motion. Finally he showed that the results were in perfect agreement also when the temperature was changed, the system assuming a new equilibrium position, as anticipated on the basis of the kinetic theory. His method yielded a precise value for Avogadro's number, i.e.

$$N = 6.82 \times 10^{23}.$$ (8)

Moreover, on substituting this value in the Clausius relation (3) Perrin obtained the diameters of the molecules: Helium, 1.7×10^{-8} cm; Argon, 2.7×10^{-8} cm; and Mercury, 2.8×10^{-8} cm.

7.2. THE LAWS OF BROWNIAN MOTION

The first theoretical studies of Brownian motion were made by Einstein (1905) and Smoluchowski (1906). Both of them chose the same parameter to characterize Brownian motion. Perrin mentioned the different methods which had been used to obtain the true speed of the particles, but none of them was fast enough to give an idea of their real motion.

Considering the particles as being identical, Perrin applied to them the Maxwell distribution function, denoting ξ^2 as the mean square of the displacement. He showed that this quantity is proportional to the duration τ of the motion, and the motion can be characterized by the parameter ξ^2/τ. This conclusion was reached under the assumption that τ is small, but long enough to assure the independence of the motions. Following Einstein, he showed that τ had to be larger than 10^{-5} sec for particles of diameter $\approx 10^{-4}$ cm in water.

Taking $\varphi(x, t)$ as the density of particles in the emulsion Perrin derived, with the help of Maxwell's distribution function, the differential equation

$$\varphi'_t = \tfrac{1}{2} \frac{\xi^2}{\tau} \varphi''_{x, x} \tag{9}$$

By introducing the Einstein diffusion coefficient D, he obtained the diffusion equation

$$\varphi'_t = D\varphi''_{x, x}, \quad \text{with} \quad D = \tfrac{1}{2} \frac{\xi^2}{\tau}. \tag{10}$$

Perrin then showed the possibility, already considered by Einstein, of relating D to the Avogadro number, N. In order to do that Perrin introduced a force, F, having the same value in each section of the cylinder containing the emulsion. The stability condition is attained when $F = 6\pi\zeta av$, where a is the radius of the particles, v their speed under the action of the force, F, and ζ is the viscosity. If the density of the particles is stationary, the flow under the action of the force F and that due to diffusion will be equal. Writing $nF \, dh = -dp$, where h is the height of the cylinder and dp, the difference of osmotic pressure between the two faces of a slide of thickness dh, Perrin obtained

$$D = \frac{RT}{N} \cdot \frac{1}{6\pi a\zeta}. \tag{11}$$

With the definition of D in Equation (10),

$$\frac{\xi^2}{\tau} = \frac{RT}{N} \cdot \frac{1}{3\pi a\zeta}. \tag{12}$$

Perrin gave an account of Einstein's theory of general Brownian motion. By a reasoning similar to the foregoing he required the equilibrium between the dynamical flow and the flow due to diffusion for a system varying with a parameter α, for which A^2 is the mean square deviation of the spontaneous variation of α, obtaining

$$\frac{A^2}{\tau} = \frac{RT}{N} \cdot \frac{2}{\kappa}, \tag{13}$$

where κ is a constant specific for the motion ($\kappa = 2\pi a^3 \zeta$ for the usual Brownian motion). In the case of rotational Brownian motion,

$$\frac{A^2}{\tau} = \frac{RT}{N} \cdot \frac{1}{4\pi a^3 \zeta}. \tag{14}$$

Perrin then gave an extensive report on his different experiments which provided a verification of Einstein's laws of Brownian motion. First, the direct observation with a microscope showed that the Brownian motions were perfectly irregular, and Perrin discussed other methods to verify this property.

If ε and ε' are mean displacements of particles performing Brownian motion at temperatures T and T' respectively, then according to Einstein

$$\frac{\varepsilon'}{\varepsilon} = \sqrt{\frac{T'}{T}} \sqrt{\frac{\zeta}{\zeta'}}. \tag{15}$$

Sedding had shown experimentally that for ultramicroscopic particles $\varepsilon'/\varepsilon = 2.2$, while the theoretical calculation gave 2.05. Perrin thought that the discrepancy was due to the effect of viscosity, and reported on some of his own measurements with particles of known diameters. The Avogadro constant, obtained by different methods, varied between 5.0×10^{23} to 8.0×10^{23}, but from his precise absolute measurements he obtained the value $N = 6.88 \times 10^{23}$.

With the help of observations on large particles of diameter of the order of 13 μ, Perrin was able to measure the rotational part of the Brownian motion. He found that, under his experimental conditions, the angle of rotation, which should be 14°, was 14.5°, a result in complete agreement with the theory.

Perrin pointed out that, for the sake of completeness, he had still to verify Einstein's relation

$$D = \frac{RT}{N} \cdot \frac{1}{6\pi a \zeta},$$

or the fact that the expression $RT/D6\pi a\zeta$ was the same for each emulsion.

Perrin gave an account of Einstein's calculation of the diffusion of sugar in water. He pointed out that Einstein's first formula[1] contained an error but Bancelin, working with Perrin, had discovered the error[2] which Einstein corrected[3]. Taking this correction into account, Einstein found that $N = 6.5 \times 10^{23}$. Perrin also discussed another experiment he had performed on the diffusion of large particles. In this experiment, performed by L. Brillouin, they first determined the coefficient of diffusion D by counting the number of particles collected on a glass plate which had been introduced into the solution. For the Avogadro number they thus obtained $N = 6.9 \times 10^{23}$ ($\pm 3\%$).

The conclusion was that the laws of perfect gases apply in all details to emulsions, thereby providing a secure experimental basis for the molecular kinetic theory and showing that the domain of its applicability is vast indeed.

Perrin proceeded to discuss other phenomena from which the existence of a molecular reality could be inferred and elements of this level of the structure of matter understood.

7.3. FLUCTUATIONS

Among the phenomena which lead to conclusions about the existence and structure of molecular reality, Perrin spoke of fluctuations as 'regimes of permanently varying inequalities in the properties of the microscopic portions of matter in equilibrium'. It was Smoluchowski who first drew attention to the problem of fluctuations by studying the variation of the density of a fluid in equilibrium.[4]

Just as in the case of temperature, the density of a fluid in equilibrium must vary continuously from place to place, and the 'accidental condensation' is given by $\gamma = (n - n_0)/n_0$, where n_0 is the concentration of particles for perfect equilibrium and n is the measured concentration. The probability of an accidental condensation, the value of which lies between γ and $\gamma + d\gamma$ is given by

$$W(\gamma)\,d\gamma = \sqrt{\frac{n_0}{2\pi}}\,e^{-n_0\gamma^2/2}\,d\gamma. \tag{16}$$

From this, by simple integration, Smoluchowski obtained the absolute value of the average condensation,

$$\int_0^\infty \gamma \sqrt{\frac{n_0}{2\pi}}\,e^{-n_0\gamma^2/2}\,d\gamma = \sqrt{\frac{2}{\pi} \cdot \frac{1}{n_0}}. \tag{17}$$

These fluctuations, which are very small in gases, can be easily verified for emulsions. Perrin recalled that Smoluchowski had also calculated the fluctuations around the critical point, and was able to give a reasonable explanation for critical opalescence in fluids.

Let a fluid, of gram-molecule M, be in thermodynamic equilibrium, and let v_0 be its volume for a uniform distribution under pressure p_0, then the mean square of the condensation is found to be

$$\overline{\gamma^2} = -\frac{RT}{N} \cdot \frac{1}{\varphi v_0 \dfrac{\partial p}{\partial v_0}}, \tag{18}$$

where

$$\varphi = \frac{n_0 M}{N}\,v_0.$$

Since at the critical point $\partial p/\partial v_0$ and $\partial^2 p/\partial v_0^2$ vanish, the condensation probability becomes

$$W(\gamma)\,d\gamma = k e^{-B\gamma^4}\,d\gamma, \tag{19}$$

with

$$-B = \frac{n_0 M}{RT}\,\frac{v_0^4}{2\cdot 3\cdot 4}\,\frac{\partial^3 p}{\partial v^3}.$$

Calculating $\partial^3 p/\partial v^3$ from the Van der Waals equation of state, one obtains $B = \frac{9}{14}\,n_0$, and the average value of the condensation becomes $1.13\,n_0^{1/4}$. In a cube containing hundred millions of molecules, the mean square condensation will be of the order of one percent. The length of the side of such a cube is of the order of magnitude of the wavelength of visible light, hence the relatively large fluctuations can cause opalescence.

Perrin pointed out that Smoluchowski's theory had been completed by Keesom, and was found to be in complete agreement with the experimental results of Kamerlingh Onnes.[5] This work was based on the theory of diffused light developed by Rayleigh and Lorentz by considering an ensemble of particles. Perrin calculated the intensity of light diffused by one centimetre cube of the fluid to be,

$$i = \frac{\pi^2}{18\lambda^4} \frac{RT}{N} (\mu_0^2 - 1)^2 (\mu_0^2 + 2)^2 \frac{1}{- v_0 \dfrac{\partial p}{\partial v_0}}, \tag{20}$$

where μ_0 is the index of refraction of the medium. Using this formula, the experimental results gave the value of Avogadro's number as 7.5×10^{23}.

Shortly after Keesom, Einstein, who did not know of Keesom's work at the time, established the theory of critical opalescence based on electromagnetic theory. He obtained the same result as Keesom.[6]

The blue colour of the sky, as is well-known from Rayleigh's work, can be explained by the diffusion of sunlight in the air. It is then possible to apply to this phenomenon the same formula for the intensity as given above and, by comparison of the light in the direction of the sun and in a direction at an angle with it, to obtain again a value for N. Various computations of N were made by using this method, one of the first being due to Lord Kelvin from the summit of Mont Rose in which the value of N was found to be between 3.0×10^{23} and 15.0×10^{23}. Bauer and Moulin[7] had made some spectro-photometric measurements of the sky from Mont Blanc, and obtained $N = 4.5 \times 10^{23}$ to 17.5×10^{23}.

Finally, Perrin pointed out that a phenomenon similar to the Brownian motion had been observed in liquid crystals. Lehmann had observed that uniaxial crystals could, between two analysers, still exhibit quickly disappearing spots of light similar to continuous sparking. Maugin had related this phenomenon to the Brownian motion.

Having reviewed the principal experiments and theories related to Brownian motion, Perrin proceeded to examine further evidence for the existence of molecules by analyzing results related to the electric charge. It is interesting to follow Perrin in this maze of experiments, all of which tend to the same goal – the determination of the existence of an elementary charge.

First on Perrin's list we find Helmholtz who was led, by a careful study of electrolysis, to the existence of an elementary electric charge. Townsend followed him in proving that the same elementary charge could be found in an ionized gas. Moreau made the same verification for the ions produced by flames. Another interesting way of obtaining the elementary charge was given by Maurice de Broglie, who combined the methods of charge displacement employed by Townsend and the Brownian motion. These experiments were performed with charged smoke granules, whose displacement was observed under the influence of an electric field. Finally, Pierre Weiss obtained similar results using ultramicroscopic particles.

After the experiments of M. de Broglie, numerous other proofs of the elementary electric charge were obtained, based on the capture of ions by dust particles or by liquid drops. The fundamental experiment along this line was that of C. T. R. Wilson (1897) who showed that droplets, obtained by adiabatic expansion in a cavity saturated with water vapor, are formed on the ions. Townsend and J. J. Thomson used this principle to determine the elementary charge by a determination of the number of ions and the number of drops. Thomson's method was improved by H. A. Wilson. All these methods were based on some kind of averaging over a number of particles depending on the counting of a certain number of drops or particles.

The real improvement in the measurements came with the study of individual charged particles. The beginning of these studies was made by H. A. Wilson, who computed the motion of a droplet under the action of an electric field and gravity. But Wilson made use of his equations for a cloud of droplets, where they were not fully justified. Millikan was the first to observe experimentally the motion of single drops, and to determine the charge on them with the help of the equation established by Wilson. An experiment performed by Ehrenhaft did not give the expected result. Perrin pointed out that the results of Ehrenhaft were based on particles which were more like sponges than spheres. Since Wilson's formula was based on Stokes' law, it was not surprising to find a great discrepancy between the results of Ehrenhaft and those of Millikan (who used spheres in his application of Stokes' law). Perrin described Millikan's experiments, and showed how it was possible to follow a single droplet ascending under the action of the electric field or falling under the gravitational field.

Taking into account the effect of Brownian motion one could obtain, as Fletcher did in Millikan's laboratory[8], the value of Ne. Using Townsend's equation,

$$Ne = 2RT \frac{\tau}{\xi^2} \frac{u}{H}, \tag{21}$$

Fletcher obtained, from 1700 observations on nine drops, the value $Ne = 28.8 \times 10^{13}$, a number which was found to be in agreement with the one given by electrolysis to within 0.5 per cent.

However, Perrin expressed doubts with regard to the exact value of the elementary charge thus obtained. The use of Stokes' law, according to him, was difficult to justify in a satisfactory manner. Perrin gave an account of the experiments which had been performed in his own laboratory to test the value of the coefficients used in Stokes' law.

After his review of the experiments which confirmed the existence of an elementary charge, Perrin added another class of experiments to his wide picture, i.e. those related to the genesis and fission of atoms. The central work in this field was due to Rutherford, who had shown that the α-particles, emitted in an 'explosion' during the fission of certain atoms, were ionized atoms of helium. Later on Crookes observed that these particles could be detected by counting the scintillations which they produced upon impact on phosphorescent materials.

The problem then was to determine the exact charge of α-particles. Regener[9] was the first to relate the number of emitted particles, counted from the scintillations produced by them and the saturation current produced in air by the same particles.

From this he arrived at a charge of the order of 8×10^{-10} e.s.u. This method was improved by Rutherford and Geiger[10] with the introduction of the so-called Geiger tube. However, the results so obtained, within experimental error, were the same. Another method was based on an absolute measurement of the charge by means of a Faraday cylinder. This method, used by Perrin, gave a value of the elementary charge of 4.65×10^{-10} e.s.u.

The counting of the number of emitted particles having been made possible, it was of interest to know the mass of the helium atom produced by the fission. Following the measurements of J. Dewar[11], Boltwood and Rutherford[12], Madame Curie performed a very precise measurement[13] with the help of which she obtained for N, the Avogadro number, the value 6.5×10^{23}.

Perrin described two other methods for obtaining N, one based on the measurement of the period of disintegration, giving $N = 7.1 \times 10^{23}$, and the other based on a measurement of the speed and energy of the α-particles compared with the amount of heat dissipated by the radium from which the particles were produced, giving the result $N = 6.2 \times 10^{23}$. Leaving the domain of radioactive disintegration, Perrin added to his general picture a brief review of Planck's calculation of N based on the black-body radiation formula, from which a value of $N = 6.16 \times 10^{23}$ had been obtained.

7.4. PERRIN'S RESULTS

Perrin's results could be summarized as follows:

Observed phenomenon		$N/10^{23}$
Viscosity of gases:	Volume of liquid	4.5
	Van der Waals equation	6.2
Brownian motion:	Distribution of particles	6.83
	Translational disturbance	6.88
	Rotational disturbance	6.5
	Diffusion	6.9
Random distribution of molecules:	Critical opalescence	7.5
	Blue of the sky	6.0
Charge of particles (in a gas)		6.4
Radioactivity:	Charge of α-particles	6.25
	Mass of helium	6.4
	Disintegration of radium	7.1
	Radiated energy	6.0
Black-body spectrum		6.4

Jean Perrin, in his masterly report to the first Solvay Conference, had covered every aspect then known that could shed light on the existence of molecules, and the discussion of his report that followed could therefore deal only with some minor points. Einstein did take the opportunity of giving an account of the work done by Weiss at Prague, pointing out the errors in Ehrenhaft's results.

REFERENCES

1. A. Einstein, *Ann. Physik* (1905).
2. M. Bancelin, *Comptes rendus* (1911).
3. A. Einstein, *Ann. Physik* (1911).
4. M. v. Smoluchowski, *Boltzmann-Festschrift*, p. 626 (1904).
5. H. Kamerlingh Onnes and W. H. Keesom, *Comm. fr. the Phys. Lab. Leyden* (1908), p. 104b; W. H. Keesom, *Ann. Phys.* **35**, 591 (1911). The work of Keesom was brought to the attention of Perrin by Einstein, and it was Einstein who developed the theory of Smoluchowski.
6. A. Einstein, *Ann. Physik*, **16**, 1275 (1910).
7. E. Bauer and M. Moulin, *Comptes rendus* (1910).
8. H. Fletcher, *Phys. Rev.* **33**, 81–110 (1911).
9. E. Regener, *Verh. Deut. Phys. Ges.* **10**, 78 (1908).
10. E. Rutherford and H. Geiger, *Proc. Roy. Soc.* **81**, 141 and 162 (1908); *Radium*, **5**, 257 (1908).
11. J. Dewar, *Trans. Roy. Soc.* **83**, 280 (1908), **85**, 410, (1910).
12. B. B. Boltwood and E. Rutherford, *Akad. Wiss., Vienna* (March 1911).
13. M. Curie, *Comptes rendus* (1910).

8. APPLICATION OF THE QUANTUM THEORY TO PHYSICO-CHEMICAL PROBLEMS

(W. Nernst)

In his report to the first Solvay Conference, Walther Nernst considered the advantages of applying quantum theory to several physico-chemical problems which had been treated by means of the molecular kinetic theory earlier.

Consider a monatomic isotropic substance (a solid) and assume that the motions of ions about their equilibrium positions are harmonic. For the purpose of calculating the heat capacity, Nernst assumed that the energy of the solid is given by the energy of the oscillators, i.e. by the sum of the kinetic and potential energies, which, in this case, are equal. Assuming further that the solid is surrounded by a gas, thermal equilibrium will be realized when the kinetic energy of the gas and that of the solid will be the same, leading to a specific heat equal to R for the solid.

Nernst reported that his experiments on many substances had shown that at low temperatures the specific heat decreased very rapidly, a contradiction already pointed out by Lord Rayleigh. The thermal equilibrium between the gas and the solid is determined by the same classical law, provided one does not take into account the theory of quanta.

Let N_0 be the number of atoms contained in a gram-atom, and let $E_0 = kT/N_0$ be the mean energy of the circular oscillation. According to Maxwell's distribution, $N = N_0 (1 - e^{-E/E_0})$. If the exchange of energy between the gas and the solid were forbidden, except by quanta of energy $\varepsilon = (R/N_0) \beta v = hv$, then the total heat capacity of a gram-atom will be given by the discrete sum,

$$\begin{aligned} W &= 3[\varepsilon(N_2 - N_1) + 2\varepsilon(N_3 - N_2) + \cdots] \\ &= 3\varepsilon N_0 (e^{-\varepsilon/N_0} - e^{-2\varepsilon/E_0}) + \cdots \\ &= \frac{3\varepsilon N_0}{e^{\varepsilon/E_0} - 1} = 3R \frac{\beta v}{e^{\beta v/kT} - 1}. \end{aligned} \tag{1}$$

After a simple derivation, Nernst obtained the Einstein formula,

$$\frac{\mathrm{d}W}{\mathrm{d}T} = 3R \frac{e^{\beta v/T} (\beta v/T)^2}{(e^{\beta v/T} - 1)^2} \tag{2}$$

for the atomic specific heat.

Nernst gave a long review of the different experimental results obtained by him and his collaborators on the specific heats of various substances. He showed that Einstein's predictions were in the right direction, but not in full agreement with the observations. The difficulties, according to Nernst, arose for different reasons, the chief being that in Einstein's formula it had been assumed that the absorption bands were all of the same type, an assumption which was found to be wrong from experiments done in Nernst's laboratory. Nernst then discussed a new formula for the specific heat derived phenomenologically by Lindemann and himself,

$$c_v = \tfrac{3}{2}R \left[\frac{(\beta v/T)^2 \, e^{\beta v/T}}{(e^{\beta v/T} - 1)^2} + \frac{(\beta v/2T)^2 \, e^{\beta v/T}}{(e^{\beta v/T} - 1)^2} \right], \tag{3}$$

which, like the Einstein formula, depends on v only. Nernst cited numerous results which showed excellent agreement with his formula, and the next step for him was to justify it.

By integrating c_v, the total heat capacity was found to be,

$$W = \tfrac{3}{2}R \left(\frac{\beta v}{e^{\beta v/kT} - 1} + \frac{\beta v/2}{e^{\beta v/2T} - 1} \right). \tag{4}$$

The only interpretation which seemed reasonable to Nernst was the following: Since both terms in the expression for W have the same value at high temperatures, and since the law of Dulong and Petit is based on the fact that at high temperatures the potential and kinetic energies are the same, Nernst suggested that the two terms vary differently at low temperatures. He assumed that,

$$W_1 = \tfrac{3}{2}R \frac{\beta v/2}{e^{\beta v/2T} - 1} \tag{5a}$$

is the potential energy, and

$$W_2 = \tfrac{3}{2}R \frac{\beta v}{e^{\beta v/T} - 1} \tag{5b}$$

is the kinetic energy.

This interpretation, however, gave rise to great difficulties, of which the principal one was related to the thermal conductivity. According to Nernst's theory the latter should become very small at low temperatures, just the opposite of what was observed.

Nernst then reported a formula, due to Lindemann, relating the oscillation frequency to the melting point T_s. The formula was based on the assumption that, at the melting point, the amplitude of atomic oscillations becomes appreciably equal to atomic size, and, if V is the atomic volume, Lindemann's formula could be written as,

$$v = 2.8 \times 10^{12} \sqrt{\frac{T_s}{mV^{2/3}}} \tag{6}$$

where the frequency v was obtained from Nernst's formula (3) for the specific heat c_v. He concluded that the agreement between the measured and computed frequencies was very good.

Finally, without giving details, Nernst reported some general results on thermal conductivity, electrical conductivity, Peltier effect and vapour pressure.

DISCUSSION

In the discussion of Nernst's report, Lorentz pointed out that his decomposition of the oscillations into circular vibrations was rather arbitrary. By calculating the mean energy of a three-dimensional oscillator, Einstein pointed out that Planck's theory, in its present state, was in contradiction for systems with more than one degree of freedom.

There was a lengthy discussion of Nernst and Lindemann's formula for the specific heat, in which Einstein, Langevin, Lindemann, Rubens, Poincaré, Rutherford, Nernst, Madame Curie, Kamerlingh Onnes, and Lorentz took part. However, the discussion concentrated on isolated technical points and did not contribute to the larger conceptual picture. Einstein did make a penetrating remark about Nernst's heat theorem and the problem of the specific heat. He said that, 'In my opinion, Nernst's theorem cannot be derived from the fact that the specific heat disappears near absolute zero, although its legitimacy becomes more probable. The question that must be answered is whether, sufficiently close to the absolute zero, a system may go from a state A to a state B in a reversible isothermal manner *without the exchange of heat*. From this it does not follow that the molecular agitation is feeble, because the passage from A to B can only be effected by utilizing the little thermal agitation that remains, and could correspond to an exchange of heat with the outside that is considerable compared to the energy of thermal agitation present. In this case, the passage of the system from the state A to the state B would be completely impossible at absolute zero. Nernst's theorem enunciates the quite plausible hypothesis, that the change from A to B is always possible in a purely static manner from the point of view of molecular mechanics.'

Einstein also expressed the point of view that the large thermal conductivity of insulators could not be understood either on the basis of ordinary mechanics or with the help of the hypothesis of energy quanta.

9. ELECTRICAL RESISTANCE

(H. Kamerlingh Onnes)

Kamerlingh Onnes, in his brief report, pointed out that the limit of resistivity in many metals was due to the remaining impurities. He had found that in many different materials the resistivity was independent of the temperature around the temperature of liquid helium, and assumed that this unexpected behaviour could be interpreted with the help of the theory of quanta.

10. THE QUANTUM OF ACTION AND NON-PERIODIC MOLECULAR PHENOMENA

(A. Sommerfeld)

Arnold Sommerfeld began his report by comparing the theory of quanta with a mechanics in which only periodic phenomena were considered. However, the way of generalizing it had already been shown by Planck with his notion of the quantum of action. In order to generalize Planck's theory, Sommerfeld assumed that during the exchanges of energy with matter the duration of the process was shorter for larger energies; conversely, the time of exchange for a small quantity of energy could be large. Sommerfeld noted: 'It seems necessary to admit that the time required for matter to absorb or give up a certain amount of energy is the smaller as the energy is larger, in such a way that the product of the energy and the time – or rather, according to the precise definition that we shall give, the integral of the energy with respect to the time – is determined by the magnitude h.'[1]

Sommerfeld illustrated his point of view by recalling that the cathode rays generated hard or soft Röntgen rays depending on their speed, and the speed of the cathode rays was related to the duration of the stopping of cathode particles. The same was true for the process of β-emission, and that the more or less hard γ-rays were simultaneously emitted.

Mathematically, Sommerfeld wrote his assumption in the form

$$\int_0^\tau H \, dt = \frac{h}{2\pi},$$ (1)

where H is the Hamiltonian of the system and τ is the duration of action for the exchange of energy. He showed, following a derivation given by Planck, that his action principle remains valid in relativity, though H now took the form $H = c^2 m_0 \sqrt{[1 - (v/c)^2]}$. Sommerfeld discussed the action principle in some detail and gave examples.

10.1. Röntgen Rays

Immediately after their discovery by Röntgen, Schuster and Stokes in England and Wiechert in Germany developed a theory according to which the Röntgen [X-] rays were produced by the electromagnetic pulses which were created by the stopping of the particles emitted by the cathode. This theory could explain a great many phenomena but suffered from one important restriction: it could explain only the polarized part of the Röntgen rays. The non-polarized part, which was most important in heavy metals, could be considered as a fluorescence of the anti-cathode. Sommerfeld explained that the two phenomena were different in nature – the polarized part being electromagnetic in the classical sense, while the other was related to some intra-atomic interactions.

Sommerfeld gave a brief account of the theory of electromagnetic pulses. What happens to the trajectory of the particle from a point O_1 at impact to the point O_2 at its stopping? At time t the pulse is situated between two spheres of radii $r_1 = ct$ and $r_2 = c(t - \tau)$, where τ is the time of stopping. The emission is thus non-symmetrical, leading to the polarization. He showed that if E_r is the total energy of the Röntgen rays emitted, and E_{pol} the part of that energy due to polarized rays, then $E_{pol}/E_r = \frac{1}{4}$.

With E_k as the energy of the cathode rays, the experiments had yielded $E_r/E_k = 1.07 \times$
$\times 10^{-3}$ or $E_{pol}/E_k = \frac{1}{6} \times 10^{-3}$.

Assuming that the potential energy of the electron (i.e. the affinity of the electron
for the atom) is negligible in comparison with its kinetic energy T, Equation (1)
becomes

$$\int_0^\tau T \, dt = \frac{h}{2\pi}. \tag{2}$$

Considering a uniform decrease of the speed, i.e., $\dot{v} = \beta c/\tau$, $\beta = v/c$, Sommerfeld
obtained

$$E_k \tau = \frac{3h}{2\pi}. \tag{3}$$

This result could also be written in the form,

$$\frac{E_{pol}}{E_k} = \frac{e^2}{9hc} \frac{\beta^2}{\sqrt{(1 - \beta^2)}}. \tag{4}$$

Taking for E_{pol}/E_k the values determined experimentally, and $e = \sqrt{4\pi} \, 4.7 \times 10^{-10}$,
Sommerfeld obtained

$$h = 10.6 \times 10^{-27} \tag{5}$$

in relatively good agreement with Planck's value. The speed of the electron being too
small to be relativistic, Sommerfeld showed that the effect of a relativistic treatment
would be negligible.

Sommerfeld found that the width of a pulse,

$$l = \beta \frac{\lambda}{2} = \frac{\beta}{2} \cdot \frac{3ch}{2\pi E_k},$$

was of the order of 3×10^{-10} cm, a result which demonstrated that the particles were
stopped by a single collision with the molecules. He pointed out that l was independent
of the substance, an interesting fact which had been ascertained at his institute in
Munich. He recalled that in 1907 W. Wien and J. Stark had attempted independently
to calculate λ by replacing E_k by $h\nu = hc/\lambda$, i.e. by considering E_k as a quantum of
frequency ν. Their result, $\lambda = ch/E_k$, was equivalent to Sommerfeld's formula to within
an unimportant factor.

10.2. γ-RAYS

The similarity between the cathode rays and β-emission was striking, and clearly the
theory of the electromagnetic pulses was, according to Sommerfeld, still valid, together with the necessary modification due to an acceleration (not a stopping) of the
relativistic electron. The energy E_γ of the accompanying γ-ray is

$$E_\gamma = \frac{e^2 \dot{v}}{6\pi c^2} \frac{\beta}{\sqrt{(1 - \beta^2)}}. \tag{6}$$

With this formula Sommerfeld obtained a (pear-shaped) distribution of the energy situated inside a cone along the axis defined by β, but never in contact with this axis. This theory seemed to be in agreement with the measurements of Edgar Meyer.[2]

The kinetic energy of the β-particle is given by the formula $E_\beta = (m - m_0)c^2$, where $m = m_0(1 - \beta^2)^{-1}$. The total energy is given by taking into account the potential energy of the Poincaré stresses. Sommerfeld wrote,

$$E = E_\beta + E_0 = m_0 c^2 (1 - \beta^2)^{-1}. \tag{7}$$

If τ is again the time during which the particle is accelerated, then by a reasoning similar to the case of Röntgen rays, Sommerfeld obtained

$$E_\beta \tau = \frac{h}{\pi b}, \quad \text{where} \quad b = 1 + \frac{m_0 c^2}{E_\beta} \left[\frac{(1 - \beta^2)^{-1}}{2\beta} \log \frac{1 + \beta}{1 - \beta} - 1 \right]. \tag{8}$$

The larger the energy E_β, the shorter would be the duration τ of the emission.

As he had done for Röntgen rays, Sommerfeld derived a formula for E_γ / E_β from which he obtained a value for h. The calculations for different materials gave him the values of h as oscillating between 0.6×10^{-27} and 2×10^{-27}.

Sommerfeld analyzed the different possible causes of errors. These seemed to be related to the inaccuracy of the experimental results, the assumption of a constant acceleration of the β-particles during the time τ of the emission, and the assumption that $H = E_\beta$ is the kinetic energy only. He noted that, in another calculation, when he took E as the total energy, he had obtained $h \cong 12 \times 10^{-27}$.

10.3. THE PHOTOELECTRIC EFFECT

Sommerfeld discussed his method of treating the photoelectric effect, which was different from the theories of Lenard and Einstein. He considered the electron as bound to an atom by a quasi-elastic force of the type, $-fx$, and subject to an external force, $eF = eE \cos nt$, due to the incident electromagnetic wave ($n = 2\pi\nu$, ν being the frequency of light). The motion of the electron is then given by,

$$m\ddot{x} + f(x) = eF = eE \cos nt, \tag{9}$$

and its kinetic and potential energies are, respectively,

$$T = \frac{m}{2} \dot{x}^2 \quad \text{and} \quad U = \frac{f}{2} x^2. \tag{10}$$

On the basis of Sommerfeld's action principle an electron would be released when the action integral is satisfied,

$$W = \int_0^\tau (T - U)\, dt = \frac{h}{2\pi}. \tag{11}$$

On integration by parts,

$$W = \frac{m}{2} x\dot{x} - \frac{e}{2} \int_0^\tau xF \, dt \, ;$$

x and $\dot{x}$ oscillate rapidly with a slowly varying amplitude, and W shall be equal to $h/2\pi$ at a maximum, i.e. for $dW/dt = 0$. Equivalently, this gave Sommerfeld the condition that

$$\dot{x}^2 = \frac{f}{m} x^2 \, ,$$

and he obtained

$$T = hv_0 + \frac{en_0}{2} \int_0^\tau xF \, dt, \tag{12}$$

where $n_0 = \sqrt{(f/m)}$ and v_0 is the eigen-frequency of the system $(n_0 = 2\pi v_0)$.

For $v = v_0$, the virial term on the right hand side of the above condition could be shown to be zero, a result which had already been obtained by Einstein and, with Einstein, Sommerfeld concluded that 'the emission of the photoelectric particles is independent of the intensity of the incident radiation and is determined by its frequency in a universal manner.'

Integrating the differential equation for the motion of the electron, in the case where n is only slightly different from n_0, Sommerfeld obtained,

$$T = hv_0 - \frac{(n_0 eE)^2}{2m(n^2 - n_0)^2} (\varepsilon - \sin \varepsilon), \tag{13}$$

with $\varepsilon = (n - n_0)\tau$. Since ε is a function of τ, Sommerfeld introduced a new relation between τ and T by noting that for $t = \tau$, $T = \bar{U}$, where $\bar{U}$ is the average value of U during the last oscillation, and obtained

$$T = \frac{(n_0 eE)^2}{2m(n^2 - n_0^2)} [1 - \cos \varepsilon]. \tag{14}$$

The comparison of the two expressions, (13) and (14), for T gave,

$$1 - \cos \varepsilon + \varepsilon - \sin \varepsilon = 2mv_0 h \left[\frac{n^2 - n_0^2}{n_0 eE} \right]^2. \tag{15}$$

By plotting this curve, Sommerfeld showed that for $n > n_0$ a photoelectric effect would take place, but that it would disappear for $n < n_0$. This result corresponded to the law given by Stokes for fluorescence. For $\varepsilon = 0$, $T = hv_0$, as already shown, but the maximum of the curve was found to be at $\varepsilon = -\pi/2$, for which

$$T = hv_0 \cdot \frac{2}{4 - \pi} = 2.3 \ hv. \tag{16}$$

Comparing with the experimental results of Wright[3] on aluminium, Sommerfeld showed that his theory was in agreement with these, while Einstein's theory of the photoelectric effect was not. In his view, the measurements of Wright led to the conclusions that: (1) the maximum of the photoelectric energy did not vary linearly with the frequency; (2) the photoelectric effect was selective and was influenced by the eigen-frequencies of the atom. Sommerfeld argued in favour of the second conclusion, and showed that a similar conclusion could be drawn from a formula derived by Lindemann[4] which showed complete agreement with the results of Pohl and Pringsheim[5] for Rb, K, Na and Ba.

He examined the factor 2π which occurred in his general assumption, Equation (1). This choice had been made in order to recover Einstein's results. He thought, however, that this choice could not really be justified, and that other possibilities should be considered. One of these would be to recover Einstein's law for the easiest emission, i.e., for the smallest value of τ. In this case, the result of his calculation gave $\int_0^\tau H \, d\tau = \frac{1}{4} h$. Sommerfeld discussed the difficulties of his theory, especially the prediction of a time τ much larger than what had been known from experimental results.

10.4. THEORY OF THE IONIZATION POTENTIAL

The ionization in a gas and in photoelectric effect presented certain similarities. In both cases the phenomenon was the liberation of an electron by the action of an electromagnetic field, the difference being in the nature of the electromagnetic field itself. Sommerfeld reported on his method of calculating the ionization potential, based on his theory of the action integral. For helium, he obtained the ionization potential $V = 9.3$ volts, his result being of the same order of magnitude as Townsend's[6], but somewhat smaller (Townsend's $V = 14.5$ volts). He assumed that the discrepancy was due to the fact that he had neglected the potential energy or affinity of the electron in the atom. Using Townsend's result, he calculated the affinity U.

10.5. ENERGY QUANTA AND PLANCK'S EMISSION QUANTA

In the last part of his report, Sommerfeld compared Planck's theory of black-body radiation and Einstein's explanation of the photoelectric effect, using light quanta, with his own theory of the element of action. For Sommerfeld, the important difference lay in the fact that his theory was 'reconcilable with classical electrodynamics', whereas the others were in complete disagreement with it. To emphasize this point, he recalled that the times τ, computed in the different cases he had considered, had all come from electrodynamic calculations. He showed that, in the theory of Röntgen rays, Planck's assumptions would lead to $E\tau = h$, whereas his theory gave $E_k \tau = 3h/2\pi$. These results were in complete disagreement, since E/E_k was found to be of the order of 10^{-3}.

DISCUSSION

Einstein immediately noticed that a difficulty existed in the physical interpretation given by Sommerfeld. The function $T - U$ could not be supposed to be zero for a free particle, because the existence of a free particle shared in the corresponding parts of the elements of action, and this depended upon the speed of the frame of reference.

Henri Poincaré posed a series of questions on the application of Sommerfeld's ideas of the element of action to systems of several particles. One problem was to determine as to which part of the interaction potential would be attributed to a certain particle when the latter was considered separately. Poincaré gave an example showing that Sommerfeld's principle, if applied to a molecule, considered as a gun, would imply that the recoil time of the gun was longer than the time required by the bullet to take on its speed. Sommerfeld's answer was essentially that his action principle was not general enough, and his intention was to modify it when it would become necessary.

Lorentz pointed out some of the difficulties that would arise in the case of the interacting particles if they were treated on the basis of Sommerfeld's action principle.

M. Brillouin noted that the problem would arise not only for the potential energy (as mentioned by Poincaré) but also for the kinetic energy, which was different in different coordinate systems.

Langevin discussed the difference between Planck's principle, relating the quantum of action to an extension in phase space, i.e. a statistical problem, and Sommerfeld's action principle which was a dynamical principle.

Answering a question raised by Rutherford concerning the stopping of cathode particles by a single atom, Madame Curie gave certain results which showed that cathode rays were progressively slowed down by the matter through which they passed. She noted that it could be taken for granted that some particles were stopped only after a long range through matter, accompanied by a series of deviations in their trajectories, but she could not ascertain whether certain particles would be stopped by a single collision.

Following a remark of Lorentz showing the weakness of another model of cathode rays presented by Wien, Einstein pointed out that Planck's quantum theory was able to yield, for Röntgen rays, results of the type obtained by Sommerfeld. His conclusion was based on a direct quantization of the kinetic energy of the electron and on the assumption that in a collision an electron emits radiation such that the energy comprised within the interval $d\nu$ is

$$\frac{1}{3}\frac{e^2\nu^2}{c^3}\,d\nu.$$

Sensing a weakness in Einstein's argument, Lorentz pointed out that, 'Einstein has decomposed any arbitrary motion of a particle into a Fourier series, every term of which corresponds to a given frequency ν. If I understand it correctly, in his way of looking at it there would be a radiation corresponding to one of these terms if the value of its proper quantum of energy $h\nu$ were lower than the total amount of energy available.'

Einstein agreed that Lorentz' objection touched on a weak point of the theory. He thought that in the 'primitive' form of the theory of quanta, only a single quantum of a given frequency would be emitted in a collision, in such a way that the result of the integration would give an average value as if it were for a large number of collisions. This way of looking at things, Einstein thought, was certainly artificial, and it showed the weak side of the theory of monochromatic energy quanta. According to Sommerfeld's idea, the frequencies, for which $\nu > T/h$, may not be emitted by the collision of the electron, because the collision was not instantaneous. Hence the higher terms in the Fourier series would not occur in the field of emission, because they also did not occur in the Fourier series which expresses the acceleration during the collision. Einstein thought that Sommerfeld's idea had the great advantage of preserving Maxwell's equations in the calculation of the field of emission; it, however, created other great difficulties which could not be ignored.

Madame Curie thought that there were reasons in favour of an electron possessing a certain speed before being ejected out of the atom. She spoke of a model of the atom, similar to J. J. Thomson's, in which the electron would already have a certain speed. But Poincaré wondered why these electrons would then not produce electromagnetic radiation.

There followed a long discussion between Madame Curie and Rutherford on the process of the creation of γ-rays in β-decay. She did not believe that it was experimentally established that the stopping of the β-particles was accompanied by γ-rays. Rutherford emphasized the complexity of β-radiation, and also pointed out that experiments had indeed proved that γ-rays were produced by β-rays.

Lorentz wondered why the action in Sommerfeld's principle, which should change continuously, had a positive sign. From the discussion on this point it emerged that the sign had been chosen in such a way that Stokes' law, or Einstein's law, would be satisfied.

REFERENCES

1. This was the first statement of the dependence of the time-energy product on Planck's constant.
2. Edgar Meyer, *Sitzber. Berliner Akad.* **32**, 647 (1910).
3. Wright, *Physik. Z.* **12**, 338 (1911).
4. F. A. Lindemann, *Berliner Phys. Ges.* **13**, 482 (1911).
5. R. Pohl and E. Pringsheim, *Berliner Phys. Ges.* **13**, 474 (1911).
6. J. S. E. Townsend, *The Theory of Ionisation of Gases by Collision*, London, pp. 24–27.

11. KINETIC THEORY OF MAGNETISM AND THE MAGNETONS

(P. Langevin)

The kinetic theory of para- and ferro-magnetism allows one to calculate, based on experimental results, the molecular magnetic moments. The simplest possible case is that of dilute paramagnetic substances, paramagnetic gases like oxygen or dilute solutions of paramagnetic salts. For such substances, Curie's empirical law gives the susceptibility as being inversely proportional to the absolute temperature,

$$\chi_m = \frac{C_m}{T}, \tag{1}$$

where C_m is the Curie constant per gram-molecule.

In his report to the first Solvay Conference, Paul Langevin sought, first of all, to derive Curie's law for paramagnetic susceptibility on the basis of the kinetic theory. In order to do so, Langevin considered the substance placed in an external magnetic field, H, and assumed that each molecule had a magnetic moment μ.

Let I_0 and I be the resultant magnetic moments when all the molecular moments are aligned along an axis at the temperature T. With the help of the Maxwell–Boltzmann distribution, Langevin obtained,

$$\frac{I}{I_0} = \frac{\text{ch}\, a}{\text{sh}\, a} - \frac{1}{a} \tag{2}$$

with $a = I_0 H/RT$. If a is small compared to unity, he got, after expansion, $I/I_0 = a/3$, and finally

$$\chi_m = \frac{I_0^2}{3RT} \tag{3}$$

in agreement with the Curie law, the Curie constant C_m being given by $C_m = I_0^2/3R$. Taking for C_m the value determined experimentally by Curie, Langevin obtained for I_0 a number of the order of one for saturated iron.

As a second step, Langevin introduced Weiss' hypothesis concerning the interaction between the molecules in the form of a second magnetic field, proportional to the magnetization and dependent on a coefficient N, different for each substance. With this assumption he again obtained,

$$\frac{I}{I_0} = \frac{\text{ch}\, a}{\text{sh}\, a} - \frac{1}{a}, \tag{2a}$$

with $a = I_0/RT\,(H+NI)$, so that a could now become much larger. A graphical solution of this equation showed that the saturation would be realized nearly completely at very low temperatures. It was at these low temperatures that Kamerlingh Onnes and Weiss had made their measurements of the molecular magnetic moments.

Another characteristic of Langevin's solution was the existence of spontaneous magnetization for $RT/NI_0^2 < \frac{1}{3}$. The temperature $\theta = NI_0^2/3R$, called the Curie temperature, corresponded to the disappearance of this phenomenon.

Taking into account the fact that a is always small, Langevin obtained

$$I\,(T-\theta) = C_m H. \tag{4}$$

As before, it was possible to take the experimental value for C_m and to calculate I_0 from $I_0 = \sqrt{3RC_m}$.

The discovery of the magneton by Weiss arose from the calculation of I_0. Weiss had noticed that C_m was a constant for a large domain of the temperature, but then suffered discontinuous changes, its values for these different domains being integers. Langevin found that by taking the value 1123.5 for the magneton, he was able to show that, to within a good approximation, it was a multiple of C_m.

Finally, by using Sommerfeld's action principle for the motion of the electron around the atom, a motion which is assumed to give the magnetic moment, and by taking a potential $U\,(r) = 1/n \cdot A/r^n$, Langevin obtained

$$I_0 = \frac{m}{Me} \cdot \frac{h}{8\pi} \frac{n}{n+2} \tag{5}$$

for the magnetic moment of a gram-molecule. With $n=1$, $h=6.5 \times 10^{-27}$, $M=7 \times 10^{23}$ (Avogadro's number), and $e/m = 1.77 \times 10^7$, Langevin found $I_0 = 1080$ c.g.s. units, while Weiss had obtained experimentally, $I = 1123.5$ for the saturation of Ni at low temperatures, thus showing a rather good agreement.

DISCUSSION

In the discussion, Wien wondered why Langevin had applied Sommerfeld's theory instead of Planck's, since the motion was periodic. Langevin showed that with Planck's assumptions the results would be in disagreement with the experiments. Langevin and Poincaré also observed, in answer to Hasenöhrl, that for the phenomena under consideration Maxwell's distribution was still valid.

12. THE PROBLEM OF THE SPECIFIC HEATS

(A. Einstein)

Albert Einstein, in his report, pointed out that it had been in the domain of specific heats, particularly in the exact derivation of the specific heat of a monatomic gas from the equation of state, that the kinetic theory had obtained its most beautiful confirmation. And, again, it was in the domain of specific heats that serious insufficiency of the application of the kinetic molecular mechanics had been noticed.

Einstein's first paper on the quantum theory of specific heat had been written in

November 1906 and published in 1907.[1] By the late fall of 1911 when Einstein gave, at the first Solvay Conference, his report on the current status of the problem of specific heats, his theory of 1907 had obtained considerable experimental support, which had largely come from the work done in Walther Nernst's laboratory in Berlin. Incidentally, it was this experimental verification of Einstein's ideas in the crucial domain of specific heat, one of Lord Kelvin's dark clouds which had hovered over the molecular kinetic theory, that had made Nernst into an enthusiastic supporter of quantum theory.

12.1. SPECIFIC HEAT OF MONATOMIC SOLIDS

According to the molecular kinetic theory, the atomic heat of a substance (or 'molecular heat' in the case of polyatomic bodies) is defined as the product of its specific heat and its atomic (or molecular) weight, i.e. it is that amount of heat which must be supplied to a gram-atom (or gram-molecule) of the substance in order to raise its temperature by one degree. The thermal content of a monatomic solid, say a crystal, is nothing more than the energy of the elastic vibrations of its atoms, which are arranged in the form of a space-lattice, about their positions of equilibrium. If classical statistics is applied to these vibrations, in particular the law of equipartition of kinetic energy, one arrives at the following conclusion: The mean kinetic energy of an atom vibrating in space, i.e. with three degrees of freedom, is $\frac{3}{2} kT$, and its mean potential energy is $3 kT$.[2] Considering now a gram-atom of the substance, the mean energy is given by

$$\bar{E} = 3 kT N = 3 RT, \tag{1}$$

where N is Avogadro's number and R is the gas constant, and T, the absolute temperature. The atomic heat of the body at constant volume becomes,

$$C_v = \frac{d\bar{E}}{dT} = 3R = 5.94 \text{ cal/deg}. \tag{2}$$

This is the law of Dulong and Petit [3], according to which the atomic heat (at constant volume) of monatomic solid bodies has the value 5.94 cal/deg, independently of the temperature. This law is obeyed by many elements more or less closely. However, there are elements which, far from following this rule, show systematic differences, especially at low temperatures. In fact, as early as 1875 F. H. Weber[4] found that the atomic heat of diamond at $-50\,°C$ is about 0.75 cal/deg. The atomic heats of other elements (boron, beryllium, silicon) were also found to be much too small at ordinary temperatures. It appeared that the defect from Dulong and Petit's normal value occurs generally at low temperatures, and becomes more pronounced the lower the temperature. The classical molecular kinetic theory offered no solution of these low values of the atomic heat.

    Einstein was the first to recognize[1] that quantum theory would solve the problem which had arisen in the molecular kinetic theory of the specific heats. Just as in the theory of radiation, the method of classical statistics leads of necessity to a wrong law in the field of atomic heats. Hence, in this case also, the law of the equipartition of energy had to be abandoned.

Einstein had reworked Planck's derivation of the expression for the average energy of one of the oscillators that absorb and emit electromagnetic radiation. Einstein began his 1907 paper with a new derivation of the equation for the average energy, going back to the fundamentals of statistical mechanics which he had developed independently a few years earlier.[4a] He again showed that a treatment based on classical methods gave the equipartition result for the average energy $\bar{U}$ of an oscillator,

$$\bar{U} = \frac{R}{N} T. \tag{3}$$

This result was based on the classical assumption that equal regions of phase space should be given equal weights in the process of averaging. In Einstein's interpretation of the probabilities used in statistical mechanics, this assumption meant that the system spent equal fractions of any long time interval in regions of equal phase volume. Now in order to avoid the equipartition result and to obtain Planck's expression for the average energy of an oscillator, this assumption had to be replaced by another – i.e., only those regions of phase space in which the energy took on discrete values 0, ε, 2ε, ..., $n\varepsilon$, ... were to have non-zero weights, and these integral multiples of the unit energy ε of the oscillator were to be weighted equally. On the basis of this assumption, the average energy of the oscillator obtained the value

$$\bar{U} = \varepsilon \left[e^{N\varepsilon/RT} - 1 \right]^{-1}. \tag{4}$$

If $\varepsilon = (R/N)\beta v$, where $\beta = h/k$, and v the frequency of the oscillator, then Equation (4) could be written as,

$$\bar{U} = \left(\frac{R}{N_0} \right) \beta v \left[e^{\beta v/kT} - 1 \right]^{-1}. \tag{5}$$

Then, using Planck's result based on classical electromagnetic theory, the frequency spectrum of the black-body radiation could be written as

$$\varrho (v, T) = \left(\frac{8\pi v^2}{c^3} \right) \bar{U}, \tag{6}$$

where c is the velocity of light.

In Planck's notation Equation (6), with Equation (5), gives the usual form of the Planck distribution law for black-body radiation,

$$\varrho (v, T) = \left(\frac{8\pi v^2}{c^3} \right) (hv) \left[e^{(hv/kT)} - 1 \right]^{-1}. \tag{7}$$

At this point Einstein sought to modify kinetic theory to bring it in accord with the radiation law, raising a major point of principle. Einstein noted: 'While up to now molecular motions have been supposed to be subject to the same laws that hold for the motions of the bodies we perceive directly (except that we also add the postulate of complete reversibility) we must now assume that, for ions which can vibrate at a definite frequency and which make possible the exchange of energy between radiation and matter, the manifold of possible states must be narrower than it is for the bodies in our direct experience. We must in fact assume that the mechanism of energy transfer is such that the energy can assume only the values 0, hv, $2hv$, ..., nhv, ...'[1]

This was not all, and Einstein proceeded to state: 'I now believe that we should not be satisfied with this result. For the following question forces itself upon us: If the elementary oscillators that are used in the theory of the energy exchange between radiation and matter cannot be interpreted in the sense of the present kinetic-molecular theory, must we not also modify the theory for the other oscillators that are used in the molecular theory of heat? There is no doubt about the answer, in my opinion. If Planck's theory of radiation strikes to the heart of the matter, then we must also expect to find contradictions between the present kinetic-molecular theory and experiment in other areas of the theory of heat, contradictions that can be resolved by the route just traced. In my opinion this is actually the case, as I try to show in what follows.'[1]

Einstein had fully recognized that classical theory was inadequate for dealing with the problems that had arisen, and had placed himself on record that there had to be a quantum *theory* which would deal with the properties of both matter and radiation. The contradictions which Einstein had immediately in mind concerned the violations of the equipartition theorem that were exhibited in the specific heats of solids and the Dulong and Petit's law.

By 1906 there were good reasons to believe that atoms had an internal structure and that they contained electrons, in some way. Einstein noted in particular Paul Drude's work on dispersion which showed that, while the infrared-absorption frequencies could be assigned to ionic vibrations, ultraviolet-absorption frequencies appeared to be associated with electronic vibrations. But if this were indeed so then once again the equipartition theorem would demand too much, since it would require a full contribution of $(R/N)\,T$ from each electronic vibration, and the heat capacity would have to be far greater than the value given by Dulong and Petit's law.

Einstein now proceeded to resolve the contradictions with one stroke. He decided that if 'Planck's theory strikes to the heart of the matter,' and the quantum hypothesis possesses universality, then the average energy of any oscillator is not given by the equipartition value, $(R/N)\,T$, Equation (3), but by Equation (4), and in this case the energy and the specific heat depend on the frequencies of the atomic vibrations in the solid. Einstein took all atomic vibrations to be independent and of the same frequency v, recognizing explicitly that in doing so he was making an oversimplification.

The energy content of the gram-atom will therefore be,

$$\bar{E} = 3R\beta v\left[e^{(\beta v/T)} - 1\right]^{-1}$$
$$= \frac{3Nhv}{e^{hv/kT} - 1}, \tag{8}$$

from which is obtained the atomic heat at constant volume, Einstein's formula, by differentiation,

$$C_v = \frac{d\bar{E}}{dT} = 3R \cdot \frac{x^2 e^x}{(e^x - 1)^2},\tag{9}$$

where

$$x = \frac{hv}{kT}.$$

According to this, the atomic heat of monatomic solid bodies is not a constant which is independent of the temperature, as Dulong and Petit's law requires, but is a function of v/T, and is therefore a function of the temperature for a body with fixed v. Its form is such that for $T=0$ (i.e., $x=\infty$) the atomic heat is itself zero. and then increases gradually with increasing temperature, approaching asymptotically at high temperatures (i.e., with small x) the classical value $3R$. Dulong and Petit's law is thus true only in the limit of small values of hv/kT, i.e. for low frequencies of atomic vibration, or high temperatures, exactly as is the case with Rayleigh's law of radiation. The departures from Dulong and Petit's law, in passing from high to low temperatures, become marked the sooner the greater the frequency of the atoms. In the case of diamond, for instance, the specific heat approaches the Dulong–Petit value when it is heated to temperatures of over $1000\,°C$, and falls to almost a tenth of that value when it is cooled to $-50\,°C$.

12.2. NERNST'S THEOREM AND ATOMIC HEATS

The absolute temperature scale had been introduced by Lord Kelvin, and the absolute zero of this scale found a physical significance in a thermodynamical law proposed by Walther Nernst in December 1905. Nernst's Heat Theorem, often called the Third Law of Thermodynamics, established an essential relation between the thermal behaviour of matter at temperatures near absolute zero and important problems of physical and chemical interest. It states: If we regard a system of condensed (i.e. liquid or solid) bodies, which passes isothermally at temperature T from one state to another, and if A is the maximum work which can be gained from this reaction, then $dA/dT=0$ for the limit $T=0$. That is, in the immediate neighbourhood of absolute zero, the maximum work which can be gained is independent of the temperature. By applying the two laws of thermodynamics, this statement can also be formulated by saying that in the neighbourhood of the absolute zero all processes develop without change of entropy or, in the words of Planck, that at the absolute zero of temperature the entropy of every chemically homogeneous body is equal to zero.

The consequence of Nernst's heat theorem is that in the immediate neighbourhood of absolute zero, the atomic heat of condensed systems remains unchanged during any transformation. Again, in the words of Planck, not only the difference of the atomic heats (before and after the reaction) is to assume the value zero at absolute zero, but also each atomic heat itself is to do the same. Thus it follows from Nernst's theorem, in agreement with the requirement of quantum theory, that the atomic heats of solid bodies disappear at absolute zero.

Walther Nernst sought to develop new methods for the determination of the specific heat at definite temperatures, which at low temperatures was a particularly delicate task. Nernst and his collaborators studied a wide variety of elements and compounds from room temperature down to liquid air temperatures, all showing a marked decrease in specific heat as the temperature was lowered. By February 1910 Nernst began to report on his results [5] and remarked that 'one gets the impression that the specific heats are converging to zero as required by Einstein's theory'. Nernst reported qualitative agreement with Einstein's equation, and announced that his co-workers, F. A. Lindemann and A. Magnus, were examining the degree to which quantitative agreement was also present.

In a lecture to the French Physical Society just a few weeks later, Nernst reported on this quantitative agreement as well, and described Einstein's equation in connection with the Dulong and Petit law, the 'old enigma'.[6] Nernst mentioned that the data obtained thus far agreed with Einstein's formula for the specific heat. Einstein's theory also gave an intrinsic interest to specific heat measurements in connection with Nernst's own heat theorem. By April 1910 Nernst was convinced of the importance of Einstein's result, and soon he would turn his attention to the quantum theory on which it was based.

The only unknown parameter in Einstein's Equation (9) was the frequency v, and once this was fixed the value of the specific heat was determined for all temperatures. Einstein had himself argued[1] that this vibrational frequency must be identical with the optical absorption frequency as determined by the method of 'residual rays'. At that time (1907) Einstein had not given a general method of relating the vibrational frequency, which determines the thermal behaviour, to other measurable properties of the solid.

This frequency v could be determined by several independent methods. One way that was always possible was the following: For a given substance, choose an experimentally well-known value of the atomic heat, C_v^*, which corresponds to a definite temperature T^*. From Equation (9) it follows then that

$$\frac{x^2 e^x}{(e^x - 1)^2} = \frac{C_v^*}{3R}, \qquad (10)$$

from which $x = hv/kT^*$ can be determined, and thereby v. From the v thus determined, the whole C_v curve can be calculated for all temperatures, and compared with experiment.

Besides this 'empirical' method of determining v, there are a number of more 'theoretical' methods which do not require the use of the values of the atomic heat. Einstein, in 1911, discovered an important connection between the frequency v and the elastic properties of the body.[7] This connection is brought out by imagining the atoms of a body arranged upon a space-lattice, as in a crystal, and supposing that a certain atom is arbitrarily disturbed from its position of rest. This atom, when released, will execute vibrations about its position of equilibrium. If these vibrations are assumed to be simply periodic, i.e. monochromatic [that this assumption was an oversimplification and essentially inadmissible was recognized by Einstein], then the frequency v is greater the smaller the atomic mass, and the greater must be the force which restores this atom to its equilibrium position. The more compressible the body is, the stronger is the restoring force. Thus v must be the greater, the smaller the atomic weight and the compressibility of the substance. From these considerations, Einstein [7] obtained the formula,

$$v = \frac{2.8 \times 10^7}{A^{1/3} \varrho^{1/6} \kappa^{1/2}}, \tag{11}$$

where A = atomic weight, ϱ = density, and κ = the compressibility of the body.

In June 1910, F. A. Lindemann, Nernst's collaborator, found another interesting relation, connecting v with the melting point, by working out the idea that the amplitude of vibration of the atom at the melting point is of the order of magnitude of the distances between the atoms. [8] If T_s is the absolute melting point, then it follows that,

$$v = 2.8 \times 10^{12} \cdot \frac{T_s^{1/2} \varrho^{1/3}}{A^{5/6}}. \tag{12}$$

Another formula deduced by E. Grüneisen [9] showed,

$$v = 2.91 \times 10^{11} \cdot A^{-5/6} [C_v^{1/2} \alpha^{-1/2} \varrho^{1/3}]_0, \tag{13}$$

where C_v is the atomic heat at constant volume, α is the coefficient of thermal expansion, and the index zero means that the value of $C_v^{1/2} \alpha^{1/2} \varrho^{1/3}$ at absolute zero is to be used.

The abnormal behaviour of diamond with respect to its atomic heat can be seen from formulas (12) and (13). Diamond has a high melting point, very low compressibility, and a low atomic weight. Its v is thus comparatively large and, from the above considerations, it follows that its atomic heat falls below the Dulong–Petit value at comparatively high temperatures.

12.3. Proposed Improvements of Einstein's Theory of Atomic Heats

The experiments of Nernst and his collaborators proved quite convincingly that the atomic heat of all solid bodies tends towards a zero value as the temperature falls. In general, the courses of these decreasing values showed a notable agreement with Einstein's Equation (9). At low temperatures, however, systematic discrepancies were found in all cases, and the observed atomic heats fall off much more slowly than Einstein's formula required.

W. Nernst and F. A. Lindemann [10] tried to take these discrepancies into account by constructing an empirical formula, and this actually expressed the observations much more accurately than did Einstein's formula. The Nernst–Lindemann formula stated:

$$C_v = \frac{3R}{2} \left\{ \frac{xe^x}{(e^x - 1)^2} + \frac{(x/2)^2 e^{x/2}}{(e^{x/2} - 1)^2} \right\}, \tag{14}$$

where $x = hv/kT$. Equation (14) obtains meaning if it is assumed that one half of all the atoms vibrate with frequency ν, and the other half vibrate with the frequency $\nu/2$. This assumption at least recognized the fact that the 'monochromatic' theory of atomic heats, which had assumed a single fixed ν for all atoms, was only an exaggerated idealization of the real state of affairs. Einstein himself had recognized this limitation and drawn attention to the need of amending his theory.[7]

If the body consists of N atoms, it possesses in general $3N$ natural frequencies, of which the slowest are phonons or sound waves, while the quickest fall in the infra-red. The most general possible movement of each atom then consists in a superposition of all these natural frequencies. Since each natural frequency represents a linear, i.e. simple periodic, motion, exactly like the motion of a Planck oscillator, one can allot to each natural frequency of period ν the theoretical quantum amount $h\nu/(e^{h\nu/kT}-1)$ as if the natural period were identical with a linear oscillator. The total mean energy then becomes

$$\bar{E} = \sum_{i=1}^{3N} \frac{h\nu_i}{e^{h\nu_i/kT} - 1}, \tag{15}$$

in which the summation is carried over all $3N$ natural frequencies $\nu_1, \nu_2, \nu_3, \ldots \nu_N$, i.e. over the entire elastic spectrum of the substance. And, by differentiation with respect to T one obtains the atomic heat,

$$C_v = \frac{\mathrm{d}\bar{E}}{\mathrm{d}T} = k \sum_{i=1}^{3N} \frac{x_i^2 e^{x_i}}{(e^{x_i} - 1)^2}, \tag{16}$$

where $x_i = h\nu_i/kT$.

12.4. Einstein's Review of the Specific Heat Problem

During the summer of 1911 Einstein sought to remove the restriction from his model that the vibrations in a crystal were monochromatic. He sought to construct a model of interacting atoms, taking into account the rapid transfer of vibrational energy from one atom to its neighbours, but he did not achieve any positive result. While thus 'tormenting himself' with this calculation, he received from Walther Nernst the proof-sheets of the first report of the Nernst–Lindemann formula.[10] Einstein immediately saw that they had obtained a valuable empirical equation, but he was not convinced of the theoretical foundation which the authors had proposed for it – that is, that the spectrum contained equal numbers of vibrations at the two frequencies, ν and $\nu/2$. Einstein was sure that the true spectrum was more complex, even though he himself could not see his way to determining it.

In his report to the first Solvay Conference Einstein gave a review of the status of the theory of specific heats. The divergent views held by Nernst and Einstein on the significance of the Nernst–Lindemann formula were discussed in detail. After his review of the problem of specific heats, Einstein dealt with the question of how mechanics had to be modified in order to take into account the new elements brought into the picture by Planck's theory of black-body radiation and the understanding of thermal properties of matter.

Einstein recalled that classically the probability $\mathrm{d}W$ for a linear oscillator to have its energy between E and $E+dE$ at temperature T is given by

$$dW = Ce^{-E/kT} \, dE, \tag{17}$$

while according to Planck the probability must be $W=Ce^{-E/kT}$ for values of E that are multiples of $h\nu$, and for all other values of the energy one must put $W=0$. The average energy of the oscillator thus becomes,

$$\bar{E} = \frac{\sum EW}{\sum W} = \frac{0 \cdot e^{-0/kT} + h\nu e^{-h\nu/kT} + 2h\nu e^{-2h\nu/kT} + \cdots}{e^{-0/kT} + e^{-h\nu/kT} + \cdots}, \tag{18}$$

and $\bar{E}$ then has the value given by Planck's formula. This result leads to certain difficulties and Einstein, treating the case of diamond, showed contradictions with experimental results on thermal conductivity.

Einstein attempted to find the behaviour of the statistical law from the thermal properties of crystals. If one starts with the entropy, $S=k \log W + \text{const.}$, and takes a body of specific heat C in contact with an infinite reservoir of temperature T, then at some instant the energy will pass from E to $E+\varepsilon$, and the temperature will pass from T to $T+\tau$, and then

$$dS = \frac{C \, d\tau}{T + \tau} - \frac{C \, d\tau}{T},$$

or,

$$S = -\frac{\varepsilon^2}{2CT^2},$$

and, by Boltzmann's theorem,

$$W = \text{const.} \times e^{-\varepsilon/2kCT^2}. \tag{19}$$

For a body characterized by a frequency ν, and having n gram-atoms, Einstein obtained

$$C = 3nR \frac{(h\nu/kT)^2 \, e^{h\nu/kT}}{(e^{h\nu/kT} - 1)^2}. \tag{20}$$

With

$$E = 3nN \frac{h\nu}{e^{h\nu/kT} - 1},$$

he obtained

$$\overline{\left(\frac{\varepsilon}{E}\right)^2} = \frac{h\nu}{E} + \frac{1}{3Nn} = \frac{1}{Z_q} + \frac{1}{Z_f}, \tag{21}$$

where $Z_q = E/h\nu =$ the average number of Planck's quanta of energy present in the body; and $Z_f = 3nN =$ the total number of degrees of freedom of all the atoms of the system. For an infinite system $1/Z_f$ is zero, and

$$\sqrt{\overline{\left(\frac{\varepsilon}{E}\right)^2}} = \frac{1}{\sqrt{Z_q}},$$

and Einstein showed that this factor is much too small in the case of diamond.

He summarized his result as follows: 'When a body exchanges energy by means of a quasi-periodic mechanism of frequency v, the statistical properties of the phenomenon are determined as if the energy were transported in whole quanta of magnitude hv. Although we do not know the mechanism by which this property can be explained, we must in any case admit that the disappearance of the periodic energy must take place in quanta of magnitude hv.'

Einstein also pointed out how difficult it is to understand that a phenomenon like the photoelectric effect should be independent of the intensity of light but depend only on the frequency. In the same context, Einstein wondered why the energy should be related to the frequency.

By means of a rough approximation, Einstein then deduced the results obtained by Nernst and Sommerfeld. He analyzed the rotation of diatomic molecules, considering them as being characterized by a unique frequency v. Assuming that the motion, in the first approximation, could be considered as a linear oscillator, he obtained for the energy

$$E = \tfrac{1}{2} I (2\pi v)^2, \tag{22}$$

where I is the moment of inertia of the system.

Einstein finally considered a monatomic gas placed in a field of black-body radiation. He assumed the molecules to be charged so that they can absorb and emit radiation. Using Kirchhoff's and Planck's laws he showed that during collisions the relation $hv < ZE$ (where $Z \approx 1$) has to be observed. If the collisions were rapid, Maxwell's equations would imply very high frequency. The collisions must therefore be smooth, and one must have $h = E\tau Z$ (with $Z \approx 1$). In this way Einstein recovered Sommerfeld's results.

During the discussion of his report, Einstein emphasized that in its present form the theory of quanta was useful but not satisfactory. Recalling the report of Lorentz at the Conference, he mentioned that classical mechanics was also insufficient. He therefore asked as to what could still be considered valid. For Einstein, the fundamental concepts were the conservation of energy and Boltzmann's theorem relating the entropy and probability.

Einstein described how Boltzmann's principle had to be understood. He defined the probability of a complexion as the ratio τ_1 / T, where T is a (long) duration of observation and τ_1 the time during which the system is in the complexion Z_1. The system, initially in a given complexion, evolves to complexions having higher probability. Since for an ensemble of systems, $S_{\text{total}} = \sum S$ and $W_{\text{total}} = \prod W$, Einstein deduced that $S = k \log W + \text{const}$. He illustrated this result by means of different examples.

12.5. COMPLETION OF EINSTEIN'S THEORY OF SPECIFIC HEAT

The central problem of Einstein's theory of specific heat consisted in calculating the 'elastic spectrum' of a given body, i.e. in determining the position of the natural periods for any body. The theory was worked out from two sides. P. Debye[11] considered an elastic continuum as an approximation to the actual atomically constructed body. M. Born and Th. von Kármán[12], on the other hand, replaced the crystal of limited size by one of infinite dimensions. The problem of solving the elastic spectrum was worked out differently in the two cases.

The Debye theory rested upon the classical theory of elasticity, treating bodies as structureless continua, and leaving out the crystalline and atomic structure of the body. From this, Debye derived the important law: the number $Z(v)\,dv$ of all those natural periods whose frequency lies between v and $v+dv$ is given by

$$Z(v)\,dv = 4\pi V\left(\frac{1}{c_l^3} + \frac{2}{c_t^3}\right) v^2\,dv, \tag{23}$$

where V is the volume of the body, c_l and c_t are the velocities of the longitudinal and transverse waves in the body respectively. Now, in a continuum the number of natural frequencies is infinite. Since the body consists of N atoms, it may not possess more than $3N$ natural frequencies. In order to achieve this, Debye made the following hypothesis: Instead of calculating the elastic spectrum of the real body of N atoms, he replaced it by that of the continuum as an approximation, cutting it off arbitrarily at the $3N$-th natural period. Debye thus obtained the greatest frequency v_m, the upper limit of the spectrum from the condition

$$\int_0^{v_m} Z(v)\,dv = \frac{4\pi V}{3}\left(\frac{1}{c_l^3} + \frac{2}{c_t^3}\right) v_m^2 = 3N, \tag{24}$$

from which

$$v_m = \left[\frac{9N}{4\pi V\left(\dfrac{1}{c_l^3} + \dfrac{2}{c_t^3}\right)}\right]^{1/3}. \tag{25}$$

The atomic heat of the body, from Equation (16), thus turns out to be,

$$C_v = k \int_0^{v_m} \frac{\left(\dfrac{hv}{kT}\right)^2 \cdot e^{hv/kT}}{(e^{hv/kT} - 1)^2} Z(v)\,dv, \tag{26}$$

and can be written as

$$C_v = \frac{9R}{x_m^3} \int_0^{x_m} \frac{x^4 e^x\,dx}{(e^x - 1)^2}, \tag{27}$$

where

$$x_m = \frac{hv_m}{kT} = \frac{\Theta}{T}.$$

The atomic heat is thus only a function of the magnitude x_m, i.e. it depends only on the ratio Θ/T. Debye's result may be expressed as follows: Considering the temperature T as a multiple of a temperature Θ, which is characteristic of a particular body, the atomic heat can be represented for all monatomic bodies by the same curve.

For high temperatures, the Debye formula gives, as it must, the classical Dulong–Petit value, $C_v = 3R$, just as the Einstein and Nernst–Lindemann formulas do. However, it differs from the latter in falling much more slowly at low temperatures. While the atomic heats, according to both Einstein and Nernst–Lindemann formulas, fall exponentially [with $(1/T^2)\exp(-\text{const}/T)$] at low temperatures, Debye's result leads to the fundamental law that the atomic heats of all bodies at low temperatures are proportional to the third power of the absolute temperature.

Born and von Kármán, going beyond Debye, took into account the real crystalline structure of the body, i.e. the space-lattice arrangement of the atoms, giving a complete account of the theory of atomic specific heats. Debye's theory, however, in addition to the Einstein terms, provided a physically satisfactory and mathematically tractable picture.

REFERENCES

1. A. Einstein, *Ann. Physik* **22**, 180 (1907).
2. L. Boltzmann, *Wiener Ber.* **63** (11), 731 (1871) and F. Richarz, *Wied. Ann.* **67**, 702 (1899).
3. Dulong and Petit, *Ann. Chim. Phys.* **10**, 395 (1819).
4. F. H. Weber, *Poggend. Ann.* **147**, 311 (1872); **154**, 367, 553 (1875).
4a. See J. Mehra, 'Einstein and the Foundation of Statistical Mechanics', *Physica* **79A**, 447–477 (1975).
5. W. Nernst, *Sitzber. Preuss. Akad. Wiss. Berlin, Kl. Math. Phys.* (1910), p. 262.
6. W. Nernst, *J. Phys. Théor. Appl.* **9**, 721 (1910).
7. A. Einstein, *Ann. Physik* **34**, 170, 590 (1911); **35**, 679 (1911).
8. F. A. Lindemann, *Phys. Z.* **11**, 609 (1910).
9. E. Grüneisen, *Ann. Physik* **39**, 291 *et seq.* (1912).
10. W. Nernst and F. A. Lindemann, *Sitzber. Kgl. Preuss. Akad. Wiss.* (1911), p. 494; *Z. Elektrochemie*, **17**, 817 (1911).
11. P. Debye, *Ann. Phys.* **39**, 189 (1912).
12. M. Born and Th. von Kármán, *Phys. Z.* **13**, 297 (1912); **14**, 15, 65 (1913); also M. Born, *Ann. Phys.* **44**, 605 (1914).

BIBLIOGRAPHY

1. *La Théorie du Rayonnement et les Quanta*, the Proceedings of the first Solvay Conference, Gauthier-Villars, Paris, 1912.
2. *The Quantum Theory*, F. Reiche, Methuen & Co., London, 1930, Chapter IV.
3. M. J. Klein, 'Einstein, Specific Heats and the Early Quantum Theory', *Science* **148**, 173 (1965).

SECOND SOLVAY CONFERENCE 1913

VERSCHAFFELT   LAUE   RUBENS   GOLDSCHMIDT   HERZEN   LINDEMANN   de BROGLIE   POPE   GRUNEISEN   HOSTELET

HASENOHRL   JEANS   BRAGG   Mme CURIE   SOMMERFELD   EINSTEIN   KNUDSEN   LANGEVIN

NERNST   RUTHERFORD   WIEN   J.J. THOMSON   WARBURG   LORENTZ   BRILLOUIN   BARLOW   KAMERLINGH ONNES   WOOD   GOUY   WEISS

3

The Structure of Matter*

1. INTRODUCTION

The first Solvay Conference on Physics had been immensely successful, both in bringing into focus the profound problems which the quantum theory of Planck and Einstein had tried to explain and in providing an impetus for the understanding of new physical phenomena. Following the initiative and advice of H. A. Lorentz, Ernest Solvay established, on 1 May 1912, an International Institute of Physics. This foundation was created initially for a period of thirty years, and Solvay endowed it with a capital of one million Belgian francs. The funds were to be used for the encouragement of research on a deeper understanding of natural phenomena, with grants and subsidies going to promising research workers. Moreover, periodic conferences with select participants, like the first one, would be organized at Brussels under the direction of the administrative council and the scientific commission of the new institute. The International Institute of Physics was initially assigned quarters at the Institute of Physiology (in Parc Léopold), also founded by Ernest Solvay.

The Scientific Council of the second Solvay Conference on Physics, which took place in Brussels from 27 to 31 October 1913, consisted of H. A. Lorentz (Haarlem), President, Madame M. Curie (Paris), M. Brillouin (Paris), H. Kamerlingh Onnes (Leyden), M. Knudsen (Copenhagen), W. Nernst (Berlin), E. Rutherford (Manchester), and E. Warburg (Berlin). Among the other invited participants were: G. Gouy, P. Langevin (France); W. Barlow, W. H. Bragg, J. H. Jeans, W. J. Pope, and J. J. Thomson (Great Britain); E. Grüneisen, H. Rubens, A. Sommerfeld, W. Voigt, and W. Wien (Germany); A. Einstein, M. von Laue and P. Weiss (Switzerland); F. Hasenöhrl (Austria); and R. W. Wood (U.S.A.). R. B. Goldschmidt (Brussels), Secretary, M. de Broglie (Paris), and F. A. Lindemann (Berlin), acted as the Scientific Secretaries.

The theme of the second Solvay Conference was 'The Structure of Matter'. Two years had gone by since the first Conference, a period during which enormous progress had taken place in physics – progress, not all of which would come into focus at the second Solvay Conference.

Laue's discovery, in 1912, of the diffraction of Röntgen rays in crystals had provided most important new information, and opened up new avenues for the investigation of the structure of matter. This discovery removed all doubts about the necessity of

---

* *La Structure de la Matière*, Rapports et Discussions du Conseil de Physique tenu à Bruxelles du 27 au 31 Octobre 1913, Gauthier-Villars, Paris, 1921.

ascribing wave properties to X-rays. The corpuscular features of this radiation, in its interaction with matter, as stressed by William Bragg, had been strikingly illustrated by Wilson's cloud chamber photographs showing the tracks of high speed electrons liberated by the absorption of the radiation in gases. Laue's discovery provided direct incentive for the investigations of crystalline structures by William and Lawrence Bragg. The Braggs analyzed the reflection of monochromatic radiation from the various sequences of parallel plane configurations of atoms in crystal lattices, and they were able both to determine the wavelength of the radiation and deduce the type of symmetry of the lattice. The discussion of these developments formed the main topic of the conference.

The first report at the conference was given by J. J. Thomson concerning his ideas regarding the electronic constitution of atoms. Thomson, without departing from classical physical principles, had been able, at least in a qualitative way, to explore many general properties of matter. It is remarkable to note that the uniqueness of Rutherford's discovery of the atomic nucleus for such exploration was not yet generally appreciated by physicists at that time. At the second Solvay Conference the only reference to this discovery was made by Rutherford himself, who, in the discussion following Thomson's report, insisted on the abundance and accuracy of the experimental evidence underlying the nuclear model of the atom.

Niels Bohr, on the basis of Rutherford's nuclear model of the atom, had achieved fundamental theoretical progress. In fact, Bohr's first paper on the quantum theory of the atomic constitution had been published a few months before the second Solvay Conference. In this paper, Bohr had taken the initial steps of using the Rutherford atomic model for the explanation of specific properties of the elements, depending on the binding of the electrons surrounding the nucleus. The Rutherford–Bohr model gave, with great precision, the numerical values of the series of spectral lines of hydrogen and helium. Although the introduction of emission quanta for the explanation of spectral lines in Bohr's theory appeared to be quite arbitrary, and the combination of the quantum hypothesis with classical mechanics for the calculation of orbits seemed to be more 'bold than logical', the results spoke eloquently that the direction taken by Bohr's theory was correct.

The second Solvay Conference did not, however, deal with Bohr's theory, and Bohr was not present at it. Laue and Voigt were present, and they represented the tendencies of the physics of crystal structure. Laue discussed his theory of the interference of X-rays by three-dimensional crystal lattices, to which Sommerfeld had made important contributions. William Bragg pointed out how the theory of Laue, Friedrich and Knipping could find a simple explanation in the selective diffraction of X-rays by crystals, which led to the spectrometry of these rays and opened the path to new progress.

R. W. Wood had made important discoveries in optics, including the spectra of complicated resonances which defied all explanation at the time, and he gave a report on these resonances. They would ultimately be explained by Bohr's theory of atomic levels.

Madame Curie discussed the consequences of the fundamental law of radioactive transformations.

The second Solvay Conference completed its task on the eve of the great discovery of Moseley that would extend the domain of X-ray spectroscopy. Before he was killed at Dardanelles in World War I, Moseley had extended Bohr's theory to the emission and absorption of X-ray lines, and his work had given precision to the notion of energy levels in the atom.

Because of the intervention of the War, the proceedings of the second Solvay Conference were not published until 1921.

2. STRUCTURE OF THE ATOM

(J. J. Thomson)

J. J. Thomson first recalled that the molecular theory of matter had shown that matter could always be divided into a finite number of chemical elements, the usual elements of Mendeleyev's periodic table. This number is, however, very large and one felt that a sub-atomic theory was desirable. Such a theory was already suggested by the properties of the atoms themselves; as, for instance, by the existence of the electron, which is the same for hydrogen or for lead.

Thomson mentioned that it had been possible, with the help of X-rays, to determine the number of electrons in atoms by studying their scattering by charged particles. Barkla's measurements[1] had shown that for light gases the ratio of the diffused energy to the incident energy was independent of the 'quality' of the rays. He had also found that this ratio was proportional to the density of gases. From his formula for the scattering of X-rays Barkla had deduced that each molecule of air must have about 15 charged corpuscles.

Similar measurements were also performed by Crowther[2]. Crowther had also used the method of the scattering of β-rays through a thin sheet of matter. The β-rays are supposed to be deviated by the positive charge of the atoms, and the result would depend on the distribution of the positive charges in the atom. With the assumption of a uniform distribution of charges in a sphere, Crowther found a result three times larger than with other methods.

The evidence of the existence of positively charged particles in atoms could be found in the emission of α-particles. Thomson spoke about experiments in which the impact of cathode rays on metals produced some helium. Hydrogen was also present, but this was probably due to the fact that hydrogen remained present even in the best vacuum.

Thomson then stated that for each element the atomic weight could be written in the form $4n$, $4n+1$, $4n+2$, or $4n+3$, where n is an integer. He showed that for elements below the atomic weight of 40 (with a few exceptions), the atomic weights have the form $4n$ or $4n+3$, but this regularity no longer held for elements with atomic weight above 40. Thomson thought that these changes were due to slight variations of the masses of the constituents of the atoms above the weight of 40. He gave examples where the addition of an extra particle could drastically change the configuration.

Thomson concluded that in the atom there are, on the one hand, negative charges which are all 'equal', and, on the other hand, positive charges which present themselves under a limited number of forms. One is the α-particle, i.e. a helium atom which has lost two negatively charged corpuscles, and the other is the hydrogen atom having lost one corpuscle.

Thomson then examined the problem of the distribution of positive charges inside the atom. The first point to ascertain was whether Coulomb's law was still valid for these very small corpuscles separated by distances of the order of about 10^{-8} cm.

Considering the case of a cathode ray or a β-particle passing through a gas, Thomson noted that the particle is deviated and the corpuscles are pushed out of the gas atoms along the trajectory of the particle. Thus the β-particles are deviated by the corpuscles in the atoms and transfer some of their momentum to some corpuscles in the atoms of the gas, which then leave the atoms. Thomson computed these effects, assuming the validity of Coulomb's law. He found, for instance, that a β-particle of speed 2×10^{10} cm/sec must have a free path in air, at atmospheric pressure, varying from 60 cm for an angle of deviation of 18° to 0.66 cm for an angle of 2°. He concluded that C. T. R. Wilson's cloud chamber pictures showed precisely that the droplets, which denote the path, had a small curvature and suffered from time to time a marked change of direction in perfect agreement with the theory.

Thomson calculated the number of ions produced by a β-particle per unit length of its trajectory. He reviewed the same process for cathode rays, and then spoke about some experiments of Niels Bohr [3] describing the collision between charged particles and the corpuscles in the atoms.

Thomson's calculations showed that an α-particle causing ionization must be created with a kinetic energy of 19 000 V, i.e. having a speed of 10^8 cm/sec. This result was in contradiction with experiment, since according to Townsend [4] the ionization energy of hydrogen was of the order of 70 V. Thomson invoked the very small size of the struck particles to justify his results, thinking that only a small part of the energy of the incident particles was transferred to the corpuscles.

Thomson thought that the influence of the electrons on the trajectory of α-particles could be neglected. He mentioned Rutherford's ideas and results on the nature of the positive charges. Thomson, however, did not agree with Rutherford's interpretation, and thought that the break in the trajectory of certain α-particles was due to a particular interaction between the incident α-particle and the α-particles within the atoms.

These results led Thomson to the conclusion that Coulomb's law was valid also for phenomena inside the atom. He was thus led to construct a model of the atom based on an equilibrium between the positive and negative charges, moving on a spherical trajectory, according to a law of force varying as the square of the distance. Thomson treated the two-dimensional problem and noted that if the particles were distributed in a circle, the centre of the circle being the centre of force, the stability could be achieved only for a small number of corpuscles, in fact five of them. In order to avoid this difficulty he considered that the circle was surrounded by some negative charges, a procedure which allowed him to put more corpuscles in equilibrium within the circle.

The model thus obtained a sequence of concentric rings, being strongly negative at the centre.

The problem then was to see whether this model could describe chemical and physical properties. Thomson explained that the chemical properties could be understood by considering that the bonding was due to a dipole-dipole interaction. He assumed that the negative charges were free enough, around the positive ones, to allow the formation of dipoles. Concerning the saturation of the atom, Thomson thought that the different rings could have only a finite number of corpuscles. He supposed that the inner rings had exactly eight corpuscles, while the external rings could have any number of corpuscles from zero to eight. With this assumption he could explain the valence bonds qualitatively.

Thomson expressed doubts about Planck's theory of energy quanta. He thought that the quanta do not arise from the molecular structure of radiant energy, and claimed that the quantum property might very well be explained by means of an atomic mechanism by which the radiant energy is transformed into kinetic energy. He described the model by which such a mechanism could be obtained, envisaging that the forces acting inside the atom, produced by the interaction of electrical charges, were not the same as given by electrostatic laws. His criterion in determining the nature of these 'new forces' was to obtain an atom which had the properties of real atoms. The principal characteristics of these forces were the following: A radial repulsive force varying as the inverse cube of the distance from the centre, combined with an attractive radial force varying as the inverse square of the distance from the centre, but limited to a certain number of radial 'tubes' within the atom. In order to obtain such a model, Thomson introduced all kinds of assumptions without giving their justification. But his idea was the following: Since the conservation of energy holds (a point which was not clear with the type of forces he employed), the condition could be imposed that the work necessary for a corpuscle to get out of one of the radial tubes would be equal to the work necessary to carry the corpuscle from its position to infinity. Moreover, along the axis of one of the tubes, the equilibrium between the forces C/a^3 and A/a^2 was assumed to be stable (C and A being constants, to be determined later).

Applying Newton's law, he obtained the period of vibration T around the equilibrium position, given by

$$\frac{2\pi}{T} = \sqrt{\frac{Ce}{ma^4}}, \tag{1}$$

where m is the mass of the corpuscle and e its charge. The work, W, done by the repulsive force to carry a corpuscle from $r=a$ to $r=\infty$ is given by $Ce/2a^2$. Thus $W = n\pi\sqrt{Cem}$, where $n=1/T$. He chose C in such a way that $\pi\sqrt{Cem}=h$. With $h=6.5\times10^{-27}$, $e=4.7\times10^{-10}$, $e/m=5.3\times10^{17}$, C turns out to be 10^{-17}.

Thomson considered the photoelectric effect, and showed that if light of frequency n falls on an atom it will find a corpuscle having this frequency and obtain resonance with it. If the quantity of energy communicated to the corpuscle were sufficient, the latter would be liberated out of the radial tube and expelled by the repulsive force, its kinetic energy being equal to hn. With enough constants to help, Thomson thus arrived at an explanation of the photoelectric effect, and by similar arguments he also sought to explain the absorption of cathode rays.

DISCUSSION

Madame Curie opened the discussion of J. J. Thomson's report by expressing doubts about the interpretation of observed helium as arising from the α-particles emitted by a substance under the action of cathode rays. She pointed out that the quantity of helium produced decreases during the process, while one would expect a constant production according to the proposed explanation. She also noted that the energy occurring in radioactive emission was much larger than that obtained from cathode rays. Thomson could only answer that this problem was not fully understood.

Nernst recalled that the atomic masses did not vary like integers, but had rather strong discrepancies which could not be explained easily by Thomson's model. Thomson was aware that in his model not enough was known about the nature of the electrical forces inside the atoms, nor could he deal with the possibility of having other kinds of forces that would explain how the masses must change. Langevin thought that the inclusion of the inertia of energy could perhaps provide a satisfactory solution to this problem.

Rutherford noted that Thomson's result that the number of electrons in atoms was about half the atomic weight was in agreement with some other facts. He pointed out that the α-particles emitted by some radioactive substances were certainly free of all their electrons and had two positive charges.

Langevin pointed out that according to Thomson the ionizations from α and β particles were due to actions of different kinds, and these conclusions had been drawn by applying classical mechanics to intra-atomic collisions. He wondered whether these laws were still valid inside the atom.

R. W. Wood wondered why the deflections of particles were always in the same direction, leading to an extended curve. Rutherford drew attention to the assumption that the positive charges are situated in a very small atomic nucleus, and the comparison between theory and experiment showed perfect agreement. Moreover, the interaction law was of the Coulomb type only. Rutherford also pointed out that the β particles had energies which were much smaller than those of the α particles, so that the former could not enter the high field regions of the nucleus and their trajectories were determined by collisions with the external electrons only.

Langevin drew attention to the fact that the existence of β-particles, which were electrons belonging to the nucleus, could not be explained by Rutherford's model. Madame Curie emphasized this point also. For her it was natural to admit the existence of the electrons inside the nucleus. However, she proposed to make a clear distinction between the peripheral electrons and those leading to β-rays.

Rubens noted that there existed different types of bindings. In HCl, for instance, the chemical bond is different from the one in H_2 or Cl_2. He thought that the two atoms of the HCl molecule were bound together by electrical charges of the opposite kinds, and dissociation was the process by which the two atoms equalized their charges.

Weiss thought that, according to Debye's theory, the molecules having an electrostatic polarization could be recognized since their dielectric constant is a function of the temperature. Hasenöhrl wondered whether it was possible to determine the temperature at which Thomson's model would become unstable.

Lorentz raised a series of serious objections to the model proposed by Thomson. He explained that a model based on two radial forces could not generate forces which would allow the corpuscle to leave the tube. He showed that the potential energy had the value

$$\frac{Ce}{2a^2} - \frac{Ae}{a} = -\frac{Ae}{2a} \tag{2}$$

in the equilibrium position inside the cone, and $Ae/2a$ outside of it. The kinetic energy required to cross the cone would then be Ae/a, but a speed corresponding to this energy could never be attained in a radial vibration. Lorentz further pointed out that since only classical mechanics was used, it was doubtful whether Thomson's model could explain black-body radiation.

Madame Curie then made a long remark concerning the exponential law of radioactive disintegration, i.e.

$$N = N_0 e^{-\lambda t}, \tag{3}$$

where N_0 is the number of atoms of a radioactive substance present at time $t = 0$, N the number of atoms at time t, and λ the radioactive constant characteristic of the substance. This law can also be written as,

$$dN = -\lambda N_0 e^{-\lambda t}\, dt = -\lambda N\, dt, \tag{4}$$

and expresses the fact that the probability for a radioactive transformation is, at a given instant, the same for all atoms under consideration. Madame Curie pointed out that this statistical law cannot be influenced by any external forces, and is independent of the temperature. An important fact is that the probability that an atom would not disintegrate for a certain time must be independent of the time during which it had already existed. The complete absence of the influence of temperature on radio-active transformations obliged one to look for a mechanism inside the atom itself that would provide the necessary element of disorder leading to the application of a law of probability.

Madame Curie then briefly explained the theory of Debierne dealing with the existence of intra-atomic disorder created by a large number of particles. This state of matter could explain, by means of some random instabilities, the emission of α-particles. A formula found by Geiger had shown, in agreement with the model of Debierne, that the logarithm of the radioactive constant was propor-tional to the logarithm of the speed of α-particles emitted.

Rutherford expressed his agreement with the remarks of Madame Curie. He pointed out that, in accordance with the views she had expressed, chemical reactions depend only on the peripheral electrons.

REFERENCES

1. C. G. Barkla, *Phil. Mag.* 6th series, **7**, 543 (1904).
2. J. A. Crowther, *Phil. Mag.* 6th series, **14**, 653 (1907).
3. N. Bohr, *Phil. Mag.* 6th series, **25**, 10 (1913).
4. J. S. E. Townsend, *Phil. Mag.* 6th series, **6**, 598 (1903).

3. INTERFERENCE OF RÖNTGEN RAYS BY THREE-DIMENSIONAL CRYSTAL LATTICES

(M. von Laue)

Max von Laue first gave a brief historical review of crystallography. The knowledge of crystal lattices, in the previous 60 years, had been based only on some properties of order in crystals, such as cleavage planes, but this knowledge was, by no means, the result of direct measurements. Laue explained that since, in the early days, the only available electromagnetic waves were of the order of 10^{-5} cm or longer, it was impossible to study interference phenomena with crystals. This situation was changed, however, with the availability of Röntgen rays, the wavelength of which was estimated to be of the order of 10^{-9} cm. The first diffraction experiments on crystals with Röntgen rays were those of Friedrich and Knipping, and they gave simple but un-expected results.

Laue then established some basic results of the theory of diffraction of a three-dimensional lattice. He considered the incident radiation as plane waves (assuming each atom to be the centre of a spherical wave), represented at a distance r by $\psi\, e^{-ikr}/r$, where the function ψ depends on the wavelength λ and the direction cosines α_0, β_0, γ_0 and α, β, γ of the incident and diffracted rays.

Making use of the diffraction theory of a lattice, Laue found that the intensity of the diffracted rays would be proportional to

$$|\psi|^2 \frac{\sin^2 M_1 A}{\sin^2 \frac{1}{2}A_1} \cdot \frac{\sin^2 M_2 A_2}{\sin^2 \frac{1}{2}A_2} \cdot \frac{\sin^2 M_3 A_3}{\sin^2 \frac{1}{2}A_3} \qquad (1)$$

where

$$A_i = \frac{2\pi}{\lambda}\left[a_{ix}(\alpha - \alpha_0) + a_{iy}(\beta - \beta_0) + a_{iz}(\gamma - \gamma_0) \right],$$

a_{ix}, a_{iy}, a_{iz} being the components of the elementary vector a_i of the lattice. This function has narrow and high maxima for $A_i = 2\pi h_i$, or, using the unit vectors s_0 and s along the directions defined by the direction cosines, for

$$(a_i, s - s_0) = h_i \lambda, \qquad \text{where} \quad i = 1, 2, 3. \qquad (2)$$

Elsewhere the intensity is very small and will not be observed. Each of the equations (2) represents a cone, so that the only possible solutions are given by the intersections of three cones of axes a_1, a_2 and a_3 respectively. Laue then presented the Ewald construction of the maxima and introduced the so-called reciprocal lattice. He showed how these equations would lead to Bragg's law,

$$n\lambda = 2\,d\cos\varphi. \qquad (3)$$

This law had first been verified by the Braggs, and immediately afterwards by Moseley and Darwin.[1]

Laue pointed out that an X-ray source is not monochromatic; however, the X-ray spectrum is generally characterized by some monochromatic peaks, and this monochromatic radiation is the only one to be considered. He gave the experimental data of the Braggs, showing the maxima of diffraction, and reported on the effect of the continuous spectrum on diffraction phenomena.

Laue sought to give a particular form to the function ψ. He noted that ψ could not be a selective function, i.e. a function which would have strong maxima only for very special wavelengths corresponding to some resonances of the atoms. The impossibility of such a behaviour followed from experiments, showing that the rotations of the crystal lead to a change in the interference phenomena in full agreement with the law of reflection, while this assumption would lead to an important change of location of the diffraction pattern.

Ewald's assumption was the opposite. He had assumed that all the atoms resonated with the same intensity for each wavelength. This assumption sounded good at first, as it allowed one to understand certain experimental facts. By means of a rather technical calculation, however, Laue pointed out that Ewald's assumption was in contradiction with experimental results with substances having a ternary symmetry.

In the second part of his talk, Laue treated the problem of a crystal having several atoms in the unit cell. Considering the unit cell as a new entity, he introduced the changes of the diffraction patterns in the function ψ. He then analyzed cases in which the addition of atoms in the unit cell would lead to the disappearance of some of the diffracted rays. But, as he pointed out, this addition did not always lead to a change in the diffraction spectrum.

Laue discussed the possibility of defining precisely the structure of crystals by means of their X-ray spectrum. He pointed out that one's knowledge concerned only the intensity, while nothing was known about the phase differences in the spectrum. In his view, although X-rays gave a great deal of information about the structure, its full determination would not be possible from the spectrum.

After some comments on the shape of the observed points given by diffraction, the shape of which is affected by the width of the pencil of X-rays, Laue talked about the influence of thermal motions on the diffraction. Following Debye's theory[2], Laue showed that when the effect of temperature is taken into account, the intensity is multiplied by a factor K, given by

$$K = \exp\left[-\frac{4\pi^2 kT}{f\lambda^2} \{(\alpha - \alpha_0)^2 + (\beta - \beta_0)^2 + (\gamma - \gamma_0)^2\} \right], \qquad (4)$$

where T is the temperature, k the Boltzmann constant, and f a quantity measuring the elastic forces which bind the atoms to their equilibrium position. If δ is the angle between the incident and the diffracted rays, then

$$(\alpha - \alpha_0)^2 + (\beta - \beta_0)^2 + (\gamma - \gamma_0)^2 = 2(1 - \cos\delta), \qquad (5)$$

showing that K decreases with δ.

DISCUSSION

Lorentz opened the discussion of Laue's report by pointing out that the Bragg reflection, called 'apparent' by Laue, was in fact a real reflection. Laue argued that this was only partially true, for a real reflection would take place only on planes for which the density of atoms was large enough. Lorentz also noted that Laue's intensity formula led to maxima proportional to $M_1^2 M_2^2 M_3^2$, i.e. to the square of the number of atoms present, while the observed intensity was proportional only to the number of atoms. He compared this case with the optical one, where one has to take into account the fact that the width of the maximum is inversely proportional to the number of elements of the lattice.

Nernst asked if the observations made on diamond favoured the existence of the zero point energy. Einstein responded by saying that the zero point energy could hardly arise in elastic vibrations, since this would lead to the finiteness of all quantities depending on the temperature even at absolute zero. For Einstein this was in contradiction with Kamerlingh Onnes' discovery of superconductivity.

Lindemann pointed out that if the zero point energy existed, it would have to be as a consequence of Debye's formula for the specific heat. He also discussed the case of diamond, mentioned in Laue's report, for which a diffraction pattern could be observed all round the crystal. However, the same phenomenon was not observed for the pyrites, substances having the same constitution as diamond. Lindemann explained that this discrepancy was certainly due to the great difference between the interatomic distances of the two substances.

Maurice de Broglie reported on his experiments connected with these questions. He had found that in magnetite the diffraction pattern was not sensitive to the presence of a magnetic field.

Sommerfeld made a remark on the nature of the function ψ and certain assumptions made by Laue concerning it. Lorentz gave the outline of a calculation of the intensity of X-rays when the source is considered to be situated at a finite distance, and showed that the result is quite different from a source situated at infinity.

Einstein, Wien, Nernst, and Lorentz then discussed the question whether the zero point energy arose from an oscillation of the atoms or some other cause.

REFERENCES

1. W. L. Bragg, *Proc. Camb. Phil. Soc.* **17**, 43 (1913);
    W. H. Bragg and W. L. Bragg, *Proc. Roy. Soc.* A **138**, 428 (1913);
    H. G. J. Moseley and C. G. Darwin, *Phil. Mag.* 6th series, **26**, 210 (1913).
2. P. Debye, *Verh. Deut. Phys. Ges.* **15**, 678 and 738 (1913).

4. REFLECTION OF X-RAYS AND THE X-RAY SPECTROMETER

(W. H. Bragg)

In the first part of his report, William Bragg discussed how the reflection of X-rays could be described and understood. He then described the X-ray spectrometer which had been used for his measurements. In his spectrometer the crystal took the place of the prism, and detection was made by means of an ionization chamber. He reported on the spectra of osmium, platinum, and iridium, whose chief characteristics were the existence of a continuous spectrum particular to the substance, but also showing a family resemblance.

Bragg also discussed the measurements of the absorption coefficients of different materials. An important characteristic of absorption is the transparence of a substance to its own spectrum. As a result of this absorption some great intensity differences could be observed in the diffraction.

Bragg then reported on the exact measurement of interatomic distances. He had first of all to perform an absolute measurement for a given crystal which could be used as a reference for measurements relative to other crystals. The absolute measurement was obtained by a series of experiments on different alkali halides, the structure of which changes with the dimensions of the ions. Considering that the distances in alkali halides were well defined, the wavelengths of the X-rays emitted by platinum and palladium could then be measured. Taking into account the differences in the relative intensities of the X-rays and the diffraction patterns, Bragg gave an account of the structures he had determined for different substances.

DISCUSSION

In the discussion following Bragg's report, Wood described an experiment in which he had produced X-rays between two very small spheres of platinum, 1 to 2 mm apart. The X-rays produced by this procedure were both very hard and of great radiation density.

Rutherford commented on the nature of X-rays. He thought that the radiation found by Bragg, Moseley and Darwin, with a platinum anticathode certainly arose from an L-series and had consequently a complex character. He raised the question whether the spectrum could be most effectively considered as a superposition of a continuum and a discrete spectrum or as a superposition of a large number of characteristic radiations.

Madame Curie wondered whether the lack of certain points in the data could be explained by the absorption phenomena described by Bragg. Bragg did not think so in view of the rather large domain of the wavelengths of platinum.

Sommerfeld, in a long technical digression, discussed how the considerations of Laue concerning the function ψ [introduced in Laue's report] were wrong.

Rutherford then gave an account of the different experiments which had been performed to understand the spectrum of β- and γ-rays, and pointed out that the spectrum of the β-radiation consisted of at least fifty frequencies. He explained that this large spectrum was the result of the interaction of β-particles with peripheral electrons in atoms. He reported on another experiment by which he had tried to show that the interaction of β-particles with the peripheral electrons could be the origin of γ-rays. His studies had indicated that each substance was characterized by certain types of γ-rays, having a precise absorption, and he had studied the frequency distribution of the emitted γ-rays by the reflection method.

Einstein proposed that use should be made of the theory of quanta, and gave the formula $E_n + + h\nu_n =$ the energy of radioactive transformation, where E_n is the energy of the β-particle and ν_n the frequency of the corresponding ray emitted.

Nernst finally gave a résumé of the situation, pointing out the importance of Bragg's work. He emphasized the differences of structure between ionic substances like KCl and others such as $CaCO_3$.

5. CRYSTALLINE STRUCTURE AND CHEMICAL CONSTITUTION

(W. Barlow and W. J. Pope)

The report on the relation between crystalline structure and chemical constitution was presented by W. Barlow and W. J. Pope. This subject could be discussed along two principal lines: (1) the research of how the different possible geometrical arrangements between the constituents of crystals are obtained, and (2) the examination of the nature of the constituents that can be grouped together and the atomic forces which produce these arrangements. The authors thought that the answers to the first part were well established, while the second had been only recently studied and partially understood. They noted that the physical, geometrical and mechanical properties were in full agreement with the definition that a crystal is an homogeneous construction, i.e. a construction whose parts are uniformly repeated throughout the system. Barlow and Pope reviewed the different steps that were involved in the discovery of crystalline structures.

Bravais, in 1850, in a study of the arrangements of polyhedra, had introduced the so-called Bravais lattices of 14 kinds, corresponding by their symmetries to seven principal crystalline systems. They are obtained as the set of points defined by the centres of polyhedra, the arrangement of which is such that each one is surrounded in a similar fashion by the other one.

In 1870, Sohncke generalized the work of Bravais by removing the conditions relative to the orientations of the polyhedra. This led to the fact that all symmetric Bravais units could be geometrically decomposed into smaller units, without axial symmetry. This allowed one to construct 65 types of homogeneous arrangements, called the 'Sohncke systems'.

Fedoroff, Schönflies and Barlow enlarged Sohncke's definition by taking into account a new type of symmetry, the reflection of images. They were thus led to 230 types of systems.

Barlow and Pope then discussed the nature of interatomic forces. They assumed that each ion acts as the centre of action of a repulsive and an attractive force. The crystal was assumed to be stable. The affinity, in general, led to minor modifications only.

The requirement of the stability of the crystal introduced a new assumption. Barlow and Pope showed that, for a crystal with a single type of atoms, the replacement of ions by spheres fulfilled the equilibrium condition for cubic and hexagonal systems, systems for which the volume occupied by spheres is the largest. They presented a table giving the number of elements in the different systems characterized by an increasing number of atoms per molecule. It became clear that with the increasing complexity of the molecules, the crystalline systems tend to take on a lower symmetry.

In confirmation of their theory of stability, Barlow and Pope presented some calculations showing that non-cubic or hexagonal structures of simple elements could be explained by noting that the difference between the structures of these elements and the cubic or hexagonal ones was very small, and could therefore be understood by means of molecular affinity. They also discussed some more complicated cases, as well as the changes of structures which appear at high temperatures.

Comparing their results with those of Bragg and Laue, Barlow and Pope concluded that X-ray methods also led one to admit that the atoms keep their individual character and that they are bound to a definite portion of the molecular volume. The comparison of the results obtained by their method with those obtained by X-rays showed good agreement in general. The study of crystalline structure by means of X-ray data was still in the early phase of its development, and they were convinced that in future studies both methods would be found to be in full agreement.

Barlow and Pope described in detail numerous examples showing the analogies between classes of substances crystallizing in the same structure or having the same type of valences.

6. CRYSTAL STRUCTURE AND THE ANISOTROPY OF MOLECULES

(M. Brillouin)

Brillouin, in his report, first sought to make a clear distinction between two ways of looking at the problem of the relation between crystalline structure and the properties of molecules. The physical chemists had at first looked at it as a geometrical problem, in which the important parameters are the volume and shape of the molecules. But the molecules themselves have no individuality and the crystal is considered as a conglomeration of spheres. In the other mode of reasoning, the molecule keeps its individuality. Brillouin remarked: 'It is the mutual actions at a distance that keep the molecules separated, and the general characteristics which the first mode of reasoning attributes to the ensemble of atomic volumes can be used to represent the mutual dynamical field of action of the molecules.'

Many distinctions exist between the two modes of representation. An important one is that, with some exceptions, the interaction can be described in the first case in spaces of dimensions higher than three. The analytic difficulties in this case are great, but it has the advantage of sufficient generality.

Brillouin's purpose was to draw some conclusions, from a dynamical point of view, about the properties of molecules from properties of polymorphic substances. The physical properties such as the electrical or magnetic effects, or the propagation of light, were related linearly to the external field, and from them one could obtain information about the molecules. Brillouin chose to use these relations for his studies. He noted that the mechanical properties were not suitable for his purpose because the elastic properties of the medium, as considered by Voigt, were defined by forces extending to many molecules and their elementary law was unknown. The only properties which could be used were those given by dielectric, magnetic or optical behaviour of the substance, for which the external field acted on each molecule individually. Brillouin recalled the formulas which had been obtained by Poisson, Mossotti and Clausius, and rederived by H. A. Lorentz under more general assumptions.[1] For instance, the expressions

$$\frac{\kappa - 1}{\kappa + 2} \cdot \frac{1}{d} \quad \text{and} \quad \frac{\mu - 1}{\mu + 2} \cdot \frac{1}{d} \tag{1}$$

combine the density d and the specific inductive capacity κ, or the magnetic permeability μ, and remain constant whatever be the state of the body.

For a discussion of the relation between birefringence, crystallographic parameters and chemical composition, Brillouin cited Mallard's *Traité de Cristallographie* (1884), which had concluded that birefringence was principally due to the anisotropy of the molecule. Brillouin also hypothesized that "the orientation of the molecule with respect to the lattice alone has an influence [on birefringence], while the symmetry of the lattice and the magnitude of crystallographic parameters have no influence."

Making use of a linear relation between the field $\mathbf{E}$ and momentum $\mathbf{M}^p$ of the p-th molecule, Brillouin wrote for the average total momentum of the crystalline element,

$$\sum_p \mathbf{M}^p = \sum_p K^p \mathbf{E} = n\hat{K}\mathbf{E}, \tag{2}$$

where

$$\hat{K} = \frac{1}{n}\sum_p K^p.$$

Using the invariance of the trace of K^p, he obtained,

$$\text{Tr}\,\hat{K} = \sum_{i=1}^{3} K_i, \tag{3}$$

where K_i are the eigenvalues of K^p (the same for all p). He could now use Lorentz' formula,

$$\frac{4\pi}{3} N\hat{K}_{ii} = \frac{K_i - 1}{K_i + 2}, \tag{4}$$

and since the $\hat{K}_{ii}$ were experimentally defined, the K_i could be determined. By considering the possible orientations of the molecules, Brillouin related the $\hat{K}_{ii}$ to the K_i.

Brillouin applied his theory to $CaCO_3$ in its rhomboidal (spar) and orthorhomboidal (aragonite) forms, showing that his criterion was satisfied in the case of the optical properties, thus allowing him to draw conclusions about the orientation of the molecules.

DISCUSSION

After Brillouin's talk the discussion, however, centred on the report of Barlow and Pope. Lorentz, Lindemann, Voigt, Kamerlingh Onnes, Brillouin, Hasenöhrl, Langevin, Weiss, Barlow and Pope took part in the discussion. The main points of discussion were questions of the relation between the volume of the spheres and the attractive and repulsive dynamical forces. Brillouin acknowledged that if he had known earlier about the work of Bragg and Laue his own report would have been considerably abridged; as it was, several of his considerations had become irrelevant.

REFERENCE

1. H. A. Lorentz, 'La théorie électromagnétique de Maxwell et son application aux corps mouvant,' *Arch. neerl.* **25**, 363 (1892).

7. RELATION BETWEEN PYRO-ELECTRICITY AND TEMPERATURE

(W. Voigt)

Voigt gave a brief report on the relation between the pyro-electricity of certain crystals and temperature, based on researches at his institute in Göttingen. He thought that the results were of interest not only for crystals but also for the quantum hypothesis. It had been found that in spite of the great differences in the absolute values of pyro-electric excitation, which varied in the ratio 1:30, all the substances examined showed the same behaviour at low temperatures, tending to a zero excitation as the temperature tended toward absolute zero. This behaviour was similar to the one observed by Nernst and his collaborators in the case of specific heat. At higher temperatures, the pyro-electric phenomena also showed the same behaviour as specific heat; the excitation tended to a limiting value. The case of tourmaline was of special interest, as it seemed to have two domains of proper vibrations.

8. MOLECULAR THEORY OF SOLIDS

(E. Grüneisen)

The goal of the molecular theory of solids, as stated by E. Grüneisen in his report, was to describe a system of atoms and electrons that would reproduce the properties of solid bodies. The basic condition for achieving it was to know the structure of atoms and the mutual interactions between atoms and electrons. Grüneisen considered only those problems in which the motion of the electrons could be neglected, because by so doing the difficulties of the atomic and electronic interaction could be avoided.

Grüneisen first reviewed the problem of the specific heat of solids. He recalled how the law of Dulong and Petit could be derived from the equipartition theorem, and mentioned the considerations of Benedicks[1] and Richarz[2] in an attempt to get, by classical means, the correction to the Dulong–Petit law for low temperatures. Einstein[3] had applied Planck's theory of energy quanta to the oscillators formed by the ions in a crystal. Under the assumption of a monochromatic vibration of the oscillators, Einstein had obtained,

$$C_v = 3R \frac{x^2 e^x}{(e^x - 1)^2},$$ (1)

with $x = hv/kT$, v being the characteristic frequency. Nernst and his collaborators had shown that this theory was still not satisfactory, and Einstein recognized that the monochromatic assumption had to be changed. This problem was attacked by Debye[4], and by Born and von Kármán[5]. In both approaches, the authors considered that the system, which classically has $3N$ eigenfrequencies (N being the number of atoms), has an energy given by a sum of the factors of the type $hv/(e^{hv/kT} - 1)$ for all the eigenfrequencies v.

It remained to calculate the probability $f(v)$ of the frequency v. Debye determined $f(v)$ by solving the differential equation of the classical elasticity theory. He treated the elastic vibrations of a homogeneous sphere, and obtained

$$f(v) = 3v^2 VF,$$ (2)

where V is the volume and $F = 4\pi/c_m^3$, c_m being the average of the longitudinal and transverse speeds of the elastic vibrations. Introducing a cut-off frequency, v_m, by imposing a condition on the number of modes, Debye obtained

$$C_v = 3Nk \left[\frac{12}{x^3} \int_0^x \frac{\xi^3 \, d\xi}{e^\xi - 1} - \frac{3x}{e^x - 1} \right],$$ (3)

where $\xi = hv/kT$ and $x = hv_m/kT$, a formula which was found to be in good agreement with the experiments.

The theory of Born and von Kármán deduced the function $f(v)$ from an analysis of the harmonic motion of ions around their equilibrium positions. The difference between the two methods was important for the non-isotropic system for which the number of propagation speeds is different.

Grüneisen then derived, following Ornstein, Debye, and Ratnowsky[6], the equation of state of a solid. Considering the entropy, S_N, as the entropy of $3N$ Planck resonators, Grüneisen obtained

$$S_N = 9Nk \left(\frac{kT}{hv_m} \right)^3 \int_0^{hv_m/kT} \left[\frac{\xi e^\xi}{e^\xi - 1} - \log(e^\xi - 1) \right] \xi^2 \, d\xi,$$ (4)

showing that the entropy is a 'universal function' of v_m/T.

Now the energy of a crystal is given by $U_N = E_N + F(v)$, where E_N is the energy of the oscillators, and $F(v)$ is the energy of the crystal when all the ions are at rest. In

Debye's calculation, E_N was a function of v_m/T, so that

$$S_N = f\left(\frac{E_N}{v_m}\right) = f\left[\frac{U_N - F(v)}{v_m}\right].$$ (5)

Using the thermodynamic relations

$$\left(\frac{\partial S_N}{\partial U_N}\right) = \frac{1}{T} \quad \text{and} \quad \left(\frac{\partial S_N}{\partial V}\right)_{U_N} = \frac{p}{T},$$ (6)

Grüneisen obtained

$$[p + F'(v)]\, V = -\frac{d \log v_m}{d \log V} \cdot E_N,$$ (7)

where E_N is given by the Debye calculation,

$$E_N = 9NkT\left(\frac{kT}{hv_m}\right)^2 \int\limits_0^{hv_m/kT} \frac{\xi^3\, d\xi}{e^\xi - 1}.$$ (8)

Grüneisen pointed out that, in accordance with the observation that the modulus of elasticity decreases under strong loads, it was possible to put the relative variation of elastic constants proportional to the relative deformation. This allowed him to put,

$$\frac{dv_m}{v_m} = -\gamma \frac{dV}{V},$$ (9)

and he finally obtained

$$[p + F'(V)]\, V = \gamma E_N.$$ (10)

He gave an account of the Mie–Grüneisen theory [7], which sought to replace $F'(v)$ and γ by relations computed from interatomic forces. They had assumed that the potential of these forces was of the type

$$F(V) = -\frac{A}{V} + \frac{B}{V^m}.$$ (11)

They calculated γ for a monochromatic system, and obtained the equation of state,

$$\left[p + \frac{A}{V^2} - m\frac{B}{V^{m+1}}\right] V = \frac{3m + 2}{6} E.$$ (12)

Grüneisen did not consider this equation satisfactory because of the lack of a polar character in the forces. He deduced certain thermodynamic relations from the above equation of state, first showing that

$$-\frac{V\left(\frac{\partial V}{\partial T}\right)_p}{C_v\left(\frac{\partial V}{\partial p}\right)_T} = -\frac{V\left(\frac{\partial V}{\partial T}\right)_p}{C_p\left(\frac{\partial V}{\partial p}\right)_s} = \gamma,$$ (13)

where S denotes an adiabatic change. A method of computing γ and m for different metals was thus available. γ was found to be of the same order of magnitude for a whole series of metals, but not at all a constant, and m varied accordingly.

Turning to the ratio of the specific heats,

$$\frac{C_p}{C_v} = 1 + \gamma \frac{T}{V}\left(\frac{\partial V}{\partial T}\right)_p = 1 - \frac{T}{v}\left(\frac{\partial v}{\partial T}\right)_p , \tag{14}$$

and supposing that the pressure is infinitesimally small, Grüneisen introduced the energy

$$Q_0 = \left(\frac{C_v}{\dfrac{1}{V}\dfrac{\partial V}{\partial T}}\right)_{T=0} , \tag{15}$$

and obtained

$$\frac{V_T - V_0}{V_0} = \frac{E_T}{Q_0 - \frac{1}{2}(m+2)E_T} . \tag{16}$$

Q_0 is always large compared to E_T, and the result could be stated as follows: The change of the relative volume between the temperatures 0 and T increases approximately in the same ratio as the vibration energy at the temperature T. At high temperatures, $v_T - v_0$ increases more rapidly than E_T.

Grüneisen gave results which showed good agreement between experiments and theory for the thermal expansion. He also developed similar relations for the adiabatic and isothermal changes of state. He pointed out that the introduction of the interatomic forces allowed him to compute the heat of sublimation at $T=0$, showing very good agreement with experiments.

Grüneisen gave numerous examples of crystals for which neither the Debye nor the Einstein formula for the specific heats was satisfactory. He quickly recognized that the difficulty arose from the distribution $f(v)$ of the frequencies. A calculation by Born and von Kármán for a linear chain, with two different masses, had shown the necessity of two bands of frequencies. Grüneisen reviewed some experimental facts exhibiting the generality of this phenomenon.

DISCUSSION

Wien opened the discussion with a remark on the contribution of electrons to the specific heat. He sought to justify his theory of conductivity, towards which Lorentz had been unsympathetic. Wien's assumption was that the number of electrons N and their average velocity u were both functions of the temperature, but in such a way that N/u remained constant.

Lindemann pointed out the difficulties which existed in the theory of conductivity, the chief being that the constant of the Wiedemann–Franz law was always much too small.

In Nernst's view, it was wrong to think that electrons could have any sensible influence on the specific heat, since the specific heats of metallic and non-metallic substances were not essentially different.

Lorentz thought that in the formula for electrical conductivity, $\sigma = e^2 LN/2mu$, it had been well-established that the velocity u was a function of the temperature, and that the important quantities that would have to be changed in accordance with the new theory would be N and the mean free path L. He commented on the nature of the forces used in the Mie–Grüneisen theory, and showed that some different results might be obtained by changing the exponent of the repulsive force. He thought it would be better to consider the force as an unknown function, the first and second derivatives of which were defined by the relations given by Grüneisen.

Nernst gave a proof of the validity of his heat theorem, which had been discussed at the first Solvay Conference. Nernst now showed, with the help of an example taken from thermal expansion, that without his theorem the Carnot cycle would not be satisfied.

After a remark of Lorentz, in which he also gave a proof of Nernst's theorem, Einstein showed that in his view one part of the Carnot cycle used in the proof was not correct. Einstein proposed to consider the impossibility of attaining the absolute zero of temperature as a postulate, and pointed out some of its consequences. Nernst protested against Einstein's point of view, and insisted that many thermodynamical proofs were based on a cycle which could not be practically realized.

REFERENCES

1. C. Benedicks, *Ann. Phys.* **42**, 133 (1913).
2. F. Richarz, *Z. Anorg. Chem.* **58**, 356 (1908); **59**, 146.
3. See Einstein's report on the current status of the problem of specific heats at the first Solvay Conference, Chapter 2, Section 12.
4. P. Debye, *Ann. Phys.* **39**, 789 (1912).
5. M. Born and Th. von Kármán, *Phys. Z.* **13**, 297 (1912); **14**, 15 and 65 (1913).
6. L. S. Ornstein, *Proc. Acad. Amsterdam*, **14**, 983 (1912);
    P. Debye, *Göttinger Wolfskehl Vortrag* (1913);
    S. Ratnowsky, *Ann. Phys.* **38**, 637 (1912); *Verh. Deut. Phys. Ges.* **15**, 75 (1913).
7. G. Mie, *Ann. Physik*, **11**, 657 (1903).
    E. Grüneisen, *Verh. Deut. Phys. Ges.* **13**, 836 (1911); *Ann. Phys.* **39**, 257 (1912).

9. RESONANCE RADIATION AND ITS SPECTRUM

(R. W. Wood)

R. W. Wood gave a brief account of the experiments he had performed on the effect of gaseous vapours on radiant energy. The question was what happens to the energy absorbed by the medium.

Wood considered the case of mercury vapour which had a strong absorption line at a wavelength of 2536 Å. When a pencil of radiation, the wavelength of which was the same as that of the absorbed ray, was passed through a balloon filled with mercury vapour, the latter appeared full of diffused light, and that in fact only a very thin ray was absorbed.

Now if the atoms which had absorbed the radiation were to emit at the same wavelength, the emission would be of a very thin ray, thinner than any other emission known at that time. Wood called the radiation diffused by the molecules as the resonance radiation. He noted that he had experimentally determined that the reason why the secondary radiation was so strong was because there was no real absorption. He pointed out that Huygens' principle, which at first seemed suitable for the explanation of these phenomena, had encountered difficulties.

Wood then described experiments leading to resonance radiation in sodium vapour, whose spectrum was found to have thousands of very thin rays.

THIRD SOLVAY CONFERENCE 1921

W.L. BRAGG   W.J. DE HAAS   C.G. BARKLA   M. SIEGBAHN   L. BRILLOUIN

E. VAN AUBEL   E. HERZEN   P. EHRENFEST   J.E. VERSCHAFFELT

M. KNUDSEN   J. PERRIN   P. LANGEVIN   O.W. RICHARDSON   J. LARMOR   H. KAMERLINGH ONNES   P. ZEEMAN   M. DE BROGLIE

A.A. MICHELSON   P. WEISS   M. BRILLOUIN   E. SOLVAY   H.A. LORENTZ   E. RUTHERFORD   R.A. MILLIKAN   Madame CURIE

4

Atoms and Electrons*

1. INTRODUCTION

International scientific collaboration had been almost completely stopped during World War I. The work of the Solvay Institutes, including the holding of scientific conferences, had also been halted. The Solvay Conferences were not resumed until the spring of 1921. The third Solvay Conference on Physics took place in Brussels from 1 to 6 April 1921. Its President was H. A. Lorentz, who now lived in retirement in Haarlem, Holland, and the participants were: G. G. Barkla (Edinburgh), W. L. Bragg (Manchester), M. and L. Brillouin (Paris), M. de Broglie (Paris), Madame M. Curie (Paris), P. Ehrenfest (Leyden), W. J. de Haas (Delft), H. Kamerlingh Onnes (Leyden), M. Knudsen (Copenhagen), Secretary of the Scientific Committee, P. Langevin (Paris), J. Larmor (Cambridge), R. A. Millikan (Chicago), J. Perrin (Paris), O. W. Richardson (London), E. Rutherford (Cambridge), M. Siegbahn (Lund), E. van Aubel (Ghent), P. Weiss (Strasbourg), and P. Zeeman (Amsterdam). The secretaries of the Conference were J. E. Verschaffelt (Haarlem), M. de Broglie, W. L. Bragg, and L. Brillouin.

A. A. Michelson (Chicago), travelling through Europe at the time, was invited to attend the third Solvay Conference. Among the invited participants, those who could not attend were: W. H. Bragg (London), N. Bohr (Copenhagen), A. Einstein (Berlin), and J. H. Jeans (Dorking).

The theme of the third Solvay Conference was 'Atoms and Electrons'. Central to this subject was Rutherford's nuclear model of the atom and Niels Bohr's atomic theory. The Conference had to deal with two different types of questions: first, to discuss the justification for the new atomic model and, second, whether phenomena like the photoelectric effect or superconductivity could be satisfactorily explained by the new tools.

The Conference was opened by Lorentz with a survey of the principles of classical electron theory. This theory had provided an explanation of the essential features of the Zeeman effect, pointing directly to the motions of electrons in the atom as the origin of spectra.

Rutherford gave a detailed account of the numerous phenomena which had received, since the second Solvay Conference, such a convincing interpretation with the help of his atomic model. This model had provided an immediate understanding of the essential features of radioactive transformations and of the existence of isotopes.

---

* *Atomes et Électrons*, Rapports et Discussions du Conseil de Physique tenu à Bruxelles du 1er au 6 Avril 1921, Gauthier-Villars, Paris, 1923.

The application of quantum theory to the electron binding in the atom had also made considerable progress. The classification of stationary quantum states by the use of invariant action integrals had, in the hands of Sommerfeld and his collaborators, led to an explanation of many details in the structure of spectra, especially of the Stark effect. The discovery of the latter had definitely excluded the possibility of tracing the appearance of line spectra to harmonic vibrations of the electrons in the atom.

The continued study of high frequency and optical spectra by Siegbahn, Catalan and others, in the few years following the third Solvay Conference, made it possible to arrive at a detailed picture of the shell-structure of the electron distribution in the ground state of the atom, which clearly reflected the periodicity features of Mendeleyev's table. Such advances implied the clarification of several significant points, leading to Pauli's discovery of the principle of mutual exclusion of equivalent quantum states. The discovery of the intrinsic spin of the electron involved a departure from central symmetry in the states of electron binding, and accounted for the anomalous Zeeman effect on the basis of the Bohr–Rutherford atomic model.

Such developments were still in the future at the third Solvay Conference, where reports were given on the recent experimental progress concerning characteristic features of the interaction between radiation and matter. Maurice de Broglie discussed the effects encountered in his experiments with X-rays, which revealed a relationship between absorption and emission processes similar to that exhibited by spectra in the optical region. Millikan reported about his continued, systematic investigations on the photoelectric effect, which led to improvement in the accuracy of the experimental determination of Planck's constant.

Einstein had already made a fundamental contribution to the foundation of quantum theory by showing how Planck's radiation formula could be derived simply by means of the same assumptions which had proved fruitful for the explanation of spectral regularities, and which had found striking support in the investigations by Franck and Hertz on the excitation of atoms by electron bombardment. Einstein's formulation of the general probability laws for the occurrence of spontaneous radiative transitions, and his analysis of the conservation of energy and momentum in the emission and absorption processes, was to prove fundamental for future developments.

By the time of the third Solvay Conference in 1921, some progress had been made in atomic theory by using general arguments to ensure the validity of thermodynamical principles and the asymptotic approach of the description of the classical physical theories in the limit of large quantum numbers. One of these general arguments was Ehrenfest's principle of adiabatic invariance of stationary states. This requirement had arisen from Bohr's formulation of the correspondence principle which, from the beginning, had served as a guide in a qualitative exploration of many different atomic phenomena. The aim of the correspondence principle was to let a statistical account of the individual quantum processes appear as a rational generalization of the deterministic description of classical physics.

Bohr had been invited to give a general survey of the recent developments of quantum theory, but he was prevented by illness from taking part in the conference.

Ehrenfest undertook to present Bohr's paper, to which he added a summary of the essential points of the correspondence argument.

We shall now give an account of the individual reports presented at the third Solvay Conference.

2. NOTES ON THE THEORY OF ELECTRONS

(H. A. Lorentz)

Lorentz opened the scientific discussions at the third Solvay Conference on Physics by presenting a résumé of the theory of electrons. He first showed that Maxwell's equations of electrodynamics were in accord with the principle of special relativity. The momentum of an electron is given by

$$\frac{mv}{\sqrt{(1 - v^2/c^2)}}, \tag{1a}$$

and its energy by

$$\frac{mc^2}{\sqrt{(1 - v^2/c^2)}}, \tag{1b}$$

where m is the (constant) mass of the electron, v its velocity, and c is the velocity of light.

In the direction of translation, the shape of the electron sustains a deformation inversely proportional to $(1 - v^2/c^2)^{1/2}$. This property, first discovered for the electron, in fact holds in general, and Lorentz gave its proof. Thus he considered an electron as a sphere of radius R, carrying a charge e uniformly distributed on the surface. The calculation of the momentum and energy of such an electron yields

$$\frac{e^2 v}{6\pi c^2 R} \left(1 - \frac{v^2}{c^2}\right)^{-1/2} \tag{2a}$$

for the momentum, and

$$\frac{e^2}{6\pi R} \left(1 - \frac{v^2}{c^2}\right)^{-1/2} - \frac{e^2}{24\pi R} \left(1 - \frac{v^2}{c^2}\right)^{1/2} \tag{2b}$$

for the energy.

Lorentz obtained the stresses as being proportional to $e^2/32\pi^2 R^4$ inside the sphere. He showed that in order to remain consistent with the relativistic transformations, one had to put

$$m = \frac{e^2}{8\pi^2 c^2 R} + \tfrac{4}{3}\pi R^3 \frac{\varepsilon}{c^2}, \tag{3}$$

where ε is the energy per unit volume. According to Poincaré's well-known hypothesis, $\varepsilon = e^2/32\pi^2 R^4$. Thus $m = e^2/6\pi c^2 R$, leading to the conclusion that inside a moving electron there would be neither momentum nor a current of energy, and the normal stress and energy density would remain constant.

Lorentz pointed out that, because of Poincaré's hypothesis, the equilibrium between the Maxwell stress and a constant external pressure would not be stable. In order to avoid this difficulty, he assumed that the electron (at rest) is rigid, so that he could put $\varepsilon = 0$, and obtain $m = e^2/8\pi c^2 R$. This led to the strange consequence that inside the electron there would be a negative amount of momentum and energy.

Lorentz then treated the motion of a system of charged particles under the influence of their mutual electrostatic interactions. If k/r^2 is the force between two particles, k being a constant with a positive or negative value depending upon whether the interaction is repulsive or attractive, then the potential energy is given by $U = \sum k/r$. By the virial theorem, the time average of the kinetic energy, $\bar{T}$, is related to the average potential energy, $\bar{U}$, as

$$\bar{U} = -2\bar{T}. \tag{4}$$

As an example, Lorentz considered the nucleus of oxygen. As conceived by Rutherford, it consists of four particles A of charge $+2$, one particle B of charge $+2$, and two electrons of charge -1. From the virial theorem Lorentz obtained the inequality

$$\frac{16}{r_4} + \frac{4}{r_5} > \frac{24}{r_1} + \frac{16}{r_2} + \frac{1}{r_3}, \tag{5}$$

where $r_1 =$ the average distance between the two A's, $r_2 =$ the average distance between A and B, $r_3 =$ the average distance between the two electrons, $r_4 =$ the average distance between an electron and an A, and $r_5 =$ the average distance between an electron and B. Lorentz did not believe that this inequality could be verified. He generalized the virial theorem to a more general system, but could not find any useful consequences of the new theorem.

Considering the case of an atom in a varying magnetic field, the nucleus assumed to be at rest, Lorentz, following Larmor, showed that the effect of a magnetic field could be assimilated in the rotation of the frames of reference. He applied this result to the case of a magnetic field directed along the Z-axis, and derived the expression for the magnetic moment due to the field $\mathbf{h}$.

By considering the effect of a uniform rotation of an electron around a given axis, Lorentz showed that its energy could be expressed as $(e^2 R/36\pi c^2)\, \omega^2$, as if the rotating particle had a moment of inertia equal to $e^2 R/18\pi c^2$. He then considered an electron, placed in an electromagnetic field, and having a translational velocity v and angular velocity ω. Taking into account the forces due to the field and the motion of the electron, he showed that for a uniform and constant magnetic field the axis of rotation of the electron would have a precession given by the equation

$$-\frac{e^2 R}{18\pi c^2}\, \dot{\omega}_x + \frac{eR^2}{3c}\, (\omega_y h_z - \omega_z h_y) = 0, \tag{6}$$

i.e. the axis will describe a cone with a speed of rotation equal to $-(6\pi c R/e)\, \mathbf{h}$.

Finally Lorentz pointed out that in classical electromagnetic theory an accelerated electron must radiate, a fact which causes serious difficulties for the theory of electrons in atoms. He hoped, however, that Maxwell's equations would still be preserved in this domain.

DISCUSSION

Rutherford asked if the radius of the electron could be determined by the classical electron theory. Lorentz pointed out that what one could say was that $R \geqq e^2/6\pi c^2 m$. Lorentz also mentioned that the assumption of the rigidity of the electron, which he had used in his report, was in contradiction with general relativity.

Rutherford thought that the inverse square law would certainly not be valid inside the nucleus. Lorentz agreed; however, his purpose in assuming this was to point out certain difficulties of the classical theory.

Larmor wondered whether the electrons could be regarded as discontinuities in the ether. Lorentz thought that so much evidence about the existence of electrons as particles was available from experiments that it was difficult to have doubts about them. He also thought that the notion of the ether was becoming meaningless.

Léon Brillouin presented, in some detail, Einstein's theory[1] dealing with the mechanism of absorption and emission of radiation by a Bohr atom.

Rutherford regarded it as a matter of the highest importance that something definite should be determined about the form and dimensions of the electron. He recalled A. H. Compton's proposal about an electron in the form of an annular ring. Compton had attempted to evaluate the dimensions of such an electron by taking into account the variations in the diffraction of X-rays of very different frequencies. Rutherford noted Compton's conclusion that the dimensions of the electron were larger than the wavelengths of penetrating X-rays; however, his own preference was for the old spherical model of the electron, with well-defined mass and radius. He also thought that experiments on the scattering of α-particles by nuclei had shown that the electron had dimensions smaller than those estimated by Compton, who thought that its radius was of the order of 10^{-11} cm. Rutherford had also concluded from experiments at the Cavendish Laboratory that the electron possessed properties other than merely an electric charge. Thus the observation of the trajectory of β-particles had shown that the electron must have a magnetic moment.

REFERENCE

1. A. Einstein, 'Zur Quantentheorie der Strahlung', *Physik. Z.* **18**, 121 (1917).

3. STRUCTURE OF THE ATOM

(E. Rutherford)

Rutherford recalled that before the discovery of the electron the ideas concerning a possible structure of the atom had been necessarily vague and uncertain, but still it had already been thought that the atom consisted of a combination of positively and negatively charged particles. The discovery of the independent existence of the negatively charged electron, with its small mass, and the experimental proof that it should be a constituent part of the atom, had given a strong impulse to the study of the structure of the atom. One of the first suggestions of this kind was made by Lord Kelvin[1] in a paper entitled 'Aepinus atomized', in which he considered the equilibrium of a certain number of 'electrions', or electrons[2] as they were now called, placed inside a sphere containing a positive charge of uniform density. This model was taken up and developed in detail by J. J. Thomson.[3] Thomson's model had the great advantage that in it the problem of the distribution, motion, vibration, and radiation of the electrons could be treated by means of classical electrodynamics.

If it were admitted that the negatively charged electron was a constituent of all atoms, the following problems had to be considered: (1) the number and distribution of the electrons, and the variation of their number with the mass of the atom; (2) the nature and distribution of positive charge; and (3) the nature and distribution of the mass of the atom.

Initially the fundamental difficulties concerned the distribution of the positive and negative charges inside the atom. Based on the results of Kaufmann's experiments, indicating that the mass of the electron was probably of electromagnetic origin, one envisaged the possibility of the atomic mass arising from the electrons alone as constituents. The atom of hydrogen would thus contain 1860 electrons!

Rutherford discussed the various methods proposed by J. J. Thomson for estimating the number of electrons in hydrogen atoms, based on: (1) X-ray diffraction by matter; (2) absorption of cathode rays; (3) dispersion of light; and (4) the scattering of high speed charged particles in their passage through matter. These methods, especially (1) and (4), had shown that the number of electrons in an atom was proportional to its mass. The experiments on the scattering of high energy β-particles, done by Crowther[4] had indicated that the number of electrons was two to three times the atomic weight of the element. From precise X-ray diffraction experiments Barkla[5] found that, for light elements, the number of electrons was about half the atomic weight. These estimates were based on Thomson's atomic model which, Rutherford said, had clarified numerous important points of atomic structure and indicated the means of attacking this fundamental problem of physics.

Rutherford then discussed the nuclear model of the atom which he had himself proposed previously.[6] He recalled the principal facts which had led to the discovery of this model. In the case of a beam of α-particles striking a thin sheet of gold or some other metal, the number of α-particles, scattered per unit surface area through an angle φ, was found to be proportional to: (1) $\operatorname{cosec}^4 \frac{1}{2}\varphi$, (2) the square of the charge Ne of the nucleus, (3) $1/E^2$, where E is the energy of the α-particle, (4) and the number of atoms per unit volume of the target. Geiger and Marsden[6a] and, later on, C. G. Darwin[7] confirmed the dependence of the scattering on velocity, varying as $1/v^4$.

The exact charge of the nucleus was determined with precision by Moseley with the help of X-ray spectra.[8] Moseley's work confirmed Van den Broek's assumption[9], according to which the nuclear charge was equal to the atomic number of the element.

The experiments of Chadwick[10] had led to the conclusion that the force around the nucleus varied according to the law $1/\gamma^p$, where p was found to lie between 1.97 and 2.03; that is, it followed the inverse square law to within experimental error.

Rutherford concluded that there existed very good experimental evidence to suggest that the atom contains a small nucleus of positive charge equal to the atomic number.

Rutherford pointed out that the emission of an α-particle from the atom led to a change of four units in the mass and two units in the charge, while in the emission of β-particles the mass remained almost constant but the charge changed by one unit, in perfect agreement with Moseley's work. Moreover, Mendeleyev's periodic table was now fully understandable.

Bohr's theory had accounted for the constitution and origin of spectra in terms of the energy levels of the peripheral electrons in atoms. The theory of Bohr, Sommerfeld, and Epstein also accounted for the fine structure of spectral lines as well as the Stark and Zeeman effects in the case of simple atoms. Rutherford compared the advantages of these studies over the qualitative considerations of Langmuir.

Rutherford discussed the question of the determination of the diameter of the nucleus. If the α-particles interacted with the nucleus through Coulomb interactions only, then the radius of the nucleus would have to be less than 7×10^{-12} cm, and evidence existed of the fact that very close to the nucleus a much stronger force acted on the α-particles.

The assumption that there were no electrons in the nucleus was in good agreement with Chadwick's work. This, however, did cause difficulties for the understanding of γ-rays, which could no longer be attributed to the K radiation, and must originate from the nucleus. But if this was true of some of the γ-rays, it seemed to be difficult for some others that were very soft.

In discussing the scattering of α- and β-particles by nuclei, Rutherford noted that a careful study of Wilson's photographs of the β-trajectories had led to the conclusion that the scatterings of α- and β-particles were quite different in nature. Compton[11] had proposed that the helical trajectory of β-particles could be the result of a magnetic moment of the electron.

In the collisions of α-particles with hydrogen the trajectory of the hydrogen nucleus after collision, according to Marsden[12], was about four times the trajectory of the α-particles. Rutherford reported that a homogeneous beam of α-particles, passing through a thin column of hydrogen, produced a number of H atoms (nuclei) which was 30 times greater than what was expected theoretically, and had a velocity distribution completely different from the case in which the nuclei had been considered as point charges. The number of H atoms decreased as the velocity of the α-particles diminished.

Rutherford did not believe that the ordinary laws of force between charged particles had been confirmed, and one was led to think that there must exist forces of a type other than the electromagnetic between α and H when they are at a distance smaller than 3×10^{-13} cm from each other.

The α-particle emission had been used by Rutherford to split the nuclei of certain atoms. He reported on experiments which had shown that atoms, presumably of hydrogen, were ejected from different substances subjected to α-radiation. This type of reaction was possible for elements having an atomic weight that could be expressed as $4n+p$, with $p=2$ or 3 but not 0. This fact was interpreted by Rutherford as demonstrating the impossibility of an α-particle being able to eject a hydrogen atom from a compound nucleus consisting of α-particles only.

Rutherford now used, for the first time in his report, the word 'proton' to describe the nucleus of hydrogen.

Continuing his review of the phenomena which take place in the nuclei of elements, Rutherford spoke about isotopes. Soddy had tried to separate radium and mesothorium chemically, and had found that this was impossible; he had concluded that there could exist different substances having the same physical and chemical properties, but differing in their radioactive properties, and had called such elements 'isotopes'.[13] Rutherford explained the reason for the existence of this phenomenon: this was because two nuclei had the same charge but different masses, and thus had the same chemical properties but different radioactive ones. He gave a list of radioactive families and their different isotopes. In the experimental determination of the atomic weight of lead, which had resulted from radioactive disintegration, the predicted isotopes had been readily found.

Going over from radioactive elements to non-radioactive ones, Rutherford reported on the attempt of J. J. Thomson and Aston to determine the isotopic mass of different elements by fractional distillation or by the diffusion method. He gave an account of Aston's mass spectrograph, which had provided a good means for the separation of different isotopes.

Finally Rutherford made some remarks about the structure of the nucleus. Aston's results had shown that the nucleus was formed of α-particles, but also of protons. The α-particle was formed from four positive charges and two negative ones. More generally, an atom of mass A and atomic number N would be formed of $A-N$ negative charges, and the ratio of the number of negative to positive charges was given by $n/A = 1 - N/A$ (where $n = A - N$). Rutherford pointed out that the stability of the nucleus could not be explained without the introduction of a new type of force between the constituents of the nucleus, and suggested that perhaps some magnetic forces might be involved.

DISCUSSION

Jean Perrin noted with astonishment that in Rutherford's experiments on the collision of α-particles with the nuclei of different elements, the hydrogen atoms emitted from nuclei, especially of Al and P, could have an energy greater than that of the incident α-particles and be emitted in all directions. Perrin thought that in this case one had to abandon the notion of a simple collision, and 'consider that in spite of its great velocity the α-particle, because of the very strong electrical repulsion, would arrive considerably slowed down in the vicinity of the nucleus. At this moment a "transmutation" takes place, consisting probably of an intra-nuclear rearrangement, with a possible capture of the incident α-particle – because one does not know what really happens – and the emission of the nucleus of hydrogen, as well as other emissions. In this manner of looking at things, there is no reason why the emitted hydrogen nuclei should "remember" the direction of the initial collision, nor why their energy should be less than that of the incident projectile – because, after all, this energy comes from the electrical intranuclear energy.'

Rutherford agreed that it was possible that the α-particle could make a temporary combination with the nucleus.

De Haas raised the question why the positive and negative charges were of equal magnitude. Rutherford answered that the original idea that negative electricity signified a lack of positive electricity was due to Benjamin Franklin, but now one thought that in neutral atoms the two types of electrical charges had to neutralize each other exactly. Millikan pointed out that it had been established experimentally that the positive and negative charges were of equal magnitudes. De Haas persisted in asking whether there was any fundamental reason for this equality. Lorentz did not believe that it was possible to establish it on the basis of general principles, and Rutherford thought that there was no means of explaining the equality of the positive and negative charge except to take it as an empirical fact.

Langevin thought that it was a remarkable consequence of Aston's experiments that if the atomic weight of oxygen were taken to be 16 or of helium to be 4, then the atomic weights of all other isotopes turn out to be integral numbers. For instance, the atomic weights of the two isotopes of neon are exactly 20 and 22. It was also remarkable that the atomic weight of hydrogen remained 1.008.

Perrin thought that if Prout's hypothesis was correct, that all atoms were formed from the atoms of hydrogen, then one could calculate, using Einstein's formula for the mass of energy, how much hydrogen would be necessary to form a helium atom, and how much mass will be lost in the form of energy. If this energy were to be emitted as quanta, then Perrin found that it would have a frequency of 6×10^{21}, and lie beyond the X-ray region. Perrin also thought that this transformation of mass would release such vast amounts of energy that this idea could explain the creation of solar energy, in a way that no theory, including those of Helmholtz and Kelvin, could explain.

Perrin conjectured that the nebula from which the sun was created, by condensation, was initially a cloud of hydrogen, which became transformed successively into helium and other heavier atoms. The transmutation of hydrogen into helium would account for almost '90 billion years' of the energy output from the sun at its present rate.

Rutherford agreed that it was indeed true that an enormous quantity of energy would be released by the combination of the nuclei of hydrogen to form helium. Here one had indeed a source of energy much larger than was possible according to the ideas of Kelvin and Helmholtz. Rutherford then made a most remarkable statement: 'It has occurred to me,' he said, 'that the hydrogen of the nebulae might consist of particles which one might call the "neutrons", which would be formed by a positive nucleus with an electron at a very short distance. These neutrons would hardly exercise any force in penetrating into matter. They will serve as intermediaries in the assemblage of nuclei of elements of higher atomic weights. It is [otherwise] difficult to understand how the positively charged particles could penetrate into nuclei against the forces of repulsion, unless they had enormously high velocities.'

Knudsen wanted to know what the difference was between this 'neutron' and the atom of hydrogen. 'The electron,' answered Rutherford, 'is much closer to the nucleus in the former [neutron] than in the latter [hydrogen nucleus].'

With respect to the structure of the nucleus, Lorentz thought that in the case of lighter elements, as Rutherford had pointed out, there existed a well defined structure analogous to what one would imagine for a 'crystalline molecule'. In the case of heavy nuclei, however, there would exist a great complexity of internal motions, including large velocities of constituent particles. These irregularities and complexities of motions in the case of heavier nuclei should allow one to understand the law of radioactive transformations.

Finally Madame Curie argued that only the presence of electrostatic forces could explain the speeds of β-particles, but these forces were not compatible with the stability of the nucleus. Such a stability required attractive forces of another type.

REFERENCES

1. Lord Kelvin, *Phil. Mag.* 6th series, **3**, 257 (1902).
2. The name 'electron' was first suggested by G. Johnstone Stoney in a paper in the *Phil. Mag.* **38**, 418 (1894). After referring to a paper which he had read at the Belfast meeting of the British Association in 1874, entitled 'On the Physical Units of Nature', [in which he had expressed Faraday's law of electrolysis as follows: 'For each chemical bond which is ruptured within an electrolyte a certain quantity of electricity traverses the electrolyte which is the same in all cases.'] Stoney continued: 'In this paper an estimate was made of the actual amount of this most remarkable fundamental unit of electricity, for which I have since ventured to suggest the name *electron*. According to this determination the electron = a twentiethet [that is, 10^{-20}] of the quantity of electricity which was at that time called the ampere, viz.: the quantity of electricity which passes each second in a current of one ampere, using this term here in its modern acceptation. This quantity of electricity is the same as three eleventhets [3×10^{-11}] of the C.G.S. electrostatic unit of quantity.'
3. J. J. Thomson, *Phil. Mag.* **7**, 237 (1904).
4. J. A. Crowther, *Proc. Roy. Soc.* **84**, 226 (1910).
5. C. G. Barkla, *Phil. Mag.* **7**, 543 (1904); **21**, 648 (1911).
6. E. Rutherford, *Phil. Mag.* **21**, 669 (1911); **27**, 488 (1914).
6a. H. Geiger and E. Marsden, *Phil. Mag.* **25**, 604 (1913).
7. C. G. Darwin, *Phil. Mag.* **27**, 499 (1914).

8. H. G. J. Moseley, *Phil. Mag.* **26**, 1024 (1913); **27**, 103 (1914).
9. A. Van den Broek, *Phys. Z.* **14**, 32 (1913).
10. J. Chadwick, *Phil. Mag.* **40**, 734 (1920).
11. A. H. Compton, *Phil. Mag.* **41**, 279 (1921).
12. E. Marsden, *Phil. Mag.* **27**, 824 (1914).
13. F. Soddy, *The Chemistry of Radio-elements*, Part II, Longmans, Green & Co. (1914).

4. THE RELATION $h\nu = \varepsilon$ AND PHOTOELECTRIC PHENOMENA

(M. de Broglie)

Maurice de Broglie presented a general outline of photoelectric phenomena. These had been among the first to be interpreted by means of quantum theory. The emission, from an illuminated metal, of corpuscles whose energy depends only on the frequency of the incident radiation, and the most remarkable fact that the electrons, emitted from a body under the influence of X-rays, have an energy of the same magnitude as the cathode electrons in the radiation generating tube – such experimental facts provided immediate support for the Planck–Einstein theory of quanta. De Broglie, in his report, proposed to distinguish between high frequency radiation such as X-rays (which give a pure radiation 'characteristic of the profound architecture' of the atom) and the spectrum of visible light. He thought that the difference between these two kinds of phenomena was quite remarkable. Also, although not completely satisfactory, the Rutherford–Bohr model of the atom was, according to de Broglie, sufficient to give an account of phenomena in the atomic domain.

In the first part of his report, M. de Broglie discussed the photoelectric effect and, in the second, the inverse process, i.e. the production of radiation by the impact of charged particles on matter. Among the subjects he treated, especially from the experimental point of view, were: excitation of characteristic radiation, or fluorescence radiation; secondary β-radiation in X-rays, or the photoelectric effect at high frequency; spectrum of secondary radiation; photoelectric effect with visible and ultraviolet light; production of radiation by impact of electrons against matter (inverse photoelectric effect); ionization potential and resonance.

For de Broglie, the most interesting fact was the existence of the photo-electron, implying a process for the exchange of energy that could not be explained without the assumption of the corpuscular nature of light. He discussed Millikan's careful experiments of 1916 on the verification of Einstein's photoelectric effect equation.

In the discussion of de Broglie's report, Rutherford pointed out that the K electrons were produced at a potential difference of 10^5 V, while the β-electrons had energies of more than 2×10^6 V, and this seemed to indicate that β-rays did not arise from the K electrons. What de Broglie's experiments had shown was that X-rays of frequency ν led to a flow of electrons whose energies were given by $h(\nu - \nu_K)$, $h(\nu - \nu_L)\ldots$, where ν_K, ν_L, etc., were the frequencies of the various atomic shells.

The discussion of de Broglie's report was far-ranging and almost all participants in the Conference engaged in it.

5. PARAMAGNETISM AT LOW TEMPERATURES; SUPERCONDUCTORS AND THE RUTHERFORD–BOHR ATOMIC MODEL

(H. Kamerlingh Onnes)

It was intended that reports on these two subjects would be presented by H. Kamerlingh Onnes and P. Langevin together at the third Solvay Conference. Langevin's sickness made this collaboration impossible, and Kamerlingh Onnes presented them on the basis of the notes that had been prepared earlier.

Kamerlingh Onnes also had had the benefit of discussions on these subjects with some members of the Commission of the Institut International du Froid, who met at Leyden in the last week of October and beginning of November, 1920. Among those who had been present at these discussions were Ehrenfest, Einstein, Kuenen, Langevin, Lorentz, Kamerlingh Onnes and Weiss.

5.1. PARAMAGNETISM

In the first report, Kamerlingh Onnes discussed paramagnetism. He reviewed the experimental facts which had demonstrated the importance of low temperatures for the study of magnetism. He noted that the precision of measurements increases significantly at low temperatures, and that for substances obeying Curie's law various effects were intensified 15 to 20 times compared to that at room temperature. Moreover, at low temperatures (about 2°K) thermal motion ceases for all practical purposes, and the phenomena being studied present a more simplified structure. Studies at low temperature were particularly of interest in connection with the kinetic theory of magnetism, established by Langevin[1], since all quantities of interest could be expressed in terms of an 'effective field', H/T, which would be larger the lower the temperature. On the other hand, the discrepancy between theory and experiment had become greater at low temperatures, leading to the question of the role of quanta in the use of statistical mechanics.

Kamerlingh Onnes discussed specific cases of magnetization. Gadolinium sulphate, for instance, obeyed Curie's law down to 4.25°K, up to the very limit of the accuracy of measurement. However, if a discrepancy existed, it would not be considerable even at 2°K. The fundamental problem was to understand how a crystal could obey Langevin's law down to such low temperatures, especially since the law itself was derived on the basis of the rotation of gaseous molecules.

While at low densities, oxygen gas obeyed Curie's law, there were departures for liquid oxygen. In the gaseous state one could still study it with the help of its equation of state, taking into account the mutual interaction of two individual molecules. The question remained whether the departure from Curie's law in the gaseous and liquid states was due to quanta or to Van der Waals' cohesion. Kamerlingh Onnes also considered the case of manganese sulphate and solid oxygen, both of which showed similar properties in that they lost their paramagnetic character at very low temperatures.

Kamerlingh Onnes denoted as normal paramagnetic substances those which obeyed Curie's law down to very low temperatures. He pointed out, following Weiss, that the influence of crystalline forces, giving the potential energy as a function of the direction, should maintain the orientation in spite of the thermal motions. He gave the results of his experiments on gadolinium sulphate, the measurements on which became uncertain below 2°K. They did show, however, that the departure from Curie's law near absolute zero was very small. The calculations seemed to indicate that the elementary magnets of gadolinium had 39.2 magnetons, but Kamerlingh Onnes was convinced that the small difference between this number and the integer 39 was due to secondary effects and that Weiss' law had been verified.

In experiments on gaseous and liquid oxygen, Kamerlingh Onnes had found that for some dilution of oxygen with nitrogen the relation between the susceptibility χ and the temperature T was still a straight line, intersecting the X-axis closer to the origin for smaller concentration. He introduced the anomaly Δ of a paramagnetic substance, i.e. the correction which must be applied to the absolute temperature in such a way that with this correction the temperature would follow the Curie law. In the case of oxygen this anomaly was found to be proportional to the density. The deviation from the Curie law could be represented as the effect of a Weiss molecular field, i.e. a field proportional to the magnetization. This field was so preponderant for a large class of substances and a large domain of temperatures that Kamerlingh Onnes called it *the anomaly of the first class*, denoting as a *cryogenic* anomaly one which arose at very low temperatures.

DISCUSSION

In the discussion, P. Weiss reported on a calculation which he had made using Langevin's law in the case of solids. When the mutual interaction of the molecules, without dependence of the orientation, was taken into account, he found the Curie law. If, however, he took into account the orientation effects produced by a molecular field, he obtained,

$$\frac{1}{\chi} = \frac{1}{C}(T - \Theta)$$

where $\Theta = -\Delta$ in the notation of Kamerlingh Onnes. Θ has usually been called the Curie point, and not *the anomaly of the first class*.

Weiss also pointed out that the molecular field was not of a magnetic nature. It could in fact be computed, and was found to be a thousand times smaller than if it were magnetic.

REFERENCE

1.   P. Langevin, *Ann. Chimie Phys.* (1905), 70.

5.2. SUPERCONDUCTORS AND THE RUTHERFORD–BOHR MODEL

One of the great triumphs of the electron theory of metals had been the derivation of the Wiedemann–Franz law. Unfortunately the theory ran into difficulties in the explanation of the passage of electricity in metals at low temperatures, especially the phenomenon of superconductivity.

Kamerlingh Onnes reported on the experiments he had undertaken on supercon-ductivity. He first discussed the general behaviour of metals at low temperatures. This behaviour was characterized by a minimum resistivity independent of the tem-perature and, according to the degree of purity, by an additive resistivity. As he had mentioned at the first Solvay Conference in 1911, mercury went through an abrupt change at 4.19°K. Since that time it had been found that radium G, thallium and tin were also superconductors.

Kamerlingh Onnes wondered why the other metals were not superconductors. Was it a question of purity? It was true that impurities did have a great influence on resistivity, although thallium remained a superconductor even when not entirely pure. He pointed out that it had not been possible to render other metals superconducting even when the temperature was reduced down to 1.5°K, which was the limit at which the measurements of resistance became extremely difficult. Whether one could conclude that the resistivity of superconductors was zero in an absolute sense was still in ques-tion. He thought that there remained a 'micro-residual' resistivity.

The magnetic field exercises an important effect on superconductivity. Kamerlingh Onnes reported on experiments showing that there was a threshold value of the mag-netic field at which superconductivity would disappear, and this effect was very sensi-tive to the temperature. Another important fact was the existence of a maximal cur-rent above which the superconductors recovered their usual resistivity. He also men-tioned the possibility of the existence of a persistent current in a ring at a sufficiently low temperature.

Kamerlingh Onnes had investigated in detail the properties of superconductors that were affected by a change of temperature. The only property which seemed to be so affected was thermal conductivity, leading to such changes in the Wiedemann–Franz law that it would no longer apply to superconductors. This abrupt change had a character similar to the Curie point for ferromagnets. J. J. Thomson had employed a similar analogy under the assumption of a molecular field, and considered align-ments of electrical doublets in order to explain the phenomenon of metallic con-duction.[1]

Kamerlingh Onnes examined the notions of mean free path and collisions and decided that they were meaningless for the phenomena arising in superconductors. He discussed the extent to which the Rutherford–Bohr atomic model could help in understanding superconductivity, and put his considerations in the form of certain questions, these being:

(1) Since the atoms described by the Rutherford–Bohr model constitute a metal, what happens to their peripheral electrons? Do they all, or some of them, lose their kinetic energy?

(2) How many kinds of electrons are to be distinguished in a metal? For instance, free electrons or others more or less strongly bound. What, according to the theory of quanta, is the statistics that applies to their motions?

(3) Are the motions of conduction electrons of adjacent atoms coherent?

(4) What is the mechanism by which the conduction electrons in ordinary conductors

transmit the energy, accumulated from an external electromotive force, among the degrees of freedom of thermal agitation?

(5) Does the velocity of the conduction electrons play a role in these phenomena?

(6) What is the mechanism by which the atoms acquire the superconducting property? How do they form superconducting filaments – i.e., macroscopic routes along which the conduction electrons can just slide along without transmitting any energy to the thermal degrees of freedom of the surrounding atoms? And how are the conduction electrons guided to follow this route amidst atoms in thermal motion?

(7) Up to what limit does the superconducting state resist the change of [physical] conditions?

(8) What is the reason for the equivalence of the roles of a magnetic field and temperature in the destruction of a stable adiabatic flux of electrons amidst an agglomeration of atoms, a large number of which are in thermal motion?

Kamerlingh Onnes formulated these questions in the belief that they were beyond the capability of the classical electron theory to explain, and probably the Rutherford–Bohr atomic model and the theory of quanta could deal with them.

In the discussion of Kamerlingh Onnes' report on superconductivity, Langevin pointed out that the critical current, i_c, at which the superconducting phenomenon stopped was also the current which generated the critical magnetic field H_c. Langevin also pointed out certain difficulties relating to the charge distribution in the system.

REFERENCE

1. J. J. Thomson, *Phil. Mag.* 6th series, **30**, 192 (1915).

6. INTENSITY OF THE REFLECTION OF X-RAYS BY DIAMOND

(W. H. Bragg)

In a very brief report, Bragg gave some results he had obtained on the reflection of X-rays by diamond. The principal result was that the analysis of intensity showed the effect of the tetravalent bond of the carbon atom, the intensity being sensitive to the privileged directions in the atom.

7. THE ANGULAR MOMENTUM IN A MAGNETIC BODY

(W. J. de Haas)

Following the notes supplied by Einstein on the theory of the Einstein–de Haas effect, de Haas derived the relation between the magnetic moment **m** of an atom and its angular momentum **M**,

$$\frac{|\mathbf{M}|}{|\mathbf{m}|} = \frac{2mc}{e}, \tag{1}$$

where m is the mass and e the charge of the electron, and c is the speed of light. The derivation was based on a direct calculation of the vector potential. De Haas noted that Richardson had already derived this result in essentially the same form as Einstein.

Richardson had pointed out that: 'By virtue of the law of conservation of momentum, the momentum thus created must be counter-balanced by an equal moment with respect to the same axis. One imagines that such a reaction might act either on the electromagnetic system producing the field or on the atoms of the magnetic material. It seems highly probable that the angular momentum thus created will be compensated by the movement of a single piece of a magnetic material.'[1]

This effect, i.e. the rotational motion of a bar because of magnetization, was discovered by Einstein and de Haas in 1915. At about the same time S. J. Barnett[2] discovered the inverse effect, i.e. the magnetization of a bar by means of rotational motion.

De Haas described the different methods employed to obtain this effect. He showed that a rod of iron, hung by a wire in such a way that it was well balanced, would obtain, when magnetized, a strong rotational impulse. De Haas noted that there might be several reasons why the rod would acquire motion: (1) the Einstein–Richardson effect, (2) the effect of the Earth's magnetic field on the horizontal component of magnetization, (3) a reaction due to the reciprocal action between the magnetic field and the field of the rod. De Haas pointed out how one could seek to eliminate the perturbations (2) and (3) in order to observe the 'Einstein–Richardson' effect.

In the discussion which followed, Richardson gave a report on the results of his experiments (and those of J. Q. Stewart at Princeton in 1915) on the same effects. The resonance methods yield the ratio of the angular momentum to the magnetic moment, which is too small by a factor of $\frac{1}{2}$. This discrepancy was not explained. Larmor wondered whether it might be possible to have a motion of the nucleus in a direction opposite to that of the electron.

REFERENCES

1. O. W. Richardson, *Phys. Rev.* **26**, 243 (1908).
2. S. J. Barnett, *Phys. Rev.* **6**, 239 (1915); **10**, 7 (1917).

8. APPLICATION OF THE THEORY OF QUANTA TO ATOMIC PROBLEMS

(N. Bohr)

It had been planned in the programme of the third Solvay Conference on Physics that Niels Bohr would give a report consisting of two parts: the first part dealing with the general principles which formed the basis of the application of the theory of quanta to atomic problems, and the second on the special applications of these principles to the problems of the arrangement and motion of electrons inside the atom. Unfortunately, sickness prevented Niels Bohr from attending the Conference. Only the first part of his report was ready, and it was presented at the third Solvay Conference by Paul Ehrenfest. Bohr had also been invited by David Hilbert and the Wolfskehl

Committee to lecture on atomic theory in Göttingen during June of 1921. He had to postpone his visit to Göttingen also, because of sickness, until June 1922. His lectures on atomic theory at Göttingen came to be called 'The Bohr Festival', and became the starting point of Heisenberg and Pauli's profound involvement in the problems of quantum theory.

In the report which Bohr had prepared for the Solvay Conference, and which Ehrenfest presented on his behalf, Bohr first recalled that the model of the atom was very similar to the solar system in which the planets revolve around the sun as the nucleus, with forces proportional to the inverse square of the distance. However, the orbits of the different planets are not completely determined by their masses and that of the sun; they depend essentially on the conditions which existed when the solar system was formed, that is on its prehistory.

In order to give an account of the well-known physical and chemical properties of elements, one is forced to admit that the motion of the particles in the atom, at least in its normal state, is completely determined by the values of the masses and charges of the constituent particles. This intrinsic stability of the atoms is best seen by considering the processes which give rise to the emission spectrum characteristic of the elements.

Bohr pointed out that it was impossible, on the basis of the classical electromagnetic theory, to explain either the stability of the normal state of the atom or the constitution of radiation produced by it. For a rational interpretation of atomic properties, the basis had to be found in the theory of quanta. The fundamental postulate of these considerations was the following: An atomic system, which emits a spectrum of discrete lines, can exist in a certain number of distinct states, called *stationary states*. The atom can exist in one of these states without emitting radiation. The emission of radiation takes place only by a complete transition between two stationary states, and the emitted radiation consists of simple harmonic waves. The frequency of the emitted radiation is related to the difference of energy of two stationary states, and is given by

$$hv = E' - E'', \tag{1}$$

where h is Planck's constant.

In the classical electron theory there would have to be a continuous emission of radiation from the atom. Even if this were ignored, the influence of the constant field was important and could not be ignored, and it meant that the electron had to be considered as moving in the atom according to Coulomb's law. The effect of external forces could not be considered in the same manner as in the classical case, but had to be understood in terms of the possible orbits or stationary states.

Bohr then discussed two phenomena: the collisions of electrons with atoms and the absorption and dispersion of electromagnetic radiation by gases. He pointed out that for the understanding of these kinds of phenomena, which are characterized by an interaction with the atom that is of very short duration in comparison with the periods of electrons in the atom, the notion of stationary states was very important. On the other hand, phenomena in which the interaction takes place in a period much longer than the periods of the electrons could, in Bohr's view, be treated very well by the classical theory.

Bohr had extended Ehrenfest's adiabatic hypothesis to what he now called the *principle of the mechanical transformability of stationary states*, and applied it to an atomic system of one degree of freedom, containing one charged particle executing oscillatory motion. The position and momentum of the particle are denoted by the generalized coordinates p and q respectively. Using the fact that $J = \int p \, dq$ is an adiabatic invariant, Bohr put $\mathscr{E} = nh\omega$, or $J = nh$ (in agreement with Planck's assumption), where n is an integer and h is Planck's constant. For the three-dimensional oscillator, $J_1 = n_1 h$, $J_2 = n_2 h$, $J_3 = n_3 h$. For the class of motions for which a separation of variables can be performed, i.e. for a class including all the multi-periodic motions, such a generalization holds valid.

Bohr then introduced the notion of degeneracy. When certain frequencies of the multi-periodic motions are such that

$$m_1\omega_1 + m_2\omega_2 + m_3\omega_3 = 0, \tag{2}$$

the conditions on the action variables can be replaced by

$$J = \int (p_1 \, dq_1 + p_2 \, dq_2 + p_3 \, dq_3) = nh. \tag{3}$$

The action of an external field would be more important for the degenerate case, because the change in the orbits may be very serious.

Bohr explained that the spectrum of an atom must be considered as arising from different monochromatic lines, from a superposition of individual processes. This was in agreement with the Ritz Combination Principle of spectral lines.

Bohr mentioned the difficulties encountered in generalizing the atomic system to more than one electron. The mechanical problem was then much more difficult, and a quantity analogous to J could not be defined in general, exceptions being the multi-periodic systems.

9. THE CORRESPONDENCE PRINCIPLE

(P. Ehrenfest)

Ehrenfest mentioned certain facets of the Rutherford–Bohr model: (1) In the classical model of Rutherford, a hydrogen atom would emit a continuous spectrum; and the electron, revolving around the nucleus would, because of its uninterrupted radiation, spiral into the nucleus. (2) Bohr had submitted the motions in the Rutherford model to a quantum condition. In determining this condition, he was guided by the discreteness of the spectral series, the Ritz combination principle, and the Planck–Einstein equation $\mathscr{E} = h\nu$. (3) Bohr sought, as much as possible, that his atomic model should conform to the classical rules (including the principle of inertia, Coulomb's law, etc.). He sought to establish a general correspondence between the electron motions in the atom and the radiation emitted by it. (4) In order to find this correspondence Bohr was guided by the following heuristic principle: When the quantum numbers in the atomic system become larger and larger, the radiation emitted tends asymptotically to that which the system would emit if it followed classical rules.

Ehrenfest explained that in all cases that could be treated [by means of the separation of variables], the coordinates x, y, z of the different electrons could be expressed as a trigonometric series, the different terms of which are defined by frequencies of the type, $p_1\omega_1 + \cdots p_k\omega_k$, where p_i are integers and ω_i are a set of fundamental frequencies. The angular momenta [the action variables] $J_1, J_2, ..., J_k$, corresponding to the angular variables $\omega_1 t, ..., \omega_k t$, are time-independent. Moreover, no radiation can take place with a continuous spectrum, as would be the case classically.

In order to formulate the correspondence principle, Bohr used a kind of averaging over the transitions. Let $n'_1, ..., n'_k$ and $n''_1, ..., n''_k$ be two sets of integers representing the system in two stationary states. A transition is then characterized by the number

$$\underset{n' \to n''}{N} = \frac{1}{2\pi} \left[(n'_1 - n''_1)\, \omega_1 + \cdots + (n'_k - n''_k)\, \omega_k \right]. \tag{1}$$

Bohr introduced a linear interpolation by means of

$$\frac{2\pi J_1}{h} = n''_1 + \lambda (n'_1 - n''_1), ..., \frac{2\pi J_k}{h} = n''_k + \lambda (n'_k - n''_k), \tag{2}$$

where λ varies from 0 to 1. The average is then defined by

$$\int_0^1 \underset{n' \to n''}{N} (\lambda)\, \mathrm{d}\lambda = \underset{n' \to n''}{\bar{N}}, \tag{3}$$

and from this Bohr deduced that

$$\underset{n' \to n''}{\nu} = \underset{n' \to n''}{\bar{N}}, \tag{4}$$

where ν is the transition frequency. Thus, for large values of n' and n'' one gets the classical result asymptotically. Ehrenfest also pointed out that Bohr had introduced certain selection rules among the possible transitions.

DISCUSSION

The discussion of Bohr's report, and Ehrenfest's complementary report on the correspondence principle, covered a wide variety of questions in which the participants were Lorentz, Ehrenfest, W. H. Bragg, Langevin, Rutherford, Zeeman, M. de Broglie, and Millikan.

Millikan gave a report on the arrangement and motion of electrons in the atoms, and mentioned the results of experiments he had performed on the collision of α-particles with some atoms. Except in the case of helium, the α-particle could not eject more than one electron, whatever the substance or the speed of α-particle considered. In helium, in one-sixth of the number of collisions at the maximum ionization speed, two electrons were ejected, thus showing, in view of the speed of the α-particle, that two electrons could not always be in symmetrical positions in the same orbit. It also showed that the electrons acted completely independently.

FOURTH SOLVAY CONFERENCE 1924

P. DEBYE   L. BRILLOUIN   E. HENRIOT   Th. DE DONDER   H.E.G. BAUER   E. HERZEN   Aug. PICCARD   E. SCHRÖDINGER   P.W. BRIDGMAN   J. VERSCHAFFELT

A. JOFFÉ   O.W. RICHARDSON   W. BRONIEWSKI   W. ROSENHAIN   P. LANGEVIN   G. de HEVESY

E. RUTHERFORD   Madame CURIE   E.H. HALL   H.A. LORENTZ   W.H. BRAGG   M. BRILLOUIN   W.H. KEESOM   E. VAN AUBEL

5

The Electrical Conductivity of Metals*

1. INTRODUCTION

The fourth Solvay Conference on Physics took place in Brussels from 24 to 29 April 1924. The general theme of the conference was 'The Electrical Conductivity of Metals and Related Problems'. H. A. Lorentz was again the President of the Conference, and the participants were: Madame Curie (Paris), H. E. G. Bauer (Strasbourg), W. H. Bragg (London), P. W. Bridgman (Cambridge, Mass., U.S.A.), M. and L. Brillouin (Paris), W. Broniewski (Warsaw), P. Debye (Zurich), E. H. Hall (Cambridge, Mass., U.S.A.), G. de Hevesy (Copenhagen), A. Joffé (Leningrad), H. Kamerlingh Onnes (Leyden), H. W. Keesom (Leyden), M. Knudsen (Copenhagen), P. Langevin (Paris), F. A. Lindemann (Oxford), O. W. Richardson (London), W. Rosenhain (Teddington, U.K.), E. Rutherford (Cambridge), E. Schrödinger (Zurich), E. van Aubel (Ghent). The Scientific Council consisted of Lorentz (President), Madame Curie, W. H. Bragg, M. Brillouin, Kamerlingh Onnes, Knudsen, Langevin, Rutherford and van Aubel. J. E. Verschaffelt (Ghent) acted as the Secretary of the Conference. The Scientific Council invited Th. de Donder, E. Henriot and A. Piccard, of the University of Brussels to participate in the Conference.

The fourth Solvay Conference was devoted to the problems of metallic conduction, and the discussions illustrated the fact that appropriate methods had not yet been developed for a comprehensive description of the properties of matter.

Lorentz gave a survey of the classical physical principles by which the problem of metallic conduction had been treated until then. In a series of important papers, Lorentz had himself pursued the consequences of the assumption that the electrons in metals behave like a gas obeying Maxwell's velocity distribution law. In spite of the initial success of such considerations, serious doubts about the adequacy of the underlying assumptions had gradually arisen, and these difficulties were brought out during discussions at the Conference. Reports on the experimental progress were given by Bridgman, Kamerlingh Onnes, Rosenhain and Hall. The theoretical aspects of metallic conduction were also discussed, especially by Richardson, who sought to apply quantum theory to the problem along the lines utilized in atomic problems. The principal phenomena of immediate concern were the Wiedemann–Franz law, Ohm's law, heat conductivity, and superconductivity. The Fermi statistics, essential for the understanding of metallic conduction, was still non-existent, and a satisfactory theory of superconductivity was still over three decades away.

---

* *Conductibilité Électrique des Métaux et Problèmes Connexes*, Rapports et Discussions du Quatrième Conseil de Physique tenu à Bruxelles du 24 au 29 Avril 1924, Gauthier-Villars, Paris, 1927.

We shall now give a résumé of the reports presented at the fourth Solvay Conference.

2. APPLICATION OF ELECTRON THEORY TO THE PROPERTIES OF METALS

(H. A. Lorentz)

Lorentz devoted himself to the discussion of fundamental problems of the electrical and thermal phenomena in metals. He first considered the Drude theory (1900) in which the ions are fixed and the electrons are free to move into the intermolecular interstices. Each electron has, on the average, an energy $\frac{3}{2}kT$ and a mean free path l. The coefficient of thermal conductivity is then given by

$$s = \tfrac{1}{2}kNlv, \tag{1}$$

where N represents the number of free electrons per unit volume, and v the speed of an electron corresponding to the kinetic energy $\frac{3}{2}kT$.

In order to calculate the electrical conductivity, Lorentz considered the electron as being under the action of an electric field E. The time between two collisions being l/v, the electric field imparts to the electron a speed $-(eE/m) \times (l/v)$ in the direction of E. Taking into account the collisions at the beginning and the end of the mean free paths, he obtained the electrical conductivity,

$$\sigma = \frac{Ne^2l}{2mv} = \frac{Ne^2lv}{6kT}. \tag{2}$$

The ratio of the two conductivities is given by

$$\frac{s}{\sigma} = 3\left(\frac{k}{e}\right)^2 T. \tag{3}$$

Lorentz noted that this expression was in good agreement with the measurements (of Jaeger and Diesselhorst).

Lorentz then presented the statistical theory of metallic conductivity which he had himself pursued earlier. He considered the ions as rigid spheres with which the electrons have only elastic collisions. He neglected the collisions between the electrons and assumed that the action of external forces imparts a very small change to the velocity in comparison with the velocity due to thermal motion.

Let ξ, η, ζ be the components of the velocity v of an electron, and $f(\xi, \eta, \zeta) \, d\lambda \, dS$, the number of electrons in the elementary volume dS whose velocities are comprised in the extension $d\lambda$ around ξ, η, ζ. The number of electrons per unit volume is then,

$$N = \int f(\xi, \eta, \zeta) \, d\lambda, \tag{4}$$

and the 'electron current' would be,

$$P = \int \xi f(\xi, \eta, \zeta) \, d\lambda, \tag{5}$$

across a plane perpendicular to their direction of motion. Also the 'energy current', due to the transport of kinetic energy, in the same direction would be

$$W = \tfrac{1}{2} m \int \xi v^2 f(\xi, \eta, \zeta) \, d\lambda. \tag{6}$$

Lorentz calculated the electrical conductivity by assuming Maxwell's distribution for the function f, that is

$$f(\xi, \eta, \zeta) = A e^{-mv^2/kT} + \psi(\xi, \eta, \zeta). \tag{7}$$

Assuming a uniform temperature and a constant field E, he obtained the electrical conductivity as,

$$\sigma = \tfrac{1}{3} \left(\frac{8m}{\pi k T} \right)^{1/2} \frac{e^2}{m} Nl, \tag{8}$$

and the thermal conductivity as,

$$s = \tfrac{2}{3} \left(\frac{8k^3 T}{\pi m} \right)^{1/2} Nl. \tag{9}$$

The ratio of the two conductivities is then,

$$\frac{s}{\sigma} = 2 \left(\frac{k}{e} \right)^2 T, \tag{10}$$

different by a factor of $\tfrac{2}{3}$ from the previous formula of Drude, and in worse agreement with measurements than the earlier result (3).

On the basis of his theory, Lorentz showed that the potential difference between two extremities of a wire depends only on the temperature at which they are maintained. He also showed that for two metals joined together by two weldings, having the temperatures T and T' respectively, the electromotive force developed would be equal to

$$F = \frac{k}{e} \int_{T}^{T'} \ln \frac{N_2}{N_1} \, dT, \tag{11}$$

a relation which involves only the temperatures of the weldings. He then used the preceding relations to calculate the Peltier and Kelvin effects.

Lorentz pointed out that it was possible to remove some of the restrictions in his model, but serious difficulties soon arise, and one does not have any more insight than before. Lorentz mentioned the theories of Bohr and Richardson, which seemed to be in better accord with the observations. They had assumed that the atoms in the metal, instead of behaving like elastic spheres, exert a force of repulsion on the electrons inversely proportional to a certain power of the distance. Lorentz thought that the only merit of the theories which he had discussed was that they had brought to attention certain details and questions which would otherwise be ignored.

Lorentz briefly reviewed the attempts of Kamerlingh Onnes, Nernst, Lindemann, Wien and Keesom at introducing quanta into the theory of metals. The principal change which Wien had obtained was that the speed of the electrons was found to be independent of the temperature. These researches had left no doubt as to what the modern picture of a metal must be. The atoms must be considered as composed of a positively charged nucleus and a certain number of electrons orbiting around it. In the metal, the nuclei had to be thought of as being arranged on a lattice.

In considering the motion of electrons in metals Lorentz noted that an electron having a quantized motion could be of no interest for the purpose of electrical conductivity. A simple calculation showed that in a wire of cross-section 1 mm^2, through which a current of 1 ampere passes, the average speed of the electron would be around $2.5 \times \times 10^{-4}$ cm/sec; that is, an electron would be able to travel through several interatomic distances in one second.

Ohm's law shows that, on the average, the action of a constant electric field leads to a constant mean velocity of the electrons, implying that by means of very complicated processes the action of nuclei on the electrons gives rise to a force proportional to the velocity v. Lorentz wrote this resultant force of resistance per unit volume as being equal to $-N\chi v$. For the stationary state,

$$- N\chi v - NeE = 0, \qquad \text{or} \quad v = - eE/\chi, \tag{12}$$

and the coefficient of electrical conductivity becomes, $\sigma = Ne^2/\chi$.

The implication in the motion of electrons in a metal is not that they remain continually free throughout forever. An electron could be ejected from its orbit in the atom and be captured by the nucleus of another atom; there could be free motion between successive captures by atoms, and so on.

Superconductivity, in the theory of Kamerlingh Onnes, was due to the existence of chains or filaments of atoms along which the electrons can move freely. Lorentz pointed out that when an electron passes through an atom, it communicates to the latter a momentum. The problem was then to determine whether the irregularity of electronic motions would create an agitation of the nuclei.

Lorentz then treated the problem of the emission of electrons by incandescent metals, and pointed out the importance of a thermodynamical treatment. By considering two parallel metallic plates, joined by a wire of the same metal, he calculated the distribution of the electrons between the two plates for a given temperature, and derived the Richardson formula,

$$\frac{\mathrm{d}}{\mathrm{d}T} \ln N_e = \frac{\varepsilon}{kT^2}, \tag{13}$$

where ε is a constant related to the geometry of the arrangement. Lorentz rederived this formula by making use of the Carnot–Clausius principle, and showed that the thermionic current per unit surface area is given by,

$$i = CT^2 e^{-\eta/kT} \tag{14}$$

where C and η are two constants for the metal. Making use of this formula and some preceding results, Lorentz showed that the potential difference between a Faraday cylinder C and a metal plate M is given by

$$\varphi_{2e} - \varphi_{1e} = \frac{1}{e}(\eta_1 - \eta_2). \tag{15}$$

If an electromotive force F is applied between M and C, the above equation becomes,

$$\varphi_{2e} - \varphi_{1e} = \frac{1}{e}(\eta_1 - \eta_2) - F. \tag{16}$$

In the photoelectric effect, the available energy to expel an electron from the metal is equal to $h\nu$, and the electron leaves the metal with an energy equal to $h\nu - \eta_1$. If F is chosen such that the electron's speed is zero, then

$$F = \frac{1}{e}(h\nu - \eta_2), \tag{17}$$

which is independent of η_1, i.e. of the metal considered, a result which had also been found in the experiments of Millikan.

Lorentz then discussed the action of a magnetic field on an electron in a conductor (Hall effect). He considered the special case of a very thin metallic sphere suspended from a wire in such a way that it could rotate around a diameter. The force F will create a moment of force, and an equilibrium will be achieved between this moment and the torsional moment of the wire. Lorentz showed that in the case of a small Hall effect a stationary state could be attained. He also studied the case of a perfectly conducting sphere, and showed that the force F, due to the field H, could not be counterbalanced by a distribution of electric charges.

Lorentz compared his results with those of Kamerlingh Onnes. His main conclusion was that the assumption concerning the freedom of motion of the electrons was wrong, and that the electrons in the superconducting state must follow some very particular orbits in which they could move without resistance.

Finally, Lorentz considered the optical properties of metals. He noted that light would produce in a metal an electric field, $\mathbf{E} = \mathbf{a}e^{int}$ (with frequency n). This field would produce a current $\mathbf{C} = (\alpha + i\beta)\mathbf{E}$. The coefficients α and β could be determined by reflection and polarization experiments, giving a relation between the conductivity and the optical properties.

DISCUSSION

Langevin pointed out that the statistical theory of Lorentz had the advantage of taking into account the distribution of velocities in a rigorous manner.

Lorentz showed that this theory could very well be adapted to the question of heat dissipation by considering that the atoms were subject to movements through collisions with the electrons.

Lindemann pointed out that in order to obtain agreement with the observed specific heat Lorentz had to assume that the number of electrons was small in comparison with the number of atoms. He noted that the theory of dissociation gave this number by means of the relation $n = n_0 e^{-q/kT}$, where q is the heat of dissociation. Lorentz recognized this shortcoming in his derivation of the specific heat.

Joffé raised a question about the potential introduced by Lorentz at the surface of the conductor. He showed that for liquid metals this force would push positive ions out of the liquid. Lorentz answered by showing that the non-existence of this potential would also lead to contradictions.

3. CONDUCTION PHENOMENA IN METALS AND THEIR THEORETICAL EXPLANATION

(P. W. Bridgman)

P. W. Bridgman presented an account of the principal experimental results concerning electrical conduction in metals, and gave a brief review of the theories which sought to account for them.

Bridgman first recalled that resistivity, as a function of the atomic number, did not have any periodic character. However, when atomic resistivity, i.e. resistivity divided by the density and multiplied by the atomic weight, was considered, then one obtained a periodic function showing that the electropositive metals were much better conductors than other elements. The relation between resistivity and temperature was, for nearly all conductors, linear in the first approximation but very sensitive to impurities. Bridgman also reviewed the effects of hydrostatic pressure and showed that in a majority of cases the pressure led to a decrease of resistivity. The effect of applying tension to the metal was similar to the effect of pressure.

In the case of a metallic crystal, the resistivity was seen to be a function varying with the direction. An abrupt change takes place in the resistivity of a metal when it goes from the solid to the liquid state; the phase in which the specific volume is larger also has the larger specific resistance. Bridgman pointed out the differences in the electrical behaviour between liquids and solids under a high pressure or temperature variation.

In the case of alloys, Bridgman pointed out that when the alloy was a simple mixing of small crystals of two types, the result was an average resistivity of the constituents. However, for alloys in which there was an interpenetration of the lattices, a small amount of one of the constituent metals mixed with the other could change the resistivity completely.

Bridgman reviewed the various theoretical contributions to the understanding of conductivity since the first theory of Drude; the principal ones being due to Lorentz, Bohr, H. A. Wilson, Swann, Nicholson, Livens, Debye, and Gans.[1] It was Debye who made the fundamental point that atoms in a crystal are placed in such a compact way that the thermal motion could change the mean free path of electrons by a fraction only, and the notion that the electrons in a metal could be regarded as a perfect gas was highly improbable.

The physical picture which Drude's theory had given rise to was that in a metal there was a swarm of electrons, whose number was comparable to that of the atoms; and the electrons moved around in interatomic spaces with the energy derived from equipartition, having collisions more often with the atoms than among themselves. The physical objections to this picture arose because of the inherent difficulty of the specific heat and the assumption that the number of electrons in the metal was of the same order as the number of atoms. The equipartition principle was, of course, necessary for providing agreement with the Wiedemann–Franz law. Another difficulty arose from the fact that if a body had as many electrons as atoms, there ought to be a lot more radiation from the metal, even at ordinary temperature, than there actually was. Moreover, such a physical picture could not explain the increase of resistivity with temperature, nor its decrease with pressure, nor the phenomenon of superconductivity.

Bridgman noted that all the proposed formulas were based on the idea of the mean free path, and led to a conductivity of the form

$$\chi = 2\sqrt{\frac{2}{3\pi}} \cdot \frac{ne^2 l}{mv}, \tag{1}$$

where n is the number of free electrons per unit volume, e the charge, m the mass, l the mean free path, and v the mean speed of the electron. The various modifications consisted in the treatment of n, l or v, leading to some changes in the coefficient.

In J. J. Thomson's theory[2] the atoms had been assumed to act as randomly oriented dipoles. At low temperatures these dipoles were aligned, thus allowing for superconductivity. But Thomson's theory was subject to many objections. Wien[3], in his theory, had rejected the postulate of equipartition for the electrons, and admitted a speed of the electrons independent of the temperature. Wien's theory could explain superconductivity qualitatively, but could not explain the variation of the resistivity with temperature or the Wiedemann–Franz law. Wien's theory was generalized by Grüneisen[4] to include the effect of the variation of resistivity with pressure.

Lindemann[5], Stark[6] and Borelius[7] introduced the notion that the electrons in the metal are distributed not as a perfect gas but rather as a perfect solid – i.e. they are distributed on spatial lattices that penetrate atomic lattices. The main difficulties encountered by such models were of a geometrical nature, due to the fact that it is generally not possible to have an electron lattice compatible with the lattice of atoms.

Bridgman also commented on another theory of J. J. Thomson[8], employing a 'reticular structure', as well as on the theories of Benedicks[9] and Waterman[10], all of which had encountered difficulties.

Bridgman then presented a long report on his own theory[11] of conductivity, which was based on the idea of the mean free path of the electrons in metals. He introduced the notion that the number of free electrons per unit volume is small in comparison with the number of atoms. Since the equipartition theorem was admitted, Bridgman could evidently explain Ohm's law and the Wiedemann–Franz relation. The other modifications in his theory arose from the calculation of the mean free path l and its variations with temperature or pressure.

DISCUSSION

Lindemann pointed out that the theory of an electron gas could not give any explanation of the anisotropy of the conductivity of crystals.

Keesom raised the question of conduction through a crystalline lattice. He pointed out that he had found diffraction rings for liquid potassium and sodium, and in this regard he had not found differences between conducting and non-conducting liquids.

Again, concerning the relation between conductivity and the existence of a crystalline state, Debye reported on the experiments he had performed (with Scherrer) on concentrated solutions in order to verify the existence of some very small crystals; the result was negative.

Madame Curie pointed out that the resistivity of a metal was not changed by its bombardment with α-particles.

Joffé mentioned that the results of X-ray analyses did not favour the existence of electron lattices.

Debye conjectured that it might be possible that the electrons, because of their mutual repulsions, formed a kind of solid. Although this solid would not have a regular lattice, it could still perform oscillations; however, because of the high frequencies of the vibrations of such a system, they would have only a small effect on the specific heat.

Richardson wondered whether the dissociation formula, $n = n_0 e^{-Q/RT}$, remained valid in quantum theory. Lindemann pointed out that this formula was a result of the relations $dQ/T = dS$ and $S = k \log W$, which were always valid.

Others who took part in the discussion of Bridgman's report were Lorentz, Rutherford, W. H. Bragg, Langevin, de Hevesy, Schrödinger, Rosenhain, Bridgman, Hall, Verschaffelt, and L. Brillouin.

REFERENCES

1. H. A. Lorentz, *Arch. Néerl.* (2), **10**, 336 (1905); *Proc. Amst. Acad.* **8**, 438, 585, 684 (1905); *Theory of Electrons*, Teubner (1907).
   H. A. Wilson, *Phil. Mag.* **20**, 835 (1910).
   W. F. G. Swann, *Phil. Mag.* **27**, 441 (1914).
   J. W. Nicholson, *Phil. Mag.* **22**, 245 (1911).
   S. H. Livens, *Phil. Mag.* **29**, 173, 425 (1915); **30**, 112, 287, 549 (1915).
   P. Debye, *Ann. Phys.* **33**, 441 (1910).
   N. Bohr, Diss. Copenhagen (1911).
   R. Gans, *Ann. Phys.* **20**, 293 (1906).
2. J. J. Thomson, *Phil Mag.* **30**, 192 (1915).
3. W. Wien, *Berl. Ber.* (1913), 84.
4. E. Grüneisen, *Verh. Deut. Phys. Ges.* **15**, 186 (1913).
5. F. A. Lindemann, *Phil. Mag.* **29**, 127 (1915).
6. J. Stark, *Jahrb. Rad.* **9**, 188 (1912).
7. G. Borelius, *Ann. Phys.* **56**, 388 (1918); **57**, 231, 278 (1918); **58**, 489 (1919).
8. J. J. Thomson, *Phil. Mag.* **44**, 657 (1922).
9. C. Benedicks, *Jahrb. Rad.* **13**, 351 (1916).
10. A. T. Waterman, *Phys. Rev.* **22**, 259 (1923).
11. P. W. Bridgman, *Phys. Rev.* **9**, 269 (1917); **17**, 161 (1921); **19**, 114 (1922); *Proc. Amer. Acad.* **59**, 119 (1923).

4. YET ANOTHER THEORY OF METALLIC CONDUCTIVITY

(O. W. Richardson)

Richardson proposed to present a theory which, he thought, would account for the essential characteristics of electronic conductors. He listed these characteristics as follows: (1) the difference between metallic and non-metallic conductors; (2) the rapid exponential increase in the conductivity of non-metallic conductors with the temperature; (3) in metallic conductors (a) the existence, or the possibility, of superconductivity, (b) the dilemma, that although proof exists that electricity and heat are conducted by electrons, the latter do not contribute at all to the specific heat of the solid, and (c) the law of Wiedemann and Franz.

Richardson claimed that his theory did not require the use of mean free paths, doublets, chains, or moving lattices; rather it did not call for any of these notions explicitly as did other theories of metallic conductivity. It also did not pretend to be either a mechanical theory or to give a detailed picture of the mode of transport of the electrons. In this sense, Richardson thought, his theory was in the same situation as the current theories of photoelectric effect or Bohr's theory of spectra.

Richardson had assumed that the electrons in atoms occupy well defined stationary states. One could consider the latter as the quantized stationary states of Bohr. The energy of the electrons in these states was determined by the structure of the atom, and it was independent of the temperature of the substance. The only energy which had to be considered as a function of the temperature, in the first approximation, was that of the ensemble of atoms, and this was in agreement with the results of Debye's theory of specific heats.

In Richardson's model[1], superconductivity was explained by assuming that the Bohr orbits in solids would become tangential to each other as the temperature would drop in the neighbourhood of absolute zero, and the electrons could move freely along these orbits. Richardson argued that when a current was established in a superconductor the electrons, which previously had a choice of different paths, would now be constrained to follow that path for which the velocity does not change sign.

At higher temperatures, Richardson assumed that the orbits were no longer in contact and the passage of an electron from one orbit to another could take place through radiative processes alone, given by $h\nu = w_2 - w_1$, where w_1 and w_2 are the energies before and after the transition in the atoms. Richardson obtained the transformation velocity of the transition from one state to another as the function $F_1[(w_2 - w_1), T]e^{-w_1/kT}$.

In the case of non-metallic conductors, even close to absolute zero, the distance between the orbits remains large, and the exchange of electrons would take place through radiative processes only. If w_0 were the necessary energy for such a process at $T = 0$, then the transformation speed would be given by $e^{-w_0/kT} \times$ a slowly varying function of T.

Richardson calculated the conductivity of metals by considering that the orbits of the electrons make an angle θ with the direction of the field, thereby introducing an additional term in the radiative energy. He put

$$h\nu = w_2 - w_1 - Xe\,\Delta\cos\theta \tag{1}$$

where X is the intensity of the field, and Δ the distance between the centres of two orbits. Taking an average over the possible angles and their probability at temperature T, Richardson obtained the specific conductivity as

$$\sigma = \frac{Ne^2}{3}\frac{\Delta}{kT}F(\bar{w}, T). \tag{2}$$

By analogy, the thermal conductivity was obtained by considering a transition from one orbit to another, but with W_1 changed by the heat flux,

$$w_1 \rightarrow w_1 + c\,\frac{\partial T}{\partial x}\,\Delta \cos\theta, \tag{3}$$

where c is the atomic specific heat at constant volume. The expression for the thermal conductivity, in Richardson's calculation, turned out to be

$$\chi = \frac{Nc}{3}\,\Delta F(\bar{w},\,T) \tag{4}$$

The ratio of the two conductivities follows the Wiedemann–Franz law,

$$\frac{\chi}{\sigma} = \frac{ckT}{e^2} \tag{5}$$

At ordinary temperatures the atomic specific heat is $c=3k$, and the above ratio takes the classical Wiedemann–Franz expression,

$$\frac{\chi}{\sigma} = \frac{3k^2 T}{e^2} \tag{6}$$

DISCUSSION

The discussion which followed Richardson's report did not provide much elaboration and clarification of his model, though some difficulties were pointed out. One conceptual difficulty was that Richardson had taken W_1 and W_2 as the energies of thermal agitation of the atoms, rather than the energies of the electron in its orbits. Also, in his theory, the distinction between metals and conductors in general was not made clear. Lindemann pointed out that according to Richardson's theory diamond should be a very good conductor.

Among those who took part in the discussion were Lindemann, Debye, Lorentz, Langevin, L. Brillouin, Schrödinger, Bridgman, Bauer, Rutherford, Joffé, and Richardson.

REFERENCE

1. O. W. Richardson, *Proc. Roy. Soc.* A **105**, 401 (1924).

5. THE INTERNAL STRUCTURE OF ALLOYS

(W. Rosenhain)

Rosenhain described various types of alloys and pointed out that alloys, more than pure metals, were of interest to metallurgists.

In view of the crystalline character of the alloys, one could think that they were formed by the fragments of crystals of constituent metals, but this point of view was not supported by X-ray analysis. Another possibility was to consider that the constituents, melted to form an alloy, took interstitial positions, occupying places in the lattice that were previously empty. A third possibility was that the melted constituents in the alloy took substitutional positions, i.e. their atoms exchanged positions with one another in the lattice. Rosenhain pointed out that experimental results on the mixing of pure metals favoured this last possibility as the only admissible one. He thought that this rule may not apply to non-metallic constituents; as, for instance, hydrogen atoms might very well occupy interstitial positions in the metals.

In the third case, since the 'foreign' atom might be bigger or smaller than the atoms of the metal, a kind of distortion of the lattice must take place around such an atom. Apart from this geometrical change in the lattice, it was important that a lattice could not be lengthened or compressed beyond a certain limit without rupture.

A lattice could be deformed in three ways. First, by thermal dilatation; second, by the introduction of 'foreign' atoms; and third, by applying mechanical tensions. When a lattice is stretched to the limit by thermal dilatation, it results in melting; the lattice breaks down, and the crystal melts. As the lattice tends to the limit of disruption by the introduction of 'foreign' atoms, the system 'dislocates' itself to form another crystalline phase. A change of phase also occurs when the elastic limit is passed under a mechanical dislocation of the lattice.

Rosenhain considered the question of 'solid solubility', and pointed out that the number of 'foreign' atoms which could be introduced in the lattice was entirely determined by the limit parameter, i.e. the energy produced by the deformations through the introduction of 'foreign' atoms was limited. Consequently the atoms whose solubility was limited would be the ones to produce the largest distortion.

A deformation of the lattice would have an effect on the mechanical properties of the substance. Assuming that deformations are most easily produced when the atomic planes can slip on one another, one finds that the geometrical deformation produced by the addition of foreign atoms would lead to an increase in the hardness of the metal, in perfect agreement with experiments.

Rosenhain noted that a geometrical deformation produced by foreign atoms, if they are larger than the atoms of the metal, will lower the melting point, since a certain energy has already been stored by the deformation. He pointed out that if the foreign atoms would produce a contraction, then the melting point would be increased, in complete agreement with experiments on the palladium-silver alloy.

It is easy to produce a gliding of atomic planes close to the melting point, and easier still to produce a gliding of an array of atoms. A simple consequence of this fact, according to Rosenhain, was that a metal could be 'diffused' easily into another one at such a temperature. He also mentioned the auto-diffusion experiments of Gróh and Hevesy[1] in which the radioactive isotopes of lead were supposed to 'diffuse' into a piece of lead, but the result was negative. Rosenhain explained that this was because the atoms were too similar, so that no tension leading to the gliding of planes had been created.

In the discussion of Rosenhain's report, Schrödinger wondered whether the deformations of the lattice could be seen from a study of the X-ray diffraction intensity patterns. Joffé gave several examples of where the gliding of atomic planes had been observed. He pointed out that when the elastic limit had been reached, a crystal presented the structure of an amorphous material at many places.

REFERENCE

1. J. Gróh and G. de Hevesy, *Ann. Phys.* **65**, 216 (1921).

6. ELECTRICAL RESISTANCE AND THE EXPANSION OF METALS

(W. Broniewski)

The resistivity of metals is a function of numerous factors, only one of which varies with the temperature; this is the free space between the molecules, and it is proportional to electrical resistance. This proportionality can be verified from the melting of a number of metals. Broniewski gave the formula,

$$W_T = R_T = - \text{ const. } \ln\left(1 - \frac{T}{KF}\right), \tag{1}$$

which expresses the variation of the free space (W_T) between the molecules, or the variation of electrical resistance (R_T), with the temperature. This formula was established on the basis of certain hypotheses concerning the intermolecular forces. In Equation (1), F is the absolute melting point, and K is a constant; for most metals $K=2$, for lead $K=3$, and for metals of the iron group, $K=1.25$. Broniewski mentioned that this formula had been verified for several metals.

7. THE ELECTRICAL CONDUCTIVITY OF CRYSTALS

(A. Joffé)

Joffé recalled that the fundamental laws and methods of measurement of the electrical conductivity of crystals had been established thirty-five years previously in the classical researches of Jacques and Pierre Curie. In his own work, first with Röntgen, and later on in Russia, Joffé had used the Curies' method and confirmed their observations.

Joffé noted that a puzzling point was the fact that, for a constant potential difference V across a crystal, the current diminished. His explanation, based on experimental results, was that the crystal became polarized, and if one took into account the electromotive force P due to polarization, Ohm's law was still valid in the form

$$\frac{V - P}{I} = R = \text{const.} \tag{1}$$

Experimental investigation of the variation of the resistance R with temperature T had shown that the conductivity σ could be described by

$$\ln \sigma = \frac{A}{T} + B, \tag{2}$$

where $A = -5 \times 10^3$ was about the same for all substances.

The polarization is different for different crystals, with two main types. The first type is represented by quartz, and it was found experimentally that polarization extended throughout the crystal. Joffé found its explanation in terms of a dissociation mechanism; he was also able to explain ionic mobility in good agreement with experiments. He had found that α-, β-, γ- and X-rays also influence conductivity, and that the current increases with the intensity of radiation until a saturation point is reached, and then it remains constant.

Calcite belongs to the second type of crystals. In this case, the polarization appeared only very close to the cathode, in a layer of about 10^{-4} cm. The explanation which Joffé found in this case was that the positive ions are nearly fixed, while the negative ions have very little mobility. He noticed that even at room temperature the degree of dissociation of calcite was large, a fact confirmed by the absence of an increase of conductivity by heating.

It was found that once the polarization was produced it did not remain constant, but decreased by diffusion at a rate which increased with the temperature.

The effect of radiation was, however, a different matter. The resistivity decreased after irradiation, but the crystal remained phosphorescent for ten days. During this period the conductivity attained its maximum and then decreased slowly.

Joffé discussed the behaviour of rock salt, and its conductivity on irradiation with X-rays. Visible light seemed to affect strongly the conductivity of rock salt. He also reported on the techniques of preparing very pure crystals, and pointed out that impurities change the resistivity considerably. Joffé commented on the electrolysis in crystals that also allowed one to study the effect of impurities on the mechanism of conductivity.

The discussion of Joffé's report consisted in seeking the clarification of various points of detail. Among those who took part in this were Langevin, Rutherford, Richardson, Lindemann, Schrödinger, Hevesy, Debye, Madame Curie, Bragg, and Bauer.

8. NEW EXPERIMENTS WITH SUPERCONDUCTORS

(H. Kamerlingh Onnes)

In his report Kamerlingh Onnes revisited the different experiments and questions that he had treated in the third Solvay Conference, and dealt with the progress which had been made since then.

He reported on an experiment that had been performed to measure the degree of variation of persistent currents. The measurements were made by observation of the relative motion of two rings of lead in the superconducting state. He found that the resistance of lead in this state could not be more than a fraction of 10^{-12} of its value at 0 °C. Since his report three years previously, the measurement of micro-residual resistivity had improved more than ten times.

Another experiment had shown clearly that the electrons in superconductors, which move without any resistance, are 'guided' in their trajectories as if 'they slide along

128   CHAPTER FIVE

their fixed tubular filaments in the superconductor'. The experiment was similar to the one he had performed before, except that the ring was now replaced by a sphere. The results of this experiment were in agreement with the idea of the 'filament' that he had expressed before.

The next question was related to the existence of a threshold field above which the resistivity was the ordinary one. At the third Solvay Conference, Kamerlingh Onnes had proposed as a working assumption that

$$H_T = H_0 - c_{HT}T, \tag{1}$$

i.e. the effect of temperature was equivalent to the action of a field. Now he pointed out that the transition with the magnetic field was not a discontinuity, but a very rapid transition. Experiments on lead, indium and tin showed that c_{HT} could be considered as a constant independent of the material. He hoped that the constant, c_{HT}, would be explained by quantum theory.

Langevin had proposed that the discontinuity could be the result of a change of phase in the material. However, X-ray analysis by Keesom had shown that the structure did not change at that point.

Kamerlingh Onnes discussed the question whether superconductors formed a special class of substances. Since the occasion of his report at the third Solvay Conference the only new substance which was found to have a superconducting state was indium. However, Au, Cd, and Ga, had also been studied very close to the superconducting state at low temperatures. The question of impurities was still very important, and a definite answer about the class of superconducting materials could be made only when really pure crystals had been obtained.

Kamerlingh Onnes had examined the superconductors for some general properties which could be studied as a function of the position of elements in the periodic table. He found that they are situated at a well defined place in the periodic table. He studied the structure of superconductors on the basis of Bohr's theory of the formation of the periodic table of elements, and concluded that there had to be shells of 18 electrons, and even those of 32 electrons, in order to have superconductivity. Following his idea that the electrons move along trajectories in which they are guided by superconducting filaments, Kamerlingh Onnes tried to explain the existence of this class of materials.

DISCUSSION

Madame Curie wondered whether the current in superconductors was produced spontaneously. Keesom thought that it could only be produced by induction, but Lorentz pointed out that Brownian motion could induce a spontaneous current for which the magnetic energy would have a value $\frac{1}{2}kT$.

Bridgman raised the question why the electrons must travel along some fixed channels. Lorentz described mathematically the system which had been used by Kamerlingh Onnes and showed that the result could be explained only by assuming the stationarity of the current distribution on the sphere used in the experiment.

Bridgman sought to give an explanation of the threshold field by assuming that the magnetic permeability was different in the superconducting state than in the normal one. The transition could then be explained by a thermodynamical cycle. However, Lorentz and Keesom did not think that thermodynamics could be applied to this case.

Keesom mentioned the experiments he had performed to show that the crystal lattice remained identical in the superconducting and normal states. X-ray analysis had also shown that the diffraction pattern remained the same.

Bragg and Langevin pointed out that the variations of intensity could be important and could signify a change in the electronic structure. Debye disagreed with this interpretation, and pointed out that a change in the electronic configuration would imply a change in the interatomic forces, and therefore a change in the lattice positions.

Piccard wondered whether lightning during a thunderstorm was perhaps a superconducting phenomenon at normal temperature.

9. METALLIC CONDUCTION AND THE TRANSVERSE EFFECTS OF THE MAGNETIC FIELD

(E. H. Hall)

Hall maintained that the electric current had to be considered as a fluid and the notion of a free electron in a metal could not be abandoned. He divided the electrons into two classes: *free electrons*, and *associated electrons*, the latter being bound to the ions.

The application of the law of mass action to a metallic bar submitted to a temperature gradient, together with the idea of convection, could be used to explain conduction phenomena. The electrons from the hottest extremity travel to the other end, as a gas would do, and return as associated electrons through a mechanism of collisions with the ions.

Hall mentioned the transverse effect of a magnetic field[1] and the orientation of ions in an electric field, and calculated the charge displacement for copper. He found that the ionic conductivity of copper was larger than its total conductivity.

He discussed the role of ions in Hall effect. Taking into account the polarization of ions, he pointed out that the field, the lines of force of which were parallel to $\mathscr{H}$, had two effects: one on the electrons and the other on the ions, and the important point was to know whether the effect of polarization would be large enough to overcome the effect on the electrons, thus leading to a positive Hall effect.

In order to study the influence of the magnetic field on free electrons, Hall assumed the existence of a mean free path l and derived a formula for the conductivity which, up to a constant factor, was the same as given by Drude. The Drude formula always leads to Ohm's law, which was an essential assumption in the derivation of the Hall effect.

He calculated the coefficient R_l of the Hall effect due to free electrons. It turned out to be $R_l = 6.28\ l/kT^{0.5}$, where l is the mean free path and k the total conductivity. He then determined the coefficient R for the total Hall effect which, based on an analogy with the coupling of two electromotive forces, turned out to be $R = R_a(k_a/k) + R_l(k_l/k)$, where $k = k_a + k_l =$ total conductivity.

Hall took into account the temperature gradient which, together with the Hall potential, gives rise to an effect known as the Ettingshausen effect. The converse effect, i.e. the creation of a transverse potential gradient in a metallic piece carrying a heat flow through a magnetic field, is called the Nernst effect.

With his argument based on the idea of 'convection', Hall again separated the contributions due to associated and free electrons respectively, obtaining

$$Q = Q_a \frac{k_a}{k} + Q_l \frac{k_l}{k},$$ (1)

where Q is called the Nernst coefficient.

Finally, Hall discussed the Righi–Leduc effect. This effect is the existence of the gradient of transverse temperature which accompanies the Nernst effect, and is very similar to the Ettinghausen effect. He concluded his report by giving the calculations of all these effects for gold.

DISCUSSION

In the discussion following Hall's report, Lorentz mentioned that the dualistic theory, in which both positive and negative charges were supposed to contribute to the electric current, always encountered great difficulties. The double sign found in the Hall effect suggested very strongly the existence of two different kinds of current, and Lorentz thought that Hall had solved the problem admirably. Lorentz attempted to give the various coefficients occurring in Hall's calculations, a physical meaning. Bridgman expressed these coefficients thermodynamically. Lorentz also gave a derivation of the relations between these coefficients, based on 'energy balance' and the second law of thermodynamics. He thought that there were inherent difficulties in the application of the second law of thermodynamics to irreversible processes.

REFERENCE

1. E. H. Hall, 'Thermo-Electric Action and Thermal Conduction in Metals', *Proc. Nat. Acad. Sci.* 7, No. 3 (March 1921); 'An Electron Theory of Electric Conduction in Metals', *Proc. Nat. Acad. Sci.* 8, No. 10 (October 1922).

10. PROPAGATION OF RADIATION MOMENTUM

(A. Joffé and N. Dobronravoff)

Joffé presented a brief report on an experiment which sought to prove Einstein's assumption about the concentration of energy in light quanta. The results were in favour of this assumption.

FIFTH SOLVAY CONFERENCE 1927

A. PICCARD E. HENRIOT P. EHRENFEST Ed. HERZEN Th. DE DONDER E. SCHRÖDINGER E. VERSCHAFFELT W. PAULI W. HEISENBERG R.H. FOWLER L. BRILLOUIN

P. DEBYE M. KNUDSEN W.L. BRAGG H.A. KRAMERS P.A.M. DIRAC A.H. COMPTON L. de BROGLIE M. BORN N. BOHR

I. LANGMUIR M. PLANCK Mme CURIE H.A. LORENTZ A. EINSTEIN P. LANGEVIN Ch.E. GUYE C.T.R. WILSON O.W. RICHARDSON

Absents : Sir W.H. BRAGG, H. DESLANDRES et E. VAN AUBEL

6

Electrons and Photons*

1. INTRODUCTION

The fifth Solvay Conference on Physics took place in Brussels from 24 to 29 October 1927, and its general theme was 'Electrons and Photons'.

H. A. Lorentz was again the President of the Conference. This was the last Solvay Conference on which Lorentz would preside; he died a few months later in Haarlem, Holland, on 4 February 1928. Under Lorentz' guidance, the Solvay Conferences had developed a unique and distinguished style of presentation and discussion of problems at the frontiers of physics at a given time, and this tradition would continue in the following decades.

Among the participants in the fifth Solvay Conference were: N. Bohr (Copenhagen), M. Born (Göttingen), W. L. Bragg (Manchester), L. Brillouin (Paris), A. H. Compton (Chicago), L. de Broglie (Paris), P. Debye (Leipzig), P. A. M. Dirac (Cambridge), P. Ehrenfest (Leyden), R. H. Fowler (Cambridge), W. Heisenberg (Copenhagen), H. A. Kramers (Utrecht), I. Langmuir (Schenectady, N.Y., U.S.A.), W. Pauli (Hamburg), M. Planck (Berlin), E. Schrödinger (Zurich), C. T. R. Wilson (Cambridge).

The Scientific Committee consisted of: H. A. Lorentz (President), W. H. Bragg (London), Madame M. Curie (Paris), A. Einstein (Berlin), C. E. Guye (Geneva), M. Knudsen (Copenhagen), Secretary, P. Langevin (Paris), O. W. Richardson (London), and E. Van Aubel (Ghent).

J. E. Verschaffelt (Ghent) served as the Scientific Secretary of the Conference. Th. De Donder, E. Henriot and A. Piccard of the University of Brussels were invited by the Scientific Committee to attend the conference. E. Herzen represented the Solvay family at the Conference.

By the time of the fourth Solvay Conference in 1924 it had become quite evident that, in dealing with complicated atomic phenomena, even such limited use of mechanical pictures as had been retained in the approach based on Bohr's correspondence principle could not be upheld. Progress had been initiated, even then, that would be of great importance for the subsequent development. For instance, A. H. Compton, in 1923, had discovered the change in frequency of X-rays by scattering from free electrons (the Compton effect). Compton, as well as Debye, emphasized the importance of this discovery in support of Einstein's conception of the light-quantum or photon,

---

* *Électrons et Photons*, Rapports et Discussions du Cinquième Conseil de Physique tenu à Bruxelles du 24 au 29 Octobre 1927, Gauthier-Villars, Paris, 1928.

notwithstanding the increased difficulties of picturing the correlation between the processes of absorption and emission of photons by the electron in the simple manner which Bohr had used for the interpretation of atomic spectra.

Within a year of Compton's work, and almost concurrently with the fourth Solvay Conference, a new light was shed on such problems by Louis de Broglie's comparison of particle motion and wave propagation. De Broglie's ideas were to find striking confirmation in the experiments by Davisson and Germer and George Thomson on the diffraction of electrons in crystals, and they proved to be of fundamental importance for the establishment of the wave equation by Schrödinger. The Schrödinger equation would become one of the most powerful tools for the solution and understanding of a wide variety of atomic problems.

Another approach to the fundamental problem of quantum physics had been initiated by Kramers who, a few weeks before the fourth Solvay Conference, had succeeded in developing a general theory of the dispersion of radiation by atomic systems. The treatment of dispersion had been, from the very beginning, an essential part of the classical approach to the problems of radiation. Lorentz had repeatedly called attention to the lack of such guidance in quantum theory. With the help of arguments based on Bohr's correspondence principle, Kramers showed how a direct connection could be established between the dispersion effects and Einstein's formulation of the probabilities of individual, spontaneous and induced radiative processes.

Kramers and Heisenberg further developed dispersion theory to include new effects originating in the perturbation of the states of atomic systems produced by electromagnetic fields. Heisenberg found a basis, in this development in dispersion theory, for his discovery of the basic formalism of quantum mechanics; in it all reference to classical pictures beyond the asymptotic correspondence was completely eliminated. The work of Born, Heisenberg and Jordan, and independently Dirac's, soon gave a general formulation of Heisenberg's ideas. The classical kinematic and dynamical variables were now replaced by symbolic operators obeying a non-commutative algebra involving Planck's constant.

Dirac and Jordan developed the transformation theory along the lines of Hamilton's original treatment of problems in classical mechanics, which elucidated the relationship between the different approaches of Heisenberg and Schrödinger to the problems of quantum theory. The basis of the interpretation of the new formalisms thus became clarified, helping to understand the apparent contrast between the superposition principle in wave mechanics and the postulate of the individuality of the elementary quantum processes. By using the amplitudes and phase of the constituent harmonic components, Dirac developed a quantum theory of radiation, in which he incorporated Einstein's photon concept in a consistent manner.

One of the principal themes of discussion at the fifth Solvay Conference was the renunciation of pictorial deterministic description implied in the methods of the new mechanics. An important question, as Bohr noted, was 'to what extent the wave mechanics indicated possibilities of a less radical departure from ordinary physical description than had been hitherto envisaged in all attempts at solving the paradoxes to which the discovery of the quantum of action had from the beginning given rise ?'[1] Born's successful treatment of collision problems had emphasized the essentially statistical character of the interpretation of the new theory. The symbolic character of the whole conception appeared most strikingly in the necessity of replacing the ordinary three-dimensional space coordinates by a representation of the state of a system, containing several particles, as wave function in a configuration space, with as many coordinates as the total number of degrees of freedom of the system. This point was particularly important in the treatment of systems involving particles of the same mass, charge and spin, revealing in the case of such 'identical' particles a limitation of the individuality implied in classical corpuscular concepts. Pauli's exclusion principle took into account such novel features in the case of electrons, while Bose had dealt with them for photons even earlier by using a statistics which represented a departure from the Boltzmann method of counting the complexions of a many-particle system.

Heisenberg's explanation of the helium spectrum in 1926 was a decisive contribution to the treatment of atoms with more than one electron. This problem had remained for many years an obstacle for the quantum theory of atomic structure. Within a year Heitler and London gave an analogous treatment of the electronic structure of the hydrogen molecule, which provided a clue to the understanding of nonpolar chemical bonds.

Similar considerations of the proton wave function of the rotating hydrogen molecule led to the assignment of a spin to the proton. This, in turn, led to an understanding of the separation between ortho- and para-states, which explained the anomalies in the specific heat of hydrogen at low temperatures.

Only a few months before the fifth Solvay Conference Heisenberg formulated the uncertainty principle, which elucidated the physical content of quantum mechanics more than any other earlier consideration. In September, 1927, Bohr stated his principle of complementarity in a lecture at the Como Congress, honouring the hundredth anniversary of Volta's death.

All this revolutionary development formed the background of the fifth Solvay Conference. The conference opened with reports by Lawrence Bragg and Arthur Compton about the rich new experimental evidence regarding scattering of high frequency radiation by electrons exhibiting widely different features when firmly bound in crystalline structures and when practically free in atoms of light gases. De Broglie, Born and Heisenberg, and Schrödinger then gave reports about the great advances that had been made in the consistent formulation of quantum theory. Niels Bohr gave a report on the epistemological problems facing quantum physics.

All the principal architects of the old and new quantum theory, such as Planck, Einstein, Bohr, Ehrenfest, Debye, Born, Heisenberg, Kramers, Compton, de Broglie, Pauli and Schrödinger, were present at the conference. Among the prominent ones, those missing were Sommerfeld and Jordan. We shall now give a résumé of the individual reports presented at the conference.

REFERENCE

1. N. Bohr, in *La Théorie Quantique des Champs*, Douzième Conseil de Physique, 1961, R. Stoops Editeur, Brussels, p. 24,

2. REFLECTION OF X-RAYS

(W. L. Bragg)

X-ray diffraction and reflection from a crystal follows the laws of classical optics. In many cases, rough information on the intensity of X-rays is generally sufficient to understand the structure of the crystal under study. A quantitative theory of the intensity was, however, very difficult to establish, though it would give information on the nature of atoms and molecules in the crystal, and Bragg's purpose was to present such a theory in his report.

Lawrence Bragg pointed out that the observations with a microscope or with X-rays, although similar in some ways, exhibited fundamental differences. The principal difference was that in the case of X-rays no phase relations could be measured, thus forbidding the formation of an image. The only quantities which could be measured were the intensity and the directions of the diffracted rays.

It was important to recognize that, in the case of X-rays, all investigations had to be made in the immediate vicinity of the very limit of the resolving power of the instruments. The wavelength region in question was between 0.6 Å and 1.5 Å. In this interval, the wavelengths were sufficiently small if one wanted to study the details of the crystalline structure, which extends to several angströms, but they were still too large to penetrate the structure of the atom. Crystalline analysis with X-rays could be divided into three parts: (1) the experimental measurement of the intensities of diffracted pencils; (2) the reduction of these observations, with the help of theoretical formulas, to the measurements of the amplitude of diffracted waves; (3) the deduction of the form of the structure with the help of the measurements of diffraction in different directions.

Bragg gave an account of the earlier work on X-ray diffraction by crystals. After the fundamental work of Laue[1], Darwin[2] and, independently, Ewald[3], studied the intensity of diffraction. Two types of formulas had to be distinguished for the intensities of the reflected X-rays. One was related to the 'mosaic crystal' (a crystal constituted of very small parts), for which the usual absorption formula was valid. The other formula applied to the 'ideally perfect crystal', in which the absorption did not appear in the intensity, and the reflection was perfect. Debye[4] had studied the effect of the diminution of intensity of reflection with the increase of temperature.

The first precise quantitative measurements were made by William Bragg[5]. William Bragg had also proposed, in his Bakerian lecture[6] of 1915, the use of Fourier analysis, a method which was successfully used by Duane, Havighurst and Compton. Compton[7] rederived the results of Darwin by means of another formula, and showed that electron distribution in the atoms of the crystal was of the type indicated by the Bohr model.

Lawrence Bragg presented a formula for the integrated reflection of X-rays falling on the surface of a crystal of the mosaic type,

$$\varrho = \frac{Q}{2\mu} \frac{1 + \cos^2 2\theta}{2},$$ (1)

where $\frac{1}{2}(1 + \cos^2 2\theta)$ is the polarization factor, μ the effective absorption coefficient, and

$$Q = \left(\frac{Ne^2}{mc} F\right)^2 \frac{\lambda^3}{\sin 2\theta},$$ (2)

where e and m are the electronic constants, c the speed of light, N the number of scattering centres per unit volume, θ the angle of incidence, λ the wavelength of the X-rays, and F the atomic form factor, which had to be determined.

The measurements performed on a given crystal provide a series of values of F, which contain all the information one can obtain about the crystalline or atomic structure. The graphical representation of F as a function of $(\sin\theta)/\lambda$ tends, for small values of the argument, to the number of electrons N in the atom.

Bragg gave examples of experimental form factors in good agreement with those that had been obtained from the Thomas model[8] of the atom. Another method, based on Fourier analysis, allowed a determination of the density of the scattering matter. The formula for the distribution of scattering matter in a crystal, having a centre of symmetry, had been written by Compton in the form

$$P_z = \frac{Z}{a} + \frac{2}{a} \sum_1^\infty F_n \cos \frac{2n\pi z}{a},$$ (3)

where z is measured perpendicularly to the planes at a distance a, $P_z \, dz =$ the quantity of scattering matter between the plane at z and $z + dz$, and $Z = \int_0^a P_z \, dz =$ total scattering matter per unit of crystal structure. Bragg noted that the wavelength employed did not permit the measurement of separate details smaller than itself. This also showed that the method of Fourier analysis was incomplete for finer details.

Bragg discussed the mechanism of the diffraction of X-rays by crystals, and how the measured X-ray intensities helped one to understand atomic structure. He pointed out that, with the assumption of spherical symmetry, it was possible to deduce from the form factor the distribution of the scattering matter and the dimension of the atom. It was found, however, that peaks did not appear at the same place if one used Fourier analysis or classical methods.

Finally Bragg commented on the refraction of X-rays. The index of refraction could be thought of as arising from the diffraction in the direction of propagation of coherent radiation which interferes with the primary beam. Darwin, and later on Ewald, had pointed out that, in the formula for reflection, the real angle θ would be given by

$$\theta - \theta_0 = \frac{1 - \mu}{\sin\theta \cos\theta},$$ (4)

where θ_0 is given by

$$n\lambda = 2\,\mathrm{d}\sin\theta_0\,. \tag{5}$$

This effect of refraction had been experimentally observed by Stenström, Duane, Patterson, Siegbahn, Hjalmar, and Davis.

Compton[9] discovered the total reflection of X-rays and used it to determine the refractive index, which was found to be slightly less than unity. The refraction of X-rays was also observed by Larson, Siegbahn and Waller[10] by using a glass prism.

Bragg pointed out that when the X-ray frequency was large in comparison with the characteristic frequencies of the atom, the refractive index was in good agreement with the formula,

$$1 - \mu = \frac{ne^2}{2\pi m v^2}\,, \tag{6}$$

where n is the electron density and v is the frequency of the incident radiation. The critical reflection angle was found to be given by

$$\theta = \sqrt{\frac{ne^2}{\pi m v^2}}\,. \tag{7}$$

When the X-ray frequency was of the same order of magnitude as the atomic frequencies, it was found that

$$\mu - 1 = \frac{e^2}{2\pi m}\sum_{1}^{n}\frac{n_s}{v_s^2 - v^2}\,, \tag{8}$$

and this formula agreed well with experimental results.

DISCUSSION

In the discussion following Bragg's report, those who took part included Debye, Fowler, Heisenberg, Pauli, Lorentz, Compton, Kramers, Dirac, and Born.

Kramers discussed, at length, his theory of the anomaly in the real part of the refractive index. He considered plane polarized electromagnetic waves falling on an atom. Under the action of the waves each atom acts as an oscillating dipole contributing, by Fourier expansion, a term at frequency v. Let $\mathrm{Re}(Pe^{2\pi i v t})$ be this term, where P is a complex vector proportional to E. Kramers put $P/E = f + ig$, where f and g are functions of v. He then extended v to the region of negative values by defining f as an even function, and g as an odd function. With this he could write,

$$f(v) = \frac{1}{\pi}\,(\mathrm{p.v.})\int\limits_{-\infty}^{+\infty}\frac{g(v')}{v - v'}\,\mathrm{d}v\,, \tag{9}$$

(p.v. = principal value of the integral). This formula showed that $f(v)$ was an analytic function of a complex variable v, holomorphic above the real axis, so that he could write

$$g(v) = -\frac{1}{\pi}\,(\mathrm{p.v.})\int\limits_{-\infty}^{+\infty}\frac{f(v')}{v - v'} \tag{10}$$

giving the Kramers–Kronig dispersion relation.

REFERENCES

1. M. von Laue, *Bayer. Akad. Wiss., Math. Phys. Kl.* (1912), 203.
2. C. G. Darwin, *Phil. Mag.* **27**, 315, 675 (1914).
3. P. P. Ewald, *Ann. Phys.* **54**, 519 (1918).
4. P. Debye, *Ann. Phys.* **43**, 49 (1914).
5. W. H. Bragg, *Phil. Mag.* **27**, 881 (1914).
6. W. H. Bragg, *Phil. Trans. Roy. Soc.* A **215**, 253 (1915).
7. A. H. Compton, *Phys. Rev.* **9**, 29 (1917); **10**, 95.
8. L. H. Thomas, *Proc. Camb. Phil. Soc.* **23**, 5, 542 (1927).
9. A. H. Compton, *Phil. Mag.* **45**, 291 (1923).
10. A. Larsson, M. Siegbahn, and I. Waller, *Naturwiss.* **12**, 1212 (1924).

3. DISCREPANCIES BETWEEN EXPERIMENTS AND THE ELECTROMAGNETIC THEORY OF RADIATION

(A. H. Compton)

Arthur Compton listed a number of experimental results, in the explanation of which the classical electromagnetic theory faced difficulties. These difficulties arose, especially, in the consideration of the following points:

(1) Does an ether exist? If one admits the existence of a medium for electromagnetic vibrations, great difficulties arise.

(2) How are electromagnetic waves produced? The theory would require that an atom should contain oscillators of the same frequency as the radiation emitted by it.

(3) The photoelectric effect; difficulties arising in considering this phenomenon on the basis of wave theory.

(4) X-ray diffraction and electron recoil phenomena.

(5) Individual interactions between quanta of radiation and the electrons (Compton effect).

In all these phenomena the corpuscular nature of radiation was predominant. It was not necessary to think that the wave and corpuscular concepts were in opposition. Still, the experimental difficulties of the wave theory, which Compton had listed, could be overcome by Einstein's hypothesis of light quanta of energy $h\nu$ and momentum $h\nu/c$. Following the suggestion of G. N. Lewis[1], Compton preferred to call the light quantum or the element of radiant energy by the name of 'photon'.[1a]

The Bohr–Kramers–Slater theory[2] of virtual radiation, had offered another conception of radiation; according to it an atom in an excited state would continuously emit some virtual radiation. Compton noted the fact that in this theory the energy was only statistically conserved.

The Michelson–Morley experiment had shown that it was impossible to detect any motion with respect to the ether, and one had to imagine a medium in which all electromagnetic disturbances propagate with a given speed with respect to every individual observer no matter what the latter's motion. With this fact in view, Compton noted that there remained the difficulty of imagining waves without the existence of a medium. He thought that the corpuscular point of view thus had advantages, especially concerning the transport of energy.

In the emission of radiation, one had not been able to find an oscillator (within the atom) having the same frequency as the emitted light of X-rays. In fact, these frequencies were given by the stationary states of the electrons. While the classical electromagnetic theory could not explain this phenomenon, Bohr had considered that an electron passes from a state of higher energy to another one of lower energy with the emission of a photon. Compton pointed out that the sharp limit in the continuous spectrum of X-rays could be explained only by the introduction of the photons. In the case of the absorption of radiation also, there existed the impossibility, from the electromagnetic point of view, of creating a transition from a given stationary state to another one, while the same thing could be explained with the help of photons.

Compton then considered the photoelectric effect. The idea of photons had been introduced by Einstein in order to explain the relation between photo-current and the intensity of incident light. Einstein's photoelectric formula gave,

$$ mc^2 \left(\frac{1}{\sqrt{(1 - \beta^2)}} - 1 \right) = h\nu - W_p. \tag{1} $$

Compton reviewed the different experiments which had confirmed Einstein's formula for frequencies corresponding to small energies up to those of γ-rays. He mentioned Auger's experimental result[3], according to which the photoelectrons were expelled in a direction close to the one defined by the electric vector of the incident wave.

All the experimental facts had led to the conclusion that the existence of photoelectrons could not be explained with the help of the electromagnetic wave theory. Compton also noted the idea according to which the photon possessed a vectorial property similar to the electric vector of an electromagnetic wave; i.e. when a photon traversed an atom, 'the electrons and the nucleus received impulses in a direction opposite to its direction of propagation'. Compton mentioned experimental results showing the momentum transfer due to photons. However, none of the properties of photoelectrons had been found to be compatible with the theory of virtual radiation of Bohr, Kramers, and Slater.

Compton then described some of the difficulties associated with the diffraction of X-rays, which could not be explained by the electromagnetic theory; these were the difficulties encountered with the change of wavelength of the X-rays, the intensity of the scattered rays, and the energy of the recoil electrons.

While for heavy elements the absorption could be understood by the emission of soft secondary X-rays, characteristic of the element, it had been experimentally established that the absorption for light elements could not be attributed to the same process. Gray[4] had already noted that the scattered rays had a longer wavelength than the incident one.

Compton described how the spectroscopic measurements had furnished proof of the quantitative change of the wavelength of the X-ray scattered by an electron, and how the theory could account for it.[5] The change of the wavelength arising from the impact of the photon with an electron is given by

$$\delta\lambda = \frac{h}{mc}\left(1 - \cos\varphi\right), \tag{2}$$

where φ is the angle of deviation of the photon.

The kinetic energy of the recoil of the electron is given by

$$E_{\text{kin}} = h\nu \frac{2\alpha \cos^2\theta}{(1 + \alpha)^2 - \alpha^2 \cos\theta}, \tag{3}$$

where θ is the angle made by the recoil electron with the direction of the incident photon. C. T. R. Wilson[6] and W. Bothe[7], independently of each other, had experimentally observed the recoil electrons within a few months after their theoretical prediction by Compton.

Compton discussed the impossibility of explaining these results by means of classical theories based on the Doppler effect. He noted that the theory of virtual radiation [the Bohr, Kramers, Slater theory] was not in disagreement with the experimental results, but could not explain them quantitatively.

Compton mentioned Bothe's experiment which had demonstrated that there was no coincidence between the photons emitted in different directions. This important result had shown that a quantum of energy was not suddenly emitted as a spherical wave when an atom passed from one stationary state to another.

DISCUSSION

In the discussion which followed Compton's report, Lorentz pointed out that, in principle, the notion of the ether was not in contradiction with Maxwell's equations; and, moreover, the concept of the photon was not sufficient to be able to treat diffraction phenomena.

Bragg gave the results of an experiment by Williams, showing that the average momentum of the photon was not $h\nu/c$ but equal to $1.8 \, h\nu/c$. C. T. R. Wilson agreed that his own results had confirmed this.

Born asked Compton to explain the origin of $h\nu/c$ for the momentum of the photon.* The only answer Compton had in mind was that when a quantum of radiation of energy $h\nu$ was absorbed by an atom, the momentum communicated to the atom by the radiation was $h\nu/c$. Dirac mentioned his calculation on an electron subject to an incident radiation and arbitrary forces. He had shown that the action of radiation was equal to $1/c$ times the part of the rate of variation of the energy due to the incident radiation. Born also mentioned Wentzel's work on a rigorous study of the scattering of light by atoms, in which he had noted that only in the case of very short wavelengths was the momentum equal to $h\nu/c$.

Bohr took up the question of waves versus photons. He pointed out that experiments with radiation had effectively revealed traits which could not easily be described in classical terms. Such a difficulty arose in the Compton effect itself. Various aspects of this phenomenon could very easily be described with the help of photons, but one must not forget that the change in frequency that was produced had to be measured by instruments whose function could only be interpreted in terms of wave theory. There seemed to be a logical contradiction in all this: on the one hand the description of the incident and scattered waves required that these waves be limited in space and time, while the

* Einstein, already in 1916, had explicitly used $h\nu/c$ for the momentum of a photon of frequency ν. [A. Einstein, 'Zur Quantentheorie der Strahlung', *Mitt. Phys. Ges. Zürich* **16**, 47–62 (1916).] It is remarkable that Einstein did not make any comment on this question during the discussion.

change of energy and momentum of electron was considered as an instantaneous event at a given point in space-time. This was precisely the cause of parallel difficulties in the theory of Bohr, Kramers and Slater, in which the authors had been led to the idea of rejecting completely the existence of the photons, and to admit that the laws of conservation of energy and momentum were only statistically valid.

Bohr mentioned that the experiments of Geiger and Bothe, and of Compton and Simon, had shown that the conservation laws were valid for individual processes involving the photons. However, the dilemma in which one found oneself concerning the nature of light, was only a typical example of the difficulties which one encountered in attempting to interpret atomic phenomena by means of classical concepts. One finds a similar paradox concerning the nature of material particles. According to the fundamental ideas of de Broglie, and the experiments of Davisson and Germer, the notion of waves was as indispensable for the interpretation of the properties of material particles as for light. The whole question of representation in space and time posed serious questions, and Bohr proposed to go into the details of these questions, related to the general problems of quantum theory, in his own report at the fifth Solvay Conference.

Brillouin mentioned a consideration of Auger and Francis Perrin that had shown, by a symmetry argument, that the emission of electrons (in photoelectric effect) must happen symmetrically around the axis defined by the electric field. According to Brillouin, this presented a serious difficulty in the photon theory of the photoelectric effect. Compton answered that the photon concept was different from the classical concepts in the sense that when a photoelectron was ejected, the photon was completely absorbed and no field remained. This condition implied a limitation on the possible trajectories.

Madame Curie thought that the Compton effect might have important applications in biology, and that the high voltage technique employed in the production of high frequency X-rays would find important uses for 'therapeutic' purposes.

Among others who took part in the discussion of Compton's report were Richardson, Debye, Ehrenfest, Kramers, Pauli and Schrödinger.

REFERENCES

1. G. N. Lewis, *Nature* (18 December 1926).
1a. Sir William Bragg had called Einstein's theory of light quanta as the 'theory of neutrons'.
2. N. Bohr, H. A. Kramers, and J. C. Slater, *Phil. Mag.* **47**, 785 (1924); *Z. Phys.* **24**, 69 (1924).
3. P. Auger, *Comptes rendus*, **178**, 1535 (1924).
4. J. A. Gray, *J. Franklin Inst.* (November 1920), 643.
5. A. H. Compton, *Phys. Rev.* **22**, 409, 483 (1923);
   P. Debye, *Phys. Z.* **24**, 161 (1923).
6. C. T. R. Wilson, *Proc. Roy. Soc.* **104**, 1 (1923).
7. W. Bothe, *Z. Phys.* **16**, 319 (1923).

4. THE NEW DYNAMICS OF QUANTA

(L. de Broglie)

The point of departure of de Broglie's considerations was the fact that once the existence of elementary corpuscles of matter and radiation was accepted as experimentally given, these corpuscles had to be regarded as having a periodicity. In this view, one did not conceive of a 'material point' as a static entity limited to a region of space, but as a centre of periodic phenomena which extended all around it. To these periodic phenomena, de Broglie attributed the aspect of a wave, the wave describing the material point, given by the function,

$$u(x_0, y_0, z_0, t) = f(x_0, y_0, z_0) \cos 2\pi v_0 t_0. \tag{1}$$

In another Galilean frame, the speed of the material point is $v = \beta c$, and the frequency and phase velocity of the wave are given by, respectively,

$$\nu = \frac{\nu_0}{\sqrt{(1 - \beta^2)}}, \qquad V = \frac{c^2}{v} = \frac{c}{\beta}, \tag{2}$$

where V, the phase velocity, is greater than c.

De Broglie showed that the velocity of the material point is given by the group velocity corresponding to the dispersion law obtained by defining,

$$n = \frac{c}{V} = \sqrt{\left(1 - \frac{v_0^2}{v^2}\right)}. \tag{3}$$

For the energy, de Broglie assumed the relation, $W = h\nu$; and, at rest, $m_0 c^2 = h\nu$. Writing the function representing the wave as,

$$u(x, y, z, t) = f(x, y, z, t) \cos \frac{2\pi}{h} \varphi(x, y, z, t), \tag{4}$$

he obtained,

$$W = \frac{\partial \varphi}{\partial t}, \quad \text{and} \quad p = - \operatorname{grad} \varphi, \tag{5}$$

so that φ is just the Jacobi function. He assumed that this form was still valid if the material point were submitted to the action of a potential, $F(x, y, z, t)$, and obtained the refractive index for the waves as,

$$n = \sqrt{\left\{\left(1 - \frac{F}{h\nu}\right)^2 - \frac{v_0^2}{v^2}\right\}}. \tag{6}$$

In the case of light, assuming that the photon mass is zero, de Broglie obtained the relations,

$$W = h\nu, \quad \text{and} \quad p = \frac{h\nu}{c}. \tag{7}$$

De Broglie then discussed the ideas of Schrödinger. He thought that Schrödinger's most fundamental idea was that the new mechanics had to begin with the equations of propagation that are constructed in such a way that, in each case, the phase of their sinusoidal solutions, in the approximation of geometrical optics, should be a solution of the Jacobi equation. The waves in the new mechanics are represented by the function Ψ, which could be written in the canonical form,

$$\Psi = a \cos \frac{2\pi}{h} \varphi, \tag{8}$$

where φ, in the first approximation, is a solution of Jacobi's equation. And if h tends to zero, one must recover the classical result.

De Broglie recalled that Schrödinger, in his first paper, gave the equation

$$\Delta \Psi + \frac{8\pi^2 m_0}{h^2} (E - F) \Psi = 0, \tag{9}$$

where F is the potential function of a constant field. Schrödinger had also introduced the notion of a wave packet for the description of the material point. De Broglie acknowledged the difficulty of talking about a material point in the atom when the extension of the wave packet was about as large as the atom itself.

De Broglie also discussed Schrödinger's equation for a system of interacting points, which takes the form

$$m^{1/2} \sum_{k,l} \frac{\partial}{\partial q_k} \left[m^{-1/2} m^{kl} \frac{\partial \Psi}{\partial q^l} \right] + \frac{8\pi^2}{h^2} (E - F) \Psi = 0, \tag{10}$$

where $F(q_1, \dots q_n)$ is the potential energy, and $\frac{1}{2}\sum m^{kl}p_k p_l$ is the kinetic energy.

In both cases the conditions on Ψ are that it be uniform, continuous and finite over the whole space. These conditions lead to the well-known eigenvalue problems, giving the energy of the stationary states.

De Broglie pointed out that Schrödinger had very ingeniously shown that his wave mechanics was equivalent to Heisenberg's matrix mechanics. De Broglie also discussed Klein's generalization of the wave equation for the relativistic case. Concerning Born's ideas, he remarked that Born had rejected the notion of a wavepacket and considered Ψ as giving a statistical probability for the presence of the electron.

De Broglie discussed the difficulty of having a wave description of matter, and at the same time the existence of discrete elements of energy and also how Ψ would sometimes appear as a *pilot wave* and at other times as a *probability wave*. Many people thought that it was illusory to ask about the position or velocity of an electron in the atom at a given time if it were represented by a Ψ-wave. De Broglie himself, on the other hand, was inclined to think that it was possible to attribute to particles a position and a velocity even in atomic systems in such a way that the variables of the configuration space had a precise meaning.

De Broglie then considered the case of the hydrogen atom. He noted that the use of Ψ, given by Schrödinger, in his [de Broglie's] formula for the speed and probability density would lead to a speed zero for the electron in the atom. However, with

$$\Psi_n = F(r, \theta) \cos \frac{2\pi}{h} \left(W_n t - \frac{mh}{2\pi} \alpha \right), \tag{11}$$

which is equally well a solution of the wave equation, the electron would have a uniform speed given by

$$v = \frac{1}{m_0 r} \frac{mh}{2\pi}, \tag{12}$$

which would be zero only in the state with $m=0$.

De Broglie reviewed certain experiments which favoured the new theory. For instance, hadn't one obtained the phenomena of diffraction and interference by using electrons? These phenomena depend on the wavelength associated with the motion of the electron. De Broglie's fundamental formula, $p = h\nu/V$, gave for an electron of velocity v,

$$\lambda = \frac{V}{v} = \frac{h}{p} = \frac{h\sqrt{(1 - \beta^2)}}{m_0 v}, \tag{13}$$

or, if β is not too close to unity,

$$\lambda = \frac{h}{m_0 v}. \tag{14}$$

if $\mathscr{V}$ is the potential difference in volts that could give an electron the velocity v, then the wavelength [on expressing h, e, m_0, in c.g.s. units] would be given by

$$\lambda = \frac{7.28}{v} = \frac{12.25}{\sqrt{\mathscr{V}}} \times 10^{-8}\,\text{cm}, \tag{15}$$

which would be comparable to the wavelength of X-rays.

De Broglie commented on the results obtained by E. G. Dymond in an experiment on the diffraction of electrons by the atoms of a gas, a case which had been studied theoretically by Born. The experimental result was in qualitative agreement with the predictions. He discussed the Davisson–Germer experiment on the diffraction of electrons by crystals, obtaining results in perfect agreement with the theory. Finally, de Broglie reported on the experiment of G. P. Thomson and A. Reid. In this experiment diffraction phenomena were observed; rings of different intensities were produced by highly energetic (~ 10000 volt) electrons.

DISCUSSION

Lorentz wondered how Sommerfeld's quantum conditions could be obtained in de Broglie's theory. This difficulty, in de Broglie's view, could be resolved by the consideration of pseudo-periods, which he had discussed in his thesis. If one considered a multi-periodic system, with partial periods τ_1, $\tau_2, ..., \tau_n$, one could find quasi-periods which were almost integral multiples of the partial periods, such that

$$\tau = m_1\tau_1 + \varepsilon_1 = m_2\tau_2 + \varepsilon_2 ... = m_n\tau_n + \varepsilon_n, \tag{1}$$

where $m_1, m_2, ..., m_n$, were integers and $\varepsilon_1, \varepsilon_2, ..., \varepsilon_n$, were arbitrarily small. In order that the wave should find itself again in phase after one of these quasi-periods, it was necessary that

$$\int_{\tau_1} p_1\,\mathrm{d}q_1 = n_1 h, \quad \int_{\tau_2} p_2\,\mathrm{d}q_2 = n_2 h, \quad ..., \quad \int_{\tau_n} p_n\,\mathrm{d}q_n = n_n h, \tag{2}$$

which are exactly the quantum conditions of Sommerfeld.

Pauli offered to provide the mathematical basis of de Broglie's consideration of particles in motion along definite trajectories. Pauli noted that it was based on the conservation law of electricity,

$$\frac{\partial \varrho}{\partial t} + \mathrm{div}\,s = 0, \quad \text{or,} \quad \sum_{k=1}^{4} \frac{\partial s_k}{\partial x_k} = 0, \tag{3}$$

which is a consequence of the wave equation, written in the form,

$$is_k = \psi \frac{\partial \psi^*}{\partial x_k} - \psi^* \frac{\partial \psi}{\partial x_k} + \frac{4\pi i}{h} \frac{e}{c} \Phi_k \psi \psi^*. \tag{4}$$

Now de Broglie had introduced, remarked Pauli, the real functions a and φ, instead of the complex function ψ, defined by

$$\psi = ae^{(2\pi i/h)\,\varphi}, \qquad \psi^* = ae^{-(2\pi i/h)\,\varphi}. \tag{5}$$

Substituting these expressions in the expression for s_k, one obtained,

$$s_k = \frac{4\pi a^2}{h}\left(\frac{\partial \varphi}{\partial x_k} + \frac{e}{c}\Phi_k\right). \tag{6}$$

From this the velocity vector could be defined as $v_1 = s_1/\varrho$, $v_2 = s_2/\varrho$, $v_3 = s_3/\varrho$, which were the expressions obtained by de Broglie. Pauli noted that in a field theory, where there exists a conservation principle, it was always possible to introduce a velocity.

Schrödinger pointed out that momentum density could just as well be considered instead of the velocity. Pauli thought that this would, as indicated by Schrödinger himself, lead to other trajectories. Schrödinger also mentioned that the orbits obtained by de Broglie were circular only because of the particular wave functions he had chosen. Other possible choices would lead to much more complicated orbits. Kramers, Ehrenfest, and Brillouin also took part in the discussion of de Broglie's report.

5. QUANTUM MECHANICS

(Max Born and Werner Heisenberg)

The new mechanics is based on the idea that atomic physics is essentially different from classical physics on account of the existence of discontinuities. Born and Heisenberg, in their joint report, pointed out that the new mechanics of quanta was essentially an extension of the theories of Planck, Einstein, and Bohr.

The mechanics of quanta seeks to introduce new notions by a precise analysis of what is 'essentially observable'. Quantum mechanics is characterized by two kinds of discontinuities: the existence of particles and that of stationary states; moreover, these discontinuities introduce a statistical element which is a very important part of its basis.

Born and Heisenberg presented the mathematical formulation of the quantum mechanical scheme. They pointed out that after the unsuccessful attempts to establish the radiation theory of an excited atom from a classical point of view, Heisenberg had tried to find a description based on the Ritz Combination Principle, which defined the frequencies of spectral lines by the numbers $\nu_{ik} = T_i - T_k$, the difference of the terms T_i and T_k. Since each line was characterized by an intensity and a phase, the whole set of spectral lines could be described by the scheme

$$\begin{vmatrix} q_{11}e^{2\pi i \nu_{11} t} & q_{12}e^{2\pi i \nu_{12} t} & \cdots \\ q_{21}e^{2\pi i \nu_{21} t} & q_{22}e^{2\pi i \nu_{22} t} & \cdots \\ \cdots & \cdots & \cdots \end{vmatrix}. \tag{1}$$

Using the analogy with the product of the Fourier series representing a classical quantity, Born and Heisenberg showed that the rules followed by the set of quantities in the above scheme were given by matrix theory. In analogy with the classical theory, the real quantities were represented by Hermitian matrices. The discontinuities having been thus introduced, Planck's constant h was introduced by the relation

$$pq - qp = \frac{\hbar}{i} I, \tag{2}$$

where $\hbar = h/2\pi$ and I is the unit matrix. The matrix p does not commute with the matrix q. The Hamiltonian equations, obtained by replacing each quantity by its matrix, were still valid; that is, the equation $\mathcal{H}(p, q) = W$, where $\mathcal{H}$ is the Hamiltonian and W the diagonal form of its matrix, perfectly replaced the equation of motion.

With the matrix formulation it had been possible to obtain the correct value for the angular momentum as $\hbar\sqrt{\{j(j+1)\}}$, with $j = 0, 1, 2, \ldots$, in agreement with the rules employed in the explanation of Zeeman effect [by Landé and others]. Moreover, Pauli had been able to derive the spectrum of hydrogen by using the matrix scheme.

The next step in the matrix formulation had been to find the generalization of the canonical variables and of canonical transformations. Dirac had found that the commutator of p and q was a generalization of the Poisson brackets. It was then obvious that a pair of matrices p and q satisfying the commutation relations would be conjugate, and a transformation leaving a commutator invariant would be a canonical transformation. Having noted that these transformations were of the form $S^{-1}pS$, $S^{-1}qS$ (for a non-singular matrix S), Born and Heisenberg gave the analogue of the Hamilton–Jacobi equation in the form

$$S^{-1}\mathcal{H}(p_0, q_0) S = W, \tag{3}$$

where p_0 and q_0 are certain fixed matrices. They briefly discussed perturbation theory and the notion of orthogonal transformations, and showed that the Hamilton–Jacobi equation was equivalent to the problem of finding eigenvalues and eigensolutions of the quadratic form,

$$\sum_{nm} \mathcal{H}_{nm}(q_0, p_0) \varphi_n \bar{\varphi}_m, \tag{4}$$

that is, to find the φ_n and W in the equation

$$W\varphi_n = \sum_m \mathcal{H}_{nm}\varphi_m. \tag{5}$$

The main difficulty of matrix theory was its inability to treat non-periodic quantities. Born and Heisenberg mentioned Dirac's attempt to solve the problem by his theory of q-numbers. Dirac's formalism was similar to the introduction of hypercomplex numbers. Its chief disadvantage was the necessity of having to introduce some special quantization conditions which had disappeared in the matrix theory. Another approach had been initiated by Born and Wiener with the operator theory, a method which Eckart, Dirac, Jordan, and J. von Neumann had developed further.

Born and Heisenberg discussed some elementary features of the operator theory in the Hilbert space of square integral functions. They showed that p and q, as operators, are such that

$$p = \frac{\hbar}{i} \frac{\partial}{\partial q},$$

and q satisfies $\varphi(q, W) = q\delta(W - q)$, where δ is the Dirac delta-function. The introduction of these operators in the Hamiltonian gives rise to the Schrödinger equation, which appears here as a particular case of the theory of operators,

$$\mathcal{H}\left(q, \frac{h}{2\pi i} \frac{\partial}{\partial q}\right) \varphi(q) = W\varphi(q). \tag{6}$$

The conditions of quantization are now expressed by the fact that the solutions have to belong to the above-mentioned Hilbert space.

Born and Heisenberg reiterated the fault of matrix theory that it could treat only periodic, that is, closed, systems; it could not predict whether a certain state exists, but only the possibility of its existence. They considered a special case in which a closed system was separated into two distinct parts, which would then mutually interact. Let $\mathcal{H}_1$ and $\mathcal{H}_2$ be the Hamiltonians of the two systems, and consider that $\mathcal{H}_1$ has non-diagonal terms. The exchange of energy could be interpreted in two different ways: (1) the periodic part of $\mathcal{H}_1$ has a long period, the energy coming in and disappearing according to this frequency; (2) there are discontinuities, and the transport of energy takes place from one system to the other in quantum jumps. Born and Heisenberg showed that these two points of view were in fact two different ways of expressing the same mathematical result. If $f(W_n^{(1)})$ is a function of the eigenvalues of $\mathcal{H}_1$, then the action of the coupling could be measured by

$$\overline{\delta f_n} = f[\mathcal{H}_{nn}^{(1)}] - f[W_n^{(1)}] \tag{7}$$

where $f[\mathcal{H}_{nn}^{(1)}]$ represents the time-average value of $f(\mathcal{H}^{(1)})$. But one could also show that $\overline{\delta f_n}$ can be written as

$$\overline{\delta f_n} = \sum_m [f(W_n) - f(W_m)] \, \Phi_{nm}. \tag{8}$$

Moreover, this theorem relates the transition probability to the orthogonal transformation S, thus relating the systems 1 and 2 considered without interaction and the systems with interaction. The problem of the precise moment at which the transition would take place cannot, however, be determined.

The manner in which time dependence had been introduced in matrix theory was very cumbersome. Born and Heisenberg noted that if t and W were conjugate variables, there had to be a commutation relation between them in quantum mechanics, that is

$$Wt - tW = \frac{\hbar}{i}. \tag{9}$$

In this way, W would correspond to the operator $(\hbar/i)(\partial/\partial t)$. This operator allowed them, following Dirac and Born, to introduce a time-dependent perturbation in the Hamiltonian and, thereby, to formulate time-dependent perturbation theory.

Born and Heisenberg then developed a series of propositions necessary for the mathematical formulation and interpretation of quantum mechanics. Having established these, they observed that the next step would be to state a set of axioms from which the theory could be derived. They presented a set of axioms which had been formulated by Jordan. They also pointed out certain difficulties concerning the phases of the probability amplitudes, phases which could not be defined by experiments. This difficulty was overcome by J. von Neumann, who gave an invariant definition of the spectrum of an operator without any presupposition about the eigensolutions.

Concerning the interpretation of the theory and agreement with experiment, Born and Heisenberg pointed out that the notions of position, velocity, momentum, energy, etc., were still very well defined when taken individually. However, they could not be measured simultaneously with indefinite precision. They described the analogy with optics and stated the uncertainty principle, $p_1 q_1 \gtrsim h$. Moreover, it had been possible to include the spin of the electron in the matrix formulation, in the extended form given to it by Dirac and Jordan.

For the several-body systems, they mentioned the systematic treatment by Wigner, using group theory. Wigner had shown that a system could, in general, be split into sub-systems without interactions. One of its features was the separation between symmetric and anti-symmetric partial systems. Born and Heisenberg pointed out that the systems of the symmetric type were subject to Bose–Einstein statistics while the anti-symmetric systems obeyed Fermi–Dirac statistics.

DISCUSSION

The main emphasis in the report of Born and Heisenberg was on the Göttingen–Copenhagen point of view. Dirac mentioned the correspondence between classical and quantum mechanics that appears by using action-angle variables, and commented on the essential differences between the classical and quantum descriptions of physical processes. Quantum theory, he said, describes a state by a time-dependent wave function ψ which can be expanded at a given time t in a series containing wave functions ψ_n with coefficients c_n. The wave functions ψ_n are such that they do not interfere at an instant $t > t_1$. Now Nature makes a choice sometime later and decides in favour of the state ψ_n with the probability $|c_n|^2$. This choice cannot be renounced and determines the future evolution of the state. Heisenberg opposed this point of view by asserting that there was no sense in talking about Nature making a choice, and that it was our observation that gives us the reduction to the eigenfunction. What Dirac called a 'choice of Nature', Heisenberg preferred to call 'observation', showing his predilection for the language he and Bohr had developed together.

6. WAVE MECHANICS

(E. Schrödinger)

In classical mechanics, the coordinates $q_1 \ldots q_n$ being given, the problem is to find the set of functions $q_1(t) \ldots q_n(t)$ corresponding to all the possible solutions of the dynamical system considered. In wave mechanics, the problem is to define a function ψ of the coordinates, $q_1, ..., q_n$, and sometimes of the coordinates and the time, t, in which $q_1, ..., q_n$ remain independent variables (a fact which is also true in the Hamilton–Jacobi theory).

Starting with the Hamiltonian $H = T + V$, Schrödinger introduced the quantity

$$L = T\left(q_k, \frac{h}{2\pi} \frac{\partial \psi}{\partial q_k}\right) + V\psi^2, \tag{1}$$

and showed that the variation of ψ, according to $\delta \int L \, d\tau = 0$, with the condition $\int \psi^2 \, d\tau = 1$, leads to the equation,

$$\Delta\psi + \frac{8\pi^2}{h^2}(E - V)\psi = 0, \tag{2}$$

where Δ is now a generalized Laplacian in the Riemannian sense, and E is defined only on a possible set of values called the eigenvalues. In this case, ψ is independent of t and one finds again the stationary states introduced by Bohr. Moreover, the integral, $\int q_i \psi_k \psi_{k'} \, d\tau$, is found to be identical with the matrix q of the Heisenberg–Born theory. Schrödinger pointed out that one could, in principle, treat any type of closed systems with the help of Equation (2). However, the time, t, does not play any role in these systems, and the probability of transition cannot be defined easily.

Schrödinger then introduced the time formally, and the time-dependent equation became,

$$\Delta\psi - \frac{8\pi^2}{h^2} V\psi - \frac{4\pi i}{h} \frac{\partial \psi}{\partial t} = 0, \tag{3}$$

whose general solution is given by,

$$\psi = \sum_{k=1}^{\infty} c_k \psi_k e^{2\pi i v_k t}, \qquad v_k = \frac{E_k}{h}, \tag{4}$$

and the stationary states are those for which $c_k = \delta_{k,l}$. Schrödinger showed that, with ψ obeying the relation $\int d\tau \, \psi\psi^* = 1$, the coefficient c_k must satisfy the condition, $\sum_{k=1}^{\infty} |c_k|^2 = 1$.

Schrödinger applied his ideas to radiation, which is determined by the electric moment. The latter, in wave mechanics, is given by

$$M = \int M_{\text{classical}} \psi\psi^* \, d\tau, \tag{5}$$

and is equivalent to the matrix of the electric moment in the theory of Born and Heisenberg.

Schrödinger commented on the interpretation of the wave function ψ, and then discussed the relativistic case. Equation (3) was only an approximation for the case of small velocities. Making use of de Broglie's expression for the velocity of the phase waves, he obtained the relativistic equation,

$$\Delta\psi - \frac{1}{c^2}\frac{\partial^2\psi}{\partial t^2} + \frac{4\pi i e\varphi}{hc^2}\frac{\partial\psi}{\partial t} + \frac{4\pi^2}{c^2}\left(\frac{e^2\varphi^2}{h^2} - v_0^2\right)\psi = 0, \tag{6}$$

where v_0 $(=mc^2/h)$ is the rest frequency. He discussed how this equation would be modified by the requirement of Lorentz invariance. As an example illustrating the wave mechanical treatment, Schrödinger discussed the problem of atomic systems with several electrons.

The discussion of Schrödinger's report consisted of various points of detail, and among those who took part were Bohr, Born, Fowler, De Donder, Lorentz, Heisenberg, and Schrödinger.

7. THE QUANTUM POSTULATE AND THE NEW DEVELOPMENT OF ATOMIC THEORY

(N. Bohr)

Niels Bohr gave a report on the epistemological problems confronting quantum physics.* He discussed the question of an appropriate terminology, and stressed the viewpoint of complementarity. Bohr's main argument was that the unambiguous communication of physical evidence required that the experimental arrangement as well as the recording of observations had to be expressed in a common language, 'suitably refined by the vocabulary of classical physics'. In all actual experimental work this requirement is fulfilled by using measuring instruments such as diaphragms, lenses and photographic plates, which are so large and heavy that, notwithstanding the decisive role of the quantum of action for the stability and properties of such bodies, all quantum effects could be disregarded in taking account of their positions and motions.

In classical physics one deals with an idealization, according to which all phenomena can be arbitrarily subdivided. The interaction between the measuring instruments and the object under observation can be neglected, or at any rate compensated, in classical physics. Bohr stressed that such interaction represents an integral part of the phenomena in quantum physics, for which no separate account could be given if the instruments would serve the purpose of defining the conditions under which the observations are obtained.

The recording of observations ultimately rests on the production of permanent marks on the measuring instruments, such as the spot produced on a photographic plate by the impact of a photon or an electron. The fact that such recording involves essentially irreversible physical and chemical processes does not introduce further

---

* The paper which Bohr contributed to the proceedings of the fifth Solvay Conference was the same as his contribution at the Como Congress on 16 September 1927.

complications, but only emphasizes the element of irreversibility involved in the very concept of observation. In Bohr's view, the characteristic new feature in quantum physics is merely the restricted divisibility of the phenomena, which requires a specification of all significant parts of the experimental setup for unambiguous description.

Bohr pointed out that since several different individual effects would in general be observed in one and the same experimental arrangement, the recourse to statistics in quantum physics was therefore unavoidable. 'Moreover, evidence obtained under different conditions and rejecting comprehension in a single picture must, notwithstanding any apparent contrast, be regarded as complementary in the sense that together they exhaust all well defined information about the atomic object. From this point of view, the whole purpose of the formalism of quantum theory is to derive expectations for observations obtained under given experimental conditions.'[1] Bohr emphasized that the elimination of all contradictions is secured by the mathematical consistency of the formalism; the exhaustive character of the description within its scope is indicated by its adaptability to any imaginable experimental arrangement.

DISCUSSION

Following Bohr's report on the epistemological questions of quantum theory, there took place an extensive general discussion on causality, determinism, and probability. Those who took part in this discussion were Lorentz, Brillouin, De Donder, Born, Einstein, Pauli, Dirac, Kramers, de Broglie, Heisenberg, Langevin, Fowler, Schrödinger, Ehrenfest, Richardson, and Compton.

The exchanges of views that started at the sessions were eagerly continued within smaller groups during the evenings. Bohr had extensive opportunity of long discussions with Einstein and Ehrenfest.

A note has survived (from the fifth Solvay Conference), which Paul Ehrenfest had passed on to Einstein during one of the lectures saying, 'Don't laugh! There is a special section in purgatory for professors of quantum theory, where they will be obliged to listen to lectures on classical physics ten hours every day.' To which Einstein replied, 'I laugh only at their naïveté. Who knows who would have the laugh in a few years?'

Einstein was particularly reluctant to renounce deterministic description in principle. He challenged Bohr and the other proponents of the new quantum mechanics with arguments suggesting the possibility of taking the interaction between the atomic objects and the measuring instruments more explicitly into account. Einstein was not convinced by answers which pointed out the futility of this prospect, and he returned to these problems again at the sixth Solvay Conference in 1930.

The Bohr–Einstein dialogue which began at the fifth Solvay Conference continued for over two decades. Niels Bohr presented an account of his discussions with Einstein on epistemological questions at the eighth Solvay Conference in 1948, and expanded it to be included in the seventieth birthday volume presented to Einstein in 1949.[2] It is a document of permanent interest for the interpretation of quantum theory. Since the Bohr–Einstein discussions began at the fifth Solvay Conference, we find it appropriate to include it here as an appendix.

REFERENCES

1. N. Bohr, in *La théorie quantique des champs*, Douzième Conseil de Physique, 1961, R. Stoops Editeur, Brussels, p. 27.
2. N. Bohr, 'Discussion with Einstein on Epistemological Problems in Atomic Physics', in *Albert Einstein: Philosopher-Scientist*, Library of Living Philosophers, Inc., 1949, Harper Torchbooks, 1959.

APPENDIX

Discussion with Einstein on Epistemological Problems in Atomic Physics *

The many occasions through the years on which I had the privilege to discuss with Einstein epistemological problems raised by the modern development of atomic physics have come back vividly to my mind and I have felt that I could hardly attempt anything better than to give an account of these discussions which have been of greatest value and stimulus to me. I hope also that the account may convey to wider circles an impression of how essential the open-minded exchange of ideas has been for the progress in a field where new experience has time after time demanded a reconsideration of our views.

From the very beginning the main point under debate has been the attitude to take to the departure from customary principles of natural philosophy characteristic of the novel development of physics which was initiated in the first year of this century by Planck's discovery of the universal quantum of action. This discovery, which revealed a feature of atomicity in the laws of nature going far beyond the old doctrine of the limited divisibility of matter, has indeed taught us that the classical theories of physics are idealizations which can be unambiguously applied only in the limit where all actions involved are large compared with the quantum. The question at issue has been whether the renunciation of a causal mode of description of atomic processes involved in the endeavours to cope with the situation should be regarded as a temporary departure from ideals to be ultimately revived or whether we are faced with an irrevocable step towards obtaining the proper harmony between analysis and synthesis of physical phenomena. To describe the background of our discussions and to bring out as clearly as possible the arguments for the contrasting viewpoints, I have felt it necessary to go to a certain length in recalling some main features of the development to which Einstein himself has contributed so decisively.

As is well known, it was the intimate relation, elucidated primarily by Boltzmann, between the laws of thermodynamics and the statistical regularities exhibited by mechanical systems with many degrees of freedom, which guided Planck in his ingenious treatment of the problem of thermal radiation, leading him to his fundamental discovery. While, in his work, Planck was principally concerned with considerations of essentially statistical character and with great caution refrained from definite conclusions as to the extent to which the existence of the quantum implied a departure from the foundations of mechanics and electrodynamics, Einstein's great original contribution to quantum theory (1905) was just the recognition of how physical phenomena like the photo-effect may depend directly on individual quantum effects.[1] In these very same years when, in the development of his theory of relativity, Einstein laid down a new foundation for physical science, he explored with a most daring spirit

---

* Niels Bohr in *Albert Einstein: Philosopher-Scientist*, The Library of Living Philosophers, Inc., Tudor Publishing Company, Copyright © 1949, 1951; Harper & Row, Publishers, New York, 1959.

the novel features of atomicity which pointed beyond the framework of classical physics.

With unfailing intuition Einstein thus was led step by step to the conclusion that any radiation process involves the emission or absorption of individual light quanta or 'photons' with energy and momentum

$$E = h\nu \quad \text{and} \quad P = h\sigma, \tag{1}$$

respectively, where h is Planck's constant, while ν and σ are the number of vibrations per unit time and the number of waves per unit length, respectively. Notwithstanding its fertility, the idea of the photon implied a quite unforeseen dilemma, since any simple corpuscular picture of radiation would obviously be irreconcilable with interference effects, which present so essential an aspect of radiative phenomena, and which can be described only in terms of a wave picture. The acuteness of the dilemma is stressed by the fact that the interference effects offer our only means of defining the concepts of frequency and wavelength entering into the very expressions for the energy and momentum of the photon.

In this situation, there could be no question of attempting a causal analysis of radiative phenomena, but only, by a combined use of the contrasting pictures, to estimate probabilities for the occurrence of the individual radiation processes. However, it is most important to realize that the recourse to probability laws under such circumstances is essentially different in aim from the familiar application of statistical considerations as practical means of accounting for the properties of mechanical systems of great structural complexity. In fact, in quantum physics we are presented not with intricacies of this kind, but with the inability of the classical frame of concepts to comprise the peculiar feature of indivisibility, or 'individuality', characterizing the elementary processes.

The failure of the theories of classical physics in accounting for atomic phenomena was further accentuated by the progress of our knowledge of the structure of atoms. Above all, Rutherford's discovery of the atomic nucleus (1911) revealed at once the inadequacy of classical mechanical and electromagnetic concepts to explain the inherent stability of the atom. Here again the quantum theory offered a clue for the elucidation of the situation and especially it was found possible to account for the atomic stability, as well as for the empirical laws governing the spectra of the elements, by assuming that any reaction of the atom resulting in a change of its energy involved a complete transition between two so-called stationary quantum states and that, in particular, the spectra were emitted by a step-like process in which each transition is accompanied by the emission of a monochromatic light quantum of an energy just equal to that of an Einstein photon.

These ideas, which were soon confirmed by the experiments of Franck and Hertz (1914) on the excitation of spectra by impact of electrons on atoms, involved a further renunciation of the causal mode of description, since evidently the interpretation of the spectral laws implies that an atom in an excited state in general will have the possibility of transitions with photon emission to one or another of its lower energy states. In fact, the very idea of stationary states is incompatible with any directive for the choice between such transitions and leaves room only for the notion of the relative probabilities of the individual transition processes. The only guide in estimating such probabilities was the so-called correspondence principle which originated in the search for the closest possible connection between the statistical account of atomic processes and the consequences to be expected from classical theory, which should be valid in the limit where the actions involved in all stages of the analysis of the phenomena are large compared with the universal quantum.

At that time, no general self-consistent quantum theory was yet in sight, but the prevailing attitude may perhaps be illustrated by the following passage from a lecture by the writer from 1913:[2]

I hope that I have expressed myself sufficiently clearly so that you may appreciate the extent to which these considerations conflict with the admirably consistent scheme of conceptions which has been rightly termed the classical theory of electrodynamics. On the other hand, I have tried to convey to you the impression that – just by emphasizing so strongly this conflict – it may also be possible in course of time to establish a certain coherence in the new ideas.

Important progress in the development of quantum theory was made by Einstein himself in his famous article on radiative equilibrium in 1917,[3] where he showed that Planck's law for thermal radiation could be simply deduced from assumptions conforming with the basic ideas of the quantum theory of atomic constitution. To this purpose, Einstein formulated general statistical rules regarding the occurrence of radiative transitions between stationary states, assuming not only that, when the atom is exposed to a radiation field, absorption as well as emission processes will occur with a probability per unit time proportional to the intensity of the irradiation, but that even in the absence of external disturbances spontaneous emission processes will take place with a rate corresponding to a certain *a priori* probability. Regarding the latter point, Einstein emphasized the fundamental character of the statistical description in a most suggestive way by drawing attention to the analogy between the assumptions regarding the occurrence of the spontaneous radiative transitions and the well-known laws governing transformations of radioactive substances.

In connection with a thorough examination of the exigencies of thermodynamics as regards radiation problems, Einstein stressed the dilemma still further by pointing out that the argumentation implied that any radiation process was 'unidirected' in the sense that not only is a momentum corresponding to a photon with the direction of propagation transferred to an atom in the absorption process, but that also the emitting atom will receive an equivalent impulse in the opposite direction, although there can on the wave picture be no question of a preference for a single direction in an emission process. Einstein's own attitude to such startling conclusions is expressed in a passage at the end of the article (*loc. cit.*, p. 127f.), which may be translated as follows:

These features of the elementary processes would seem to make the development of a proper quantum treatment of radiation almost unavoidable. The weakness of the theory lies in the fact that, on the one hand, no closer connection with the wave concepts is obtainable and that, on the other hand, it leaves to chance (*Zufall*) the time and the direction of the elementary processes; nevertheless, I have full confidence in the reliability of the way entered upon.

When I had the great experience of meeting Einstein for the first time during a visit to Berlin in 1920, these fundamental questions formed the theme of our conversations. The discussions, to which I have often reverted in my thoughts, added to all my admiration for Einstein a deep impression of his detached attitude. Certainly, his favoured use of such picturesque phrases as 'ghost waves (*Gespensterfelder*) guiding the photons' implied no tendency to mysticism, but illuminated rather a profound humour behind his piercing remarks. Yet, a certain difference in attitude and outlook remained, since, with his mastery for co-ordinating apparently contrasting experience without abandoning continuity and causality, Einstein was perhaps more reluctant to renounce such ideals than someone for whom renunciation in this respect appeared to be the only way open to proceed with the immediate task of co-ordinating the multifarious evidence regarding atomic phenomena, which accumulated from day to day in the exploration of this new field of knowledge.

In the following years, during which the atomic problems attracted the attention of rapidly increasing circles of physicists, the apparent contradictions inherent in quantum theory were felt ever more acutely. Illustrative of this situation is the discussion raised by the discovery of the Stern–Gerlach effect in 1922. On the one hand, this effect gave striking support to the idea of stationary states and in particular to the quantum theory of the Zeeman effect developed by Sommerfeld; on the other hand, as exposed so clearly by Einstein and Ehrenfest,[4] it presented with unsurmountable difficulties any attempt at forming a picture of the behaviour of atoms in a magnetic field. Similar paradoxes were raised by the discovery by Compton (1924) of the change in wave-length accompanying the scattering of X-rays by electrons. This phenomenon afforded, as is well known, a most direct proof of the adequacy of Einstein's view regarding the transfer of energy and momentum in radiative processes; at the same time, it was equally clear that no simple picture of a corpuscular collision could offer an exhaustive description of the phenomenon. Under the impact of such difficulties, doubts were for a time entertained even regarding the conservation of energy and momentum in the individual radiation processes;[5] a view, however, which very soon had to be abandoned in face of more refined experiments bringing out the correlation between the deflection of the photon and the corresponding electron recoil.

The way to the clarification of the situation was, indeed, first to be paved by the development of a more comprehensive quantum theory. A first step towards this goal was the recognition by de Broglie in 1925 that the wave-corpuscle duality was not confined to the properties of radiation, but was equally unavoidable in accounting for the behaviour of material particles. This idea, which was soon convincingly confirmed by experiments on electron interference phenomena, was at once greeted by Einstein, who had already envisaged the deep-going analogy between the properties of thermal radiation and of gases in the so-called degenerate state.[6] The new line was pursued with the greatest success by Schrödinger (1926) who, in particular, showed how the stationary states of atomic systems could be represented by the proper solutions of a wave-equation to the establishment of which he was led by the formal analogy, originally traced by Hamilton, between mechanical and optical problems. Still, the paradoxical aspects of quantum theory were in no way ameliorated, but even emphasized, by the apparent contradiction between the exigencies of the general super-position principle of the wave description and the feature of individuality of the elementary atomic processes.

At the same time, Heisenberg (1925) had laid the foundation of a rational quantum mechanics, which was rapidly developed through important contributions by Born and Jordan as well as by Dirac. In this theory, a formalism is introduced, in which the kinematical and dynamical variables of classical mechanics are replaced by symbols subjected to a non-commutative algebra. Notwithstanding the renunciation of orbital pictures, Hamilton's canonical equations of mechanics are kept unaltered and Planck's constant enters only in the rules of commutation

$$qp - pq = \sqrt{-1} \frac{h}{2\pi} \tag{2}$$

holding for any set of conjugate variables q and p. Through a representation of the symbols by matrices with elements referring to transitions between stationary states, a quantitative formulation of the correspondence principle became for the first time possible. It may here be recalled that an important preliminary step towards this goal was reached through the establishment, especially by contributions of Kramers, of a quantum theory of dispersion making basic use of Einstein's general rules for the probability of the occurrence of absorption and emission processes.

This formalism of quantum mechanics was soon proved by Schrödinger to give results identical with those obtainable by the mathematically often more convenient methods of wave theory, and in the following years general methods were gradually established for an essentially statistical description of atomic processes combining the features of individuality and the requirements of the superposition principle, equally characteristic of quantum theory. Among the many advances in this period, it may especially be mentioned that the formalism proved capable of incorporating the exclusion principle which governs the states of systems with several electrons, and which already before the advent of quantum mechanics had been derived by Pauli from an analysis of atomic spectra. The quantitative comprehension of a vast amount of empirical evidence could leave no doubt as to the fertility and adequacy of the quantum-mechanical formalism, but its abstract character gave rise to a widespread feeling of uneasiness. An elucidation of the situation should, indeed, demand a thorough examination of the very observational problem in atomic physics.

This phase of the development was, as is well known, initiated in 1927 by Heisenberg,[7] who pointed out that the knowledge obtainable of the state of an atomic system will always involve a peculiar 'indeterminacy'. Thus, any measurement of the position of an electron by means of some device, like a microscope, making use of high-frequency radiation, will, according to the fundamental relations (1), be connected with a momentum exchange between the electron and the measuring agency, which is the greater the more accurate a position measurement is attempted. In comparing such considerations with the exigencies of the quantum-mechanical formalism, Heisenberg called attention to the fact that the commutation rule (2) imposes a reciprocal limitation on the fixation of two conjugate variables, q and p, expressed by the relation

$$\Delta q \cdot \Delta p \approx h, \tag{3}$$

where Δq and Δp are suitably defined latitudes in the determination of these variables. In pointing to the intimate connection between the statistical description in quantum mechanics and the actual possibilities of measurement, this so-called indeterminacy relation is, as Heisenberg showed, most important for the elucidation of the paradoxes involved in the attempts of analyzing quantum effects with reference to customary physical pictures.

The new progress in atomic physics was commented upon from various sides at the International Physical Congress held in September 1927 at Como in commemoration of Volta. In a lecture on that occasion,[8] I advocated a point of view conveniently termed 'complementarity', suited to embrace the characteristic features of individuality of quantum phenomena, and at the same time to clarify the peculiar aspects of the observational problem in this field of experience. For this purpose, it is decisive to recognize that, *however far the phenomena transcend the scope of classical physical explanation, the account of all evidence must be expressed in classical terms.* The argument is simply that by the word 'experiment' we refer to a situation where we can tell others what we have done and what we have learned and that, therefore, the account of the experimental arrangement and of the results of the observations must be expressed in unambiguous language with suitable application of the terminology of classical physics.

This crucial point, which was to become a main theme of the discussions reported in the following, implies the *impossibility of any sharp separation between the behaviour of atomic objects and the interaction with the measuring instruments which serve to define the conditions under which the phenomena appear.* In fact, the individuality of the typical quantum effects finds its proper expression in the circumstance that any attempt of subdividing the phenomena will demand a change in the experimental arrangement introducing new possibilities of interaction between objects and measuring instruments which in principle cannot be controlled. Consequently, evidence obtained under different experimental conditions cannot be comprehended within a single picture, but must be regarded as *complementary* in the sense that only the totality of the phenomena exhausts the possible information about the objects.

Under these circumstances an essential element of ambiguity is involved in ascribing conventional physical attributes to atomic objects, as is at once evident in the dilemma regarding the corpuscular and wave properties of electrons and photons, where we have to do with contrasting pictures, each referring to an essential aspect of empirical evidence. An illustrative example, of how the apparent paradoxes are removed by an examination of the experimental conditions under which the complementary phenomena appear, is also given by the Compton effect, the consistent description of which at first had presented us with such acute difficulties. Thus, any arrangement suited to study the exchange of energy and momentum between the electron and the photon must involve a latitude in the space-time description of the interaction sufficient for the definition of wave-number and frequency which enter into the relation (1). Conversely, any attempt of locating the collision between the photon and the electron more accurately would, on account of the unavoidable interaction with the fixed scales and clocks defining the spacetime reference frame, exclude all closer account as regards the balance of momentum and energy.

As stressed in the lecture, an adequate tool for a complementary way of description is offered precisely by the quantum-mechanical formalism which represents a purely symbolic scheme permitting only predictions, on lines of the correspondence principle, as to results obtainable under conditions specified by means of classical concepts. It must here be remembered that even in the indeterminacy relation (3) we are dealing with an implication of the formalism which defies unambiguous expression in words suited to describe classical physical pictures. Thus, a sentence like 'we cannot know both the momentum and the position of an atomic object' raises at once questions as to the physical reality of two such attributes of the object, which can be answered only by referring to the conditions for the unambiguous use of space-time concepts, on the one hand, and dynamical conservation laws, on the other hand. While the combination of these concepts into a single picture of a causal chain of events is the essence of classical mechanics, room for regularities beyond the grasp of such a description is just afforded by the circumstance that the study of the complementary phenomena demands mutually exclusive experimental arrangements.

The necessity, in atomic physics, of a renewed examination of the foundation for the unambiguous use of elementary physical ideas recalls in some way the situation that led Einstein to his original revision of the basis for all application of space-time concepts which, by its emphasis on the primordial importance of the observational problem, has lent such unity to our world picture. Notwithstanding all novelty of approach, causal description is upheld in relativity theory within any given frame of reference, but in quantum theory the uncontrollable interaction between the objects and the measuring instruments forces us to a renunciation even in such respect. This recognition, however, in no way points to any limitation of the scope of the quantum-mechanical description, and the trend of the whole argumentation presented in the Como lecture was to show that the viewpoint of complementarity may be regarded as a rational generalization of the very ideal of causality.

At the general discussion in Como, we all missed the presence of Einstein, but soon after, in October 1927, I had the opportunity to meet him in Brussels at the Fifth Physical Conference of the Solvay Institute, which was devoted to the theme 'Electrons and Photons.' At the Solvay Meetings, Einstein had from their beginning been a most prominent figure, and several of us came to the Conference with great anticipations to learn his reaction to the latest stage of the development which, to our view, went far in clarifying the problems which he had himself from the outset elicited so ingeniously. During the discussions, where the whole subject was reviewed by contributions from many sides and where also the arguments mentioned in the preceding pages were again presented, Einstein expressed, however, a deep concern over the extent to which causal account in space and time was abandoned in quantum mechanics.

To illustrate his attitude, Einstein referred at one of the sessions[9] to the simple example, illustrated by Fig. 1, of a particle (electron or photon) penetrating through a hole or a narrow slit in a diaphragm placed at some distance before a photographic plate. On account of the diffraction of the wave connected with the motion of the particle and indicated in the figure by the thin lines, it is under such conditions not possible to predict with certainty at what point the electron will arrive at the photographic plate, but only to calculate the probability that, in an experiment, the electron will be found within any given region of the plate. The apparent difficulty, in this description, which Einstein felt so acutely, is the fact that, if in the experiment the electron is recorded at one point A of the plate, then it is out of the question of ever observing an effect of this electron at another point (B), although the laws of ordinary wave propagation offer no room for a correlation between two such events.

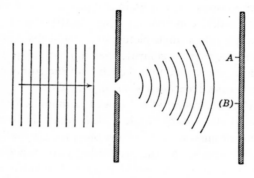

FIGURE 1

Einstein's attitude gave rise to ardent discussions within a small circle, in which Ehrenfest, who through the years had been a close friend of us both, took part in a most active and helpful way. Surely, we all recognized that, in the above example, the situation presents no analogue to the application of statistics in dealing with complicated mechanical systems, but rather recalled the background for Einstein's own early conclusions about the unidirection of individual radiation effects which contrasts so strongly with a simple wave picture (see p. 155). The discussions, however, centred on the question of whether the quantum-mechanical description exhausted the possibilities of accounting for observable phenomena or, as Einstein maintained, the analysis could be carried further and, especially, of whether a fuller description of the phenomena could be obtained by bringing into consideration the detailed balance of energy and momentum in individual processes.

To explain the trend of Einstein's arguments, it may be illustrative here to consider some simple features of the momentum and energy balance in connection with the location of a particle in space and time. For this purpose, we shall examine the simple case of a particle penetrating through a hole in a diaphragm without or with a shutter to open and close the hole, as indicated in Figs. 2a and 2b, respectively. The equidistant parallel lines to the left in the figures indicate the train of plane waves corresponding to the state of motion of a particle which, before reaching the diaphragm, has a momentum P related to the wave-number σ by the second of equations (1). In accordance with the diffraction of the waves when passing through the hole, the state of motion of the particle to the right of the diaphragm is represented by a spherical wave train with a suitably defined angular aperture θ and, in case of Fig. 2b, also with a limited radial extension. Consequently, the description of this state involves a certain latitude Δp in the momentum component of the particle parallel to the diaphragm and, in the case of a diaphragm with a shutter, an additional latitude ΔE of the kinetic energy.

Since a measure for the latitude Δq in location of the particle in the plane of the diaphragm is given by the radius a of the hole, and since $\theta \approx 1/\sigma a$, we get, using (1), just $\Delta p \approx \theta P \approx h/\Delta q$, in accordance with the indeterminacy relation (3). This result

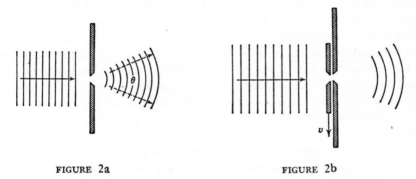

FIGURE 2a   FIGURE 2b

could, of course, also be obtained directly by noticing that, due to the limited extension of the wave-field at the place of the slit, the component of the wave-number parallel to the plane of the diaphragm will involve a latitude $\Delta \sigma \approx 1/a \approx 1/\Delta q$. Similarly, the spread of the frequencies of the harmonic components in the limited wave-train in Fig. 2b is evidently $\Delta \nu \approx 1/\Delta t$, where Δt is the time interval during which the shutter leaves the hole open and, thus, represents the latitude in time of the passage of the particle through the diaphragm. From (1), we therefore get

$$\Delta E \cdot \Delta t \approx h, \tag{4}$$

again in accordance with the relation (3) for the two conjugated variables E and t.

From the point of view of the laws of conservation, the origin of such latitudes entering into the description of the state of the particle after passing through the hole may be traced to the possibilities of momentum and energy exchange with the diaphragm or the shutter. In the reference system considered in Figs. 2a and 2b, the velocity of the diaphragm may be disregarded and only a change of momentum Δp

between the particle and the diaphragm needs to be taken into consideration. The shutter, however, which leaves the hole opened during the time Δt, moves with a considerable velocity $v \approx a/\Delta t$, and a momentum transfer Δp involves therefore an energy exchange with the particle, amounting to

$$v\,\Delta p \approx \frac{\Delta q\,\Delta p}{\Delta t} \approx \frac{h}{\Delta t},$$

being just of the same order of magnitude as the latitude ΔE given by (4) and, thus, allowing for momentum and energy balance.

The problem raised by Einstein was now to what extent a control of the momentum and energy transfer, involved in a location of the particle in space and time, can be used for a further specification of the state of the particle after passing through the hole. Here, it must be taken into consideration that the position and the motion of the diaphragm and the shutter have so far been assumed to be accurately coordinated with the space-time reference frame. This assumption implies, in the description of the state of these bodies, an essential latitude as to their momentum and energy which need not, of course, noticeably affect the velocities, if the diaphragm and the shutter are sufficiently heavy. However, as soon as we want to know the momentum and energy of these parts of the measuring arrangement with an accuracy sufficient to control the momentum and energy exchange with the particle under investigation, we shall, in accordance with the general indeterminacy relations, lose the possibility of their accurate location in space and time. We have, therefore, to examine how far this circumstance will affect the intended use of the whole arrangement and, as we shall see, this crucial point clearly brings out the complementary character of the phenomena.

Returning for a moment to the case of the simple arrangement indicated in Fig. 1, it has so far not been specified to what use it is intended. In fact, it is only on the assumption that the diaphragm and the plate have well-defined positions in space that it is impossible, within the frame of the quantum-mechanical formalism, to make more detailed predictions as to the point of the photographic plate where the particle will be recorded. If, however, we admit a sufficiently large latitude in the knowledge of the position of the diaphragm, it should, in principle, be possible to control the momentum transfer to the diaphragm and, thus, to make more detailed predictions as to the direction of the electron path from the hole to the recording point. As regards the quantum-mechanical description, we have to deal here with a two-body system consisting of the diaphragm as well as of the particle, and it is just with an explicit application of conservation laws to such a system that we are concerned in the Compton effect where, for instance, the observation of the recoil of the electron by means of a cloud chamber allows us to predict in what direction the scattered photon will eventually be observed.

The importance of considerations of this kind was, in the course of the discussions, most interestingly illuminated by the examination of an arrangement where between the diaphragm with the slit and the photographic plate is inserted another diaphragm with two parallel slits, as is shown in Fig. 3. If a parallel beam of electrons (or photons)

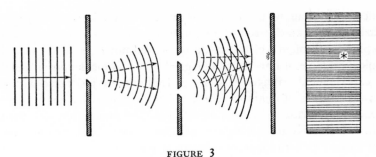

FIGURE 3

falls from the left on the first diaphragm, we shall, under usual conditions, observe on the plate an interference pattern indicated by the shading of the photographic plate shown in front view to the right of the figure. With intense beams, this pattern is built up by the accumulation of a large number of individual processes, each giving rise to a small spot on the photographic plate, and the distribution of these spots follows a simple law derivable from the wave analysis. The same distribution should also be found in the statistical account of many experiments performed with beams so faint that in a single exposure only one electron (or photon) will arrive at the photographic plate at some spot shown in the figure as a small star. Since, now, as indicated by the broken arrows, the momentum transferred to the first diaphragm ought to be different if the electron was assumed to pass through the upper or the lower slit in the second diaphragm, Einstein suggested that a control of the momentum transfer would permit a closer analysis of the phenomenon and, in particular, make it possible to decide through which of the two slits the electron had passed before arriving at the plate.

A closer examination showed, however, that the suggested control of the momentum transfer would involve a latitude in the knowledge of the position of the diaphragm which would exclude the appearance of the interference phenomena in question. In fact, if ω is the small angle between the conjectured paths of a particle passing through the upper or the lower slit, the difference of momentum transfer in these two cases will, according to (1), be equal to $h\sigma\omega$, and any control of the momentum of the diaphragm with an accuracy sufficient to measure this difference will, due to the indeterminacy relation, involve a minimum latitude of the position of the diaphragm, comparable with $1/\sigma\omega$. If, as in the figure, the diaphragm with the two slits is placed in the middle between the first diaphragm and the photographic plate, it will be seen that the number of fringes per unit length will be just equal to $\sigma\omega$ and, since an uncertainty in the position of the first diaphragm of the amount of $1/\sigma\omega$ will cause an equal uncertainty in the positions of the fringes, it follows that no interference effect can appear. The same result is easily shown to hold for any other placing of the second diaphragm between the first diaphragm and the plate, and would also be obtained if, instead of the first diaphragm, another of these three bodies were used for the control, for the purpose suggested, of the momentum transfer.

This point is of great logical consequence, since it is only the circumstance that we are presented with a choice of *either* tracing the path of a particle *or* observing interference effects, which allows us to escape from the paradoxical necessity of concluding that the behaviour of an electron or a photon should depend on the presence of a slit in the diaphragm through which it could be proved not to pass. We have here to do with a typical example of how the complementary phenomena appear under mutually exclusive experimental arrangements (see p. 158) and are just faced with the impossibility, in the analysis of quantum effects, of drawing any sharp separation between an independent behaviour of atomic objects and their interaction with the measuring instruments which serve to define the conditions under which the phenomena occur.

Our talks about the attitude to be taken in face of a novel situation as regards analysis and synthesis of experience touched naturally on many aspects of philosophical thinking, but, in spite of all divergencies of approach and opinion, a most humorous spirit animated the discussions. On his side, Einstein mockingly asked us whether we could really believe that the providential authorities took recourse to dice-playing ('... *ob der liebe Gott würfelt*'), to which I replied by pointing at the great caution, already called for by ancient thinkers, in ascribing attributes to Providence in everyday language. I remember also how at the peak of the discussion Ehrenfest, in his affectionate manner of teasing his friends, jokingly hinted at the apparent similarity between Einstein's attitude and that of the opponents of relativity theory; but instantly Ehrenfest added that he would not be able to find relief in his own mind before concord with Einstein was reached.

Einstein's concern and criticism provided a most valuable incentive for us all to reexamine the various aspects of the situation as regards the description of atomic phenomena. To me it was a welcome stimulus to clarify still further the role played by the measuring instruments and, in order to bring into strong relief the mutually exclusive character of the experimental conditions under which the complementary phenomena appear, I tried in those days to sketch various apparatus in a pseudo-realistic style of which the following figures are examples. Thus, for the study of an interference phenomenon of the type indicated in Fig. 3, it suggests itself to use an experi-

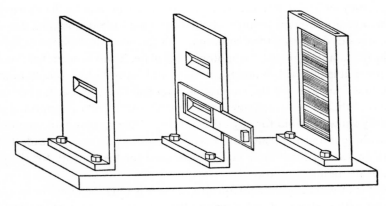

FIGURE 4

mental arrangement like that shown in Fig. 4, where the solid parts of the apparatus, serving as diaphragms and plate-holder, are firmly bolted to a common support. In such an arrangement, where the knowledge of the relative positions of the diaphragms and the photographic plate is secured by a rigid connection, it is obviously impossible to control the momentum exchanged between the particle and the separate parts of the apparatus. The only way in which, in such an arrangement, we could insure that the particle passed through one of the slits in the second diaphragm is to cover the other slit by a lid, as indicated in the figure; but if the slit is covered, there is of course no question of any interference phenomenon, and on the plate we shall simply observe a continuous distribution as in the case of the single fixed diaphragm in Fig. 1.

In the study of phenomena in the account of which we are dealing with detailed momentum balance, certain parts of the whole device must naturally be given the freedom to move independently of others. Such an apparatus is sketched in Fig. 5,

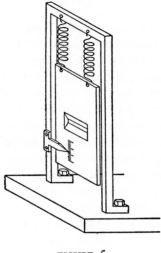

FIGURE 5

where a diaphragm with a slit is suspended by weak springs from a solid yoke bolted to the support on which also other immobile parts of the arrangement are to be fastened. The scale on the diaphragm together with the pointer on the bearings of the yoke refer to such study of the motion of the diaphragm, as may be required for an estimate of the momentum transferred to it, permitting one to draw conclusions as to the deflection suffered by the particle in passing through the slit. Since, however, any reading of the scale, in whatever way performed, will involve an uncontrollable change in the momentum of the diaphragm, there will always be, in conformity with the in-determinacy principle, a reciprocal relationship between our knowledge of the position of the slit and the accuracy of the momentum control.

In the same semi-serious style, Fig. 6 represents a part of an arrangement suited for the study of phenomena which, in contrast to those just discussed, involve time coordi-

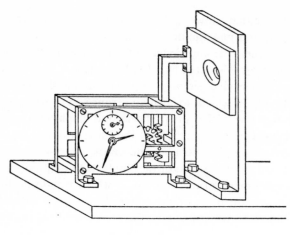

nation explicitly. It consists of a shutter rigidly connected with a robust clock resting on the support which carries a diaphragm and on which further parts of similar character, regulated by the same clockwork or by other clocks standardized relatively to it, are also to be fixed. The special aim of the figure is to underline that a clock is a piece of machinery, the working of which can completely be accounted for by ordinary mechanics and will be affected neither by reading of the position of its hands nor by the interaction between its accessories and an atomic particle. In securing the opening of the hole at a definite moment, an apparatus of this type might, for instance, be used for an accurate measurement of the time an electron or a photon takes to come from the diaphragm to some other place, but, evidently, it would leave no possibility of controlling the energy transfer to the shutter with the aim of drawing conclusions as to the energy of the particle which has passed through the diaphragm. If we are interested in such conclusions we must, of course, use an arrangement where the shutter devices can no longer serve as accurate clocks, but where the knowledge of the moment when the hole in the diaphragm is open involves a latitude connected with the accuracy of the energy measurement by the general relation (4).

The contemplation of such more or less practical arrangements and their more or less fictitious use proved most instructive in directing attention to essential features of the problems. The main point here is the distinction between the *objects* under investigation and the *measuring instruments* which serve to define, in classical terms, the conditious under which the phenomena appear. Incidentally, we may remark that, for the illustration of the preceding considerations, it is not relevant that experiments involving an accurate control of the momentum or energy transfer from atomic particles to heavy bodies like diaphragms and shutters would be very difficult to perform, if practicable at all. It is only decisive that, in contrast to the proper measuring instruments, these bodies together with the particles would in such a case constitute the system to which the quantum-mechanical formalism has to be applied. As regards the specification of the conditions for any well-defined application of the formalism, it is moreover essential that the *whole experimental arrangement* be taken into account. In fact, the introduction of any further piece of apparatus, like a mirror, in the way of a particle might imply new interference effects essentially influencing the predictions as regards the results to be eventually recorded.

The extent to which renunciation of the visualization of atomic phenomena is imposed upon us by the impossibility of their subdivision is strikingly illustrated by the following example to which Einstein very early called attention and often has reverted. If a semi-reflecting mirror is placed in the way of a photon, leaving two possibilities for its direction of propagation, the photon may either be recorded on one, and only one, of two photographic plates situated at great distances in the two directions in question, or else we may, by replacing the plates by mirrors, observe effects exhibiting an interference between the two reflected wave-trains. In any attempt of a pictorial representation of the behaviour of the photon we would, thus, meet with the difficulty: to be obliged to say, on the one hand, that the photon always chooses *one* of the two ways and, on the other hand, that it behaves as if it had passed *both* ways.

It is just arguments of this kind which recall the impossibility of subdividing quantum phenomena and reveal the ambiguity in ascribing customary physical attributes to atomic objects. In particular, it must be realized that – besides in the account of the placing and timing of the instruments forming the experimental arrangement – all unambiguous use of space-time concepts in the description of atomic phenomena is confined to the recording of observations which refer to marks on a photographic plate or to similar practically irreversible amplification effects like the building of a water drop around an ion in a cloud-chamber. Although, of course, the existence of the quantum of action is ultimately responsible for the properties of the materials of which the measuring instruments are built and on which the functioning of the recording devices depends, this circumstance is not relevant for the problems of the adequacy and completeness of the quantum-mechanical description in its aspects here discussed.

These problems were instructively commented upon from different sides at the Solvay meeting,[10] in the same session where Einstein raised his general objections. On that occasion an interesting discussion arose also about how to speak of the appearance of phenomena for which only predictions of statistical character can be made. The question was whether, as to the occurrence of individual effects, we should adopt a terminology proposed by Dirac, that we were concerned with a choice on the part of 'nature', or, as suggested by Heisenberg, we should say that we have to do with a choice on the part of the 'observer' constructing the measuring instruments and reading their recording. Any such terminology would, however, appear dubious since, on the one hand, it is hardly reasonable to endow nature with volition in the ordinary sense, while, on the other hand, it is certainly not possible for the observer to influence the events which may appear under the conditions he has arranged. To my mind, there is no other alternative than to admit that, in this field of experience, we are dealing with individual phenomena and that our possibilities of handling the measuring instruments allow us only to make a choice between the different complementary types of phenomena we want to study.

The epistemological problems touched upon here were more explicitly dealt with in my contribution to the issue of *Naturwissenschaften* in celebration of Planck's 70th birthday in 1929. In this article, a comparison was also made between the lesson derived from the discovery of the universal quantum of action and the development which has followed the discovery of the finite velocity of light and which, through Einstein's pioneer work, has so greatly clarified basic principles of natural philosophy. In relativity theory, the emphasis on the dependence of all phenomena on the reference frame opened quite new ways of tracing general physical laws of unparalleled scope. In quantum theory, it was argued, the logical comprehension of hitherto unsuspected fundamental regularities governing atomic phenomena has demanded the recognition that no sharp separation can be made between an independent behaviour of the objects and their interaction with the measuring instruments which define the reference frame.

In this respect, quantum theory presents us with a novel situation in physical science, but attention was called to the very close analogy with the situation as regards analysis and synthesis of experience, which we meet in many other fields of human knowledge and interest. As is well known, many of the difficulties in psychology originate in the different placing of the separation lines between object and subject in the analysis of various aspects of psychical experience. Actually, words like 'thoughts' and 'sentiments', equally indispensable to illustrate the variety and scope of conscious life, are used in a similar complementary way as are space-time coordination and dynamical conservation laws in atomic physics. A precise formulation of such analogies involves, of course, intricacies of terminology, and the writer's position is perhaps best indicated in a passage in the article, hinting at the mutually exclusive relationship which will always exist between the practical use of any word and attempts at its strict definition. The principal aim, however, of these considerations, which were not least inspired by the hope of influencing Einstein's attitude, was to point to perspectives of bringing general epistemological problems into relief by means of a lesson derived from the study of new, but fundamentally simple, physical experience.

At the next meeting with Einstein at the Solvay Conference in 1930, our discussions took quite a dramatic turn. As an objection to the view that a control of the interchange of momentum and energy between the objects and the measuring instruments was excluded if these instruments should serve their purpose of defining the space-time frame of the phenomena, Einstein brought forward the argument that such control should be possible when the exigencies of relativity theory were taken into consideration. In particular, the general relationship between energy and mass, expressed in Einstein's famous formula

$$E = mc^2, \tag{5}$$

should allow, by means of simple weighing, to measure the total energy of any system and, thus, in principle to control the energy transferred to it when it interacts with an atomic object.

As an arrangement suited for such purpose, Einstein proposed the device indicated

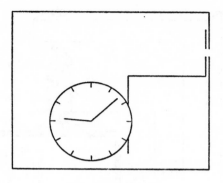

FIGURE 7

in Fig. 7, consisting of a box with a hole in its side, which could be opened or closed by a shutter moved by means of a clockwork within the box. If, in the beginning, the box contained a certain amount of radiation and the clock was set to open the shutter for a very short interval at a chosen time, it could be achieved that a single photon was released through the hole at a moment known with as great accuracy as desired. Moreover, it would apparently also be possible, by weighing the whole box before and after this event, to measure the energy of the photon with any accuracy wanted, in definite contradiction to the reciprocal indeterminacy of time and energy quantities in quantum mechanics.

This argument amounted to a serious challenge and gave rise to a thorough examination of the whole problem. At the outcome of the discussion to which Einstein himself contributed effectively, it became clear, however, that the argument could not be upheld. In fact, in the consideration of the problem, it was found necessary to look closer into the consequences of the identification of inertial and gravitational mass implied in the application of relation (5). Especially, it was essential to take into account the relationship between the rate of a clock and its position in a gravitational field – well known from the red-shift of the lines in the sun's spectrum – following from Einstein's principle of equivalence between gravity effects and the phenomena observed in accelerated reference frames.

Our discussion concentrated on the possible application of an apparatus incorporating Einstein's device and drawn in Fig. 8 in the same pseudo-realistic style as some of the preceding figures. The box, of which a section is shown in order to exhibit its interior, is suspended in a spring-balance and is furnished with a pointer to read its position on a scale fixed to the balance support. The weighing of the box may thus be performed with any given accuracy Δm by adjusting the balance to its zero position by means of suitable loads. The essential point is now that any determination of this position with a given accuracy Δq will involve a minimum latitude Δp in the control of the momentum of the box connected with Δq by the relation (3). This latitude must obviously again be smaller than the total impulse which, during the whole interval T of the balancing procedure, can be given by the gravitational field to a body with a

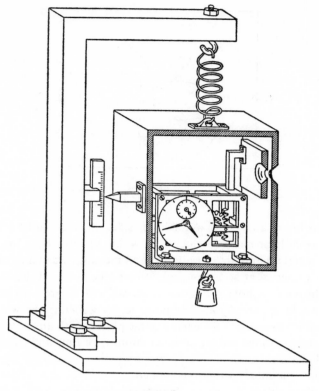

FIGURE 8

mass Δm, or

$$\Delta p \approx \frac{h}{\Delta q} < T \cdot g \cdot \Delta m, \tag{6}$$

where g is the gravity constant. The greater the accuracy of the reading q of the pointer, the longer must, consequently, be the balancing interval T, if a given accuracy Δm of the weighing of the box with its content shall be obtained.

Now, according to general relativity theory, a clock, when displaced in the direction of the gravitational force by an amount of Δq, will change its rate in such a way that its reading in the course of a time interval T will differ by an amount ΔT given by the relation

$$\frac{\Delta T}{T} = \frac{1}{c^2}\, g\, \Delta q. \tag{7}$$

By comparing (6) and (7) we see, therefore, that after the weighing procedure there will in our knowledge of the adjustment of the clock be a latitude

$$\Delta T > \frac{h}{c^2\,\Delta m}.$$

Together with the formula (5), this relation again leads to

$$\Delta T \cdot \Delta E > h,$$

in accordance with the indeterminacy principle. Consequently, a use of the apparatus as a means of accurately measuring the energy of the photon will prevent us from controlling the moment of its escape.

The discussion, so illustrative of the power and consistency of relativistic arguments, thus emphasized once more the necessity of distinguishing, in the study of atomic phenomena, between the proper measuring instruments which serve to define the reference frame and those parts which are to be regarded as objects under investigation and in the account of which quantum effects cannot be disregarded. Notwithstanding the most suggestive confirmation of the soundness and wide scope of the quantum-mechanical way of description, Einstein nevertheless, in a following conversation with me, expressed a feeling of disquietude as regards the apparent lack of firmly laid down principles for the explanation of nature, in which all could agree. From my viewpoint, however, I could only answer that, in dealing with the task of bringing order into an entirely new field of experience, we could hardly trust in any accustomed principles, however broad, apart from the demand of avoiding logical inconsistencies and, in this respect, the mathematical formalism of quantum mechanics should surely meet all requirements.

The Solvay meeting in 1930 was the last occasion where, in common discussions with Einstein, we could benefit from the stimulating and mediating influence of Ehrenfest, but shortly before his deeply deplored death in 1933 he told me that Einstein was far from satisfied and with his usual acuteness had discerned new aspects of the situation which strengthened his critical attitude. In fact, by further examining the possibilities for the application of a balance arrangement, Einstein had perceived alternative procedures which, even if they did not allow the use he originally intended, might seem to enhance the paradoxes beyond the possibilities of logical solution. Thus, Einstein had pointed out that, after a preliminary weighing of the box with the clock and the subsequent escape of the photon, one was still left with the choice of either repeating the weighing or opening the box and comparing the reading of the clock with the standard time scale. Consequently, we are at this stage still free to choose whether we want to draw conclusions either about the energy of the photon or about the moment when it left the box. Without in any way interfering with the photon between its escape and its later interaction with other suitable measuring instruments, we are, thus, able to make accurate predictions pertaining *either* to the moment of its arrival *or* to the amount of energy liberated by its absorption. Since, however, according to the quantum-mechanical formalism, the specification of the state of an isolated particle cannot involve both a well-defined connection with the time scale and an accurate fixation of the energy, it might thus appear as if this formalism did not offer the means of an adequate description.

Once more Einstein's searching spirit had elicited a peculiar aspect of the situation in quantum theory, which in a most striking manner illustrated how far we have here transcended customary explanation of natural phenomena. Still, I could not agree with the trend of his remarks as reported by Ehrenfest. In my opinion, there could be no other way to deem a logically consistent mathematical formalism as inadequate than by demonstrating the departure of its consequences from experience or by proving that its predictions did not exhaust the possibilities of observation, and Einstein's argumentation could be directed to neither of these ends. In fact, we must realize that in the problem in question we are not dealing with a *single* specified experimental arrangement, but are referring to *two* different, mutually exclusive arrangements. In the one, the balance together with another piece of apparatus like a spectrometer is used for the study of the energy transfer by a photon; in the other, a shutter regulated by a standardized clock together with another apparatus of similar kind, accurately timed relatively to the clock, is used for the study of the time of propagation of a photon over a given distance. In both these cases, as also assumed by Einstein, the observable effects are expected to be in complete conformity with the predictions of the theory.

The problem again emphasizes the necessity of considering the *whole* experimental arrangement, the specification of which is imperative for any well-defined application of the quantum-mechanical formalism. Incidentally, it may be added that paradoxes of the kind contemplated by Einstein are encountered also in such simple arrangements as sketched in Fig. 5. In fact, after a preliminary measurement of the momentum of the diaphragm, we are in principle offered the choice, when an electron or photon has passed through the slit, either to repeat the momentum measurement or to control the position of the diaphragm and, thus, to make predictions pertaining to alternative subsequent observations. It may also be added that it obviously can make no difference, as regards observable effects obtainable by a definite experimental arrangement, whether our plans of constructing or handling the instruments are fixed beforehand or whether we prefer to postpone the completion of our planning until a later moment when the particle is already on its way from one instrument to another.

In the quantum-mechanical description our freedom of constructing and handling the experimental arrangement finds its proper expression in the possibility of choosing the classically defined parameters entering in any proper application of the formalism. Indeed, in all such respects quantum mechanics exhibits a correspondence with the state of affairs familiar from classical physics, which is as close as possible when considering the individuality inherent in the quantum phenomena. Just in helping to bring out this point so clearly, Einstein's concern had therefore again been a most welcome incitement to explore the essential aspects of the situation.

The next Solvay meeting in 1933 was devoted to the problems of the structure and properties of atomic nuclei, in which field such great advances were made just in that period owing to the experimental discoveries as well as to new fruitful applications of quantum mechanics. It need in this connection hardly be recalled that just the evidence obtained by the study of artificial nuclear transformations gave a most direct test of Einstein's fundamental law regarding the equivalence of mass and energy, which was to prove an evermore important guide for researches in nuclear physics. It may also be mentioned how Einstein's intuitive recognition of the intimate relationship between the law of radioactive transformations and the probability rules governing individual radiation effects (see p. 155) was confirmed by the quantum-mechanical explanation of spontaneous nuclear disintegrations. In fact, we are here dealing with a typical example of the statistical mode of description, and the complementary relationship between energy-momentum conservation and time-space coordination is most strikingly exhibited in the well-known paradox of particle penetration through potential barriers.

Einstein himself did not attend this meeting, which took place at a time darkened by the tragic developments in the political world which were to influence his fate so deeply and add so greatly to his burdens in the service of humanity. A few months earlier, on a visit to Princeton where Einstein was then guest of the newly founded Institute for Advanced Study to which he soon after became permanently attached, I had, however, opportunity to talk with him again about the epistemological aspects of atomic physics, but the difference between our ways of approach and expression still presented obstacles to mutual understanding. While, so far, relatively few persons had taken part in the discussions reported in this article, Einstein's critical attitude towards the views on quantum theory adhered to by many physicists was soon after brought to public attention through a paper[11] with the title 'Can Quantum-Mechanical Description of Physical Reality Be Considered Complete?', published in 1935 by Einstein, Podolsky and Rosen.

The argumentation in this paper is based on a criterion which the authors express in the following sentence: 'If, without in any way disturbing a system, we can predict with certainty (i.e., with probability equal to unity) the value of a physical quantity, then there exists an element of physical reality corresponding to this physical quantity.' By an elegant exposition of the consequences of the quantum-mechanical formalism as regards the representation of a state of a system, consisting of two parts which have been in interaction for a limited time interval, it is next shown that different quantities, the fixation of which cannot be combined in the representation of one of the partial systems, can nevertheless be predicted by measurements pertaining to the other partial system. According to their criterion, the authors therefore conclude that quantum mechanics does not 'provide a complete description of the physical reality', and they express their belief that it should be possible to develop a more adequate account of the phenomena.

Due to the lucidity and apparently incontestable character of the argument, the paper of Einstein, Podolsky and Rosen created a stir among physicists and has played a large role in general philosophical discussion. Certainly the issue is of a very subtle character and suited to emphasize how far, in quantum theory, we are beyond the reach of pictorial visualization. It will be seen, however, that we are here dealing with problems of just the same kind as those raised by Einstein in previous discussions, and, in an article which appeared a few months later,[12] I tried to show that from the point of view of complementarity the apparent inconsistencies were completely removed. The trend of the argumentation was in substance the same as that exposed in the foregoing pages, but the aim of recalling the way in which the situation was discussed at that time may be an apology for citing certain passages from my article.

Thus, after referring to the conclusions derived by Einstein, Podolsky and Rosen on the basis of their criterion, I wrote:

Such an argumentation, however, would hardly seem suited to affect the soundness of quantum-mechanical description, which is based on a coherent mathematical formalism covering automatically any procedure of measurement like that indicated. The apparent contradiction in fact discloses only an essential inadequacy of the customary viewpoint of natural philosophy for a rational account of physical phenomena of the type with which we are concerned in quantum mechanics. Indeed the *finite interaction between object and measuring agencies* conditioned by the very existence of the quantum of action entails – because of the impossibility of controlling the reaction of the object on the measuring instruments, if these are to serve their purpose – the necessity of a final renunciation of the classical ideal of causality and a radical revision of our attitude towards the problem of physical reality. In fact, as we shall see, a criterion of reality like that proposed by the named authors contains – however cautious its formulation may appear – an essential ambiguity when it is applied to the actual problems with which we are here concerned.

As regards the special problem treated by Einstein, Podolsky and Rosen, it was next shown that the consequences of the formalism as regards the representation of the state of a system consisting of two interacting atomic objects correspond to the simple arguments mentioned in the preceding in connection with the discussion of the experimental arrangements suited for the study of complementary phenomena. In fact, although any pair q and p of conjugate space and momentum variables obeys the rule of non-commutative multiplication expressed by (2), and can thus only be fixed with reciprocal latitudes given by (3), the difference $q_1 - q_2$ between two space-coordinates referring to the constituents of the system will commute with the sum $p_1 + p_2$ of the corresponding momentum components, as follows directly from the commutability of q_1 with p_2 and q_2 with p_1. Both $q_1 - q_2$ and $p_1 + p_2$ can, therefore, be accurately fixed in a state of the complex system and, consequently, we can predict the values of either q_1 or p_1 if either q_2 or p_2, respectively, is determined by direct measurements. If, for the two parts of the system, we take a particle and a diaphragm, like that sketched in Fig. 5, we see that the possibilities of specifying the state of the particle by measurements on the diaphragm just correspond to the situation described on p. 165 and further discussed on p. 172, where it was mentioned that, after the particle has passed through the diaphragm, we have in principle the choice of measuring either the position of the diaphragm or its momentum and, in each case, making predictions as to subsequent observations pertaining to the particle. As repeatedly stressed, the principal point here is that such measurements demand mutually exclusive experimental arrangements.

The argumentation of the article was summarized in the following passage:

From our point of view we now see that the wording of the above-mentioned criterion of physical reality proposed by Einstein, Podolsky and Rosen contains an ambiguity as regards the meaning of the expression 'without in any way disturbing a system'. Of course there is in a case like that just considered no question of a mechanical disturbance of the system under investigation during the last critical stage of the measuring procedure. But even at this stage there is essentially the question of *an influence on the very conditions which define the possible types of predictions regarding the future behaviour of the system.* Since these conditions constitute an inherent element of the description of any phenomenon to which the term 'physical reality' can be properly attached, we see that the argumentation of the mentioned authors does not justify their conclusion that quantum-mechanical description, as appears from the preceding discussion, may be characterized as a rational utilization of all possibilities of unambiguous interpretation of measurements, compatible with the finite and uncontrollable interaction between the objects and the measuring instruments in the field of quantum theory. In fact, it is only the mutual exclusion of any two experimental procedures, permitting the unambiguous definition of complementary physical quantities, which provides room for new physical laws, the coexistence of which might at first sight appear irreconcilable with the basic principles of science. It is just this entirely new situation as regards the description of physical phenomena that the notion of *complementarity* aims at characterizing.

Rereading these passages, I am deeply aware of the inefficiency of expression which must have made it very difficult to appreciate the trend of the argumentation aiming to bring out the essential ambiguity involved in a reference to physical attributes of objects when dealing with phenomena where no sharp distinction can be made between the behaviour of the objects themselves and their interaction with the measuring instruments. I hope, however, that the present account of the discussions with Einstein in the foregoing years, which contributed so greatly to make us familiar with the situation in quantum physics, may give a clearer impression of the necessity of a radical revision of basic principles for physical explanation in order to restore logical order in this field of experience.

Einstein's own views at that time are presented in an article 'Physics and Reality', published in 1936 in the *Journal of the Franklin Institute*.[13] Starting from a most illuminating exposition of the gradual development of the fundamental principles in the theories of classical physics and their relation to the problem of physical reality, Einstein here argues that the quantum-mechanical description is to be considered merely as a means of accounting for the average behaviour of a large number of atomic systems, and his attitude to the belief that it should offer an exhaustive description of the individual phenomena is expressed in the following words: 'To believe this is logically possible without contradiction; but it is so very contrary to my scientific instinct that I cannot forgo the search for a more complete conception.'

Even if such an attitude might seem well balanced in itself, it nevertheless implies a rejection of the whole argumentation exposed in the preceding, aiming to show that, in quantum mechanics, we are not dealing with an arbitrary renunciation of a more detailed analysis of atomic phenomena, but with a recognition that such an analysis is *in principle* excluded. The peculiar individuality of the quantum effects presents us, as regards the comprehension of well-defined evidence, with a novel situation unforeseen in classical physics and irreconcilable with conventional ideas suited for our orientation and adjustment to ordinary experience. It is in this respect that quantum theory has called for a renewed revision of the foundation for the unambiguous use of elementary concepts as a further step in the development which, since the advent of relativity theory, has been so characteristic of modern science.

In the following years, the more philosophical aspects of the situation in atomic physics aroused the interest of ever larger circles and were, in particular, discussed at the Second International Congress for the Unity of Science in Copenhagen in July

1936. In a lecture on this occasion,[14] I tried especially to stress the analogy in epistemological respects between the limitation imposed on the causal description in atomic physics and situations met with in other fields of knowledge. A principal purpose of such parallels was to call attention to the necessity in many domains of general human interest of facing problems of a similar kind as those which had arisen in quantum theory and thereby to give a more familiar background for the apparently extravagant way of expression which physicists have developed to cope with their acute difficulties.

Besides the complementary features conspicuous in psychology and already touched upon (see p. 168), examples of such relationships can also be traced in biology, especially as regards the comparison between mechanistic and vitalistic viewpoints. Just with respect to the observational problem, this last question had previously been the subject of an address to the International Congress on Light Therapy held in Copenhagen in 1932,[15] where it was incidentally pointed out that even the psycho-physical parallelism as envisaged by Leibniz and Spinoza has obtained a wider scope through the development of atomic physics, which forces us to an attitude towards the problem of explanation recalling ancient wisdom, that when searching for harmony in life one must never forget that in the drama of existence we are ourselves both actors and spectators.

Utterances of this kind would naturally in many minds evoke the impression of an underlying mysticism foreign to the spirit of science; at the above-mentioned Congress in 1936 I therefore tried to clear up such misunderstandings and to explain that the only question was an endeavour to clarify the conditions, in each field of knowledge, for the analysis and synthesis of experience.[14] Yet, I am afraid that I had in this respect only little success in convincing my listeners, for whom the dissent among the physicists themselves was naturally a cause of scepticism about the necessity of going so far in renouncing customary demands as regards the explanation of natural phenomena. Not least through a new discussion with Einstein in Princeton in 1937, where we did not get beyond a humourous contest concerning which side Spinoza would have taken if he had lived to see the development of our days, I was strongly reminded of the importance of utmost caution in all questions of terminology and dialectics.

These aspects of the situation were especially discussed at a meeting in Warsaw in 1938, arranged by the International Institute of Intellectual Co-operation of the League of Nations.[16] The preceding years had seen great progress in quantum physics owing to a number of fundamental discoveries regarding the constitution and properties of atomic nuclei as well as important developments of the mathematical formalism taking the requirements of relativity theory into account. In the last respect, Dirac's ingenious quantum theory of the electron offered a most striking illustration of the power and fertility of the general quantum-mechanical way of description. In the phenomena of creation and annihilation of electron pairs we have in fact to do with new fundamental features of atomicity, which are intimately connected with the non-classical aspects of quantum statistics expressed in the exclusion principle, and which have demanded a still more far-reaching renunciation of explanation in terms of a pictorial representation.

Meanwhile, the discussion of the epistemological problems in atomic physics attracted as much attention as ever and, in commenting on Einstein's views as regards the incompleteness of the quantum-mechanical mode of description, I entered more directly on questions of terminology. In this connection I warned especially against phrases, often found in the physical literature, such as 'disturbing of phenomena by observation' or 'creating physical attributes to atomic objects by measurements'. Such phrases, which may serve to remind of the apparent paradoxes in quantum theory, are at the same time apt to cause confusion, since words like 'phenomena' and 'observations', just as 'attributes' and 'measurements', are used in a way hardly compatible with common language and practical definition.

As a more appropriate way of expression I advocated the application of the word *phenomenon* exclusively to refer to the observations obtained under specified circumstances, including an account of the whole experimental arrangement. In such terminology, the observational problem is free of any special intricacy since, in actual experiments, all observations are expressed by unambiguous statements referring, for instance, to the registration of the point at which an electron arrives at a photographic plate. Moreover, speaking in such a way is just suited to emphasize that the appropriate physical interpretation of the symbolic quantum-mechanical formalism amounts only to predictions, of determinate or statistical character, pertaining to individual phenomena appearing under conditions defined by classical physical concepts.

Notwithstanding all differences between the physical problems which have given rise to the development of relativity theory and quantum theory, respectively, a comparison of purely logical aspects of relativistic and complementary argumentation reveals striking similarities as regards the renunciation of the absolute significance of conventional physical attributes of objects. Also, the neglect of the atomic constitution of the measuring instruments themselves, in the account of actual experience, is equally characteristic of the applications of relativity and quantum theory. Thus, the smallness of the quantum of action compared with the actions involved in usual experience, including the arranging and handling of physical apparatus, is as essential in atomic physics as is the enormous number of atoms composing the world in the general theory of relativity which, as is often pointed out, demands that dimensions of apparatus for measuring angles can be made small compared with the radius of curvature of space.

In the Warsaw lecture, I commented upon the use of not directly visualizable symbolism in relativity and quantum theory in the following way:

Even the formalisms, which in both theories within their scope offer adequate means of comprehending all conceivable experience, exhibit deepgoing analogies. In fact, the astounding simplicity of the generalization of classical physical theories, which are obtained by the use of multidimensional geometry and non-commutative algebra, respectively, rests in both cases essentially on the introduction of the conventional symbol $\sqrt{-1}$. The abstract character of the formalisms concerned is indeed, on closer examination, as typical of relativity theory as it is of quantum mechanics, and it is in this respect purely a matter of tradition if the former theory is considered as a completion of classical physics rather than as a first fundamental step in the thoroughgoing revision of our conceptual means of comparing observations, which the modern development of physics has forced upon us.

It is, of course, true that in atomic physics we are confronted with a number of unsolved fundamental problems, especially as regards the intimate relationship between the elementary unit of electric charge and the universal quantum of action; but these problems are no more connected with the epistemological points here discussed than is the adequacy of relativistic argumentation with the issue of thus far unsolved problems of cosmology. Both in relativity and in quantum theory we are concerned with new aspects of scientific analysis and synthesis and, in this connection, it is interesting to note that, even in the great epoch of critical philosophy in the former century, there was only question to what extent *a priori* arguments could be given for the adequacy of space-time coordination and causal connection of experience, but never question of rational generalizations or inherent limitations of such categories of human thinking.

Although in more recent years I have had several occasions of meeting Einstein, the continued discussions, from which I always have received new impulses, have so far not led to a common view about the epistemological problems in atomic physics, and our opposing views are perhaps most clearly stated in a recent issue of *Dialectica*,[17] bringing a general discussion of these problems. Realizing, however, the many obstacles for mutual understanding as regards a matter where approach and background must influence everyone's attitude, I have welcomed this opportunity of a broader exposition of the development by which, to my mind, a veritable crisis in physical science has been overcome. The lesson we have hereby received would seem to have brought us a decisive step further in the never-ending struggle for harmony between content and form, and taught us once again that no content can be grasped without a formal frame and that any form, however useful it has hitherto proved, may be found to be too narrow to comprehend new experience.

Surely, in a situation like this, where it has been difficult to reach mutual understanding not only between philosophers and physicists but even between physicists of different schools, the difficulties have their root not seldom in the preference for a certain use of language suggesting itself from the different lines of approach. In the Institute in Copenhagen, where through those years a number of young physicists from various countries came together for discussions, we used, when in trouble, often to comfort ourselves with jokes, among them the old saying of the two kinds of truth. To the one kind belong statements so simple and clear that the opposite assertion obviously could not be defended. The other kind, the so-called 'deep truths,' are statements in which the opposite also contains deep truth. Now, the development in a new field will usually pass through stages in which chaos becomes gradually replaced by order; but it is not least in the intermediate stage where deep truth prevails that the work is really exciting and inspires the imagination to search for a firmer hold. For such endeavours of seeking the proper balance between seriousness and humour, Einstein's own personality stands as a great example and, when expressing my belief that through a singularly fruitful cooperation of a whole generation of physicists we are nearing the goal where logical order to a large extent allows us to avoid deep truth, I hope that it will be taken in his spirit and may serve as an apology for several utterances in the preceding pages.

The discussions with Einstein which have formed the theme of this article have extended over many years which have witnessed great progress in the field of atomic physics. Whether our actual meetings have been of short or long duration, they have always left a deep and lasting impression on my mind, and when writing this report I have, so-to-speak, been arguing with Einstein all the time, even in discussing topics apparently far removed from the special problems under debate at our meetings. As regards the account of the conversations I am, of course, aware that I am relying only on my own memory, just as I am prepared for the possibility that many features of the development of quantum theory, in which Einstein has played so large a part, may appear to himself in a different light. I trust, however, that I have not failed in conveying a proper impression of how much it has meant to me to be able to benefit from the inspiration which we all derive from every contact with Einstein.

REFERENCES

1. A. Einstein, *Ann. Phys.* **17**, 132 (1905).
2. N. Bohr, *Fysisk Tidsskrift* **12**, 97 (1914). (English version in *The Theory of Spectra and Atomic Constitution*, Cambridge University Press, 1922.)
3. A. Einstein, *Physik. Z.* **18**, 121 (1917).
4. A. Einstein and P. Ehrenfest, *Z. Phys.* **11**, 31 (1922).
5. N. Bohr, H. A. Kramers and J. C. Slater, *Phil. Mag.* **47**, 785 (1924).
6. A. Einstein, *Berl. Ber.* (1924) 261; (1925) 3 and 18.
7. W. Heisenberg, *Z. Phys.* **43**, 172 (1927).
8. Atti del Congresso Internazionale dei Fisici, Como, Settembre 1927 (reprinted in *Nature*, **121**, 78 and 580 (1928)).
9. Institut International de Physique Solvay, *Rapport et discussions* du 5e Conseil, Paris (1928) 253ff.
10. *Ibid.*, 248ff.
11. A. Einstein, B. Podolsky, and N. Rosen, *Phys. Rev.*, **47**, 777 (1935).
12. N. Bohr, *Phys. Rev.* **48**, 696 (1935).
13. A. Einstein, *J. Franklin Inst.* **221**, 349 (1936).
14. N. Bohr, *Philosophy of Science*, **4**, 289 (1937).
15. IIe Congrès international de la Lumière, Copenhagen 1932.
16. *New Theories in Physics* (Paris 1938), 11.
17. N. Bohr, *Dialectica*, **1**, 312 (1948).

Centenaire de Fresnel

Liste des participants.

Ch. H.

2	H. A. Lorentz	1	— —
3	J. E. Verschaffelt	2	2
	Ch. Lefebure	3	
4	P. Langevin	4	—
5	O. W. Richardson	5	1
6	A. Einstein	6	1
8 7	Kramers	7	1
9	C. T. R. Wilson	8	1
N. Bohr 10	N. Bohr	9	1
R. H. Fowler 11	R. H. Fowler	10	1
15	A. H. Compton (with ... and ...)	11 12 13 14 15	2 2
16	W. Pauli jr.	16	1
17	W. Heisenberg	17	1
18	P. Debye		
19	L. Brillouin	18 suppl.	—
20	Louis de Broglie	19	—
21	M. de Donder	20	1
22	C. E. Guye	21 Benn 20	—
		21 nombre	
23	W. L. Bragg	22 / 22	1
24	M. Born	23	2

The ceremony, in honour of the hundredth anniversary of Augustin Fresnel's death, took place at the Académie des Sciences in Paris during the fifth Solvay Conference, and its participants were invited. Those who wished to attend signed their names: H. A. Lorentz, J. E. Verschaffelt, Ch. Lefébure, P. Langevin, O. W. Richardson, A. Einstein, H. A. Kramers, C. T. R. Wilson, N. Bohr, R. H. Fowler, A. H. Compton, W. Pauli jr, W. Heisenberg, P. Debye, L. Brillouin, Louis de Broglie, Th. De Donder, C. E. Guye, W. L. Bragg, M. Born.

SIXTH SOLVAY CONFERENCE 1930

A. PICCARD   W. GERLACH   C. DARWIN   P.A. DIRAC   H.A. KRAMERS   J.H. VAN VLECK   W. HEISENBERG
                                                        H. BAUER   P. KAPITZA   L. BRILLOUIN   P. DEBYE   W. PAULI   J. DORFMAN   E. FERMI
E. HENRIOT   MANNEBACK   Mme CURIE   P. LANGEVIN   A. EINSTEIN   O. RICHARDSON   B. CABRERA   N. BOHR   W.J. DE HAAS
        J. VERSCHAFFELT   A. COTTON   J. ERRERA   O. STERN
E. HERZEN   Absents : Ch.E. GUYE et M. KNUDSEN
        Th. DE DONDER   P. ZEEMAN   P. WEISS   A. SOMMERFELD

7

Magnetism*

1. INTRODUCTION

The sixth Solvay Conference on Physics took place in Brussels from 20 to 25 October 1930. The general theme of the Conference was 'Magnetism'.

After the death of Lorentz, Paul Langevin had become President of the Scientific Council of the International Institute of Physics, founded by Ernest Solvay. The members of the Scientific Council for the sixth Solvay Conference were: N. Bohr (Copenhagen), B. Cabrera (Madrid), Madame Curie (Paris), Th. De Donder (Brussels), A. Einstein (Berlin), Ch. E. Guye (Geneva), M. Knudsen, Secretary (Copenhagen), and O. W. Richardson (London).

Among the invited participants in the sixth Solvay Conference were: A. Cotton (Paris), C. G. Darwin (Edinburgh), P. Debye (Leipzig), W. J. de Haas (Leyden), P. A. M. Dirac (Cambridge), J. Dorfman (Leningrad), E. Fermi (Rome), W. Gerlach (Munich), W. Heisenberg (Leipzig), P. Kapitza (Cambridge), H. A. Kramers (Utrecht), W. Pauli (Zurich), A. Sommerfeld (Munich), O. Stern (Hamburg), J. H. Van Vleck (Madison, Wisconsin, U.S.A.), P. Weiss (Strasbourg), and P. Zeeman (Amsterdam).

The Scientific Council also invited E. Henriot and A. Piccard, both of Brussels, to attend the Conference. E. Herzen attended the Conference as a representative of the Solvay family.

J. E. Verschaffelt (Ghent), E. Bauer (Paris), L. Brillouin (Paris), J. Errera (Brussels, and C. Manneback (Brussels) acted as the Scientific Secretaries of the Conference.

Langevin opened the conference with a tribute to the memory of H. A. Lorentz who had been associated with the Institute and the Solvay Conferences since their inception, and had guided their activities with his initiative, wisdom and leadership.

The subject of the conference was the 'Magnetic Properties of Matter'. Langevin himself had made important contributions to this field. Through the studies of Weiss and his school, experimental knowledge in magnetism had increased substantially, and the advent of quantum mechanics had enriched the theoretical understanding of the magnetic properties of matter. The subject represented a new frontier of research in physics, and was thus ripe for discussion at a Solvay Conference.

The first report, on magnetism and spectroscopy, was given by Sommerfeld. He discussed in particular the knowledge of angular momenta and magnetic moments, which had been derived from the investigations of the electronic constitution of

---

* *Le Magnétisme*, Rapports et Discussions du Sixième Conseil de Physique tenu à Bruxelles du 20 au 25 octobre 1930, Gauthier-Villars, Paris, 1932.

atoms, resulting in the explanation of the periodic table. J. H. Van Vleck reported on the latest results and their theoretical interpretation in the peculiar variation of the magnetic moments within the family of the rare earths. E. Fermi discussed the magnetic moments of atomic nuclei, in which, as first pointed out by Pauli, the hyperfine structure of spectral lines was to be found.

Cabrera and Weiss gave general surveys of the rapidly increasing experimental evidence about the magnetic properties of matter. They discussed the equation of state of ferromagnetic materials, comprising the abrupt change of their properties at definite temperatures like the Curie point. Attempts had been made earlier to correlate such effects, especially by Weiss' introduction of an internal magnetic field associated with the ferromagnetic state. However, a clue to the understanding of these phenomena was found by Heisenberg's comparison of the alignment of the electron spins in ferromagnetic substances with the quantum statistics governing the symmetry properties of the wave functions responsible for the chemical bonds in Heitler and London's theory of molecular formation.

Pauli, in his report, gave a comprehensive theoretical treatment of magnetic phenomena. He also discussed the problems raised by Dirac's quantum theory of the electron, which allowed for the incorporation of the intrinsic spin and magnetic moment of the electron in a natural manner. The question was raised as to how far could such quantities as the spin and magnetic moment be considered measurable in the same sense as the mass and charge of the electron, which can be defined on the basis of phenomena that can be understood in classical terms. The concept of spin, just as the quantum of action, referred to phenomena that could not be analyzed classically. It is an abstract notion which permits a generalized formulation of the conservation of angular momentum. Pauli's report discussed these matters in details, including the impossibility of measuring the magnetic moment of a free electron.

The reports of Cotton and Kapitza dealt with the development of experimental techniques for the investigation of magnetic phenomena. Kapitza had succeeded in producing magnetic fields of previously unsurpassed strength within limited spatial extensions and time intervals. Cotton had designed huge permanent magnets providing constant fields. Madame Curie pointed out how such magnets could be used for the investigation of radioactive phenomena, such as Rosenblum's work on the fine structure of α-ray spectra.

By the time the sixth Solvay Conference took place numerous problems had been understood, which had been beyond one's grasp at the fourth Conference on the electrical conduction in metals in 1924. Sommerfeld had, in 1927, obtained important results in the theory of metallic conductivity by replacing the Maxwell velocity distribution of the electrons by a Fermi distribution. Bloch had developed a detailed theory of metallic conduction, including the temperature dependence of these phenomena, on the basis of wave mechanics. The theory still failed to account for superconductivity, and it would be only in the 1950s that a deeper understanding of this phenomenon would come, based on the interactions of many-body systems.

The Bohr–Einstein dialogue on epistemological questions, which had started at the fifth Solvay Conference in 1927, was resumed at the 1930 meeting. This was the last meeting that Einstein attended; the seventh Solvay Conference took place in 1933, but Einstein emigrated to the United States that year.

We shall now give résumés of the various reports presented at the conference.

2. MAGNETISM AND SPECTROSCOPY

(A. Sommerfeld)

Let **M** be the angular momentum of an electron rotating around a nucleus, and **μ** its magnetic moment, then

$$\mathbf{\mu} = \frac{e}{2m} \mathbf{M}. \tag{1}$$

Sommerfeld expressed μ as a multiple of $\hbar$, and wrote the elementary magnetic moment as

$$\mu = \frac{e}{2m} \hbar.$$

He introduced the Bohr magneton,

$$\mu_B = \frac{eN}{2m} \hbar,$$

where N is the Avogadro number. Sommerfeld noted that the idea of the magneton had been introduced by Bohr in a discussion with him. Langevin had previously introduced a similar notion at the first Solvay Conference, and Gans had made an equivalent proposition at the Karlsruhe conference of 1911. However, Sommerfeld maintained, the initial idea of a magneton was due to P. Weiss, who had $\mu_W = 1125$ gauss/cm, so that $\mu_B/\mu_W = 4.97$.

Sommerfeld recalled Pauli's note[1] of 1920 in which he had made use of the quantization in space to understand the normal Zeeman effect in hydrogen. The average of $\cos^2 \theta$, which had the value $\frac{1}{3}$ in Langevin's theory, became equal to $\frac{1}{3}(k+1)(k+\frac{1}{2})/k^2$ in Pauli's case, where k is an integral quantum number arising from the projection of **M** along the magnetic field **H**. The calculation of the magnetic susceptibility by this method gave,

$$\chi = \frac{p^2 \mu_W^2}{3RT} = \frac{k^2 \mu_B^2}{RT} \overline{\cos^2 \theta}, \tag{2}$$

where p is the number of Weiss magnetons and $p\mu_W$ is the molecular moment. Comparison with experiments for NO and O_2 gave good agreement. Epstein and Gerlach had applied Pauli's results to paramagnetic atomic ions with success, with a few exceptions.

In the case of the anomalous Zeeman effect Sommerfeld introduced the quantum number j, corresponding to the total angular momentum. j could be considered as the result of the compound motion of the electron in the orbit plus its proper rotation.

Now the magnetic moment μ is related to the Bohr magneton by the formula $\mu = jg\mu_B$, where g is the Landé factor. The magnetization coefficient χ is again given by $p^2 \mu_W^2 / 3RT$ or $\frac{1}{3} j(j+1) g^2 \mu_B^2$ leading to

$$p\mu_W = \sqrt{j(j+1)}\, g\mu_B. \tag{3}$$

This analysis showed that the number of magnetons increases linearly until about the first half of the iron group, and then decreases irregularly.

Sommerfeld pointed out that the quantitative difference between Pauli's theory and his own for the anomalous Zeeman effect was small. He examined the rare earths for which the calculation of the 'spectroscopic magnetons' led to a perfect agreement.

In the series of the rare earths, the groups $N\,(n=4)$ is completed with the f electrons, i.e. those with azimuthal quantum number $l=3$. From Pauli's exclusion principle, the number of f electrons is equal to $2(2l+1)=14$. Sommerfeld introduced two quantities, z, the number of f electrons contained in each ion, and z', the number of electrons necessary to fill a group. For these quantities, he had the relation $z+z'=2(2l+1)=14$. Using Hund's rule, Sommerfeld expressed the number of magnetons in terms of z and z'. The agreement with experiments was satisfactory, except in the case of Sm (samarium) and Eu (europium). Van Vleck treated these exceptions in more detail in his report.

The difficulties in the iron group arose from the fact that the multiplets are too narrow, $h\,\Delta v \ll kT$, while for the rare earths $h\,\Delta v \gg kT$. This important difference implied an influence of the temperature.

Assuming that each level is represented by a number of ions proportional to $(2j+1)e^{-hv_j/kT}$, Sommerfeld obtained

$$p = 4.97 \sqrt{\left(\frac{\sum N_j j(j+1)\, g^2}{\sum N_j} \right)}, \tag{4}$$

where the sum extends from $j=l-s$ to $j=l+s$. For large multiplets this formula is the same as for the rare earths, but in the case of narrow multiplets Sommerfeld obtained,

$$p = 4.97 \sqrt{\left(\frac{\sum (2j+1)\, j(j+1)\, g^2}{\sum (2j+1)} \right)}. \tag{5}$$

This formula did not fit very well with the experimental results even when the coupling between l and s disappeared.

Sommerfeld then discussed the case of diatomic molecules, which Van Vleck had treated quantum mechanically, but which Sommerfeld translated into the language of the old quantum theory.

In the second part of his report, Sommerfeld rederived the results of Van Vleck in the framework of the Dirac equation. He wrote the Dirac equation in the form,

$$\Lambda u = 0, \tag{6}$$

with

$$\Lambda = \sum_{1}^{4} \alpha_k \left(\frac{\partial}{\partial x_k} + i\Phi_k \right) + B, \tag{6a}$$

and

$$\Phi_k = \frac{2\pi}{hc} (\mathbf{A}, iV), \tag{6b}$$

where V is the scalar potential and, $\mathbf{A}$, the vector potential, $B = m_0 c^2/\hbar c$, and α_k are operators satisfying the commutation relations

$$\alpha_i \alpha_k + \alpha_k \alpha_i = 2\delta_{i,k}. \tag{7}$$

Sommerfeld wrote the solution of Equation (6) in the form

$$u = \Psi e^{i\Omega t}, \tag{8}$$

and developed Ψ and Ω in powers of H (the uniform magnetic field along the Z-axis). The problem is then solved by perturbation theory, and the magnetic moment of the atom in the direction of H is found in the form

$$\mu_H = \hbar c \left(M_{kk} - 2cH \sum_{l}' \frac{M_{lk}M_{kl}}{\omega_k - \omega_l} \right), \tag{9}$$

where the M's are matrix elements of the perturbation function,

$$M_{lk} = \int \bar{\psi}_l s \psi_k \, d\tau,$$

$$M_{kl} = \int \bar{\psi}_k s \psi_l \, d\tau, \tag{10}$$

$$M_{kk} = \int \bar{\psi}_k s \psi_k \, d\tau,$$

where

$$s = \frac{ie}{2\hbar c} (\alpha_2 x - \alpha_1 y),$$

and $\psi_l, \bar{\psi}_l$ are the stationary states without the magnetic field H.

Sommerfeld distinguished between two cases: (a) broad multiplets, i.e., n, l, j fixed, but m varying, (b) narrow multiplets, i.e., n, l fixed, but m, j varying. The magnetic moment of the gram atom, χH, is then equal to

$$\chi H = N \frac{\sum\limits_{f} \mu H e^{-E/kT}}{\sum\limits_{f} e^{-E/kT}}, \tag{11}$$

where N is the Avogadro number, and $f = m$ in case (a), and $f = m, j$ in case (b). Also, $\chi = \chi_{\text{dia.}} + \chi_{\text{par.}}$, is the sum of dia- and para-magnetic susceptibilities.

In the case of a broad multiplet, j remains fixed and corresponds to the lowest state of the multiplet. Sommerfeld obtained, in agreement with previous results,

$$\chi_{\text{par.}} = \frac{j(j+1)}{3RT} g^2 \mu_B^2. \tag{12}$$

In the case of a narrow multiplet, the summation is on j and m; and for a doublet, with $j_1 = l + \frac{1}{2}$ and $j_2 = l - \frac{1}{2}$, Sommerfeld got,

$$\chi_{\text{par.}} = \frac{\mu_B^2}{3RT} [3 + l(l+1)]. \tag{13}$$

Generalizing this formula by replacing the factor 3 by $4s(s+1)$, with $s = \frac{1}{2}$, he obtained the number of Weiss magnetons as

$$p = 4.97 \sqrt{\{4s(s+1) + l(l+1)\}}. \tag{14}$$

This formula had been obtained for the case of a small magnetic field, but was in agreement with the one for a strong field in old quantum theory. Sommerfeld pointed out that the calculation for a strong field could also be performed in quantum mechanics, leading to the same result. The origin of the discrepancy corresponds to the mutual interaction between neighbouring levels, which is automatically introduced in quantum mechanics.

DISCUSSION

Kramers pointed out that the Hund rules seemed to be very good in the case of rare earths, but might present some discrepancies for other substances. Sommerfeld agreed that in the theory of the 'number of spectroscopic magnetons', one could calculate paramagnetism if one knew the spectroscopic nature of the fundamental term considered. Whether this fundamental term is always determined by Hund's rule is another question. For neutral atoms, there are certainly exceptions to this rule, as for instance Cr. However, for the ions, especially trivalent ions of the rare earths, Sommerfeld knew of no exceptions.

Kramers then remarked that the measurements of de Haas and Gorter were in good agreement with Hund's theory, except at low temperatures. Van Vleck pointed out that this disagreement at low temperatures was quite normal. Hund's theory was for free ions, so that for solids it was valid so long as the interatomic forces, which tend to orient the ions, were weak compared to kT. Sommerfeld pointed out that Hund's rules had been proved theoretically by Wigner with the help of group theory, and later on by Slater without it.

Weiss noted that the linear character of the law $\chi(T-\theta) = C$, which could be written as

$$\frac{1}{\chi T} = \frac{1}{C} - \frac{n}{T},$$

with $n = \theta/C$, was in good agreement with experiments. A theory giving an explanation of this phenomenon must not only show that in the expansion in terms of $1/T$, the term $1/T^2$ has the right constant, but also explain this linearity.

After a remark of Van Vleck on the influence of an external field on the electronic motion, Kramers gave a résumé of the problem under consideration. He pointed out that the problem of the action of a weak external electric field, with potential $w = ax^2 + by^2 + cz^2$, on a free electron, was identical to the problem of the quantization of the asymmetric top. When the external field is strong, one can calculate by supposing that the spin vector, s, is independent of l, the orbital angular momentum vector; the effect of the external field, in the first approximation, is to split the orbital angular momentum l into $2l+1$ levels.

Dirac made a comment on Sommerfeld's quantum mechanical calculation. He pointed out that Sommerfeld's conclusion, showing that the same result would follow for the weak and strong fields, could be easily understood. Both depend on the quantity $\Sigma_{l,k} M_{lk} M_{kl}$, which is the trace of M^2. But the trace is independent of the representation, so that whether this quantity is calculated with j or l diagonal does not make any difference.

REFERENCE

1. W. Pauli, *Phys. Z.* **21**, 615 (1920).

3. THE SUSCEPTIBILITY OF SAMARIUM AND EUROPIUM IONS

(J. H. Van Vleck)

In order to calculate that part of the susceptibility which is independent of the intensity of the magnetic field, it is necessary to know the energies of the stationary states, W_n, to an approximation of the second order in H. Van Vleck put

$$W_n = W_n^{(0)} + H W_n^{(1)} + H^{(2)} W_n^{(2)},\qquad (1)$$

and for the magnetic moment,

$$M_n = -\frac{\partial W_n}{\partial H} = -W_n^{(1)} - 2H W_n^{(2)}.\qquad (2)$$

From this he obtained the susceptibility as,

$$\chi = \frac{N \sum_n M_n e^{-W_n/kT}}{H \sum e^{-W_n/kT}}.\qquad (3)$$

Expanding the exponential in powers of H, he obtained that part of the susceptibility which is independent of the field,

$$\chi = \frac{N \sum_n \left(\dfrac{(W_n^{(1)})^2}{kT} - 2W_n^{(2)} \right) e^{-W_n^{(0)}/kT}}{\sum e^{-W_n^{(0)}/kT}}.\qquad (4)$$

$W^{(2)}$ is, in general, small and negligible, but in the case of samarium and europium it is the dominant term. Expanding W_n with the help of the Russell–Saunders coupling and, disregarding the diamagnetic part, Van Vleck obtained the expression for χ. Comparing with the expressions for χ in the cases of broad and narrow multiplets, Van Vleck showed that in the case of samarium and europium the multiplets were of an intermediate type, and the theoretical formula agreed with experimental results. He also showed that the values for the rare earth and iron series were nearly the same with and without corrections.

4. MAGNETIC MOMENTS OF NUCLEI

(E. Fermi)

It was well known that the electrons have a magnetic moment, and Fermi thought that there was good reason to believe that the protons have a magnetic moment also. The difficulty was in understanding why the magnetic moment was so weak. On the basis of Dirac's theory, the electron has a magnetic moment of one Bohr magneton when free or weakly coupled. Following Darwin, Fermi gave a calculation based on the Dirac equation showing that the magnetic moment, for orbits of the order of 10^{-12} to 10^{-13} cm, is about one-hundredth to one-thousandth of the Bohr magneton.

In analogy with the electron, Fermi calculated the magnetic moment of the proton with the formula

$$\mu_p = \frac{eh}{4\pi m_p c},$$ (1)

where m_p is the proton mass. This moment is about 1840 times smaller than that of the electron.

Considering that the hyperfine structure is the result of the magnetic moment of the nucleus alone, Fermi sought to draw some conclusions in the case of an atom with a single electron in the closed shell. He considered the effect on different terms. The S (or $^2S_{1/2}$) term is split into two terms having an energy difference

$$\delta\left(^2S_{1/2}\right) = \frac{2i+1}{i}\frac{8\pi}{3}\mu\mu_0\psi^2(0),$$ (2)

where μ_0 ($= eh/4\pi m_0 c$) is the Bohr magneton, μ ($= 2i(Z/M)(eh/4\pi m_p c)$), the nuclear magnetic moment, $i\hbar$, the mechanical moment, and $\psi(0)$, the value of the electron s-wave function at the position of the nucleus.

For the P terms ($^2P_{1/2}$), he also got a splitting with an energy difference,

$$\delta\left(^2P_{1/2}\right) = \frac{2i+1}{i}\tfrac{3}{8}\mu\mu_0\int\tfrac{1}{3}\psi\,d\tau.$$ (3)

Fermi pointed out that it was very difficult to determine the value of $\psi(0)$. However, he computed $\psi(0)$ by means of a statistical method, with an error of about 30% in the value of ψ_0^2, and found that in the case of thallium the value of the hyperfine structure for $7^2S_{1/2}$ was four times smaller than for $6^2P_{1/2}$. Fermi thought that this difference could not be explained by nuclear magnetic moment alone. The reason which, he thought, could contribute to the discrepancy would be that the forces between the electron and the nucleus were not of the Coulomb type, and that there was a resonance between the electrons within and those outside the nucleus.

DISCUSSION

Pauli pointed out that the hyperfine structure of Li was particularly interesting, and it seemed to be incomprehensible, especially the structure of the fundamental ortho-term of the ion Li$^+$ for which the theory gives triplets. Pauli explained that the rule which was valid for atoms could not be applied to the nuclei. He discussed how one could interpret the hyperfine structure of the isotopes of Li.

Zeeman mentioned that the value of the nuclear moment I could be determined by the splitting in an intense field, leading to a set of $2I+1$ components. Thus the value of the nuclear moment could be established with great certainty.

Among others who took part in the discussion were Kramers, Richardson, Langevin, Dirac and Heisenberg.

5. EXPERIMENTAL STUDY OF PARAMAGNETISM; THE MAGNETON

(B. Cabrera)

Cabrera first recalled the definition of magnetization, $\mathbf{I}$, and specific magnetization, $\mathbf{j}$. If d$\mathbf{M}$ is the magnetic moment of a body placed in a magnetic field H, then these quantities are defined by the relations d$\mathbf{M}=\mathbf{I}\,\mathrm{d}v$, and d$\mathbf{M}=\mathbf{j}\,\mathrm{d}m$, d$v$ and dm being the volume and mass elements of the body.

The susceptibility, k, and the specific susceptibility or the coefficient of magnetization, χ, are defined by,

$$\mathbf{I} = k\mathbf{H}, \quad \text{and} \quad \mathbf{j} = \chi\mathbf{H} \tag{1}$$

There are three kinds of substances – diamagnetic, paramagnetic, and ferromagnetic – for which, χ is negative, positive but independent of H, or positive and function of H, respectively. The coefficients of susceptibility can be obtained by a measurement of the force exerted by the field on the body. Along the X-axis this force is given by

$$\mathrm{d}X = \frac{k_{ap}}{2}\frac{\mathrm{d}H^2}{\mathrm{d}x}\,\mathrm{d}v, \tag{2}$$

where $k_{ap}=k-k_0$, and k_0 is the susceptibility of the medium.

Cabrera described the problem of making measurements. The first difficulty concerned the shape of the magnet, which had to be chosen in such a way that dH^2/dx would be constant over a large volume. The measurement was then performed by measuring the moment of a wire or rod, but the accuracy could not exceed one in one thousand. He also described an apparatus for the measurements on liquids.

Experimentally, the condition for mixing had been obtained as

$$\chi = m_1\chi_1 + m_2\chi_2 + \cdots, \tag{3}$$

with $m_1+m_2+\cdots=1$, where χ_1, χ_2, ..., are the susceptibilities of the different bodies and $m_1, m_2...$, are the proportions in which they enter the mixing. The value of χ generally changes with the orientation of the body, and Cabrera described methods for the measurement of χ along different magnetic axes.

The influence of temperature on paramagnetism may be studied with the help of the Curie law which paramagnetic substances follow,

$$\chi T = C. \tag{4}$$

Assuming that each molecule has a magnetic moment μ, Langevin had proved that for a gram-molecule of the gas,

$$\mathbf{J}_m = \mu N L(a), \tag{5}$$

where N is Avogadro's number, $L(a) = \coth a - 1/a$ and $a = \mu H/kT$. Thus, in the first approximation, Langevin's law yields

$$\chi_m = \frac{\mu_m^2}{3RT}, \tag{6}$$

where $\mu_m = N\mu$. Cabrera pointed out that, but for the constant, the quantum mechanical formulation also leads to the Curie law, and that it no longer had to be considered as being merely an empirical law.

Cabrera discussed various procedures which had been used for the quantization of Langevin's law. For instance, Sommerfeld had replaced $L(a)$ by $L_j(a)$, where

$$L_j(a) = \frac{2j+1}{j} \coth \frac{2j+1}{j} a - \frac{1}{2j} \coth \frac{a}{2j}, \tag{7}$$

and as $J \to \infty$, $L_j(a) \to L(a)$. However, the results were not as good as with the classical expression. The main difficulty with $L_j(a)$ arose from the assumption that the magnets could take any orientation, and this assumption did not seem to be realistic.

The sulphate of Gd-octohydrate, which had been studied by H. R. Woltjer and Kamerlingh Onnes, obeyed Langevin's law, Equation (5), for all values of a between 1 and 5.6. Unfortunately, this was the only case which indicated the saturation predicted by Langevin's law. In all the other cases known at that time, only the linear part of the curve could be reached, satisfying the Curie–Weiss law,

$$\chi_a(T + \Delta) = C_a, \tag{8}$$

where C_a is a constant depending on the cation of the paramagnetic substance, and Δ is the correction introduced by Weiss to the Curie law. Cabrera mentioned the causes for the existence of Δ, but as Weiss had already recognized, Δ is too big to be the direct effect of a magnetic field.

The work of Cabrera and Palacios had given an explanation of the molecular field, by assuming that the orientation of the magnetic axis is defined by a position energy ε_i, arising from the internal structure and bonds. They obtained, in the first approximation,

$$\chi_a = \frac{\mu^2 N}{3k} \frac{1 - l^2}{T + \Delta}, \tag{9}$$

with

$$\Delta = \frac{2l(lk - p)}{k(1 - l^2)},$$

where

$$l = \sum_1 \varrho_i \cos\theta_i, \qquad p = \sum_i \varepsilon_i \cos\theta_i, \qquad q_i = \sum \varrho_i \varepsilon_i.$$

ε_i is the energy for the position i of the magnetic axis, defined by the angle θ_i, and ϱ_i is the statistical weight corresponding to it.

Even when approximations to higher orders were taken into account, the series of chlorides of Ru, Rh, Pd, Os, Ir, and Pt showed a behaviour which could not be explained by the simple laws of Curie and Weiss. Moreover, their behaviour was not of the type of ferromagnetic substances Fe, Co, Ni, and empirically they seemed to follow the law

$$(\chi + k)(T + \Delta) = C. \tag{10}$$

Cabrera noted that paramagnetism decreases with increasing atomic weight until Pt^{++}, which is diamagnetic. In fact diamagnetism was always present, and the susceptibility had to be written in the form $\chi = \chi_d + \chi_p$ (with d and p for dia- and paramagnetism respectively). Thus the Curie–Weiss law takes the form

$$(\chi - \chi_d)(T + \Delta) = C, \tag{11}$$

which is of the type given above, in which k can have both signs. This fact had been interpreted as showing the existence of a paramagnetism independent of T.

Cabrera pointed out that the Curie constant C_a was independent of the atoms or groups of atoms to which the cation, responsible for the diamagnetism, was bound. However, this was not always the case, and C_a would take different values if the magnetic moment was different in different groups of atoms. It might even happen that C_a would change for a given temperature in the same substance.

Weiss had discovered a relation for the variation of C_a, based on Langevin's theory, in which $C_a = \mu_0^2/3R$. It was,

$$\sqrt{C_a} = nK, \tag{12}$$

where n is an integer, and K, a constant. This relation is satisfied by many substances and shows, through a precise calculation of K, the validity of the concept of the Weiss magneton.

The Bohr magneton is $\mu_B = Neh/4\pi mc = 4.957\ \mu_W$. Cabrera pointed out that, according to quantum mechanics,

$$\mu = g\sqrt{j(j+1)}\ \mu_B = 4.957\ g\sqrt{j(j+1)}\ \mu_W, \tag{13}$$

showing that the coefficient of μ_W is always a fraction, a result which was in contradiction with experiment. However, if one took into account the rules given by Hund for the configurations, with the corrections of Van Vleck and A. Frank, then this coefficient was nearly an integer for a number of substances, particularly in all compounds of the rare earths. There was still disagreement with the iron family.

Finally, Cabrera discussed some mineral complexes as special cases of cations. He showed that for Cr^{+++}, Fe^{+++}, Fe^{++}, Ni^{++} and Cu^{++}, etc. the general law, Equation (10), was obeyed.

DISCUSSION

In the discussion, Kapitza presented a new method for the measurement of magnetic susceptibility, which was different from Curie's and was more sensitive. Weiss mentioned the improvement in measurements brought about by A. Piccard and Quincke, and pointed out that the quantization around a privileged direction, one of the ideas in the theory of Cabrera and Palacios, had also been considered by E. Bauer.

De Haas, Langevin, Van Vleck, Sommerfeld, Heisenberg, Darwin, and Dorfman also took part in the discussion of Cabrera's report.

6. QUANTUM THEORY OF MAGNETISM: THE MAGNETIC ELECTRON

(W. Pauli)

The reports of Sommerfeld and Van Vleck had shown that a quantitative interpretation of magnetic properties could be given for paramagnetic gases and dilute salts of the rare earths. However, the knowledge of the magnetism of solid bodies was still qualitative. The discovery of electron spin had allowed one to understand the paramagnetism, independent of temperature, of certain metals, as well as ferromagnetic properties.

Pauli discussed how, by using the exclusion principle and neglecting the action of the spins on the motion of the electrons, the general form of the wave function could be obtained. The electrons are not completely independent, but follow Fermi statistics, and this statistics is a consequence of the exclusion principle.

Pauli calculated the number of stationary states for the electron in the momentum interval between p and $p + dp$ in a volume V, obtaining

$$dZ = 2V \frac{2\pi (2m_0)^{3/2}}{h^3} \sqrt{E}\, dE, \tag{1}$$

with $p = \sqrt{2m_0 E}$, m_0 being the electron mass, and E the energy. For low temperatures, the exclusion principle implies that all states, up to an energy η, are occupied. η is determined by

$$N = \int_0^{\eta} dZ = 2V \frac{2\pi (2m_0)^{3/2}}{h^3} \tfrac{2}{3}\eta^{3/2}, \tag{2}$$

while the total energy is given by

$$E = \int_0^{\eta} E\, dZ = 2V \frac{2\pi (2m_0)^{3/2}}{h^3} \tfrac{2}{5}\eta^{5/2}. \tag{3}$$

Putting $n = N/V$, and $\varepsilon = E/V$, Pauli obtained

$$\varepsilon = \tfrac{3}{5}\eta = \tfrac{3}{40} \left(\frac{6}{\pi}\right)^{3/2} \frac{h^2}{m_0} \left(\frac{n}{2}\right)^{2/3}. \tag{4}$$

For low temperatures, the condition that $kT \ll \eta$ is well satisfied at the usual density of the metal.

Pauli pointed out that the exclusion principle does not allow two electrons, corresponding to the same stationary state, to align their magnetic axes simultaneously with the external field. The two electrons can have their axes parallel only if their kinetic energies are different, such that magnetization is necessarily related to an increase of the kinetic energy of the system.

Let η_1 and η_2 be the maximum kinetic energies, parallel and antiparallel to the field, respectively. Then, if

$$\eta = \eta_1 - \mu_0 H = \eta_2 + \mu_0 H \tag{5}$$

is the energy necessary to turn around a spin, the total number of electrons can be expressed as,

$$N = V \frac{2\pi (2m_0)^{3/2}}{h^3} \tfrac{2}{3} \left[(\eta + \mu_0 H)^{3/2} + (\eta - \mu_0 H)^{3/2}\right]. \tag{6}$$

The total magnetic moment is then found to be,

$$M = \mu_0 V \frac{2\pi (2m_0)^{3/2}}{h^3} \tfrac{2}{3} \left[(\eta + \mu_0 H)^{3/2} - (\eta - \mu_0 H)^{3/2}\right]. \tag{7}$$

Assuming that, $\mu_0 H \ll \eta$, for the intensity of practically realizable fields, Pauli obtained,

$$M = \mu_0 V \frac{2\pi (2m_0)^{3/2}}{h^3} 2\mu_0 H \eta^{1/2}. \tag{8}$$

Consequently,

$$\chi = \frac{M}{HV} = \frac{9}{10} \frac{n\mu_0^2}{\varepsilon}. \tag{9}$$

Thus Pauli obtained that part of paramagnetism which is independent of the temperature.[1]

Pauli then discussed the diamagnetism of free electrons. Writing the Hamiltonian in the form,

$$\mathcal{H} = \frac{1}{2m_0} \left[\left(p_x - \frac{eH}{2c} y\right)^2 + \left(p_y + \frac{eH}{2c} x\right)^2\right], \tag{10}$$

he showed that the probability for a representative point of the position and momentum of the electron to be in a given element of phase space was given by the Maxwell distribution, which meant that the statistical distribution of the electrons was not affected by the presence of H. Then he treated the problem according to quantum mechanics. Writing the angular momentum as $P = xp_y - yp_x$, with $x^2 + y^2 = r^2$, and $p_x^2 + p_y^2 = p_r^2 + p^2/r^2$, he obtained the Hamiltonian in the form,

$$\mathcal{H} = \frac{1}{2m_0}\left(p_r^2 + \frac{p^2}{r^2}\right) + \frac{m_0}{2}\left(\frac{eH}{2m_0c}\right)^2 r^2 + \frac{eHP}{2m_0c}. \tag{11}$$

The last term is constant, and the rest is the Hamiltonian of a harmonic oscillator with the eigenfrequencies $\sigma = (1/2\pi)\,(eH/2em_0c)$, and the energy eigenvalues given by

$$E = (2n + |m| + 1)\,h\sigma, \tag{12}$$

where n is the radial quantum number, and m, the azimuthal one. Moreover, $P = m\hbar$. Also, since $\hbar\sigma = \mu_0 H$,

$$E = (2n + |m| + 1)\,\mu_0 H. \tag{13}$$

Pauli pointed out that the introduction of a cylinder of radius R, with its axis of revolution coincident with the Z-axis, generally made the calculation very complicated. Following Landau, he assumed that the radius of the trajectory corresponding to the energy kT was much smaller than R. With this assumption, Pauli obtained the partition function corresponding to different states as,

$$S = F\,\frac{4\pi m_0}{h^2}\,\mu_0 H \sum_{n=0}^{\infty} e^{-(2n+1)\,\mu_0 H/kT}, \tag{14}$$

where

$$F = \pi R^2, \qquad \text{and} \qquad \frac{eH}{hc} = \frac{4\pi m_0}{h^2}\,\mu_0 H.$$

The magnetic moment M is then given by $M = kT\,(\partial \log S/\partial H)$, and turns out to be

$$M = -\mu_0\left[\coth\frac{\mu_0 H}{kT} - \frac{kT}{\mu_0 H}\right]. \tag{15}$$

The Langevin factor now appeared with the minus sign, due to diamagnetism. Pauli introduced the Fermi statistics, and finally obtained the susceptibility as

$$\chi = \tfrac{3}{5}\,\frac{n\mu_0^2}{\varepsilon}. \tag{16}$$

The comparison with experiments showed that it was of the right order of magnitude.

Pauli treated the problem of the mutual interaction of atoms as a perturbation problem. He assumed that the electrons are all in the ground state and that there is one electron per atom. Using the exclusion principle and the permutability of the electrons, he gave in detail the general form of the wave function (the Slater determinant). He then introduced the Hamiltonian,

$$\mathcal{H} = \sum_k E_{\text{kin.}}^{(k)} + \sum_{j<k}\left(\frac{e^2}{r_{ik}} + \frac{e^2}{r_{a_i,\,a_k}}\right) - \sum \frac{e^2}{r_{a_i,\,k}}, \tag{17}$$

where the terms in e^2 represent the electron-electron interaction, the ion-ion interaction, and the electron-ion interaction, respectively.

On considering that the unperturbed system is given by,

$$\left[E_{\text{kin.}}^{(k)} - \frac{e^2}{r_{a_i, k}} \right] u_{a_i}(q_k) = E_{a_i}^0(q_k), \tag{18}$$

Pauli showed that, if the functions $u_{a_i}(q_k)$ are nearly orthogonal (i.e. overlap only slightly), one has

$$\mathcal{H}_{ik} = \int u_{a_i}(q_i) u_{a_k}(q_i) u_{a_i}(q_k) u_{a_k}(q_k) \times$$

$$\times \left(\frac{2e^2}{r_{ik}} + \frac{2e^2}{r_{a_i a_k}} - \frac{e^2}{r_{a_i, i}} - \frac{e^2}{r_{a_k, k}} - \frac{e^2}{r_{a_i, k}} - \frac{e^2}{r_{a_k, i}} \right) dv_i \, dv_k. \tag{19}$$

The eigenvalue problem could then be put into the form,

$$\mathcal{H}_I x(f_1, \ldots, f_r) - \sum_{i<k} \mathcal{H}_{ik} x(P_{ik} \mid f_1, \ldots, f_r) - Ex(f_1, \ldots, f_r) = 0, \tag{20}$$

where $\mathcal{H}_I$ is the non-perturbed energy increased by that part of the Coulomb interaction which corresponds to the interaction between the ions, and P_{ik} permutes the indices i and k.

Following Bloch, Pauli showed that the value of the average energy corresponding to a spin moment s was,

$$\bar{E}_s = \mathcal{H}_I - \frac{n(n-2) + s(s+1)}{n(2n-1)} \sum_{i<k} \mathcal{H}_{ik}, \tag{21}$$

a result which was first obtained by Heitler.

This method was used and generalized by Heisenberg for the case of ferromagnetism. In order to obtain the partition function, $S = \sum e^{-E/kT}$, Heisenberg had replaced E by $\bar{E}_s$, and introduced the magnetic field H. He obtained,

$$S = \sum_{s=0}^{\infty} \sum_{m=-s}^{+s} f(s) \, e^{-E_s/kT + m(2\mu_0 H/kT)}, \tag{22}$$

where $f(s)$ is the statistical weight for $\bar{E}_s$. Putting $J = \sum_{k \neq i} \mathcal{H}_{i,k}$, Pauli showed that, with some approximations,

$$\bar{E}_s = \text{const.} - J \frac{s^2}{2n}. \tag{23}$$

With $\alpha = \mu_0 H/kT$, and $\beta = J/2kT$, this gave,

$$S = \sum_{s=0}^{n} \sum_{m=-s}^{s} f(s) \, e^{2(\alpha m + \beta s^2/2n)}. \tag{24}$$

Heisenberg's model had used $\bar{E}_s$ as the approximation for the eigenvalues. Slater and Bloch later on gave a regular approximation method for the eigenvalues of the system, based upon which Pauli obtained the expressions for $\bar{M}$ in the case of two- and three-dimensional lattices. An important result of this theory was that ferro-magnetism could appear only in the case of a three-dimensional lattice.

Dirac had shown that the Hamiltonian could be put in the form,

$$\mathscr{H} = \mathscr{H}_I + \sum_{i<k} \mathscr{H}_{ik} \tfrac{1}{2}[1 + (\sigma_i, \sigma_k)], \tag{25}$$

where σ_i is the spin vector of the electron i. Pauli discussed the relation between this Hamiltonian and the Ising model.

Pauli reviewed briefly the different theories which had attempted to explain why Fe, Ni and Co are ferromagnetic while many other substances are not, but he did not consider any of them fully satisfactory. He gave an outline of Kramers' method for the paramagnetic rotation of the plane of polarization in crystals.

In the second part of his report, Pauli spoke about Dirac's relativistic quantum mechanics of the electron.

He first discussed the experiments which were supposed to give the momentum of free electrons, with the goal in view to emphasize Bohr's assertion that it was impossible to observe the spin of free electrons or protons by means of experiments based on the usual concepts of particle motion. One can show that since the magnetic moment of the electron is $\mu_0 = eh/4\pi m_0 c$, the necessary conditions [such that the effects of actions on this moment are not hidden by the Lorentz force] are favourable to the appearance of diffraction effects which forbid the observation of these actions. Pauli discussed the stopping of an electron moving along the Z-axis by a magnetic field applied in the opposite direction, the Stern–Gerlach experiment for free electrons, the compensation of the Lorentz force by an electric field, and the measurement of a magnetic field produced by an electron. By using Heisenberg's uncertainty principle, he showed that these experiments could not give the results expected classically.

Pauli discussed the Klein–Gordon equation and the difficulties raised by it for the definition of a relativistically invariant density. Dirac had solved this problem by introducing his equation,

$$\left(\sum_v \gamma^v p_v - im_0 c\right)\psi = 0, \tag{26}$$

where $\gamma^v \psi$ satisfy the relation

$$\gamma^\mu \gamma^v + \gamma^v \gamma^\mu = 2\delta_{\mu v}, \tag{27}$$

$\gamma^v \psi$ being an abbreviation for $\sum_\sigma \gamma^v_{\varrho\sigma}\psi_\sigma$, ($\varrho = 1, \ldots, 4$). He proved that $\varrho = \sum_\sigma \bar{\psi}_\sigma \psi_\sigma$ and $s_k = \sum_{\varrho\sigma} \bar{\psi}_\varrho \alpha^k_{\varrho\sigma}\psi_\sigma$ (with $\alpha^k = i\gamma^4\gamma^k$) are relativistically invariant. He showed that in the approximation of classical motion these equations could be reduced to

$$\left\{-(E' + e\phi_0) + \frac{m_0 v^2}{2} + \mu_0(\boldsymbol{\sigma}\cdot\mathbf{H}) + \tfrac{1}{2}\mu_0\left[\boldsymbol{\sigma}\left(\mathbf{E}\cdot\frac{\mathbf{v}}{c}\right)\right] - \tfrac{1}{2}\mu_0 i\left(\mathbf{E}\cdot\frac{\mathbf{v}}{c}\right)\right\}\psi = 0, \tag{28}$$

where $E' = E_0 - m_0 c^2$, and ϕ_0 is the electric potential. The third term represents the spin energy in the magnetic field, and the fourth term the spin energy in the electric field with the correct Thomas factor $\tfrac{1}{2}$.

Pauli then discussed the possibility of producing polarized electron waves, and described certain experiments in which the polarized electrons are produced and detected.

Finally, Pauli discussed the question of the negative energy states in Dirac's theory and the difficulties related to it.

DISCUSSION

In the discussion of Pauli's report, Dirac pointed out the difficulties of Landau's method which Pauli had employed. The Hamiltonian, in this method, was defined only up to a function of the field, and it was not clear how the energy of an electron could be defined in the magnetic field. Pauli showed that the form of the Hamiltonian was taken in such a way that adiabatic changes in the field would not modify the Hamiltonian.

Heisenberg remarked that, in certain crystals, the intensity of the magnetic field along the hexagonal axis was much weaker than in a direction perpendicular to it. This fact was, at first, in contradiction with Weiss' theory, but had been explained by Bloch and Gentile. It could be shown that this result was found only for relatively weak fields, i.e. fields which were comparable to the reciprocal actions of the spins.

Brillouin explained how his result for the one-dimensional problem of a periodic potential could be generalized. He gave a construction of the so-called Brillouin zones, and showed how the numbering of the plane waves could be reduced to a numbering in the first zone. He implicitly introduced the notions of energy bands and effective mass, and wrote the energy in the form $E = P_0 + (h/2m^*) k^2$, where m^* is a coefficient which could be positive or negative, and P_0 is the mean potential. For $m^* \cong m_0$, one obtains paramagnetism of the Pauli type, and for m^* large, paramagnetism of the Curie type.

Dorfman reported on some experiments which he had undertaken in order to determine whether the free or the bound electrons are responsible for ferromagnetism.

Kramers gave an account of his investigation on the explanation of the experiments of Becquerel and de Haas. He found that above $2\,°\mathrm{K}$ the magnetic rotation could be expressed by $\varrho = \varrho_\infty(\lambda, T)$ $\times th(n\mu_B H/kT)$.

Concerning the impossibility of defining experimentally the magnetic moment of a free electron, Bohr remarked: 'I should emphasize that the direct inobservability of the intrinsic magnetic moment of the electron does not imply that the concept of spin has lost its significance as a means of explaining the fine structure of spectral lines and the polarization phenomena of electron waves. Only the manner in which the concept of spin appears in the formalism of quantum mechanics is such that it does not lend itself to an independent interpretation based on classical notions.'

Pauli's report gave rise to a long discussion, and others who took part in it were Darwin, Dirac, Pauli, Kapitza, Debye, Cotton, Heisenberg, Weiss, Fermi, Richardson, and Van Vleck.

REFERENCE

1. W. Pauli, Z. Phys. **41**, 81 (1926).

7. THE EQUATION OF STATE OF FERROMAGNETS

(P. Weiss)

The analogy between magnetization and fluids had already been noticed by P. Curie. Weiss thought that the paramagnets were to ferromagnets what ideal gases were to fluids of high density, and he regarded the theory of the molecular field as a development of this idea. This theory takes into account paramagnetism and the mutual actions of the elements causing a moment, and Weiss examined the two cases separately.

Weiss defined, in analogy with perfect gases, that a substance is *purely paramagnetic* if its internal energy, U, is a function of T, the temperature, only. If σ is the magnetization, then $\partial U/\partial\sigma = 0$, and, as Langevin had shown, the equation of state of such a substance should be of the form

$$\sigma = f\left(\frac{H}{T}\right), \tag{1}$$

i.e., a function of H/T only. The first order expansion implies, $\sigma = c(H/T)$, i.e. Curie's law. Weiss introduced two new quantities,

$$H_m = -\frac{\partial U}{\partial\sigma}, \tag{2}$$

where H_m is the molecular field, and

$$\sigma = f\left(\frac{H + h_m}{T}\right), \tag{3}$$

h_m being the correction to the molecular field in the equation of state for substances which are not purely paramagnetic. He put $h_m = n\sigma$, n being a constant.

With the help of some thermodynamical relations, Weiss showed that $H_m = h_m$ if both are independent of T. Also, the magnetic energy then takes the form

$$U_m = -n\frac{\sigma^2}{2}. \tag{4}$$

Using Langevin's function, he finally deduced the existence of the Curie temperature, θ, and gave the general law,

$$\sigma = \frac{C}{T - \theta} H, \tag{5}$$

or the coefficient of magnetization,

$$\chi = \frac{T - \theta}{C} = \frac{T}{C} - n. \tag{6}$$

Weiss discussed the thermodynamic properties of the molecular field. The linear relation between T and χ had been observed for large intervals of temperature, for generally small values of σ. This leads to $h_m = n\sigma$, so that if the molecular field is a function of temperature, this must happen for higher order terms only. Weiss showed that if $h_m = \phi(\sigma)$, i.e. if $\partial H_m/\partial T = 0$, then for $\sigma = $ const., the relation between H and T given by the equation

$$\frac{1}{T}\frac{\partial H_m}{\partial T} = \frac{\partial^2 H}{\partial T^2} \tag{7}$$

is a straight line. This was, however, not in agreement with experiments, and Weiss explained this discrepancy by taking into account the variability of the moments with temperature. Weiss noted the existence of an anomaly in the specific heat of ferromagnets, and pointed out that the excess of energy was not equal to the magnetic energy at absolute zero, as had been generally assumed.

Weiss remarked that for the large region in which the Curie law is valid, the experimental knowledge consisted of the knowledge about the Curie constant only. He reviewed the different theoretical calculations of the Curie constant C, and discussed how the knowledge of the curve giving spontaneous magnetization at all temperatures could give σ/σ_0 and the value $a=\sigma_0 H/RT$. Moreover, σ, as a function of T, could, in general, be expressed as

$$\sigma = \sigma_0(1 - AT^2 - BT^4 - \cdots). \tag{8}$$

Weiss examined the saturation law of ferromagnetics as a function of temperature. Far from the Curie point, where the paramagnetic susceptibility is negligible, one does not observe spontaneous magnetization, but only an asymptotic approach toward it. Weiss thought that this phenomenon must arise from an interaction between elementary magnetic moments, and must be of an interatomic nature.

8. THE VOLUME ANOMALY OF FERROMAGNETS

(P. Weiss)

Historically, attention to the exceptional variations of the volume of ferromagnets had been drawn by the discovery of invar (a ferronickel alloy, Fe_2Ni) by Guillaume. At ordinary temperatures, its coefficient of expansion is twenty times smaller than that of pure iron. Weiss pointed out that this behaviour was common to all the ferromagnets, and was not just a consequence of a change in the lattice structure. Weiss reported on some experimental results, especially the work of Chevenard, exhibiting this anomaly. He then discussed Bauer's theory of the volume anomaly of ferromagnets.

From the free energy, $\psi = U - TS$, one has the relation $p = \partial\psi/\partial V$, where p is the pressure which must be applied to the substance in such a way that, at a given temperature, the volume V takes on a certain value. The free energy could be calculated from the molecular field hypothesis, i.e.

$$H_m = -\frac{\partial U}{\partial I} = NI, \tag{1}$$

where N is the constant of the molecular field and I is the magnetization. Weiss wrote, $\psi = \psi_0 + \psi_m$, where ψ_0 is the free energy due to the non-ferromagnetic part, and ψ_m is the free energy of magnetic origin, and pointed out that the volume anomaly was due to ψ_m. He found,

$$\frac{\Delta V}{V} = \frac{k}{2} NI^2, \tag{2}$$

where k is the compressibility coefficient. This showed that $\Delta V/V$ is always negative, a fact which was in contradiction with experimental results already for invar.

Weiss calculated the work done in expansion, and found it to be only a small fraction of the magnetic energy. He concluded that magnetic phenomena are not affected by this difference in thermal expansion. Weiss speculated about the various possible causes of the insufficiency of the theory.

9. GYROMAGNETIC PHENOMENA

(P. Weiss)

Weiss pointed out that there are two gyromagnetic phenomena, one inverse of the other: the rotation of the body of a substance by magnetization, and the magnetization produced by rotation. They allow one to determine the ratio of the angular momentum to the magnetic moment [of elementary magnets], thus providing information about magnetization that is independent of other sources, such as magnetic measurements and spectral analysis.

Weiss recalled some historical facts: Richardson, in 1908, had explained the rotation by magnetization as a consequence of the properties of electrons; the experiment, yielding a positive result, was performed in 1915 by Einstein and de Haas. De Haas gave a report on this subject at the third Solvay Conference in 1921. The inverse phenomenon, i.e. magnetization by rotation, was observed by Perry in 1890. It was deduced in 1909 by S. J. Barnett, who observed this phenomenon in 1914, and published it in 1915, the same year as the publication of the work by Einstein and de Haas.

As discovered by Einstein, the ratio of the angular momentum M to the magnetic moment μ is independent of any assumption about the orbit, and equal to $2m/e$. According to the Uhlenbeck and Goudsmit theory of the spinning electron, this ratio is one-half of that given by Einstein. For any spectroscopic state, in general, the ratio of the angular momentum to the magnetic moment is given by

$$\frac{1}{g}\frac{2m}{e},$$

where g is the Landé factor.

The principle of gyromagnetic measurements can be expressed by the mechanical law

$$\int C\, dt = I\, \Delta\omega + \Delta M, \tag{1}$$

where C is the moment due to external forces acting on the body, I is its moment of inertia, ω its angular velocity, and $I\omega$ its angular momentum; M is the momentum due to the carriers of magnetic moments. If all the carriers of magnetic moments are of the same type, one has

$$\int C\, dt = I\, \Delta\omega + \frac{2m}{e}\, g\, \Delta\mu. \tag{2}$$

As shown by Barnett, this equation reduces to

$$I\, \Delta\omega + \frac{2m}{e}\, g\, \Delta\mu = 0. \tag{3}$$

In Barnett's theory each carrier of magnetic moment had been considered as a gyroscope of negative charge. With this assumption, Barnett had derived the relations,

$$H\mu = M\omega \quad \text{or} \quad H = R\omega, \qquad (4)$$

where H is the magnetic field, μ the magnetic moment, and R the gyromagnetic ratio.

Weiss analyzed the results, leading to the conclusion that the paramagnetic and ferromagnetic properties were due to the spinning electron. He discussed some experiments on Dy_2O_3 for the determination of the Landé factor, showing good agreement with the theory.

DISCUSSION

The discussion of the reports given by Weiss first dealt with the problem of the specific heat of ferromagnets. In the case of Ni, the example discussed by Weiss, there was a point of discontinuity in the first derivative when the part due to the molecular field was removed. Weiss noted that the effect of residual ferromagnetism on the curve was not known.

Einstein wondered whether the discontinuity in the specific heat arose from a quantum degeneracy. Weiss thought that the drop in the specific heat at the Curie point would not be instantaneous as required by the molecular field theory.

Kapitza pointed out that gases dissolved in metals could completely change their magnetic properties, but Weiss thought that the effect of impurities could be reduced to less than one in one thousand.

Heisenberg pointed out that there did not yet exist a satisfactory calculation of the specific heat of ferromagnetic substances on the basis of quantum theory. The energy of a ferromagnetic substance seemed to depend not only on its magnetization but also explicitly on its temperature.

Pauli expressed his doubts whether in paramagnetic substances, at very low temperatures, the variation was according to $T^{3/2}$ or T^2. He thought that a more realistic variation, which rendered the high temperature behaviour more natural, was $aT^{3/2}+bT^{5/2}$.

Bauer gave a brief report on his theory of the anomalous expansion in ferromagnets. He pointed out that Weiss' molecular field theory leads to an anomaly of the thermal expansion. Bauer's assumption was to admit that the constant n of the molecular field depends on the specific volume v, and he obtained a general formula for the coefficient of expansion. The more general case of the expansion anomalies in relation to magnetostriction and compressibility at the Curie point had been treated by Verschaffelt.

10. SUPER-DIELECTRIC SUBSTANCES

(J. Dorfman)

Dorfman described the behaviour of 'super-dielectric' substances, of which he treated the salt of 'Seignette' (double tartarate of sodium and of potassium hydrate) as an example. Its behaviour was analogous to ferromagnets, but its dielectric constant was found to be a function of the intensity of the electric field. The super-dielectric property of the 'Seignette' salt diminished as the temperature was increased.

11. CONSTANT MAGNETIC FIELDS

(A. Cotton)

Cotton first described the large electromagnet at Bellevue, and gave the technical details about its construction and characteristics. This electromagnet seemed to represent the limit of what was then technically feasible, providing a magnetic field of 70000 G for a large surface. However, by the addition of elements a stronger field could be obtained.

Cotton described the different experiments which had been performed with this large electromagnet, and those that were intended. All of these experiments were made possible because of the availability of a large uniform magnetic field. For instance, Rosenblum performed very precise α-ray spectroscopy and was able to separate the different rays of thorium C. He also obtained the corresponding energies with great accuracy, using a field of 36000 G. Rutherford had obtained similar results by a completely different method based on the lengths of the trajectories in air.

Cotton discussed another phenomenon, the magnetic birefringence of pure liquids. The phase difference measuring the birefringence is proportional to the length crossed by light orthogonally to the lines of force of the field. He pointed out that optical studies of a liquid in electrostatic and magnetic fields should give interesting information on the symmetries of molecules.

The Faraday rotation is proportional to the magnetic potential difference, $\int H \, dl$, between the extremities of the tube containing the substance. In this case also, the existence of strong magnetic fields was important.

Cotton also discussed the Zeeman effect experiments. The previous measurements had never been made with fields stronger than 35000 G, and his magnet at Bellevue offered new experimental possibilities.

Among the experiments which could benefit from the powerful new magnet, Cotton noted, were the action of a magnetic field on light sources, the studies of magnetic properties of crystals, and the influence of a magnetic field on the dielectric properties of metals.

Following Cotton's report, Madame Curie commented on the advantages of a large magnetic field for the resolution of spectra and, in general, for studies in radioactivity. Cotton, Kapitza, de Haas, Zeeman, Piccard, Kramers, Debye and Madame Curie, discussed the technical aspects of the magnet.

12. INTENSE MAGNETIC FIELDS

(P. Kapitza)

Kapitza observed that by using classical techniques for obtaining magnetic fields, such as the ferrocobalts by Weiss or the electromagnet by Cotton, the limit of 100 kG could hardly be surpassed. This limitation could be overcome if the electromagnet were used for a very short time only, providing a gain in the strength of the field at the expense of duration. The sacrifice of the duration of the magnetic field would introduce some experimental difficulties, but in many cases they could be overcome. The main difficulty was the production of a strong field itself during a very short time.

Kapitza described the different methods by which he had tried to solve the problem. The magnetic field is given by $H = k \sqrt{(W/a\varrho)}$, where W is the power supplied, a the radius of the coil, ϱ the resistance of the conductor, and k a coefficient depending on the form of the coil. The production of heat was a serious problem. If a coil of

1 cm radius, supplied with 4×10^4 kW for 1 sec, developing a field of 10^6 G, were heated to $10^4\,°C$, the reduction of the duration to 10^{-2} sec would raise the temperature to only $10^2\,°C$.

In order to produce such large power for a short duration, Kapitza described three types of apparatus using electrical, chemical and mechanical means for the storage of electricity. The first employed a very large condenser. With it the greatest difficulty was a practical one, because such a condenser would have to be extremely large and would be charged to a very high voltage, giving rise to the problem of isolating it.

A practical procedure was realized by using storage batteries. With 2 to 3×10^3 kW, Kapitza obtained a field of the order of 125 kG, which could be used for studies of the Zeeman effect and the deviation of α-particles in a Wilson cloud chamber.

The difficulty of switching off the current in a very short time was such that it was not possible to increase the field any more. Kapitza overcame it by using a mechanical storage of electricity. He employed turbo-alternators with a massive rotor, using the large mass to store a large quantity of kinetic energy. With these improvements Kapitza was able to obtain fields of the order of 320 kG.

Kapitza described the technique of measuring the field, and then discussed the application of these high magnetic fields. The first studies were made on the change of the resistivity of metals in a magnetic field. It was found, in a study of 35 metals at low temperatures, that, with the exception of ferromagnets, the change of resistance was proportional to the square of the field for weak fields, but had a linear variation in more intense fields.

Kapitza also employed strong fields for the measurement of magnetic susceptibility. He described the apparatus he had employed for the measurement of magneto-striction, with which he also expected to study the deformations produced at the atomic level.

The discussion of Kapitza's report consisted of questions concerning experimental techniques and measurements. Among those who took part in it were Cotton, Kramers, Madame Curie, Gerlach, Verschaffelt, Brillouin, de Haas, Piccard, Heisenberg, and Sommerfeld.

13. THREE COMMUNICATIONS

(Th. De Donder)

De Donder gave a report on three problems, dealing with them in a very formal manner: (1) The interpretation of Planck's constant h and the Bohr magneton μ by means of Einstein's theory of gravitation; (2) Electro- and magneto-striction; (3) The generalization of the Dirac equation.

8

The Structure and Properties of Atomic Nuclei*

1. INTRODUCTION

The seventh Solvay Conference on Physics took place in Brussels from 22 to 29 October 1933. Its general theme was the 'Structure and Properties of Atomic Nuclei'. This was the last conference to take place before World War II. A conference on elementary particles and their mutual interactions had been planned for 1939, but world events intervened and it had to be cancelled.

Paul Langevin was President of the Scientific Committee whose members were: N. Bohr (Copenhagen), B. Cabrera (Madrid), P. Debye (Leipzig), Th. De Donder (Brussels), A. Einstein (Le Coq-sur-mer, Belgium), Ch. E. Guye (Geneva), A. Joffé (Leningrad), O. W. Richardson (London), and J. E. Verschaffelt (Ghent), Secretary. Einstein and Guye did not attend the Conference.

Among the invited participants were: P. M. S. Blackett (London), W. Bothe (Heidelberg), J. Chadwick (Cambridge), J. D. Cockcroft (Cambridge), Madame M. Curie (Paris), L. de Broglie (Paris), M. de Broglie (Paris), P. A. M. Dirac (Cambridge), C. D. Ellis (Cambridge), E. Fermi (Rome), G. Gamow (Leningrad), W. Heisenberg (Leipzig), F. Joliot (Paris), Madame I. Joliot-Curie (Paris), E. O. Lawrence (Berkeley, California, U.S.A.), L. Meitner (Berlin), N. F. Mott (Bristol), W. Pauli (Zurich), R. Peierls (Cambridge), F. Perrin (Paris), M. S. Rosenblum (Paris), Lord Rutherford (Cambridge), and E. Schrödinger (Berlin).

The Scientific Secretaries of the Conference were: J. E. Verschaffelt (Ghent), E. Bauer (Paris), H. A. Kramers (Utrecht), L. Rosenfeld (Liège), E. T. S. Walton (Cambridge). The Scientific Committee also invited the following to attend the Conference: E. Henriot, A. Piccard, E. Stahel, M. Cosyns, and E. Herzen, all from Brussels.

The seventh Solvay Conference, devoted to the structure and properties of atomic nuclei, took place at a time when this field was undergoing extremely rapid development. Cockcroft presented the first report at the conference. After briefly referring to the abundant evidence about nuclear disintegrations by the impact of α-particles, which had been obtained in the preceding years by Rutherford and his collaborators, Cockcroft described in detail the important new results obtained by the bombardment of nuclei with protons accelerated to great velocities with the appropriate high voltage equipment (then available).

---

* *Structure et Propriétés des Noyaux Atomiques*, Rapports et Discussions du Septième Conseil de Physique tenu à Bruxelles du 22 au 29 Octobre 1933, Gauthier-Villars, Paris, 1934.

SEVENTH SOLVAY CONFERENCE 1933

Sitting (left to right): E. Schrödinger, Madame I. Joliot, N. Bohr, A. Joffé, Madame Curie, P. Langevin, O. W. Richardson, Lord Rutherford, Th. De Donder, M. de Broglie, L. de Broglie, Lise Meitner, and J. Chadwick.

Standing (left to right): E. Henriot, F. Perrin, F. Joliot, W. Heisenberg, H. A. Kramers, E. Stahel, E. Fermi, E. T. S. Walton, P. A. M. Dirac, P. Debye, N. F. Mott, B. Cabrera, G. Gamow, W. Bothe, P. M. S. Blackett (at back), M. S. Rosenblum, J. Errera, E. Bauer, W. Pauli, J. E. Verschaffelt, M. Cosyns (at back), E. Herzen, J. D. Cockcroft, C. D. Ellis, R. Peierls, A. Piccard, E. O. Lawrence, and L. Rosenfeld.

Cockcroft and Walton's experiments on the production of high speed α-particles by the impact of protons on lithium nuclei had given the first direct experimental verification of Einstein's formula for the general relation between energy and mass, $E = mc^2$, and in the following years this formula served as a guide in nuclear research. Cockcroft also described how closely the measurements of the variations of the cross-section for the process with proton velocity confirmed the predictions of wave mechanics, to which Gamow and others had been led in connection with the theory of spontaneous α-decay [the tunnel effect]. Cockcroft also presented the evidence then available regarding artificial nuclear disintegrations, and compared the results of experiments in Cambridge with proton bombardment with those just obtained in Berkeley with deuterons accelerated in Lawrence's newly constructed cyclotron.

Another recent progress of the utmost significance was Chadwick's discovery of the neutron. It represented a really dramatic development, resulting in the confirmation of Rutherford's anticipation of a heavy neutral constituent of atomic nuclei. Rutherford[1] had first expressed this conjecture in his Bakerian Lecture on the 'Nuclear Constitution of Atoms', delivered on 3 June 1920, and again at the third Solvay Conference in spring 1921. Chadwick began his report with a description of the search in Cambridge for the anomalies in α-particle scattering, and discussed some of the most pertinent considerations regarding the part played by the neutron in nuclear structure, as well as the important role it played in inducing nuclear transformations.

The participants in the seventh Solvay Conference learned about another decisive discovery, namely artificial radioactivity, produced by controlled nuclear disintegrations. This discovery was made only a few months before the conference, and Frédéric Joliot and Irène Curie gave an account of it in their report. They gave a survey of many aspects of their fruitful researches, in which the processes of β-decay with the emission of positive as well as negative electrons had been ascertained. In the discussion following their report, Blackett told the story of the discovery of the positron by Anderson and himself in cosmic ray researches, and its interpretation in terms of Dirac's relativistic electron theory. A new stage had thus been reached in the development of quantum physics; it concerned the creation and annihilation of material particles analogous to the processes of emission and absorption of radiation in which photons are created and disappear.

Many features of radioactive processes were discussed at the seventh Solvay Conference. Gamow gave a report on the interpretation of γ-ray spectra, based on his theory of spontaneous and induced α-particle and proton emission and their relation to the fine structure in α-ray spectra. The important problem of continuous β-ray spectra also came under discussion, especially Ellis' investigations of the thermal effects produced by absorption of the emitted electrons, and it seemed irreconcilable with the detailed energy and momentum balance in the β-decay process. The evidence based on the spins of nuclei involved in the process seemed contradictory to the convervation of angular momentum. In order to avoid such difficulties, Pauli introduced the idea of the neutrino, a particle of zero rest mass and spin one-half, which is emitted in β-decay together with the electrons. This concept became the basis of a most fruitful development in elementary particle physics.

Heisenberg, in his report, dealt with the question of the structure and stability of atomic nuclei. He had recognized, on the basis of the uncertainty principle, the difficulties of the presence of electrons within atomic nuclei. Heisenberg immediately grasped the discovery of the neutron as the foundation for the new view of nuclear structure, with neutrons and protons as the proper nuclear constituents. On this basis he developed the explanation of many properties of nuclei. Heisenberg's conception implied that the phenomenon of β-decay should be considered as evidence of the creation of positive or negative electrons and neutrinos with the release of energy in the accompanying change of a neutron to a proton, and vice versa.

Soon after the seventh Solvay Conference, Fermi achieved great progress by developing his theory of β-decay, which became the starting point of subsequent important developments in the theory of weak interactions.

The early 1930s marked the beginning of the new developments in modern nuclear physics. With the concepts of proton, electron, neutron, positron, and neutrino, already formulated by 1932, and Yukawa's conception of the meson available in 1935, nuclear physics set out on the road of extraordinary conceptual and technological advance. Some of the excitement of this adventure was present at the seventh Solvay Conference, in which Rutherford took part and was a central figure. This was the last Solvay Conference he attended before his death in 1937.

We shall now give the résumés of individual reports presented at the conference.

REFERENCE

1. E. Rutherford, 'Nuclear Constitution of Atoms', *Proc. Roy. Soc.* **A 97**, 374–400 (1920).

2. DISINTEGRATION OF ELEMENTS BY ACCELERATED PROTONS

(J. D. Cockcroft)

Cockcroft discussed the earlier experiments of Rutherford, of Rutherford and Chadwick, and of Blackett with a Wilson chamber, in which α-particles were absorbed by light elements, leading to the emission of protons. The energy required for such an emission is of the order of 3 MeV. Cockcroft pointed out that although this energy is necessary, classically speaking, for a particle to pass through the potential barrier, the situation is quite different in wave mechanics. In the latter case, one has the tunnel effect, the derivation of which was given by Gamow, which allows a finite fraction of incident particles, of relatively small energy, to penetrate inside the atom.

Gamow's formula for the probability for a particle of charge $Z'e$, mass m and energy E, to penetrate a nucleus of atomic number Z and radius r_0 is given by,

$$W = \exp\left[-\frac{e^2}{\hbar} \sqrt{2m} \frac{ZZ'}{E} (2U_0 - \sin 2U_0) \right],$$ (1)

where $U_0 = \cos^{-1} \sqrt{(r_0 E / Z E^2)}$ and $\hbar = h/2\pi$. The probability W would be larger for a proton than for an α-particle. The energy of the protons had to be between 10^2 to 6×10^2 keV to penetrate the nuclei of a large number of light elements.

Cockcroft then turned to the detailed technical problem of the production of high voltages for the acceleration of particles. He described the various types of equipment used in Cambridge, as well as the Van de Graaff machine and the apparatus using Tesla induction coils. Cockcroft gave a detailed account of the production and acceleration of protons. He concluded this part of his report on the experimental setup by comparing the different arrangements, their respective advantages and disadvantages, and the domains of their applicability.

Cockcroft gave an account of his investigation of the disintegration of lithium, the technique employed being the stopping of the emitted particles in air or in a thin film of mica. He was able to confirm the reaction

$$\tfrac{7}{3}\text{Li} + \tfrac{1}{1}\text{H} = \tfrac{4}{2}\text{He} + \tfrac{4}{2}\text{He}, \tag{2}$$

the two α-particles being emitted in opposite directions as required by the law of momentum conservation. He gave an account of the work of P. I. Dee and E. T. S. Walton, who had arrived at the same result with this reaction, using a Wilson cloud chamber.

It was important to perform an exact counting of the emitted particles in order to verify Gamow's law. The agreement was better than expected. Cockcroft also gave the results of Rutherford and Oliphant on the disintegration of lithium and boron.

Cockcroft had examined the disintegrations of fluorine, beryllium and carbon, and the following reactions were found to exist:

$$\tfrac{19}{9}\text{F} + \tfrac{1}{1}\text{H} = \tfrac{16}{8}\text{O} + \tfrac{4}{2}\text{He} \tag{3a}$$

$$\tfrac{9}{4}\text{Be} + \tfrac{1}{1}\text{H} = \tfrac{6}{3}\text{Li} + \tfrac{4}{2}\text{He} \tag{3b}$$

$$\tfrac{12}{6}\text{C} + \tfrac{1}{1}\text{H} = \tfrac{9}{5}\text{B} + \tfrac{4}{2}\text{He}. \tag{3c}$$

Cockcroft and Walton had also performed experiments on heavier elements, as well as on the disintegration of elements by using deuterons. He reported the existence of the following reactions:

$$\tfrac{9}{4}\text{Be} + \tfrac{2}{1}\text{He} = \tfrac{7}{3}\text{Li} + \tfrac{4}{2}\text{He} \tag{4a}$$

$$\tfrac{14}{7}\text{N} + \tfrac{2}{1}\text{H} = \tfrac{12}{6}\text{C} + \tfrac{4}{2}\text{He} \tag{4b}$$

$$\tfrac{12}{6}\text{C} + \tfrac{2}{1}\text{H} = \tfrac{13}{6}\text{C} + \tfrac{1}{1}\text{H}. \tag{4c}$$

DISCUSSION

The discussion of Cockcroft's report was opened by Rutherford. He expressed great pleasure at the development of what he called 'modern alchemy', and mentioned some interesting new results which he and Oliphant had just obtained by the bombardment of lithium with protons and deuterons. These experiments yielded evidence about the existence of hitherto unknown isotopes of hydrogen and helium with atomic mass 3.

Lawrence described in detail the construction of his cyclotron, and gave an account of the latest investigations performed at Berkeley.

Madame Curie thought that the reaction

$$_3^7\text{Li} + {_1^1}\text{H} = 2\,{_2^4}\text{He}$$

was the first experiment which had provided the verification of Einstein's mass-energy relation.

Others who took part in the discussion were Heisenberg, Bohr, M. de Broglie, Langevin, Gamow, F. Perrin, Joliot, Walton, Meitner, and Chadwick.

3. ANOMALOUS SCATTERING OF α-PARTICLES; TRANS-MUTATION OF ELEMENTS BY α-PARTICLES; THE NEUTRON

(J. Chadwick)

3.1. ANOMALOUS SCATTERING

Chadwick discussed the extent to which the interaction between an α-particle and an atomic nucleus is given by Coulomb's law, since both quantum and classical mechanics lead to the same scattering formula. In order to explain the anomalous scattering of fast α-particles by light elements, it is necessary not only to use quantum mechanics but to admit, as Bieler[1] had done, that below a certain limit, r_0, the interaction is no longer given by Coulomb's law but by a very strong attractive force. The interaction remains of the Coulomb type, so long as $2Ze^2/\frac{1}{2}MV^2 > r_0$, for α-particles of velocity 2×10^9 cm/sec, and elements for which $Z > 29$.

Chadwick pointed out that the variation of the observed scattering from the Coulomb type was very sensitive to the scattering angle and the speed of the incident particles. Since the de Broglie wavelength is much smaller for fast particles than the corresponding radius of the nucleus, classical mechanics is no longer applicable in this domain.

Chadwick used the decomposition of the plane wave into spherical harmonics to obtain an estimate of the scattering. He explained that the collision parameter p_l, for each element of the decomposition, was of the form $p_l = l\hbar/MV$, where M and V are the mass and speed of the α-particles. The scattering anomaly was very large when all the l's, from zero to a large value of l, gave a contribution to the scattering. Chadwick showed that, for aluminium, only $l = 0$ and $l = 1$ were necessary to explain the experimental results. He also gave the results for boron, carbon and beryllium.

REFERENCE

1. Bieler, *Proc. Camb. Phil. Soc.* **21**, 686 (1923); *Proc. Roy. Soc.* **A 105**, 34 (1924).

3.2. TRANSMUTATION BY α-PARTICLES

Chadwick reported on investigations of the transmutations with α-particles. The incident α-particle is absorbed at different possible energy levels within the nucleus, the higher levels corresponding to the excited states. The particles return to the ground state through the emission of γ-rays. This picture of the process was confirmed experimentally. Chadwick reported that for thick plates a resonance effect was observed, the absorption being much stronger for certain precise values of the energy of α-particles.

Finally he gave a picture of the energy states in the nucleus, in as far as it could be described on the basis of experimental results. In the case of aluminium he had found that the particle could be at four different levels corresponding to energies from 1 to 6 MeV.

3.3. THE NEUTRON

Chadwick had reported on the 'possible existence of the neutron' in February 1932 in *Nature*[1], and he gave a complete account of his discovery in a paper on 'the existence of the neutron' which was received by the *Proceedings of the Royal Society* on 10 May 1932.[2]

In his report to the seventh Solvay Conference, Chadwick gave a brief account of how the neutrons had been 'observed' and the reactions through which their mass had been determined. The neutron, at that time, was certainly the only particle without electric charge, and it had therefore to be detected through an indirect procedure, such as the ionization of recoiling particles. The neutrons are not emitted by all the elements which are bombarded by beams of α-particles, and Chadwick assumed that this difference between elements was due to certain general rules concerning the nuclear structure.

The determination of the mass of the neutron was a difficult procedure, but it had been possible to show by means of detailed energy balance that the mass of the neutron was about the same as the proton mass.

Chadwick had found that the interaction between the neutron and the nucleus was very weak except at a very short distance, less than 10^{-12} cm, and that the scattering of neutrons was essentially the case of a collision between particles.

Chadwick discussed Massey's theory of the collisions between neutrons and atomic nuclei. The neutron was considered as a hydrogen atom in a 'zero' quantum state, the potential being given by

$$V(r) = e^2 \left(\frac{1}{r} + \frac{Z}{a_0} \right) e^{-2Zr/a_0}, \tag{1}$$

where Z is the effective nuclear charge, supposed to be large, a_0 is the radius of the first Bohr orbit in hydrogen, and a_0/Z is the 'radius' of the neutron. From experiments on the scattering of neutrons by heavy elements, Massey had deduced the maximum neutron 'radius' to be 2×10^{-13} cm. However, this model was found to be in contradiction with the experimental results on proton-neutron collisions.

The study of the collisions between neutrons and protons was considered to be most important for the knowledge of nuclear structure. It had been found experimentally that, in the centre of mass system, the collisions were approximately spherically symmetric. The collision radius was found to be of the order of 8×10^{-13} cm. The wave mechanical theory of collisions between two independent particles gave the collision cross-section as

$$Q = \frac{h^2}{\pi M^2 v^2} \sum_n (2n + 1) \sin^2 \delta_n, \tag{2a}$$

where M is the relative mass of the system, v the initial speed, and δ_n are constant phases depending on M, v, and the interaction energy, $V(r)$, between the particles. Moreover, the angular distribution in a coordinate system, in which the centre of mass is at rest, is given by

$$I(\theta) \sin \theta = \frac{h^2}{8\pi^2 M^2 v^2} \left| \sum_n (e^{2i\delta_n} - 1)(2n + 1) P_n(\cos \theta) \right|^2 \sin \theta, \qquad (2b)$$

where

$$P_0(\cos \theta) = 1, \qquad P_2(\cos \theta) = \cos \theta, \quad \text{etc.}$$

The experimental results on proton-neutron collisions had showed that only P_0 was significant in this case. Chadwick used a binding energy between neutron and proton of the order of 10^6 eV to calculate Q, the collision cross-section, and found $Q \sim 2 \times 10^{-23} \cdot v^{-2}$, where v is measured in units of 10^9 cm/sec.

Chadwick then discussed the disintegration by neutrons. It had been found that many experiments, in which a neutron was emitted by using α-particles, could now be reversed. For instance, the following reactions were observed:

$$^{14}_{7}\text{N} + ^{1}_{0}n = ^{11}_{5}\text{B} + ^{4}_{2}\text{He} \qquad (3a)$$

$$^{16}_{8}\text{O} + ^{1}_{0}n = ^{13}_{6}\text{C} + ^{4}_{2}\text{He} \qquad (3b)$$

$$^{20}_{10}\text{Ne} + ^{1}_{0}n = ^{17}_{8}\text{O} + ^{4}_{2}\text{He}. \qquad (3c)$$

Chadwick concluded his report by giving the energy balance in these disintegrations.

DISCUSSION

Mott outlined the theory of the scattering of α-particles in hydrogen. It was based on the idea that the dimensions of the nucleus are smaller than the wavelength of the proton; thus the scattering resulting from the nuclear potential is independent of the scattering angle. For the scattered intensity of protons, of velocity V in a direction θ, he found the expression $|A + Ze^2(2mv^2)^{-1} \operatorname{cosec}^2\tfrac{1}{2}\theta|^2$, where A is the constant amplitude of scattering due to nuclear forces.

Peierls pointed out that this theory was in contradiction with the experimental results of Chadwick. However, good agreement was obtained if one took into account Delbrück's assumption, that is, including a repulsion of the nucleus at very short distance in Gamow's potential.

Others who took part in the discussion of Chadwick's report included Heisenberg, Bothe, Irène Joliot-Curie, Lise Meitner, Gamow, Langevin, Madame Curie, Lawrence, Ellis, Fermi, F. Perrin, Cockcroft, and Rutherford.

REFERENCES

1. J. Chadwick, *Nature* (27 February 1932).
2. J. Chadwick, *Proc. Roy. Soc.* **A 136**, 692–708 (1932).

4. PENETRATING RADIATION FROM ATOMS UNDER THE ACTION OF α-RAYS

(F. Joliot and I. Joliot-Curie)

In 1930 Bothe and Becker discovered a new type of transformation: they found that certain light elements, under the action of α-rays, emitted a highly penetrating radiation of nuclear origin, but of feeble intensity.[1] They concluded that this radiation was of electromagnetic nature. This very weak effect was observed with Be, B, Li, F, Al, Na and Mg. The same type of experiment was also performed by Webster in the Cavendish laboratory, and he was able to show the existence of a dissymmetry with respect to the direction of the incident α-particles.

The Joliots performed experiments with a very intense source of α-rays, and observed that the new radiation was very penetrating and could expel protons and He-nuclei of very high energy from a paraffin target. Another important fact was the absence of ionization in the beam from the source, which made them think that this radiation probably consisted of high energy γ-rays. After several fruitless efforts at interpreting this radiation as γ-rays, they followed Chadwick's assumption according to which the source of this radiation were neutrons, particles of mass one and charge zero. With this assumption, the experiments could be easily explained, and are characterized by two types of phenomena: first, the radiation due to neutrons, which is dissymmetrical; and second, the γ-radiation, leading to fast electron emissions.

The Joliots gave a description of the different observed properties of neutrons. The expulsion of protons from a paraffin target had given the first experimental proof of the existence of neutrons. The Joliots gave an account of the different methods used for the detection of neutrons, the main difficulty being the separation of γ-rays from the neutrons, both of which showed similar behaviour.

The Joliots discussed the analogy which existed between the intensities of the γ-rays and the neutrons emitted from a given substance. They considered the case of radiation from beryllium when irradiated with α-particles from polonium; two types of radiation were found to be present, neutrons of 4.5 and 7.8 MeV and γ-rays of 5 MeV. They showed figures giving the variations of the neutron and γ-excitations as a function of the energy of the α-particles, and observed that the corresponding curves for neutrons and γ-rays were very similar, presenting the same extrema. The important difference was the fact that the neutron and γ-emissions did not begin at the same energy; the γ-emission began at a higher energy. The Joliots discussed the difficulties of a correct interpretation of the nuclear reaction

$$\ce{^9_4Be} + \ce{^4_2He} = \ce{^{12}_6C} + \ce{^1_0}n \nearrow + \gamma \nearrow . \tag{1}$$

Similarly they studied the reactions involving the irradiation by α-particles of B, Li, F, Na, Al, Mg and N, and concluded that in spite of certain difficulties of the interpretation of the observed radiation, their work had in general verified the formulas of the nuclear reactions involved.

The Joliots discussed the materialization and transmutation of electrons. For instance, the experiments of Anderson[2], Blackett and Occhialini[3], had shown the existence of positrons. The Joliots reported on experiments involving the irradiation of different materials by a large source of neutrons and γ-rays, which had shown that the number of positive electrons increases with the atomic number. By certain absorption measurements they were able to determine the nature of the positrons, and to conclude that the latter came from γ-rays only.

Meitner and Phillip, Anderson and Neddermeyer, and the Joliots, had independently observed the creation of pairs of electrons, having one positive and one negative electron, the positive one generally having the higher energy. Since the total energy observed in such a reaction was about 1.6 MeV, the Joliots proposed the following interpretation: a photon of large energy $h\nu$, on encountering a heavy nucleus, would transform itself into two electrons of opposite signs. The creation of two electrons requires an energy of 1.02 MeV (assuming that the rest masses of both the positive and the negative electron are the same). The surplus energy of the quantum, $h\nu - 1.02$ MeV, then appears as the kinetic energy, W, transmitted to the two electrons and, in certain cases, may also appear as the energy $h\nu'$ of a scattered quantum.

The Joliots pointed out that the presence of matter was essential in the foregoing reaction, although it does not participate in the reaction. They thought that this phenomenon (pair production) could explain the anomalous absorption of γ-rays above 10^6 eV, because in this region the Klein–Nishina formula is no longer valid! The Joliots noted that this phenomenon was the first in which the transformation of γ-rays into matter had been observed, and Madame Curie had called it the 'materialization of electrons'.

The Joliots then reported on experiments which had demonstrated that in the transmutation of elements, not only protons, neutrons and α-particles could be emitted, but also electrons of both signs, their number depending on the nature of the substance. Based on their experiments, they thought that the following reactions should take place:

$$^{10}_{5}\text{B} + {}^{4}_{2}\text{He} = {}^{13}_{6}\text{C} + {}^{1}_{0}n + e^{+} \tag{2a}$$

$$^{27}_{13}\text{Al} + {}^{4}_{2}\text{He} = {}^{30}_{14}\text{Si} + {}^{1}_{0}n + e^{+}, \tag{2b}$$

e^{+} being the positron. In the case of Be, the observed positrons could come from the process of the 'internal materialization' of the photon, as in the reaction,

$$^{9}_{4}\text{Be} + {}^{4}_{2}\text{He} = {}^{12}_{6}\text{C} + {}^{1}_{0}n + h\nu \tag{3a}$$

or

$$^{9}_{4}\text{Be} + {}^{4}_{2}\text{He} = {}^{12}_{6}\text{C} + {}^{1}_{0}n + (e^{+} + e^{-}). \tag{3b}$$

The Joliots also mentioned that they were led to the assumption that the proton had a complex structure, being formed of a neutron and a positron, and that this assumption was supported by some experimentally observed nuclear reactions.

DISCUSSION

An extensive discussion took place after the report of the Joliots, and almost all the invited members took part in it.

Lise Meitner reported on experiments on the velocity distribution of neutrons, as well as the distribution of directions of emission in the collisions between neutrons and protons.

Concerning the law of interaction between a proton and a neutron, Heisenberg pointed out that this force attains an appreciable value only if the distance between the particles is small, of the order of 10^{-12} cm.

Blackett presented an analysis of the positive electrons. He pointed out that since they could be created in pairs along with the negative electrons, they had to have a spin $\frac{1}{2}$ for the conservation of momentum. Blackett also discussed electron showers in cosmic rays. The electrons so observed had energies of 10^7 to 10^9 volts. He reported on the experimental work of Rossi, Fünfer and Gilbert, and the theoretical results of Oppenheimer and Plesset, but the complexity of the showers could not be explained on their basis.

REFERENCES

1. W. Bothe and Becker, *Z. Phys.* **66**, 289 (1930).
2. C. D. Anderson, *Phys. Rev.* **43**, 491 (1933).
3. P. M. S. Blackett and Occhialini, *Proc. Roy. Soc.* **139**, 699 (1933).

5. THEORY OF THE POSITRON

(P. A. M. Dirac)

The question of negative energy arises when one studies the motion of a particle in relativity. In non-relativistic theory the energy W is a function of the speed v or the momentum p,

$$W = \tfrac{1}{2}mv^2 = \frac{1}{2m}\,p^2, \tag{1}$$

so that W is always positive, but in relativity,

$$W^2 = m^2c^4 + c^2p^2$$

or

$$W = c\,\sqrt{(m^2c^2 + p^2)}, \tag{2}$$

so that W could be negative.

The classical assumption is that W is always positive and cannot assume negative values. In quantum theory, however, a variable may undergo discontinuous changes, in such a way that W could pass from a positive to a negative value. Although the theory could not explain the structure of the electron, and could not be used for distances of the order of the electron radius, it certainly remained correct elsewhere, and if one did not reject the theory one had to explain the negative value predicted by it.

Dirac pointed out that whether the calculation was performed in classical or in quantum mechanics, the result was that an electron of negative energy would be deviated by the field as if it were an electron of positive energy, having a positive electric charge $+e$. One was thus tempted to suggest that an electron in a negative energy state constituted a positron, but this was not acceptable because the observed positron certainly did not have a negative kinetic energy.

Dirac proposed that the use of Pauli's exclusion principle would give a better result by considering that in our world nearly all the negative energy states are occupied. The positrons are then the 'holes' in this uniform distribution, and this view allows a correct interpretation of the observed phenomena. If this assumption is correct, the mass of the positron should be equal to the mass of the electron. Also the average life-time of a proton must be 3×10^{-7} sec if it is moving slowly in the air at atmospheric pressure. In this conception the creation of electron-positron pairs is equivalent to a photoelectric effect in the uniform distribution of negative charge.

A new assumption had to be made in order to avoid an infinite electric field. Dirac assumed that fields were produced only by charges above the uniform distribution. Although such an assumption could be made without ambiguities in the absence of an external field, certain mathematical difficulties arise because of the presence of such a field.

Dirac treated the particular case of a constant electric field, weak enough to be a perturbation of the system. Using the Hartree–Fock approximation method, and assigning to each electron an eigenfunction $\psi(q)$, Dirac introduced the density matrix R by

$$(q' |R| q'') = \sum_r \bar{\psi}_r(q') \psi_r(q''). \tag{3}$$

The equation of motion for R is,

$$ih\dot{R} = HR - RH, \tag{4}$$

where H is the Hamiltonian for an electron moving in a field, and is given by

$$H = c\varrho_1(\boldsymbol{\sigma}, \mathbf{p}) + \varrho_3 mc^2 - eV, \tag{5}$$

where the ϱ's and σ are the usual spin matrices and V is the electrostatic potential.

The condition that the distribution satisfies the Pauli principle is given by $R^2 = R$. If R_0 is the distribution that produces no field, it is given by

$$R_0 = \tfrac{1}{2}\left(1 - \frac{W}{|W|}\right), \tag{6}$$

where W is the kinetic energy of an electron, given by

$$W = c\varrho_1(\boldsymbol{\sigma}, \mathbf{p}) + \varrho_3 mc^2. \tag{7}$$

Dirac considered a 'permanent' state for which the equation of motion is given by

$$0 = HR - RH. \tag{8}$$

$R = R_0$ satisfies this equation only if V=constant.

Dirac assumed V to be a quantity of the first order, and looked for a solution for R of the form $R = R_0 + R_1$, where R_1 is of the first order also. With $\gamma = W/|W|$, he obtained the equation of motion as,

$$(m^2c^2 + p^2)^{1/2} R_1 + R_1 (m^2c^2 + p^2)^{1/2} = \tfrac{1}{2}\frac{e}{c}(V - \gamma V\gamma). \tag{9}$$

The quantity of interest here is the electric density, $D(R_1)$, corresponding to the distribution R_1. This quantity is related to the trace of R_1 with respect to the spin variables. After a long calculation, involving a logarithmic divergence which was eliminated by a cut-off in the energy, Dirac obtained,

$$- e(x|D(R_1)|x) = -\frac{e^2}{\hbar c} \cdot \frac{2}{3\pi} \left(\log\frac{2P}{mc} - \frac{5}{6} \right) \varrho - \frac{4}{15\pi} \frac{e^2}{\hbar c} \left(\frac{\hbar}{mc} \right)^2 \nabla^2 \varrho, \qquad (10)$$

where P is the magnitude of the momentum, and ϱ is the electric density producing the potential V in such a way that

$$\nabla^2 V = -4\pi\varrho. \qquad (11)$$

Dirac concluded that, on the basis of his calculation, the electric charges normally observed on the electrons, protons, and other charged particles, are not the real charges carried by these particles and occurring in fundamental equations, but are smaller by a factor of 136 to 137.

DISCUSSION

Peierls gave another example in which he had treated two divergences arising in the theory of 'holes'.

Pauli was interested in the way in which the exclusion principle had been introduced in the theory of 'holes', but he pointed out the difficulties of the theory related to the fact that the vacuum would have an infinite energy.

Bohr pointed out that the Klein–Nishina formula was no longer valid in a region of the order of the electron radius. Since Dirac's work implied disagreement with this formula in this region, Bohr wondered whether it was possible to verify it experimentally. Bohr pointed out the great success that had been achieved in applying the Dirac equation to the theory of the positron, but he also noted the limitation of the theory. The theory could not be applied to distances of the dimensions of the electron itself, that is dimensions smaller than $\delta = e^2/mc^2$, or for corresponding proper times, $\tau = \delta/c = e^2/mc^3$. Bohr discussed some of the difficulties in the construction of a quantum electrodynamics, and the motion of a field in general.

6. THE ORIGIN OF γ-RAYS AND THE NUCLEAR ENERGY LEVELS

(G. Gamow)

Gamow first discussed certain aspects of the theory of the atomic nucleus based on the protons and electrons as constituents, and pointed out the contradictions to which this model gave rise. Then he discussed the new nuclear model with the protons and neutrons. On the basis of the new assumption he obtained the curve giving the binding energy with respect to the atomic mass, as well as the general shape of the nuclear potential.

The theory of nuclear structure was not yet well established. Gamow's model was based on a potential well of infinite dimensions, in which the energy levels were given by

$$J_{j+1/2} \left[\frac{2\pi\sqrt{2m}}{h} \sqrt{E}\, r_0 \right] = 0, \qquad (1)$$

where r_0 is the radius of the sphere outside which the potential is infinite, and $J_{j+1/2}$ is the Bessel function of order $j+\frac{1}{2}$. The difference $\Delta_1 E$ between the two first levels

S_1 and P_1 is given by

$$\Delta_1 E = \frac{h^2}{8mr_0^2}.$$ (2)

For $r_0 = 0.9 \times 10^{-12}$ cm, this formula gave 0.5×10^6 eV for the energies of the α-particles, and 2×10^6 eV for protons and neutrons.

Gamow discussed the Rutherford–Ellis theory of γ-emission, and pointed out the difficulties which arise in it. Then he discussed the different modes of probability transitions corresponding to dipole, quadrupole, etc., symmetries of the field. The dipole transition probability was found to be given by

$$\chi = \frac{64\pi^4 Z^2 e^2 v_{m,n}^3}{3hc^3} |r_{m,n}|^2,$$ (3)

where $v_{m,n}$ is the frequency of the γ-quantum emitted, Ze the charge of the particle emitting it, and $r_{m,n}$ the matrix element corresponding to the coordinates. Using the relation

$$\sum_m (E_m - E_n) |x_{m,n}|^2 = \frac{h^2}{8\pi^2 m}$$ (4)

from wave mechanics, Gamow found for χ the expression,

$$\chi \lesssim \frac{8\pi^2 Z^2 e^2}{3mc^3 h^2} (h v_{m,n})^2.$$ (5)

Taking $h v \sim mc^2$, he obtained, for the α-particle or the proton, the same value, $\chi \lesssim 3 \times 10^{15}$ sec^{-1}. He pointed out that a very special charge distribution in the nucleus could very well lead to an annhilation of the dipoles, and give rise to quadrupole symmetries only.

Gamow discussed internal conversion in which, instead of a γ-quantum, an electron is emitted with the same energy. If the distance between the electron and the nucleus is large in comparison with the γ-wavelength, the effect is equivalent to the ordinary photoelectric effect. If this distance is shorter than the γ-wavelength the interaction between the electron and the nucleus can be considered as being direct, without the retarded potential. In the intermediate case, the γ-wavelength and the electron-nuclear distance are of the same order. Moreover, the electrons situated close to the nucleus, in the case of heavy nuclei, have velocities comparable to that of light; they are thus ejected by γ-rays with very high velocities, and a relativistic calculation has to be made in this case.

Gamow then treated the following subjects: nuclear excitation by disintegration through α-particles, nuclear excitation by β-emission, nuclear excitation by γ-emission, and the action of γ-rays on atomic nuclei. He pointed out that for sufficiently hard γ-rays, the Klein–Nishina formula was no longer in full agreement with the experimental results, and it seemed plausible to consider that the γ-rays could have an interaction with the nucleus. Since the protons and neutrons were considered as being too heavy to allow such an interaction, the explanation was that the interaction took place between the nuclear electrons [an electron dissociated from a neutron] and the γ-rays, giving rise to a β-emission.

DISCUSSION

Rosenblum gave his results concerning the fine structure of α-rays, and pointed out that this structure is always accompanied by a corresponding γ-structure.

There was a long discussion between Fermi, Mott, Heisenberg, Bohr, Peierls and Perrin on the possibility of transitions leading to dipole radiation.

Mott discussed the problem of β-emission. The energy of an α-ray corresponds to the difference of energy between the initial and final states of the nucleus. Ellis and Mott had assumed that, in β-disintegration, this difference corresponds to the maximum of the β-spectrum. This assumption, according to Mott, was justified by the fact that the energies of the γ-rays emitted after a β-disintegration are always smaller than the maximum value in the spectrum.

Among those who took part in the extensive discussion after Gamow's report were Rosenblum, Ellis, Fermi, Mott, Rutherford, Lise Meitner, Heisenberg, Madame Curie, Perrin, and Bohr.

7. STRUCTURE OF THE NUCLEUS

(W. Heisenberg)

Heisenberg pointed out that our knowledge of nuclear structure does not seem to be in contradiction with quantum mechanics. The theory of α-disintegration developed by Gamow, Condon and Gurney, had shown that the heavy constituents of the nucleus are submitted inside it to energy conditions very similar to those which govern the motion of electrons around the nucleus. An analysis of the energies involved in nuclear reactions had shown that the speeds of these heavy particles was not relativistic, and, therefore, non-relativistic quantum mechanics could be applied.

Heisenberg recalled an argument of Bohr's relating the mass defect to the nuclear radius. If r_0 is the nuclear radius of helium and, Δp, the variation in the momentum of a proton inside its nucleus, then

$$\Delta p \sim \frac{\hbar}{r_0}, \tag{1}$$

hence, its average kinetic energy,

$$\bar{E}_{\text{kin.}} \sim \frac{1}{2M} \left(\frac{\hbar}{r_0}\right)^2. \tag{2}$$

Since the average kinetic energy is of the same order as the average potential energy, the mass defect turns out to be of the order of

$$4 \times \frac{1}{2M} (mc)^2 \left(\frac{\hbar c}{e^2}\right)^2 \sim 0.01 \, Mc^2, \tag{3}$$

which provides a good order of magnitude compared to the experimental value of $0.029 \, Mc^2$.

This argument had shown that quantum mechanics was applicable and that Coulomb forces, which would lead to a much smaller mass defect, did not play an important role inside the nucleus, at least for light nuclei.

Dealing with some of the difficulties related to the existence of electrons in the nucleus, Heisenberg pointed out that, on the basis of the Dirac theory and the uncertainty principle, a much larger mass defect should be associated with the β-emission. He discussed the idea according to which quantum mechanics would be applicable to the heavy constituents of the nucleus *only*, and pointed out that in this case the presence of the electrons would be essential for the binding energy of the nucleus.

Heisenberg discussed another assumption concerning nuclear structure. He recalled that the masses of nuclei are multiples of α-particles and not of the protons, thus inferring the presence of α-particles as constituents of the nucleus. However, a definitive answer could not be given concerning other constituents.

Heisenberg discussed Gamow's droplet model of the nucleus, which was based on the assumption of very rapidly decreasing forces between the constituents of the nucleus. This model was in agreement with the experimental fact that the average density of nuclei seemed to be independent of their dimensions, the nuclear radii increasing as the cube root of the atomic masses.

In Gamow's droplet model, in which the nuclei were assumed to consist of α-particles only, the energy ascribed to the nucleus was of the form

$$\varepsilon = - CN_\alpha + \frac{(2eN_\alpha)^2}{r}, \tag{4}$$

where N_α is the number of α-particles constituting the nucleus, and r its radius. For $r = R \sqrt[3]{N_\alpha}$, the energy becomes

$$\varepsilon = - CN_\alpha + 4 \frac{e^2}{R} (N_\alpha)^{5/3}. \tag{5}$$

The mass defect curve represented by this equation thus presents a minimum, showing that nuclei containing a large number of α-particles would disintegrate spontaneously. Gamow had noted the existence of similar relations for nuclei containing electrons in addition to α-particles. However, the hypothesis of the stability of a nucleus in β-disintegrations was violated in all these considerations, and it was necessary to examine other models of the structure of the nucleus.

Heisenberg then discussed the discovery of the neutron by J. Chadwick, and by I. Curie and F. Joliot, which made it possible to consider the nucleus as made up of α-particles, electrons, protons and neutrons. Different models had been proposed. In some, only the α-particles, protons and neutrons were included, while in others the electrons as well. These models had been advanced by F. Perrin, Iwanenko, Gapon, Bartlett, and Landé. Gamow and Perrin had included the electrons in their models, which introduced statistical difficulties. However, one had to consider the structure of the neutron, and it was assumed to be formed of a proton and an electron.

Heisenberg discussed the mutual interaction between protons and neutrons. It could either be envisaged as an ordinary force, or, by analogy with the case of molecules, it could be considered as an exchange interaction. The former hypothesis would regard the neutron as an indissoluble elementary particle. Diverse hypotheses were possible as far as the exchange interaction was concerned. One could, for instance, consider only the exchange of charge, without any modification of the spin. Or, as Majorana had done, one could envisage an interaction in which the negative charge and spin were exchanged simultaneously between the particles.

Different hypotheses were also possible regarding the mutual interaction between the neutrons. It seemed, however, that the neutron-neutron interaction within the nucleus was much smaller than the neutron-proton interaction, and one could, in the first approximation, neglect it altogether. The Coulomb forces between the protons within the nucleus could thus also be neglected.

The mathematical representation of the exchange interaction could be developed according to two different schemes: (1) One could introduce, for each nuclear particle, five coordinates: three coordinates of position $\mathbf{r}_k$, a spin variable σ_k, and a new variable ϱ_k which takes the value $+1$ or -1 depending on whether the particle is a proton or a neutron; (2) Each particle is characterized by four variables: $\mathbf{r}_k$ and σ_k, but the coordinates are designated differently for the neutrons and the protons (e.g., $\mathbf{r}_K$, σ_K, and $\mathbf{r}_k$, σ_k, respectively). The Schrödinger wave function is written, in the first case, as

$$\varphi(\mathbf{r}_1, \sigma_1, \varrho_1; \mathbf{r}_2, \sigma_2, \varrho_2, \ldots), \tag{6}$$

and in the second case, as

$$\varphi(\mathbf{r}_I, \sigma_I; \mathbf{r}_{II}, \sigma_{II}, \ldots, \mathbf{r}_1, \sigma_1; \mathbf{r}_2, \sigma_2, \ldots). \tag{7}$$

The relation between these two schemes is expressed by the equations,

$$\begin{cases} \varphi(\mathbf{r}_1, \sigma_1, +1; \mathbf{r}_2, \sigma_2, +1; \ldots) = \varphi(\mathbf{r}_I, \sigma_I; \mathbf{r}_{II}, \sigma_{II}; \ldots) \\ \varphi(\mathbf{r}_1, \sigma_1, +1; \mathbf{r}_2, \sigma_2, -1; \ldots) = \varphi(\mathbf{r}_I, \sigma_I; \mathbf{r}_1, \sigma_1; \ldots) \\ \varphi(\mathbf{r}_1, \sigma_1, -1; \mathbf{r}_2, \sigma_2, -1; \ldots) = \varphi(\mathbf{r}_1, \sigma_1; \mathbf{r}_2, \sigma_2; \ldots) \\ \cdot \quad \cdot \quad \cdot \quad \cdot \quad \cdot \quad \cdot \quad \cdot \quad \cdot \quad \cdot \quad \cdot \quad \cdot \quad \cdot \quad \cdot \quad \cdot \quad \cdot \quad \cdot \quad \cdot \quad \cdot \end{cases} \tag{8}$$

Heisenberg introduced the matrices

$$\varrho^{\xi} = \begin{vmatrix} 0 & 1 \\ 1 & 0 \end{vmatrix} \quad \text{and} \quad \varrho^{\eta} = \begin{vmatrix} 0 & -i \\ i & 0 \end{vmatrix}. \tag{9}$$

With these, the neutron-proton interaction term in the Hamiltonian, under the hypothesis of the exchange of negative charge, and the scheme (1) becomes,

$$J(rkl) \tfrac{1}{2} [\varrho_k^{\xi} \varrho_l^{\xi} + \varrho_k^{\eta} \varrho_l^{\eta}], \tag{10}$$

and with scheme (2), it becomes

$$-J(r_{Kl}) \cdot P'_{Kl}, \tag{11}$$

where P'_{Kl} represents the operator for the permutation of the variables r_K, σ_K with r_l, σ_l.

On the other hand, the exchange interaction envisaged by Majorana, in the scheme (1), leads to the term

$$J(rkl) \tfrac{1}{4} [\varrho_{k}^{\xi} \varrho_{l}^{\xi} + \varrho_{k}^{\eta} \varrho_{l}^{\eta}] [1 + (\sigma_{k} \sigma_{l})], \tag{12}$$

and, in the representation (2), it leads to

$$-J(r_{Kl}) P_{Kl}, \tag{13}$$

where P_{Kl} is the operator corresponding to a permutation of the space coordinates r_K and r_l. A basic fact used by Majorana was the 'droplet' character of the nucleus (the Gamow model). The finite radius of the protons was introduced through exchange interactions, as already used by Heitler and London for the molecules.

Heisenberg gave the mathematical justification for these results. For a nucleus composed of n_1 neutrons and n_2 protons, he wrote the Schrödinger wave function in the form

$$\Phi = \begin{vmatrix} \varphi_1(\mathbf{r}_I, \sigma_I) & \cdots & \varphi_{n_1}(\mathbf{r}_I, \sigma_I) \\ \varphi_1(\mathbf{r}_{II}, \sigma_{II}) & \cdots & \varphi_{n_1}(\mathbf{r}_{II}, \sigma_{II}) \\ \cdots & \cdots & \cdots \\ \varphi_1(\mathbf{r}_{n_1}, \sigma_{n_1}) & \cdots & \varphi_{n_1}(\mathbf{r}_{n_1}, \sigma_{n_1}) \end{vmatrix} \begin{vmatrix} \varphi_1(\mathbf{r}, \sigma_1) & \cdots & \varphi_{n_2}(\mathbf{r}_1, \sigma_1) \\ \cdots & \cdots & \cdots \\ \cdots & \cdots & \cdots \\ \varphi_1(\mathbf{r}_{n_2}, \sigma_{n_2}) & \cdots & \varphi_{n_2}(\mathbf{r}_{n_2}, \sigma_{n_2}) \end{vmatrix}. \tag{14}$$

With this, the potential energy for the exchange interaction becomes,

$$\begin{aligned} E_{\text{pot}} &= -\int \Phi^* \sum_{K, k} J(r_{K, k}) P_{K, k} \Phi \, d\omega \\ &= -\int d\mathbf{r} \, f[\varrho_N(\mathbf{r}), \varrho_P(\mathbf{r})], \end{aligned} \tag{15}$$

where f is a symmetric function of ϱ_N and ϱ_P, which tends, at large densities, to the limit $2\varrho_N J(0)$ or $2\varrho_P J(0)$, depending on whether $\varrho_N < \varrho_P$ or $\varrho_N > \varrho_P$.

Making use of the Thomas–Fermi method, in the form given by Dirac, Heisenberg obtained the kinetic energy,

$$E_{\text{kin}} = \frac{h^2}{M} \frac{4\pi}{5} \left(\frac{3}{8\pi}\right)^{5/3} \int d\mathbf{r} \, (\varrho_N^{5/3} + \varrho_P^{5/3}). \tag{16}$$

Finally,

$$E = \frac{h^2}{M} \cdot \frac{4\pi}{5} \left(\frac{3}{8\pi}\right)^{5/3} (n_1^{5/3} + n_2^{5/3}) V^{-2/3} - V f\left(\frac{n_1}{V}, \frac{n_2}{V}\right), \tag{17}$$

where V is the volume of the nucleus, which can be obtained from the condition $dE/dV = 0$. Heisenberg then calculated the function f and the Coulomb term in the energy.

Heisenberg compared the theoretical formulas with the experimental results, and obtained some general consequences for the isotope chart. Various properties of the nuclei, such as stability, abundance in nature, spin, etc., all present a certain periodicity for a mass variation of 4 and a charge variation of 2. This was confirmed by the neutron-proton nuclear model. The Gamow model, with the assumptions of Perrin (i.e. the nuclear constituents being α-particles, neutrons, protons and electrons), leads to a very strong effect of the mass on nuclear properties. Heisenberg showed that the proton-neutron model could very well reproduce the isotope chart.

With respect to α-decay, Heisenberg pointed out that the Gamow, Condon, Gurney theory had described the stability of the nucleus in a manner which was independent of the model used. He then examined β-decay. According to Gamow's theory, the nuclear stability should follow the same criterion for α- and β-disintegrations. The continuous spectrum of the β-emission had given rise to great difficulties of inter-pretation. No relation of the type of the Geiger–Nuttal law for α-particles could be given for the β-emission. Bohr had been willing to consider that energy conservation was violated in β-decay, but Pauli had introduced the concept of the neutrino to preserve energy and momentum conservation laws.

Heisenberg concluded by discussing the applications of the neutron-proton nuclear model to questions of mass defect and stability, diffusion and disintegration, in all of which his theory agreed well with the experimental results.

DISCUSSION

Pauli discussed the difficulties of interpreting the existence of a continuous β-spectrum. Bohr had suggested that the laws of conservation of energy and momentum were violated in this case, but Pauli was not willing to pay such a price. Pauli mentioned his own solution of the problem. 'In June 1931, at the occasion of a conference in Pasadena, I proposed the following interpretation: the conservation laws remain valid, and the expulsion of β-particles is accompanied by a very penetrating radiation of neutral particles, which has not yet been observed. The sum of the energies of the β-particle and the neutral particle (or neutral particles, because one does not know whether there is only one or several) emitted by the nucleus in a single process, would be equal to the energy which corresponds to the upper limit of the β-spectrum. It goes without saying that we admit not only the conservation of energy, but also the conservation of momentum, and of angular momentum, as well as of the statistics, in all elementary processes.'

Pauli described the properties of the nuclear particle which he had hypothesized, and which Fermi had named the 'neutrino'. It had to have a spin 1/2 and obeyed Fermi statistics.

Chadwick reported on experiments which had been undertaken to detect the neutrinos.

Perrin discussed his assumption that in β-decay only one neutrino was emitted. He thought that the form of the intensity distribution of the β-spectrum allowed one to make a reasonable assumption about the mass of the neutrino. Assuming that energy and momentum conservation could be applied, he found that, on the basis of a relativistic calculation, the intrinsic mass of the neutrino had to be zero.

Peierls wondered why Heisenberg regarded the neutron as an elementary particle. Heisenberg replied that this was just an impression. The mass defect of the neutron appeared to be extremely small, and if the neutron had a complex structure one would have observed some evidence in the disintegrations.

Perrin pointed out that there was a symmetry in the equations ${}_0^1n \rightleftharpoons {}_1^1\text{H} + e^-$ and ${}_1^1\text{H} \rightleftharpoons {}_0^1n + e^+$, and both had to be accepted or rejected simultaneously.

Lise Meitner, Bohr, Dirac, Fermi, Cockcroft, Debye, and De Donder also took part in the discussion after Heisenberg's report.

9

Towards the Spectrum of Elementary Particles and the Hierarchy of Interactions

1. INTRODUCTION

Between 1933, the year in which the seventh Solvay Conference on Physics took place, and 1948, when the eighth Conference was organized, there occurred many changes and developments in the physics of elementary particles. First, a number of those who had been in the forefront of the development in earlier years, such as Rutherford, retired from active work or died. Others, like Dirac, Heisenberg, and Pauli, who had done much to create quantum mechanics and quantum field theory, continued to work along the lines on which they had been successful in earlier years. Meanwhile a new generation had specialized in quantum mechanics and its extension to the problems of nuclear and particle physics, and in the 1930s and 1940s more effort went into these fields than perhaps into any other topic of physics ever before. Often, what one found were negative results and serious difficulties pointing to defects with the formalism and results of the fundamental theory.

The field of elementary particle physics became separated into more or less distinct classes of phenomena. It was found that besides the quantum theory of electro-dynamics, which had been pursued immediately after the creation of quantum mechanics, there existed different phenomena of nuclear decays of which only the α-decay could be explained reasonably by means of a semi-phenomenological approach. Within nuclear interactions, other than the electromagnetic, two classes of phenomena had to be distinguished: the strong and the weak interactions.

Not only did the theoretical physicists make progress by going into the details of the phenomena connected with the existence and interaction of elementary particles, but the experimentalists developed techniques and the machinery to create, detect, and study new particles. Some of these had properties predicted by the theory, while others brought along new puzzles which had to be understood.

During World War II, few physicists had been able to carry on their normal work. Efforts of many of them were devoted to the problems of the application of nuclear physics, and these efforts shed light only on some limited areas of particle physics. Thus the problem of nuclear forces came under better focus. A small number of physicists, however, were still able to elaborate freely on theoretical schemes which were far removed from immediate application: such as Dirac in England, Tomonaga in Japan, and Pauli in Princeton, New Jersey. The increased amount of money spent on the machinery used in nuclear physics and technology during the War also proved to be beneficial for the growth of particle physics. After the War, particle accelerators of increasing size and energy were constructed, which produced fast charged particles of energies comparable to those available in cosmic rays.

It was at this point in the development that the eighth Solvay Conference took place in 1948, the first since 1933. Before discussing the Conference, we shall give an outline of the progress that had been achieved in quantum and particle physics.

2. THE DEVELOPMENT OF QUANTUM ELECTRODYNAMICS

In dealing with quantum electrodynamics in the 1930s and 1940s, one has to bear in mind that such a field did not exist, as a special topic in particle physics. In the beginning one thought of dealing with quantum field theory, of which the interactions of charged particles via photons provided a typical example. On the other hand, it was not clear whether all interactions in particle physics could be essentially reduced to those of the electromagnetic type. The concept of the strength of the interaction developed only gradually. By then, the electromagnetic interactions represented a specific class of interactions, namely the renormalizable one, and it was not clear whether all forces in the theory of elementary particles belonged to this class. This problem has continued till today, when the unification of electromagnetic and weak interactions on the basis of a renormalizable theory preoccupies many theoreticians of particle physics.

We shall, therefore, first discuss the progress of quantum field theory in general between 1933 and 1947/48, which is essentially identical with the development of quantum electrodynamics.

2.1. BEGINNING OF QUANTUM FIELD THEORY

Quantum field theory began almost with the first papers on quantum mechanics. P. Jordan was responsible for the sections at the end of the Born–Jordan[1] and Born–Heisenberg–Jordan[2] papers, which dealt with the quantization of the electromagnetic field and derived its fluctuations.

Dirac,[3] in his celebrated paper of 1927, considered the interaction of an atom with the electromagnetic field with the goal of building a relativistic quantum theory with interaction. Dirac proceeded by first dealing with an approximation which was not strictly relativistic. He considered the interaction of a non-relativistic atom with the relativistic radiation field; and, in order to have a discrete number of degrees of freedom of the latter, he enclosed the system in a finite box, decomposing the radiation field into its Fourier coefficients. Moreover, he chose the following dynamical variables,

$$b_r = \sqrt{N_r}\, e^{-i\theta_r/\hbar} \quad \text{and} \quad b_r^+ = \sqrt{N_r}\, e^{+i\theta_r/\hbar}, \tag{1}$$

where N_r is the absolute square of the Fourier coefficient a_r, and θ_r is a phase variable conjugate to N_r. The variables satisfy the commutation relations

$$b_r b_r^+ - b_r^+ b_r = 1, \tag{2}$$

all others being zero. The N_r can be shown to take on only integral values, larger than or equal to zero, and b and b^+ can be interpreted as annihilation and creation operators of the photons. By this procedure, Dirac was able to deal with transitions of the energy of photons as well as the emission and absorption of spectral lines. In a second paper Dirac treated the theory of dispersion.[4]

A year after this first paper on 'second quantization', Jordan and Wigner[5] developed a similar scheme for Fermi fields, in which the creation and annihilation operators satisfy, instead of Equation (2), the anti-commutation relations,

$$\psi_r\psi_r^+ + \psi_r^+\psi_r = 1, \tag{3}$$

all others being zero.

2.2. QUANTUM THEORY OF FIELDS

Dirac's method of second quantization had a great impact on further development. It got Heisenberg and Pauli started on a very ambitious project, the quantum mechanics of wave fields, and they published two long papers on it in 1929 and 1930.[6] They considered an unspecified relativistic classical field, having a scalar Lagrangian density $\mathscr{L}$,

$$\mathscr{L} \equiv \mathscr{L}\left(\varphi, \psi, ..., \frac{\partial\varphi}{\partial x^\mu}, \frac{\partial\psi}{\partial x^\mu}\right), \tag{4}$$

which depends only on the fields φ and ψ and their first derivatives. The conjugate momenta were defined canonically,

$$\pi_\varphi = \frac{\partial\mathscr{L}}{\partial\dot\varphi} . \tag{5}$$

For fields and conjugate momenta, commutation or anticommutation rules were given at equal times; for instance, in the case of a scalar field φ,

$$[\pi(\mathbf{r}), \varphi(\mathbf{r}')] = - i\delta(\mathbf{r} - \mathbf{r}'), \tag{6}$$

where $\mathbf{r}$ and $\mathbf{r}'$ indicate two different space-time points and δ is the Dirac δ-function.

This procedure was manifestly non-covariant, and Heisenberg and Pauli presented a complicated proof in their first paper to prove relativistic invariance, which they simplified considerably in their second paper. What one has to show is that the classical Lagrangian *and* the commutation relations are relativistically invariant, and one does it most elegantly by introducing the relativistic generalization of the δ-function. However, only preliminary forms of this step, which was due to Schwinger much later, are to be found in the Heisenberg–Pauli papers of 1929/30.[6a]

The new formalism was applied by Heisenberg and Pauli to quantum electrodynamics, but the gauge group caused technical problems. Moreover, the negative energy solutions of the electron were not properly interpreted at that time, and the first infinities of the interacting fields were discovered when considering the self-interaction of an electron. The two papers of Heisenberg and Pauli opened the door to the difficult questions of quantum field theory which would occupy physicists for the next twenty years.

Dirac criticized the formalism of Heisenberg and Pauli, especially the complete equivalence of fields and particles which they had assumed. Dirac considered the field as something more elementary.[7] In his attempts to improve the formulation, Dirac introduced the so-called many-time formalism, i.e. wave functions containing different times,

$$\psi \equiv \psi(t_1, \mathbf{x}_1; t_2, \mathbf{x}_2; \ldots), \qquad (7)$$

and wrote the equations of motion in a manifestly covariant form. The new wave function provided a representation different from the Schrödinger and the Heisenberg representations, and Dirac's method became important in the later development of quantum electrodynamics, especially the derivations of Tomonaga, Schwinger and Feynman.

N. Bohr and L. Rosenfeld treated a problem connected with field quantization, namely the limitation in the accuracy of simultaneous measurements.[8] By that time the formal equivalence of the Heisenberg–Pauli approach to field theory, on the one hand, and Dirac's approach, on the other, had been recognized.

2.3. DIVERGENCES IN QUANTUM FIELD THEORY

The foremost problem in the 1930s became the infinities which arose in most calculations of physical quantities. Already quite early, Ehrenfest had discovered a difficulty in Dirac's paper on absorption and emission of radiation; since use had been made of a pointlike electron, the latter should obtain an infinite self-energy.[9] Heisenberg and Pauli had mentioned this fact explicitly in their 1929 paper on quantum field theory. Other infinities or divergences were discovered subsequently: for example, the logarithmically divergent charge 'renormalization'[10]; and the infinite self-energy of the photon.[11] Only the lowest order perturbation calculations yielded finite results, such as the celebrated formula of O. Klein and Y. Nishina[12] for the Compton effect. The status of the problems of infinities in quantum electrodynamics was presented several times by Weisskopf.[13] The important result of these studies was that the self-energy of the Dirac electron was found to be 'only' logarithmically divergent.[14]

Several remedies for the divergences were tried in the years after 1929. Apart from the simple and 'brutal' subtraction of infinities,[15] more subtle limits, like the 'λ-limiting' procedure, were considered.[16] Even more drastic alterations in the mathematical structure of the theory like abandoning the classical Lagrangian [nonlinear electrodynamics] or changing the structure of the space of states ['indefinite metric'] were proposed.

The only problem that could be solved was the 'infrared catastrophe', a divergence occurring because of the zero mass of the photons, which could be explained by abandoning perturbation theory.[17] Although this treatment is still discussed and criticized, one believes that the problem is now well-understood.

3. THE HIERARCHY OF INTERACTIONS

It was natural that the new scheme of quantum field theory should have been applied first to electromagnetic phenomena. In fact, one did not know of any other phenomena to apply it to. The situation changed drastically, however, in 1933/34 when Fermi developed his theory of the β-decay. Then came Yukawa's meson theory of nuclear forces (1935), and by the end of the decade physicists were convinced that a hierarchy of interactions, from the 'strong' via the electromagnetic to the 'weak' interactions, determined the processes occurring between elementary particles.

3.1. THE THEORY OF BETA-DECAY

In June 1931, Pauli had proposed that the continuous spectrum of electrons created in β-decay could be explained by introducing the neutrino. Thus the reaction, $n \rightarrow p + e^- + \bar{\nu}$, became the fundamental process in β-decay. In 1933 Joliot-Curie discovered the [artificial] positron decay, and the time was ripe to think about a mechanism for this interaction. This was done by Fermi.[18] He had noticed that the theories of light particles, at that time, did not explain the binding of particles inside the nucleus.

Fermi noted: 'The simplest way for the construction of a theory which permits a quantitative discussion of the phenomena involving nuclear electrons, seems then to examine the hypothesis that the electrons do not exist as such in the nucleus before the β-emission occurs, but that they, so to say, acquire their existence at the very moment when they are emitted; in the same manner as a quantum of light, emitted by an atom in a quantum jump, can in no way be considered as pre-existing in the atom prior to the emission process. In this theory, then, the total number of electrons and neutrinos (like the total number of light quanta in the theory of radiation) will not necessarily be constant, since there might be processes of creation or destruction of these particles.

'According to the ideas of Heisenberg, we will consider the heavy particles, neutron and proton, as two quantum states connected with two possible values of an internal coordinate, ϱ, of the heavy particle. We assign to it the value $+1$ if the particle is a neutron and -1 if the particle is a proton.

'We will then seek an expression for the energy of interaction between the light and the heavy particles which allows transitions between the values $+1$ and -1 of the coordinate ϱ, that is to say, transformation of neutrons into protons and vice-versa, in such a way, however, that the transformation of a neutron into a proton is necessarily connected with the creation of an electron which is observed as a β-particle, and of a neutrino; whereas the inverse transformation of a proton into a neutron is connected with the disappearance of an electron and a neutrino.'

The β-interaction which Fermi wrote down for the first time was, in modern notation, vector current-current coupling,

$$\mathscr{L}_i = (\bar{\psi}_p \gamma^\mu \psi_n)(\bar{\psi}_e \gamma_\mu \psi_\nu) \cdot G, \tag{8}$$

where G is the Fermi coupling constant, the factor $(\bar{\psi}_p \gamma^\mu \psi_n)$ represents the hadron [vector] current, and $(\bar{\psi}_e \gamma^\mu \psi_v)$ represents the lepton current.

Fermi's theory was developed further in the 1930s. Other possibilities of the four Fermi-couplings, like scalar or tensor coupling, replacing the γ^μ by scalar or tensor operators, were studied, but the structure of a local coupling of four Fermi fields has remained unchanged until today.

3.2. YUKAWA'S THEORY OF STRONG INTERACTIONS

Once a reasonable theory of β-decay had been proposed by Fermi, one wondered whether it could not explain all the nuclear forces. Tamm and Iwanenko viewed the nuclear forces connecting the nuclei as being created by the virtual exchange of an electron-neutrino pair.[19] The intriguing aspect of such an explanation was that the short range of nuclear forces could follow from the fact that the interaction was weak at large distances, but seemed to become very strong at zero distance.

A year later, Yukawa published a different hypothesis.[20] Yukawa, drawing an analogy from electrodynamics, considered the nuclear forces to be mediated by a boson field, i.e. by an intermediate particle. If the particle had a non-vanishing rest mass, μ, the short range of nuclear forces could be immediately accounted for, since the static solution of the scalar wave equation is given by

$$V(r) = \frac{e^{-(\mu c/\hbar)\cdot r}}{r}. \tag{9}$$

$V(r)$ describes the potential of a source, similar to that of an electric charge, where the exchange objects are scalar particles of mass μ. From the range of nuclear forces, which is of the order of 10^{-13} cm, Yukawa estimated the rest mass of the boson as being roughly equal to 200 electron masses. Hence the name 'mesotron' was given to this object.

The new theory of strong forces was not immediately accepted, but in 1937 evidence for the existence of charged particles, heavier than electrons but lighter than protons, became available in cosmic ray experiments. The 'meson' was 'found' by C. D. Anderson* and S. M. Neddermeyer in a cloud chamber.[21] Thus the 'heavy quantum' predicted by Yukawa seemed to exist, the more so since the mass and lifetime of the new particle seemed to agree with the theory. Yukawa had anticipated that the boson would decay weakly because he also hoped to account for the β-interaction by means of the same object.

Soon, however, new difficulties arose. The newly discovered 'heavy quantum' seemed to have only weak interaction with nuclei and could not be considered respon-

---

* The particle discovered by Anderson, now known as the μ-meson or muon, was referred to as the mesotron, Anderson's original term. Yukawa's original name 'heavy quantum' was felt to be too provisional and too general for permanent use. In informal usage, the Yukawa particle was at times referred to as the 'yukon', but the name did not catch on. At a meeting in E. Bretscher's house in Cambridge, H. J. Bhabha, M. H. L. Pryce, and N. Kemmer agreed to use the word 'meson' in print. C. F. Powell's use of the word 'meson' after the War led to its general use. [See N. Kemmer, 'The Impact of Yukawa's Meson Theory on Workers in Europe', *Suppl. Prog. Theor. Phys.* (1965).]

sible for the strong nuclear forces. The situation thus became very problematic again. On the one hand, there existed a theory explaining numerous features of nuclear forces in close analogy with quantum electrodynamics, that is, one had an interaction Lagrangian of the type

$$\mathscr{L}_i = g\,(\bar{\psi}_N \psi_N)\Phi_\pi,\tag{10}$$

where a scalar coupling of the nuclear density $(\bar{\psi}_N \psi)$ to a scalar meson Φ had been assumed; on the other hand, the coupling constant g appeared to be as small as the coupling constant in Fermi's theory.

In addition, it was not clear whether the intermediate quanta of Yukawa were scalar or vector particles. The spin dependence of nuclear forces appeared to require a vector meson rather than a scalar one, and Proca's theory seemed to be appropriate in this connection.[22] In 1938 Kemmer considered, for the first time, pseudoscalar fields.[23] Thus one became familiar with several kinds of massive boson fields. Since the study of nuclear forces in nuclei was the main source of information, the statements about the properties of the meson depended on complicated calculations. Perturbation theory was applied, although it could not, in any way, be justified. The alternative schemes of the 'strong coupling theory' were likewise problematic, especially when extended relativistically.[24] Thus the introduction of the meson seemed to have raised more questions than it answered. By the end of World War II, the only point on which most physicists working on these problems seemed to agree was that, to some extent, only the electromagnetic and weak interactions had been understood.

4. NEW ASPECTS IN PARTICLE THEORY

During the first decade of the existence of elementary particle theory, starting in 1930 with the prediction of the existence of the positron on the basis of Dirac's hole theory, considerable progress had been achieved in the understanding of the new phenomena in spite of the difficulties of infinities. Perhaps the most striking success was the recognition of new types of interactions between an increasing variety of fundamental particles. The so-called strong interaction, in particular, led to considerable problems, and one felt that the methods hitherto used in quantum theory and quantum electrodynamics would perhaps break down. No perturbation methods could be applied to these problems. It was soon found that even in the case of a far weaker coupling than the electromagnetic, i.e. in weak interactions, considerable troubles arose when one tried to apply higher order perturbation theory.[25] The reason is that the weak interaction includes a coupling constant G, which has a dimension of length squared if the Fermi fields are quantized according to the canonical rules. In these theories, however, the interaction term explodes at very high energies.

We shall leave the track of the quantum field theory discussed thus far, and turn to a radically different interaction scheme. It was developed by Heisenberg, beginning in 1943, and it gave the concept of an elementary particle a completely new meaning. This is the theory of the S-matrix.

4.1. THE S-MATRIX

With the insurmountable difficulties of quantum field theory in mind, Heisenberg wondered whether one should not leave the entire topic of interacting fields and try to establish a theory which uses only experimentally observable quantities.[26] Among the latter he included the energy and momentum of free particles, the probabilities of scattering, absorption and emission of elementary particles, and finally the discrete energy values of closed systems, which manifest themselves in the masses of elementary particles, for instance. In order to combine the apparently disconnected quantities, Heisenberg invented the scattering matrix.[27] By means of this tool, the scattering and eigenvalue problems may be treated alike.

The scattering matrix is the operator which connects the ingoing state of elementary particles with the outgoing state, characterized, for example, by the momenta $\mathbf{k}'_1, ..., \mathbf{k}'_n$ and $\mathbf{k}_1, ..., \mathbf{k}_m$. Thus the transition amplitude becomes,

$$\langle \mathbf{k}_1, ..., \mathbf{k}_m | S | \mathbf{k}'_1, ..., \mathbf{k}'_n \rangle . \tag{11}$$

Its main property is that it is *unitary*, or

$$S^+ S = 1 . \tag{12}$$

One can use the expression (11) for all possible transitions; in other words, once the scattering matrix is given, it is possible to derive all the transition amplitudes and, therefore, solve the problems of particle physics in principle.

On the other hand, the scattering matrix, or, more accurately, the reaction matrix, defined by

$$R = S - 1 \tag{13}$$

has a close relation to the interaction term in quantum theory, provided it exists.

4.2. ANALYTIC PROPERTIES OF THE S-MATRIX

When dealing with bound states, Heisenberg discovered the analytic properties of the scattering matrix in the complex energy plane.[28] Bound states are represented by poles on the positive imaginary axis of the momentum, as can be seen from the following consideration. Suppose the scattering process, occurring at the origin, turns an incoming spherical wave, $(1/r)e^{-ikr}$, of momentum k, into an outgoing one, $(1/r)e^{+ikr}$, then we obtain for the final wave,

$$\frac{1}{r} (e^{-ikr} - S(k) e^{ikr}) . \tag{14}$$

In this solution the last term disappears asymptotically for positive imaginary momentum; hence a singular behaviour of the S-matrix at positive imaginary momentum describes a bound state.

Møller discussed further the analytic properties of the S-matrix and related, for instance, the radioactive systems to properties of the S-matrix in a complex momentum space.[29]

Since the description by means of interacting fields had become too problematic, one sought to postulate the existence of a scattering matrix. In this manner, there arose a new picture of elementary particles. In this matrix theory only finite quantities, which are directly observable, are used. The properties of elementary particles, such as their masses and couplings, are directly reflected in the structure of the S-matrix. For instance, bound states appear as poles in the complex planes of variables. In a certain sense, the difference between elementary and composite objects disappears.

The symmetry properties of particles and their interactions can also be treated in the form of the S-matrix. If the S-matrix commutes with a symmetry operator, the corresponding conservation law is valid.

Thus it seemed very attractive to start with a scattering matrix and derive from it the properties and interactions of elementary particles. However, the problem concerning the principles from which the S-matrix should be derived remains, and the question arises whether one can ever know enough to construct *the* scattering matrix in particle physics.[30] In spite of these considerations of principle, the idea of the S-matrix found its way into particle physics and was used as an auxiliary tool in the construction of a more satisfactory quantum electrodynamics.

REFERENCES

1. M. Born and P. Jordan, *Z. Phys.* **34**, 858 (1925).
2. M. Born, W. Heisenberg, and P. Jordan, *Z. Phys.* **35**, 557 (1926).
3. P. A. M. Dirac, *Proc. Roy. Soc.* A **114**, 243 (1927).
4. P. A. M. Dirac, *Proc. Roy. Soc.* A **114**, 710 (1927).
5. P. Jordan and E. P. Wigner, *Z. Phys.* **47**, 631 (1928).
6. W. Heisenberg and W. Pauli, *Z. Phys.* **56**, 1 (1929); **59**, 168 (1930).
6a. The first invariant function was the D-function of Jordan and Pauli, *Z. Phys.* **47**, 151 (1928). The Δ-function was computed by Dirac in 1934, *Proc. Camb. Phil. Soc.* **30**, 150 (1934).
7. P. A. M. Dirac, *Proc. Roy. Soc.* A **136**, 453 (1932).
   P. A. M. Dirac, V. Fock and B. Podolsky, *Phys. Z. Sov. Union*, **2**, 468 (1932).
   P. A. M. Dirac, *Phys. Z. Sov. Union*, **3**, 64 (1933).
8. N. Bohr and L. Rosenfeld, *Kgl. Danske Vid. Selskab., Math-fys. Medd.* **12**, no. 8 (1933).
9. W. Pauli, *Naturwiss.* **21**, 841 (1933).
10. P. A. M. Dirac, *Seventh Solvay Conference*, 1933 (Gauthier Villars, Paris, 1934).
11. W. Heisenberg, *Z. Phys.* **90**, 209 (1934).
12. O. Klein and Y. Nishina, *Z. Phys.* **52**, 853 (1929).
13. V. F. Weisskopf, *Kgl. Danske Vid., Sels. Math.-fys. Medd.* **14**, no. 6 (1936); *Phys. Rev.* **56**, 72 (1939).
14. V. F. Weisskopf, *Z. Phys.* **89**, 27 (1934); **90**, 817 (1934).
15. W. Pauli and M. E. Rose, *Phys. Rev.* **49**, 462 (1936).
16. G. Wentzel, *Z. Phys.* **86**, 479, 635 (1933).
    P. A. M. Dirac, *Ann. Inst. H. Poincaré*, **9**, 13 (1939).
17. F. Bloch and A. Nordsieck, *Phys. Rev.* **52**, 54 (1937).
18. E. Fermi, *Ric. Sci.* **2**, no. 12 (1934).
19. I. Tamm and D. Iwanenko, *Nature*, **133**, 981 (1934).
20. H. Yukawa, *Proc. Phys. Math. Soc., Japan*, **17**, 48 (1935).
21. C. D. Anderson and S. M. Neddermeyer, *Phys. Rev.* **51**, 884 (1937); **54**, 88 (1938).

22. A. Proca, *J. Phys. Rad.* **7**, 347 (1936).
23. N. Kemmer, *Proc. Roy. Soc.* **A 166**, 127 (1938).
24. G. Wentzel, *Helv. Phys. Acta*, **13**, 269 (1940).
25. See W. Heisenberg, *Z. Phys.* **101**, 533 (1936).
26. W. Heisenberg, *Z. Phys.* **120**, 513, 673 (1942).
27. A similar idea was applied to certain problems of nuclear physics by J. A. Wheeler, *Phys. Rev.* **52**, 1107 (1937).
28. W. Heisenberg, *Z. Phys.* **120**, 673 (1943).
29. C. Møller, *K. Dansk. Vid. Sels. Mat.-fys. Medd.* **22**, no. 19 (1946).
30. W. Heisenberg, *Z. Phys.* **123**, 93 (1944); *Z. Naturf.* **1**, 608 (1946), and following papers.

EIGHTH SOLVAY CONFERENCE 1948

G.BALASSE L.FLAMACHE L.GROVEN O.GOCHE M.DEMEUR J.ERRERA VAN ISACKER L.VAN HOVE E.TELLER Y.GOLDSCHMIDT L.MARTON C.C.DILWORTH I.PRIGOGINE J.GEHENIAU E.HENRIOT M.VAN STYVENDAEL

P. KIPFER   P. AUGER   F. PERRIN   R. SERBER   L. ROSENFELD   B. FERRETTI   C. MOLLER   M. LEPRINCE-RINGUET

P. SCHERRER   E. STAHEL   O. KLEIN   P.M.S. BLACKETT   P.I. DEE   F. BLOCH   O.R. FRISCH   R.E. PEIERLS   H.S. BHABHA   J.R. OPPENHEIMER   G.P.S. OCCHIALINI   C.F. POWELL   H.B.G. CASIMIR   M. de HEMPTINNE

10

The Elementary Particles*

1. INTRODUCTION

Fifteen years after the seventh conference, the eighth Solvay Conference on Physics took place in Brussels from 27 September to 2 October 1948. The general theme of the Conference was 'The Elementary Particles'. Numerous techniques, which had been developed and promoted during the War, had become important for the progress of elementary particle physics, e.g. the particle accelerators and microwave devices. In addition, photographic emulsion techniques had improved so much that sensible observation of tracks of charged particles could be made and information about important properties of photographed objects could be obtained. It was at this crucial moment in the development of elementary particle physics, three years after World War II had come to an end, that a new Solvay Conference was held.

The Scientific Committee consisted of Sir Lawrence Bragg (Cambridge), President, N. Bohr (Copenhagen), Th. De Donder (Brussels), Sir O. W. Richardson (Alton, Hants.), E. Verschaffelt (The Hague), H. A. Kramers (Leyden), and E. Henriot (Brussels), Secretary. The other members of the Scientific Committee, P. Debye (Ithaca, N.Y.), A. Joffé (Leningrad), A. Einstein (Princeton, N.J.), and F. Joliot (Paris), did not attend the Conference.

Among the invited speakers were: C. F. Powell (Bristol), P. Auger (Paris), F. Bloch (Stanford, Calif.), P. M. S. Blackett (Manchester), H. J. Bhabha (Bombay), L. de Broglie (Paris) [represented at the Conference by Madame A. Tonnelat], R. E. Peierls (Birmingham), W. Heitler (Dublin), E. Teller (Chicago, Ill.), R. Serber (Berkeley, Calif.), and L. Rosenfeld (Manchester). N. Bohr, J. R. Oppenheimer (Princeton, N.J.) and W. Pauli (Zurich) commented on the current situation in particle physics.

Niels Bohr, after recalling the great stimulation which the Solvay Conferences on Physics had provided for the clarification of fundamental physical problems, discussed the notions of causality and complementarity. He summarized the epistemological situation in the following words:

'The impossibility of subdividing the individual quantum effects and of separating a behaviour of the objects from their interaction with the measuring instruments serving to define the conditions under which the phenomena appear implies an ambiguity in assigning conventional attributes to atomic objects which calls for a reconsideration of our attitude towards the problem of physical explanation. In this novel situation, even the old question of ultimate determinacy of natural phenomena

---

* *Les Particules Élémentaires*, Rapports et Discussions du Huitième Conseil de Physique tenu à l'Université libre de Bruxelles du 27 Septembre au 2 Octobre 1948, R. Stoops, Brussels, 1950.

has lost its conceptual basis, and it is against this background that the viewpoint of complementarity presents itself as a rational generalization of the very idea of causality.'

Although the 1948 Solvay Conference was devoted to the physics of elementary particles, two reports were given on different topics. First, P. M. S. Blackett discussed 'The Magnetic Field of Massive Rotating Bodies', and then E. Teller spoke 'On the Abundance and Origin of Elements'.

Blackett mentioned the hypothetical relation,

$$P = - \beta_1 \frac{G^{1/2}}{2c} \, U, \qquad (1)$$

which seems to exist between the magnetic moment P of a massive rotating body having an angular momentum, U; G and c are the gravitational constant and the velocity of light respectively, and β_1 is a constant of order 1. This relation would be satisfied approximately by the Earth, the Sun and 78 Virginis.

The understanding of this relation, Blackett pointed out, is not contained within the structure of present day physical theory and can only be achieved within the framework of a general theory embracing gravitational and electromagnetic phenomena. In this connection, he referred to a hypothesis of H. A. Wilson that a mass element moving with a velocity $\mathbf{v}$ produces a magnetic field at a distance r, given by

$$H = - \beta_1 \frac{G^{1/2}}{c} \, m \, \frac{\mathbf{v} \cdot \mathbf{r}}{r^3}, \qquad (2)$$

in analogy with the magnetic field of a moving charge.[1] This relation, although untrue for freely translating bodies, might apply to rotating bodies.

Blackett then discussed recent measurement of the magnetic field of the Earth, indicating that the Earth's crust, in which no currents can be assumed, contributes essentially to the magnetic field. Matter therefore seems to exhibit a new property, since the effect cannot be explained by the existence of magnetic materials. Blackett also described new measurements of the magnetic fields of certain stars by Babcock and suggested a value of 0.7 for the constant β_1 in Equation (1). This value seems to be confirmed for the mean field of the Sun as well.

The explanation of the new effect depends on whether one assumes that the mass *per se* in a massive body is responsible for it, or, rather, it is the number of electric charges. The second assumption would be more natural if, for a short time, the positive and negative charges are separated by rotation. It seems that the present knowledge of the constitution of celestial bodies favours this explanation.

Finally, in order to account for the definite sign in Equation (1), some asymmetry between positive and negative charges has to be introduced. According to the same equation, this difference is proportional to the square root of the gravitational constant. Hence, if Dirac's hypothesis of a decreasing value of G is correct, the rotating celestial bodies had a much larger magnetic field in earlier stages of the universe.

In the discussion of Blackett's report, more speculations about the empirical facts and theory were mentioned. Rosenfeld, in particular, recalled the unified theories of gravitation and electromagnetism. Since the time of Blackett's report, many more observations of the magnetic fields of stars and celestial objects have been made and other unified theories of electromagnetism and gravitation have been proposed. The situation, however, has not changed since 1948, and the relation (1) has remained in the realm of scientific speculation.

Teller's report also did not deal directly with particle physics, but with the question of the original formation of elements in the universe. Teller presented a paper, prepared jointly with Maria G. Mayer, summarizing the current status of the subject. Such a report was of interest, because in the previous decade considerable knowledge had been obtained about the conditions, under which nuclear reactions take place, and Teller himself was leading, at that time, a technological project based on the same type of physics.

The abundance of elements in various parts of the universe shows some striking regularities. One might consider three major examples: the Earth's crust, the meteorites, and the atmospheres of the Sun and the stars. Although the distribution of elements in the Earth's crust is most accessible, it represents a very biased sample of the more original distribution, since some of the gaseous parts have escaped the gravitational field, whereas certain metals have been concentrated in the core and others show up abundantly in the crust. Meteorites provide in many cases a far better sample for the relative cosmic abundances of elements, especially of the composition of the planet's core. The volatile elements are retained only in the atmospheres of the major planets and of the Sun or other stars.

Separating the elements into two classes, one containing elements up to iron, the other containing the heavy elements, one *finds mostly elements of the first class* in the universe, predominantly H, He, C and O. Moreover, great fluctuations of abundance occur when passing from one light element to the other, whereas the heavier elements are distributed more evenly. Turning to a detailed study of isotope abundances, further peculiarities are noted, such as the reduced amount of deuterium in star atmospheres in comparison to its occurrence on Earth, or the high amount of C^{13} in the spectra of some stars.

In order to explain these features, one must refer to the specific history of the cosmic body to be studied, as well as to processes that go on. The latter are, of course, nuclear reactions. But apart from the specific nature of these processes one finds, among the heavy elements, an increased abundance of some having a preferred nuclear structure. Nuclei containing 50 or 82 protons or neutrons occur particularly frequently.[3] One assumes that the lighter elements are formed by thermonuclear reactions, essentially by the addition of protons to already existing nuclei. Heavy nuclei, on the other hand, first contained a large excess of neutrons, and their present distribution results from stability against fission.

Teller discussed three theoretical schemes that should help to explain the origin of elements. First the neutron capture theory, according to which all nuclei are built up by successive neutron capture followed by beta-decay.[4] Then the occurrence of an element is determined by the competition between neutron capture and beta-decay of the neutrons. According to this theory, all light nuclei should eventually be trans-formed into heavier ones unless the process is stopped.

The hypothesis that elements are formed in thermodynamic equilibrium accounts for the fact that nuclei having a strong binding energy occur particularly frequently.[5] But this situation is valid only for light nuclei.

Finally, one has to assume that heavy elements are formed by nuclear fission. One starts from a primordial substance, which Gamow had called *Ylem*, containing many protons and even more neutrons [excess neutrons], which breaks up into lighter stable and quasi-stable nuclei. Teller even speculated that the *Ylem* might be a neutron fluid and that protons are produced from it by electron emission.

2. THE MESON FAMILY

The two following reports dealt with mesons, i.e., the objects predicted by Yukawa in 1935. These reports were particularly important since great confusion had arisen in the late 1930s. Let us recall the story. In 1935 Yukawa had put forward his theory of 'heavy quanta' to explain the short range character of the nuclear forces. However, Yukawa assumed at the same time that the heavy quantum, having a mass of approxi-mately 200 electrons, could decay into electrons and neutrinos, and thus be responsible for weak interactions.

The work of Anderson and Neddermeyer, during the period between 1936 and 1938, finally led to the identification of a particle with apparently the right properties, espe-cially the mass and the lifetime predicated by Yukawa.[6] The Yukawa particle should interact strongly with a nucleus. If the meson is charged negatively then it will be captured by nuclei in a state of high quantum number, which decays subsequently by γ-emission or the emission of an Auger electron. Due to its large mass, the heavy quantum would spend a large fraction of time moving within the nucleus. If it were to interact strongly, it must be absorbed before decaying. In the early experiments, such a behaviour was indeed found, but further experiments by Conversi, Pancini, and Piccioni[7] proved the opposite. Hence the Anderson–Neddermeyer particles were not the Yukawa mesons, a result which was confirmed by the fact that they could penetrate hundreds of nuclei before interacting.

Before the experiments of Conversi *et al.* brought certainty, Sakata and Inoue[8] sug-gested that the studies of cosmic rays had revealed the existence of mesons of two types. Marshak and Bethe[9] went even further. The contradiction between the apparently large cross-section for the creation of mesons in nucleon-nucleon collisions, as dem-onstrated by the large meson flux at sea level, and the subsequent very weak interac-tion of the resulting mesons suggested that the immediately created mesons are heavier and decay spontaneously with a lifetime of 10^{-8} sec into other mesons which interact only weakly with matter. While these theoretical speculations went on, new experi-ments with the recently developed 'nuclear emulsions' allowed one to isolate the decay process. It was about these observations that C. F. Powell reported at the eighth Solvay Conference.

2.1. OBSERVATIONS ON THE PROPERTIES OF MESONS IN THE COSMIC RAY RADIATION

(C. F. Powell)

Powell reported, in particular, on the most recent results of the Bristol group. He noted: 'It has been found possible to account for all the main phenomena, involving slow charged mesons with masses in the interval from 100–400 m_e, observed in photographic plates exposed to cosmic radiation, in terms of two types of particles, π and μ, which can be charged either positively or negatively.[9] A π-particle, of mass 310 m_e, decays spontaneously with the emission of a μ-particle of mass $\sim 200 \, m_e$, and a neutral meson of mass $\sim 80 \, m_e$.[10] The mean lifetime, τ_π, of both the positive and negative π-particles, is of the order of 5×10^{-9} sec. The π-particles can be produced in processes which lead to the explosive disintegration of nuclei[11], and when brought to the end of their range in photographic emulsions they are captured by atoms and produce disintegrations of both light and heavy nuclei.[12] Their modes of creation and extinction are therefore consistent with the view that they have a strong interaction with the nucleons. On the other hand, the μ-mesons are produced directly, either rarely, or not at all; and when brought to rest in silver bromide, they never, or only rarely, produce nuclear disintegration observable in the conditions of our experiment.[13] The observed behaviour of the μ-mesons is therefore consistent with the view that they have a very weak interaction with nucleons.'

Phenomenologically, the situation on the photographic plates is slightly more complicated. The ϱ-mesons are those which produce no visible secondary tracks, i.e. mainly $\mu^\pm$-particles and some $\pi^\pm$-particles where one fails to see the secondary μ^+, and also some π^- leading to disintegrations with the emission of neutrons. The mesons which produce heavy charged secondaries, hence visible tracks, were called σ-mesons; they include mainly π^-- and some μ^--particles, producing occasionally a star.

Powell reported that there were about eighty identified examples of the process in which a heavy π^+ decayed at the end of its range into a μ^+ which also stops in the emulsion.[14] From the fact that the velocity of the emitted μ^+ is fairly constant, he concluded that the masses of the two particles, m_π and m_μ are constants within narrow limits.[15] The exact values of the masses had been determined by the scattering of the particles with the constituents of the emulsion to be $m_\mu = 200 \pm 30 \, m_e$ and $m_\pi = 270 \pm 40 \, m_e$, respectively.[16]

Twenty-five examples of the production of mesons had been found until then in which the mesons reached their end of range within the photographic emulsion.[17] These events were nearly exclusively of the σ-type, hence π^--mesons, indicating that π^+-particles are emitted with higher energies since the nuclei from which they are produced repel them.

One had also determined the direction of the meson flux and found a downward flux of ϱ-mesons, i.e., of $\mu^\pm$-particles. The upward-going ϱ-events could be explained as resulting from the decay of $\pi^\pm$-mesons. The results were obtained by introducing blocks of lead close to the emulsions during the exposure.[18] It was also found that the μ-particles never produce disintegrations under the conditions of the experiment.

In an additional note Powell discussed estimates of the lifetime of π^- decaying into μ^-, and the evidence for the existence of a meson having the mass of 800 m_e.[19] The tracks of these heavier mesons show grain densities in between the $\pi^{\pm}$ and the nucleons; thus one could assume, with all reservations, the existence of the τ-particles of lifetime larger than 10^{-10} sec.

The first indication of the existence of these particles, which are now called K-mesons, was found by Leprince-Ringuet and L'Héritier while observing a charged particle, in a Wilson chamber, which collided with an electron.[20] The analysis of the kinematics of the collision led to a mass of the incident particle of 1000 m_e, but it was later criticized that one could not exclude the possibility of a proton collision with an electron.

In 1947, Rochester and Butler found two forked tracks in a cloud chamber with a magnetic field, and identified them as decays of two particles (one charged, one neutral) having a mass of about 1000 m_e.[21]

A year later Leprince-Ringuet and collaborators again found the first evidence for a very heavy meson in an emulsion, when they observed a particle track from which a slow π^- meson had emerged.[22] Nowadays it does not seem to be that this event was due to a K-meson; the first reliable report on a K-meson in nuclear emulsion was given by Brown et al.[23]

We should make a few remarks here about the 'laboratory' in which these new particles had been found, the cosmic radiation. In 1900, the experiments by C. T. R. Wilson and by Elster and Geitel showed conclusively the electrical conductivity of air.[24] While searching for the radiation that caused this conductivity, the balloon experiments of V. F. Hess and Kolhörster made it clear that a penetrating radiation, passing downwards through the Earth's atmosphere, gave rise to some of the ionization.[25] Later on this radiation was called 'cosmic radiation', and some of it consists of charged particles. It had first been assumed that most of the primary radiation consisted of electrons, but the east-west asymmetry of cosmic radiation (due to the deflection of the cosmic particles in the magnetic field of the Earth) proved that an important fraction was positively charged, and finally one found less than 1% of the primary radiation to be electrons.[26] Instead, helium and heavier nuclei were detected at altitudes of 30 000 km.

The ideas about the origin of cosmic radiation remained controversial for a long time. Nowadays one knows that an important part of the protons comes from the Sun; while others, mostly particles of higher energies, originate in the magnetic field of the galaxy. The primary spectrum of protons drops exponentially with the energy, and particles in the range of several GeV to several thousands of GeV are available. Thus the collision energy is sufficient to produce particles heavier than mesons. Indeed, in cosmic radiation, not only the positrons, $\mu^{\pm}$-particles, and $\pi^{\pm}$-mesons were first detected, but also the hyperons Λ, Σ, Ξ. The problem with cosmic rays is that one needs many exposures to observe a certain event. The detection devices, such as cloud chambers and photographic emulsions, could not be turned on specific events in cosmic ray physics, and one had to rely on accidents.

In the mid-1950s the importance of cosmic rays for the study of elementary particles and their interactions faded away as accelerators producing protons of higher and higher energies became available. The high energy machines provided an enormous advantage, since beams containing 10^{11} fast particles per pulse could be concentrated in a very small area. Accelerator techniques had started by using two simple condenser plates having a potential difference. To enhance the potential difference Van de Graff generators and tandem accelerators were developed before the age of circular accelerators started, particularly with the cyclotron, to which E. O. Lawrence had contributed greatly. In Lawrence's laboratory at Berkeley, cyclotrons and synchro-cyclotrons of increasing energy became available, and the 188-inch machine was able to accelerate protons to nearly 0.5 GeV. The future, however, belonged to synchrotrons, and one of the earliest machines, the Bevatron of Berkeley, was designed for energies sufficient to create anti-protons. The anti-proton was detected during the same year that the machine came into operation.

2.2. ARTIFICIAL MESONS

(R. Serber)

The second report on mesons at the eighth Solvay Conference was given by Serber. In the first part of his report Serber discussed the methods and pre-requisites of particle production; and, in the second part, the experimental results.

π-mesons were produced when a target was bombarded by high energy α-particles, accelerated in the Berkeley cyclotron, according to the reaction,

$$\alpha(N) \to P + \pi^-(\alpha). \tag{3}$$

The threshold for meson production is 260 MeV, and the production increases rapidly with the rising energy of the particles. One could prove that no other light particles were involved in the process (3). The π-mesons thus produced decay into μ-particles,

$$\pi \to \mu + \begin{cases} \nu \\ \gamma \end{cases}. \tag{4}$$

Serber pointed out that, 'In writing (4) we have supposed the third particle involved to be a neutrino or a gamma ray, since the evidence on the masses of the π- and μ-meson now seems to be compatible with a third particle of zero mass, and we should like to avoid the hypothesis of a new kind of neutral particle. Lattes and Gardner now give as their best value for the π mass, $M_\pi = 286 \pm 6$ electron masses. Professor Brode finds for μ-mesons of cosmic ray origin, $M_\mu = 212 \pm 4 \, m_e$.'

And, he continued: 'The ratio of these two numbers is $R = 1.35 + 0.04$, which is to be compared to the ratio $R = 1.32$ which is to be expected for zero mass of the third particle. Lattes and Gardner are now carrying out a more direct measurement of the ratio. A very preliminary value is $R = 1.36$ which, however, is based on the measurement of the masses of only two μ's, the values found being 209 and 211.'

The mean life of the π-meson is found to be roughly, $\tau = 9 \times 10^{-9}$ sec. 'It may be mentioned,' said Serber, 'that a π lifetime of such a magnitude, together with the previously mentioned fact that of the thirty π^{+}'s seen to stop all decayed, rules out the connection originally suggested by Yukawa between the beta-decay of nucleus and the beta-decay of mesons. For if beta-decay of the π-meson is a process competing with [(4)], the above evidence shows that its lifetime must be greater than 3×10^{-9} sec. On the other hand, the observed beta-decay rate would demand a meson lifetime of about 3×10^{-11} sec, a hundred times shorter than is admissible.'

The μ^{-}-meson never causes a star in the emulsions. From the capture in nuclei one can estimate, $\tau_{\pi} = 1.2 \times 10^{-8}$ sec. The decay of the μ-meson yields its spin. Serber pointed out that, 'For integral spin we would expect decay into electron and neutrino, and non-energetic electrons. However, for spin $\frac{1}{2}$, a plausible hypothesis would be

$$\mu \rightarrow e + 2\nu. \tag{5}$$

An interesting feature of this suggestion is that the energy spectrum of the electron may depart quite radically from an ordinary beta-spectrum.'

In the second part of his report Serber discussed some experimental details of determining the properties of π- and μ-particles.

DISCUSSION OF THE REPORTS OF POWELL AND SERBER

The discussions of the reports by Powell and Serber were lively and concentrated on theoretical and experimental questions. Teller mentioned the experiments of Steinberger carried out at Chicago that seemed to prefer the μ-decay, Equation [(4)]. L. Rosenfeld summed up the discussion of Serber's report by remarking that:

'The necessity of abandoning the conception of beta-decay, suggested by Yukawa, and involving the mesons as intermediary, does not, at the present stage imply any loss of simplicity.

'In fact, we have in any case four independent couplings between the various bonds of elementary particles; from this point of view it does not matter whether we assume a direct coupling between π-mesons and leptons, as in Yukawa's conception, or between nucleons and leptons, as in Fermi's original theory. In the latter case, the resulting indirect coupling between μ-mesons and leptons turns out to be weaker than that between π-mesons and μ-mesons, which agrees with the Berkeley findings.'

After Powell's report, the evidence for the heavier mesons, in particular, was discussed. The hypothesis that the charged meson is a proton was proposed, but subsequently rejected. Leprince-Ringuet gave a report on the event which he had found with L'Héritier and referred to further studies. N. Bohr then discussed particular aspects of the stopping of mesons in nuclear emulsions. 'In heavy materials,' he noted, 'the mesons have more chances to be captured before decay.' There was general agreement that these new particles had to be taken seriously, and the enlarged meson family came to be recognized by the participants at the eighth Solvay Conference.

3. DIFFERENT TYPES OF INTERACTIONS

The two reports on new kinds of mesons were followed by a talk on the experimental properties of cosmic ray products, and was presented by P. Auger.

3.1. PROPERTIES OF PARTICLES CONTAINED IN LARGE ATMOSPHERIC SHOWERS

(P. Auger)

Auger discussed in particular the properties of those particles which penetrate farther than electrons normally do. Investigations had been carried out on the absorption of these objects and their ability to produce secondaries when passing through lead.

An array of counters was placed at small distances from each other and plates of lead could be interposed in between. Two types of particles result from the showers, one being absorbed twice as much as the other one. The latter component produces many secondary electrons in lead. Auger did not think that the explanation of this effect could be given by the hypothesis that only mesons and electrons were contained in cosmic rays.

In the discussion of Auger's report, Ferretti raised questions about the reliability of the counter experiments. The question was also raised whether the 'penetrating' component was not partly produced in the layer of lead. Oppenheimer suggested that 'We go back to consider the origin of these great showers. I think it is not settled, but believe that they are not made by primary electrons. There is a great mass of evidence that when primary nuclei collide in the atmosphere they produce three components, mesons of various kinds, nucleons, and soft radiation. On purely theoretical grounds it is easier to understand the soft radiation as a decay phenomenon. Thus in an Auger shower there may be, together with the cascades, nucleons and mesons. The mesons will be of such a high energy that a part of them will not have decayed into "cold" mesons, but remain "hot", i.e. still capable of nuclear interactions.'

Blackett also made a few remarks concerning the same complexity of phenomena which are observed by counters. There occurs the creation of very energetic non-electron pairs, the creation of several particles into a small angle region as well as genuine cascades involving only electrons and photons, and finally the τ-mesons, both neutral and charged. 'The study of photographs serves to show the extraordinary complexity of shower phenomena, and to emphasize the difficulties of making clear-cut and unique deductions from counter experiments alone. The study of these shower photographs must surely begin with their classification into distinct types before any very detailed quantitative measurements are made. The first approach must thus be more "botanical" than quantitative.'

3.2. NUCLEAR FORCES

(L. Rosenfeld)

The relation of the phenomena observed in cosmic ray experiments to the 'well-known' problems of nuclear forces was discussed by L. Rosenfeld who gave an extensive report on the status of the latter. He started by outlining the main difficulties of his endeavour. 'Nuclear interactions can either be described in a purely phenomenological way by means of the concept of nuclear potential, or interpreted as due to the nuclear field, related to some kind of particles of intermediate mass. At the present stage, both descriptions involve too many uncertainties or even contradictions to allow a consistent picture of nuclear forces to be drawn. The present report will therefore be confined to a discussion of some rather disconnected aspects of the problem, which have been the subject of recent study. We shall first examine how far the field description of nuclear interactions is in harmony with the experimental facts concerning nuclear processes and with the various kinds of mesons lately discovered. We shall then inquire what inferences can be drawn about the properties of the mesons responsible for the nuclear field from the results of a more phenomenological analysis of the relevant empirical evidence.'

The concept of a field, Rosenfeld stated, is very suitable to provide an invariant description of interactions. It couples sources by transporting energy via an intermediate elementary particle. Hence in a field theory particles are coupled, and the range of interaction is connected with the mass μ of the particle transmitting the interaction according to Yukawa's formula, $V(r) = e^{-\mu c/\hbar \cdot r}/r$.

'According to this general picture,' said Rosenfeld, 'we must expect the nuclear field, of range $R \sim 10^{-13}$ cm, to be associated with particles of mass $\mu \sim 300$ to 400 times the mass of the electron. The order of magnitude of the constant g (analogous to the elementary electric charge in the theory of electromagnetic interactions), which gives a measure of the intensity of the sources of a nuclear field, can be estimated from the strength of the nuclear potential; it is expressed by the dimensionless ratio,

$$\frac{g^2}{\hbar c} \approx 0.1 .' \tag{6}$$

This field theory of the nuclear interaction would explain naturally the anomalous magnetic moments of the proton and the neutron as being due to the clouds of charged nuclear fields, although a quantitative calculation could not be made. [This is still the status in 1975!]

The relationship of the nuclear field to the newly discovered mesons is rather complex, but 'Obviously the π-mesons are strongly coupled to the nucleons and must essentially contribute to the nuclear interactions; their mass gives a quite acceptable value for the range of the forces. Precisely how the nuclear forces are brought about by the π-mesons will primarily depend on their spin, about which we have no experimental indication whatever. Much simpler course, for the time being, is to *assume that they have integral spin (0 or 1)*. It is then possible to take over the usual meson theory of nuclear interactions, as it has been developed until now. In particular, the charge independence of the nuclear interactions will then require the existence of neutral π-mesons, which would presumably be highly unstable with multiple photon emission.'

The μ-meson, on the other hand, may have half integral spin, and Rosenfeld proposed the following decay for the pions:

$$\pi^\pm \to \mu^\pm + \mu^0, \tag{7}$$

where μ^0 has a much smaller mass than $\mu^\pm$-mesons, and one must introduce a new coupling constant $\check{g}$ for this process. The lifetime of the meson is 10^{-8} sec, hence 'it is tempting to imagine that the π-mesons are coupled to pairs of μ-mesons and to pairs of leptons in *exactly* the same way; in fact, as if μ-mesons and leptons were different states of a single species of elementary particles of half integral spin.' This statement sounds like a prophecy of the universal weak interaction theory. It must, however, be viewed with some care because Rosenfeld linked it to the Yukawa theory of weak interactions, involving the pion as the 'intermediate boson'.

Concerning the decay of the μ-meson, Rosenfeld concluded from the energy spectra of decay electrons that again a neutral μ^0 participates in the interaction,

$$\mu^{\pm} \to e^{\pm} + v + \mu^0. \tag{8}$$

He noted: 'Theoretically this process would be described by introducing a direct coupling between $\mu - \mu^0$ meson pair; the coupling parameter, determined by the lifetime t_0 of about 3 sec, would be of the order of magnitude g_F (the Fermi coupling constant). The complete set of interactions between the various kinds of particles might then be symbolized by either of the following schemes, in which the dotted arrows represent indirect couplings.'

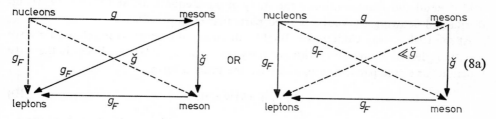

Rosenfeld then returned to the more complicated phenomenological aspects of nuclear potentials which act between the nucleons in a nucleus. This analysis is rather involved since many-body interactions come in. But even in the simple case of the deuteron, spin-orbit couplings cause non-central potentials.

The range of the nuclear potential does not allow a clear prediction of the mass of the intermediate mesons, since one may assume, according to Schwinger, two types of mesons, i.e. pseudoscalar and vector mesons, giving an effective potential in the ^1S-state proportional to

$$V(\alpha) - \frac{1}{r} \left[e^{-xr} + (2\mu^2 - \gamma^2) e^{-x_\mu r} \right] \tag{9}$$

where x is the mass of the pseudoscalar meson, x_μ the mass of the vector meson, and γ^2 the ratio of the coupling strengths.[27] But it appears, Rosenfeld concluded, that the nuclear forces cannot be well described by this theory. However, the isotopic spin structure of the nuclear interaction was not clear in 1948.

In the discussion Serber reported on experiments of high energy neutron-photon scattering, done at Berkeley by Segré, Wilson, Powell and collaborators, using 90 and 40 MeV neutrons. He concluded that, 'It seems to me that the above results point the moral that we know very little in fact about the nuclear forces, and that the meson theories predicated on the supposed properties of such forces must be taken with appropriate caution.'

4. ASPECTS OF QUANTUM FIELD THEORY

The theoretical description of phenomena connected with elementary particles made use exclusively of quantum field theory. The schemes which had been developed in the 1930s were taken up with new vigour and important extensions were studied, and some of them were discussed at the eighth Solvay Conference in 1948.

4.1. GENERAL RELATIVISTIC WAVE EQUATIONS

(H. J. Bhabha)

The increasing number of elementary constituents of matter and the possibility that these objects have spins different from 0 and $\frac{1}{2}$ forced physicists to deal quite generally with higher spin equations. As often before, work started with a paper by Dirac in 1936, who had treated the force free case and then interaction with an electromagnetic field.[28] Pauli and Fierz continued the study of these equations, and Proca (1936) and Rarita and Schwinger (1940) discussed particular examples.[29]

At the 1948 Solvay Conference, Bhabha discussed the progress in relativistic wave equations describing particles with arbitrary spin. The equations are linear in the wave function for a free particle case, and have the general form

$$(\alpha^k p_k + x\beta)\,\psi = 0. \tag{10}$$

In Equation (10), p^k is the four-momentum of the object to be described, the α's and β's are matrices and the x's represent the mass. From relativistic invariance certain properties follow for the matrices α. The latter quantities also enter into the construction of the (free) current

$$S^k = \psi^+ D\alpha^k\psi, \tag{11}$$

and the energy density

$$T^{00} = \psi^+ D\alpha^0 p^0\psi, \tag{12}$$

where ψ^+ is the hermitian conjugate to the wave function ψ, and D is a hermitian matrix.

An important question connected with the equations of higher spin particles is that of statistics. Since 1936 Pauli had devoted himself to this problem, showing finally that in the case of no interaction the integral spin equations have always to be quantized with Bose–Einstein statistics, whereas half-integral spin equations require Fermi–Dirac statistics.[30] The essential point in Pauli's arguments was the postulate that measurements at two space points with space-like distances can never disturb each other. As Bhabha mentioned in his report: 'From the experimental point of view the situation is that we have information only in three cases, proton, neutron, and electron, and in all three cases the particles have one-half spin and obey Fermi–Dirac statistics. I don't think that we know at the present time that any elementary particle has certainly an integral spin.' To which Serber replied that the arguments for spin 1 of the π-mesons were very strong.

The equations for higher spin have other consequences. First, one can derive the fact that the eigenvalues α^0 are related to the masses of the particles. If α^0 has only one eigenvalue, then $p^2 = x^2$ is single-valued; otherwise ψ describes a set of particles with different masses.

Normally one requires that the matrices α will be algebraically irreducible. As a consequence subsidiary conditions arise: for instance, the Lorentz condition in the case of spin 1 objects. The quantization of higher spin equations leads to formal problems which have been solved only partially.

4.2. Study of the General Theory of Particles with Spin by the Fusion Method

(L. de Broglie)

Bhabha's communication was followed by a report of de Broglie, which was presented by Madame M. A. Tonnelat. The 'fusion method' was developed by de Broglie since 1932 in order to obtain a wave mechanics of the photon.[31]

The essence of this method is to consider particles with spins higher than $\frac{1}{2}\hbar$ as composite objects, beginning with the photon.[32] Thus, photons should be considered as composite objects formed by the intimate fusion of two spin $\frac{1}{2}$ objects, giving a vector photon and a scalar photon. The Bose statistics then immediately follows for the composite particles.

Thus the formalism for describing the photon arises from the Dirac equation for the electron, and leads to a first order differential equation for the photon. One can then introduce interaction into the scheme, and de Broglie treated the zero-mass and non-zero-mass cases together. Since the zero-mass case gives troubles with quantization, he favoured a finite rest-mass for the photon. The divergence problems of the interacting photons are removed by introducing a non-point-like interaction and a 'smallest length'.[33] The same idea was also advocated by N. Rosen, and a bit later by A. March and W. Heisenberg.

Similarly, particles with spins higher than 1 can be composed of spin $\frac{1}{2}$ objects; for instance, gravitons with spin 2 would be formed by four spin $\frac{1}{2}$ particles. This procedure also allows one to explain naturally the resemblance between Newton's and Coulomb's laws, as Madame Tonnelat pointed out in an appendix to de Broglie's report.

In the discussion of the de Broglie–Tonnelat report Schrödinger and Casimir asked about the influence exerted by a photon with a mass ($< 10^{-44}$g according to de Broglie) on the electromagnetic phenomena, and it was found that only negligibly small effects would be caused.

4.3. Quantum Theory of Damping and Collisions of Free Mesons

(W. Heitler)

Whereas the problems discussed by Bhabha and de Broglie concerned the formal structure of quantum field theory, the reports of W. Heitler and R. E. Peierls dealt with one of the most crucial difficulties of the theory.

Heitler spoke on the quantum theory of damping, which is a heuristic attempt to eliminate the infinities of quantum field theory in a relativistically invariant manner. In this case, the problems are usually formulated in a Hamiltonian scheme, and the total Hamiltonian, H, is split up into two parts

$$H = H_0 + H_{\text{int}}.\tag{13}$$

where H_0 is the free Hamiltonian and H_{int} is the interacting part. When one calculates the transition of a state 0 to a state n in terms of a perturbation series, not only does the series diverge but so also do the individual terms, other than the first term which depends only on H_0.

Physical reasons require that the first term be the dominant one if H_{int} contains a small coupling constant, and that a few terms of the perturbation series should suffice to calculate the process. The problem then immediately becomes to render the first divergent term convergent. This is achieved by inserting into the equation for the transition element H_0, instead of the total H,

$$U_{n|0} = H^0_{n|0} - i\pi \sum_A H^0_{n|A} U_{A|0} \delta(E - E_A).\tag{14}$$

Integrating Equation (14) over dE_A gives,

$$U_{n|0} = H^0_{n|0} - i\pi \sum_A H^0_{n|A} \varrho_A U_{A|0},\tag{15}$$

where ϱ_A is the density function. The second term in Equation (15) is called the *damping term*. It is always small in electron-photon processes, since there

$$H^0_{n|A}\varrho_A \ll 1,\tag{16}$$

whereas in meson processes, $H^0 \varrho \gg 1$. One cannot, therefore, neglect the damping term in the latter.

The general solution of Equation (15) is formally expressed as

$$U = \frac{\bar{H}}{1 + i\pi\bar{H}_\varrho},\tag{17}$$

where $\bar{H} = H^0$. If the state density ϱ increases with the square of the energy ε, the cross-section becomes

$$\phi = |U|^2 \varrho \propto \frac{1}{\varrho} \propto \varepsilon^{-2},\tag{18}$$

whereas perturbation theory yields the increase with ε^2. Heitler concluded: 'The present case can also be regarded as a limiting case of the so-called strong coupling theory where a finite size of the sources is introduced and where a characteristic quantity, the spin inertia, occurs. The present theory is the limiting case of spin inertia $\to 0$, or very small.'

Heitler then applied his fundamental Equation (15) to meson collisions. The cross-section for the scattering of a meson from the nucleon decreases due to the damping. The theory also predicts that multiparticle production processes occur very rarely in high energy collisions and nearly no antiprotons will be produced. But if two nuclei collide, several mesons may come out due to the compound structure of the nuclei (plural production).

4.4. PROBLEMS OF SELF-ENERGY

(R. E. Peierls)

Since the diverging terms are neglected in the transition matrix, processes that depend only on them cannot be treated, especially those related to the self-energy, which became most important later on. Peierls treated these problems in his report. He noted that, 'It is well known that the infinite self-energy resulting from almost all forms of field theory that have been studied is one of the most important obstacles to the progress of fundamental theory. The position has remained virtually unchanged for about fifteen years in so far as there still exists no formalism which is free from singularities, compatible with the requirements of Lorentz invariance and other general laws, and which, in the domain to which ordinary non-relativistic quantum theory is applicable, represents the earlier results achieved there and confirmed by experiment.'

Peierls continued: 'Nevertheless the subject has developed in several ways. Firstly, a number of attempts to remove the difficulties have been investigated and while a few lead to mathematical problems of such complexity that no conclusions have been drawn with certainty, most of them have been shown to fail.'

The existence of many particles, instead of a very few, has dimmed the hope of having a simple theory. 'It is not known today whether a problem like the electromagnetic self-energy can be solved on its own, or whether only a formalism which accounts also for all interactions could have a chance of success. But at any rate a satisfactory solution of this self-energy, if it is possible, would not represent the last word and one is, therefore, more inclined to accept formalisms which will contain arbitrary parameters or arbitrary functions which at this stage do not come out of the theory but have to be taken from experiment. Equally it would not be surprising if the formalism would not finally settle the value of the fine structure constant or the ratio of the proton to electron mass.'

As an example, Peierls dealt briefly with the self-interaction of electrons with their own electromagnetic field, but pointed to similar problems in meson-nucleon theory. This question goes back already to the classical theory of electrons. The ordinary Coulomb attraction between two particles of opposite charge becomes arbitrarily large, if the particles approach indefinitely closely. To eliminate this difficulty Lorentz and others had considered equations in which the point charge was replaced by a charge distribution of finite radius a.

The classical Lorentz equation,

$$m \frac{d^2x}{ds^2} + \frac{2}{3}\frac{e^2}{c^3}\left[\frac{d^3x}{ds^3} - \left(\frac{d^2x}{ds^2}\right)^2 \frac{dx}{ds}\bigg/\left(1 + \left(\frac{dx}{ds}\right)^2\right)\right] = F \qquad (19)$$

gives the motion of an electron of mass m and charge e, moving under the influence of a force F on a world-line s. In this equation, terms are neglected which are small if the Fourier transform of the velocity contains only frequencies small compared to τ^{-1}. The latter quantity is 10^{23} sec if the classical electron radius e^2/mc^2 is taken.

The damping term on the right hand side of Equation (19) accounts for the emitted radiation, and this equation holds as long as the wavelength is smaller than 2×10^{-13} cm. 'Since both energy conservation and the laws of emission of radiation are well confirmed within the range of applicability of Equation [(19)], the same is evidently true of this equation itself. We have thus the difficulty in the form that infinitely small a makes the inertia term infinite, whereas a finite a makes it impossible to describe [(19)] as an approximation to a Lorentz invariant equation.'

In order to remove some of the difficulties, Dirac had proposed a new interpretation of the classical theory by redefining the field at the position of the electron.[34] Dirac used as this field's intensity,

$$A = \tfrac{1}{2}(A_{\text{ret.}} - A_{\text{adv.}}). \qquad (20)$$

Equivalent formalisms leading to Equation (19), i.e. without the terms of higher order, were devised by G. Wentzel and others.[35]

The problem with Equation (19) is that one obtains 'runaway solutions' if one starts with a finite acceleration, even when $F = 0$. The boundary condition of finite acceleration at $t = +\infty$ would exclude them; however, it seems difficult to extend such a condition to quantum theory.

Instead of making the classical equation (19) an exact equation by certain manipulations, one can rather assume that it is an approximation of a non-linear equation in the case of not too intense fields.[36] In these attempts one replaces the integrand of the action integral of the electromagnetic field,

$$\mathbf{E}^2 - \mathbf{H}^2, \qquad (21a)$$

by

$$f(\mathbf{E}^2 - \mathbf{H}^2, \mathbf{E} \cdot \mathbf{H}), \qquad (21b)$$

where $\mathbf{E}$ and $\mathbf{H}$ are the electric and magnetic fields, respectively, and f is a function such that, for weak fields $\mathbf{E}$, $\mathbf{H}$, the usual expression follows, but the field close to an electron saturates. Then the electron radius is the distance at which the field becomes critical, and the field deviates strongly from those of Maxwell's equations. While this extension leads to reasonable consequences in the classical case, nobody knows how the quantization can be achieved.

Another possibility of obtaining a finite classical field theory consists in subtracting from the Coulomb force between two particles another force which is equal to the

Coulomb force at close approach, but decreases more rapidly with distance, say like a meson force.[37] But the new fields make a negative contribution to the energy density, and a particle losing such radiation is likely to gain energy and become unstable.

McManus, on the other hand, had tried to formulate a theory of an extended electron by introducing a relativistic form factor.[38] It is, however, difficult to handle this scheme, and even more to quantize it properly.

Finally, the action-at-a-distance approach of Feynman and Wheeler, and their predecessors, avoids the problem of self-energy from the very beginning. The important point here is that the causal transmission of energy comes about by an absorption condition.[39]

To summarize, acceptable cures exist within the framework of classical theory for the divergence of self-energies. The problem is, however, how to extend these proposals to quantum field theory. An electron interacts with the Coulomb field as well as with the transverse radiation field. The latter, however, is quantized and one has to calculate it according to quantum [perturbation] theory. The second order term is

$$E_2 = \sum_{n \neq 0} \frac{|V_{0n}|^2}{E_0 - E_n}, \tag{22}$$

where E_0 is the state in question and E_n are the possible intermediate states. The sum (22) diverges quadratically in Dirac's electron theory, but only logarithmically in the hole theory.[40] The same degree of infinity applies to the static term.

A further difficulty in quantum electrodynamics appears because of the fact that, owing to the zero mass of the photons, the probability of soft photon emission becomes infinite in perturbation theory. By replacing the perturbation approximation by a more complete method, Bloch and Nordsieck had solved the problem.[41]

In order to eliminate the divergences of self-energy altogether, one has to modify quantum field theory. One method, used by Heitler [see the foregoing], was to leave out from perturbation calculation all intermediate (=bound) photons. Other solutions involved the λ-limiting procedure of Wentzel. Pais had studied the subtraction formalisms.[42] Also, the quantization of space was applied to the self-energy problem[43] and non-Hamiltonian theories had been developed.[44] Since one cannot define a state better than up to 10^{-23} sec, a Hamiltonian does not make direct sense in the quantum field theory of interaction.

In any case, Peierls concluded, some change in the fundamental equations seemed to be necessary, but it might not show up clearly in the electron-proton scattering due to the extended structure of the latter.

5. THE STATUS OF QUANTUM ELECTRODYNAMICS

In the years immediately preceding and following 1948, quantum electrodynamics had become a more complete scheme. Some of the progress was reported and discussed at the 1948 Solvay Conference.

5.1. Non-Relativistic Quantum Electrodynamics

(H. A. Kramers)

H. A. Kramers reported on non-relativistic quantum mechanics and the correspondence principle. All the important applications of classical electron theory are independent of the structure of the electrons; only the charge e and mass m appear as parameters in the equations. In passing on to a quantum theory, one has to perform the following preparations: (a) to find the simplest way in which the structure-independent part of the assertions of the electron theory can be mathematically expressed; (b) to bring in these expressions in the form of the canonical equations of motion. These questions were treated to some extent by Serpe, Opechowski and Kramers.[45] This work, although performed at the non-relativistic level, allowed an analysis of emission, absorption, and scattering of light, and also somewhat clarified the problem of divergences in quantum field theory.

Kramers considered an electron of mass, m_0, and charge, e, having the linear dimension of magnitude a, the electron radius. If the wavelengths characterizing the fields are large with respect to a, one can expect a structure-independent behaviour of the electron provided its velocity is small with respect to c. The vector potential $\mathbf{A}$ is now split into two parts,

$$\mathbf{A} = \mathbf{A}_0 + \mathbf{A}_1, \tag{23}$$

where $\mathbf{A}_1$ is due to the external field, and $\mathbf{A}_0$ is responsible for the electron current ('proper' field),

$$\mathbf{A}_0 = \theta \frac{e}{a} \frac{\partial \mathbf{r}}{\partial t} \tag{24}$$

for $r \gg a$, $(\theta \sim 1)$. The fundamental, structure-independent, equations then take the form,

$$m\ddot{\mathbf{r}} = -e \langle \mathbf{A}_1 \rangle + e\dot{\mathbf{r}} \times \langle \operatorname{rot} \mathbf{A}_1 \rangle - \frac{\partial U}{\partial \mathbf{r}}, \tag{25a}$$

and

$$\Box \mathbf{A}_1 = \dddot{\mathbf{A}}_0, \tag{25b}$$

where $\langle \ \rangle$ indicates averages over the electron, and U represents the external potential.

From these relations Kramers was able to calculate the important quantities in quantum electrodynamics. By taking the difference of the energies of an elastically bound and a free electron, he obtained,

$$\Delta \varepsilon = \tfrac{3}{2}\hbar k_1 \left\{ 1 + \frac{1}{\pi} \frac{k_1}{K} \left(\log \frac{K}{k_1} + c \right) + \cdots \right\}, \tag{26}$$

where k_1 is the frequency of the electron oscillator and C becomes zero in the local interaction limit. The correction term to the ground state of the harmonic oscillator, $\tfrac{3}{2}\hbar k_1$, is the same which appears in Bethe's formula for the Lamb shift.[46]

Thus Kramers' theory of electrodynamics permits some of the calculations which are typical of the 'consistent' quantum electrodynamics. Some of the spurious divergences due to the point model of the electron may be removed. But Kramers concluded that 'the establishment of a relativistic theory of charged particles has hitherto been, and will perhaps remain so for some time, a story of artful and ingenious guessing.' Kramers, one of the great experts on the quantum electrodynamics of 1920s and 1930s did not recognize how close the solution of the problem was, along the lines he had indicated, and to which he had contributed so much. It appears that Kramers had been guided a little too much by the correspondence principle, and when asked in the discussion he confessed that he had not attempted a relativistic theory because 'there are so many things which are not correspondence-like in the ordinary way. Secondly,' he said, 'I think there does not exist a really consistent classical relativity theory [of the electron].'

5.2. ELECTRON THEORY

(J. R. Oppenheimer)

Oppenheimer opened his report on the electron theory with the remark: 'In this report I shall try to give an account of the developments of the last year in electrodynamics.' What had happened?

The quantum field theory of Dirac, Heisenberg and Pauli, in the late 1920s, led to an understanding of the emission, absorption and scattering processes, but it also brought about the difficulty of the infinite displacement of spectral lines.[47] Pair production improved the divergence properties of the self-energy integral.[48] On the other hand, it led to infinite vacuum polarization. The vacuum polarization gives rise to an 'induced' charge, but only the sum of the 'true' [bare] and the induced charges can be measured; this is charge normalization. What was needed for further progress was the theoretical calculation of an experimentally observed effect [Lamb shift], but this experiment was not performed before 1947.[49] All earlier experiments failed to give deviations from the expectation of a suitably handled subtraction quantum electrodynamics in which infinite terms were just neglected.

Oppenheimer noted: 'The problem then is to see to what extent one can isolate, recognize and postpone the consideration of those quantities, like the electron's mass and charge, for which the present theory gives infinite results – results which, if finite, could hardly be compared with experience in a world in which arbitrary values of the ratio, $e^2/\hbar c$, cannot occur. What one can hope to compare with experience is the totality of other consequences of the coupling of charge and field, consequences of which we need to ask: does theory give for them results which are finite, unambiguous and in agreement with experiments?'

The earlier methods rested on an expansion in the coupling constant e up to the second order, on the radiative scattering corrections, on the Lamb shift, and on the electron's g-value.[50]

Oppenheimer continued: 'Such a procedure would no doubt be satisfactory, if cumbersome, were all quantities involved finite and unambiguous. In fact, since mass and charge corrections are in general represented by logarithmically divergent integrals, the above outlined procedure serves to obtain finite, but not necessarily unique or correct, reactive corrections for the behaviour of an electron in an external field; and a special tact is necessary, such as that implicit in Luttinger's derivation of the electron's anomalous gyromagnetic ratio, if results are to be, not merely plausible, but unambiguous and sound. Since, in more complex problems, and in calculations carried to higher orders in e, this straightforward procedure becomes more and more ambiguous, and the results are more dependent on the choice of Lorentz frame and of gauge, more powerful methods are required. Their development has occurred in two steps, the first largely, the second wholly, due to Schwinger.[51']

The first formal step consists in a contact transformation which describes the electron, not subject to external fields, in terms of the observable charge and mass. It yields the following results: (i) an infinite term in the electron's electromagnetic inertia; (ii) an ambiguous light quantum self-energy; (iii) an infinite vacuum polarizability, and the familiar results [Møller scattering, Compton effect, emission and absorption]. The results in radiative corrections, however, are frequently dependent on gauge and Lorentz frame. Although the fundamental equations of quantum electrodynamics are Lorentz and gauge invariant, the solutions in powers of e are not. Hence one must leave the frame and gauge uncertain until afterwards. A covariant formalism, however, was developed by Tomonaga[52] and Schwinger.[51] An important step in it is the use of the interaction representation in which the state vector is defined as

$$|\psi(t)\rangle = e^{iH_0^s/\hbar}|\phi_s(t)\rangle , \tag{27}$$

where H_0^s is the free Hamiltonian in the Schwinger representation and, $|\phi_s(t)\rangle$, the corresponding state vector. Thus the state vector is not constant as in the Heisenberg picture; it is so only if there is no coupling to fields, i.e., if the elementary charge is zero. On the other hand, the solution of the field equations is largely simplified since they involve only the free Hamiltonian. Schwinger eliminated from these equations the 'virtual' transitions by contact transformations and obtained finite and covariant results.

Oppenheimer concluded his report by listing five questions:

(1) Can the development be carried further, to higher powers of e, (a) with finite results, (b) with unique results, (c) with results in agreement with experiment?

(2) Can the procedure be freed of the expansion in e, and carried out rigorously?

(3) How general is the circumstance that the only quantities which are not, in this theory, finite, are those like the electromagnetic inertia of electrons, and the polarization effects of charge, which cannot directly be measured within the framework of the theory? Will this hold for charged particles of other spin?

(4) Can these methods be applied to the Yukawa-meson fields of nucleons? Does the resulting power series in the coupling constant converge at all? Do the corrections improve agreement with experience? Can one expect that when the coupling is large there is any valid content to the Maxwell–Yukawa analogy?

(5) In what sense, or to what extent, is electrodynamics – the theory of Dirac pairs and the electromagnetic field – 'closed'?

During the years that followed, question (1) would be answered affirmatively; questions (2) and (3) still remain open, whereas question (4) has to be answered negatively to a large extent. Oppenheimer had himself speculated in this sense in 1948. The domain of quantum electrodynamics is only weakly connected with other areas, but the hadronic phenomena are not 'closed' in the same sense.

In the discussion of Oppenheimer's report, Dirac expressed his hope that in a non-perturbative treatment the infinities might disappear. But Bohr replied that one would then lose all the successes. Only for the meson-nucleon problems did one need a completely new scheme. Heitler wondered whether the renormalization procedure implied equal charge for all elementary particles.

Finally, W. Pauli added a short communication dealing with a correction to the Compton effect in the formalism of Weisskopf, Schwinger and Tomonaga. The results confirmed the statement that quantum electrodynamics could now be considered as a good candidate for a consistent theory.

6. THE STATUS OF PARTICLE THEORY IN 1948

The reports and discussions at the eighth Solvay Conference in 1948 demonstrated some important progress in the understanding of phenomena involving elementary particles. One theory, quantum electrodynamics, seemed to be on the way to becoming a consistent scheme. On the other hand, important new discoveries had been made recently concerning the meson problem and one could hope for important progress in the near future. This generally optimistic trend was reflected also in the last general discussion at the eighth Solvay Conference. In fact, this discussion was almost entirely devoted to the tasks of the future, especially the problems of interactions other than the electromagnetic one.

6.1. Muon Problems

Oppenheimer drew attention to the decay of the μ-meson, particularly whether the 'neutral meson' in Equation (8) $[\mu^{\pm} \to e^{\pm} + \nu + \mu^0]$ is a neutrino. Møller reported on calculations which he and his collaborators had done on the comparison of nuclear β-decay and the μ-decay. The result is 'that the decay of the μ-meson is due to a similar process as the β-decay and that we have a kind of Fermi interaction between all particles of spin $\frac{1}{2}$.'

6.2. Pion Problem and Heavier Mesons

It was generally agreed [reports of Teller and Powell] that the pion decay, $\pi \to \mu + + \begin{cases} \nu \\ \gamma \end{cases}$, takes roughly 10^{-11} sec. The cross-section with nucleons is large but does not increase considerably with energy.

τ-mesons seem to occur rather frequently and a typical decay time of 10^{-8} sec was reported by Blackett.

6.3. The Situation in Atomic Physics

The eighth Solvay Conference closed with some general comments on 'the present situation in atomic physics' by Niels Bohr. He again referred to the satisfactory situation in quantum electrodynamics, but pointed out that the non-dimensional coupling constant,

$$\alpha = \frac{e^2}{\hbar c} \sim \frac{1}{137},\tag{28}$$

prevented one from calculating it within the framework of the theory itself.

The fact that nuclear constitution requires new forces foreign to electromagnetic theory causes a new situation in the understanding of microscopic phenomena, since no correspondence arguments can be applied due to the short range character of the Yukawa forces. Meson theory, Bohr said, 'is yet in a most preliminary stage and new viewpoints will obviously be demanded for the theoretical comprehension of the rapidly increasing experimental evidence about the various types of mesons and the conditions for their production.'

He continued: 'In a future, more comprehensive theory of elementary particles, the relation between the elementary unit of electric charge and the universal quantum of action may play a more fundamental role than in present theories.' Bohr pointed to a new complementarity of fields and particles which dominates the understanding of elementary particle phenomena.

The prophetic words of Bohr still ring in the air. The development of elementary particle physics has taken another path. More and more particles have been discovered, the hierarchy of interactions came to be firmly established, but only few connections between different forces have become obvious. Elementary particle physics became a highly specialized field of physics, and the thirteen years until the Solvay Conference on quantum field theory [1961] would see an enormous extension in the knowledge of a variety of phenomena, but only a little unification towards a general understanding.

REFERENCES

1. H. A. Wilson, *Proc. Roy. Soc.* A **104**, 451 (1923).
2. See the earlier summary of V. M. Goldschmidt, *Det. Norske Videnskap., Akademi i Oslo, I., Mat.-Naturv. Kl.* no. 4 (1937).
3. H. E. Suess, *Z. Naturf.* **2a**, 311, 604 (1947).
4. See for instance, R. A. Alpher, H. Bethe, and G. Gamow, *Phys. Rev.* **73**, 803 (1948).
5. S. Chandrasekhar and L. Heinrich, *Astrophys. J.* **95**, 288 (1942);
   C. F. von Weizsäcker, *Phys. Z.* **38**, 176 (1937); **39**, 633 (1938).
6. C. D. Anderson and S. H. Neddermeyer, *Phys. Rev.* **51**, 884 (1937); **54**, 88 (1938).
7. M. Conversi, E. Pancini, and O. Piccioni, *Phys. Rev.* **71**, 209 (1947).
8. S. Sakata and T. Inoue, *Prog. Theor. Phys.* **1**, 143 (1946).
9. R. Marshak and H. A. Bethe, *Phys. Rev.* **72**, 506 (1947).
10. W. L. Gardner and C. M. G. Lattes, *Science* **107**, 270 (1948).

11. C. M. G. Lattes *et al.*, *Nature*, **159**, 694 (1947); **160**, 453, 486 (1947).
12. G. P. S. Occhialini and C. F. Powell, *Nature*, **162**, 168 (1948).
13. R. H. Brown *et al.*, *Nature*, **163**, 47, 82 (1949).
14. C. F. Powell and Rosenblum, *Nature*, **161**, 473 (1948).
15. C. M. G. Lattes *et al.*, *Proc. Phys. Soc. (London)*, **61**, 173 (1948).
16. Y. Goldschmidt-Clermont, *et al.*, *Proc. Phys. Soc., (London)*, **61**, 183 (1948).
17. G. P. S. Occhialini and C. F. Powell, *Nature*, **162**, 168 (1948).
18. R. H. Brown *et al.*, *Nature*, **163**, 47, 82 (1949).
19. B. Peters and H. Bradt, Bristol Symposium, 1948.
20. L. Leprince-Ringuet and M. L'Héritier, *C. R. Acad. Sci. (Paris)*, **219**, 618 (1944).
21. D. Rochester and C. C. Butler, *Nature* **160**, 855 (1947).
22. L. Leprince-Ringuet, *et al.*, *C. R. Acad. Sci. (Paris)*, **226**, 1897 (1948).
23. R. H. Brown *et al.*, *Nature*, **163**, 82 (1949).
24. C. T. R. Wilson, *Proc. Camb. Phil. Soc.* **11**, 32 (1900); J. Elster and H. F. Geitel, *Phys. Z.* **2**, 560 (1900).
25. V. F. Hess, *Phys. Z.* **13**, 1084 (1902); **14**, 610 (1913); W. Kolhörster, *Berl. Deutsch. Phys. Ges.* **16**, 719 (1914).
26. R. Hulsizer and B. Rossi, *Phys. Rev.* **73**, 1402 (1948).
27. J. Schwinger, *Phys. Rev.* **60**, 164 (1941).
28. P. A. M. Dirac, *Proc. Roy. Soc.* A **155**, 447 (1936).
29. M. Fierz, *Helv. Phys. Acta*, **12**, 3 (1939);
    A. Proca, *J. Phys. Rad.* **7**, 347 (1936);
    W. Rarita and J. Schwinger, *Phys. Rev.* **60**, 61 (1940).
30. W. Pauli, *Phys. Rev.* **58**, 716 (1940).
31. L. de Broglie, '*Une Nouvelle Théorie de la Lumière: la mécanique ondulatoire du photon*', Paris, 1940, 1942.
32. L. de Broglie, '*Une Nouvelle Conception de la Lumière*', Actualités Scientifiques, fasc. **81**, Hermann, Paris, 1934.
33. L. de Broglie, *C. R. Acad. Sci. (Paris)*, **200**, 361 (1935).
34. P. A. M. Dirac, *Proc. Roy. Soc.* A **180**, 1 (1942).
35. G. Wentzel, *Z. Phys.* **86**, 479 (1933).
36. M. Born, *Proc. Roy. Soc.* A **143**, 410 (1934);
    M. Born and L. Infeld, *Proc. Roy. Soc.* A **144**, 425 (1935).
37. F. Bopp, *Ann. Phys.* **38**, 345 (1940); **42**, 575 (1943);
    E. C. G. Stückelberg, *Helv. Phys. Acta*, **14**, 51 (1941).
38. McManus, Birmingham University dissertation, 1947.
39. R. P. Feynman and J. A. Wheeler, *Rev. Mod. Phys.* **17**, 157 (1945).
40. V. F. Weisskopf, *Z. Phys.* **89**, 27 (1934); **90**, 817 (1934*)*.
41. F. Bloch and A. Nordsieck, *Phys. Rev.* **52**, 54 (1937).
42. A. Pais, *Verh. Kon. Acad. (Amsterdam)*, **19**, (1947).
43. H. S. Snyder, *Phys. Rev.* **71**, 38 (1947).
44. P. A. M. Dirac, *Proc. Camb. Phil. Soc.* **30**, 150 (1934);
    W. Heisenberg, *Z. Phys.* **90**, 209 (1934).
45. J. Serpe, *Physica*, (8) **2**, 226 (1941);
    W. Opechowski, *Physica*, **2**, 161 (1941);
    H. A. Kramers, *Nuovo Cim.* **15**, 108 (1938).
46. H. A. Bethe, *Phys. Rev.* **72**, 339 (1947).
47. J. R. Oppenheimer, *Phys. Rev.* **35**, 461 (1930).
48. See V. F. Weisskopf (1934), quoted above.
49. W. E. Lamb and R. C. Retherford, *Phys. Rev.* **72**, 241 (1947).
50. H. Lewis, *Phys. Rev.* **73**, 173 (1948);
    W. E. Lamb and N. Kroll, *Phys. Rev.* **75**, 388 (1949);
    V. F. Weisskopf and J. B. French, *Phys. Rev.* **75**, 1240 (1949);
    H. A. Bethe, *Phys. Rev.* **72**, 339 (1947);
    P. Luttinger, *Phys. Rev.* **74**, 893 (1948).
51. J. Schwinger, *Phys. Rev.* **74**, 1439 (1948).
52. S. Tomonaga, *Prog. Theor. Phys.* **1**, 27, 109 (1946).

APPENDIX

Even the physicists are capable of bursting into song (or doggerel) when their spirits are moved or if the occasion calls for it. Maxwell and Einstein versified now and then, with less spectacular effect than some of their other achievements. The following attempts at versification were presented at the banquet during the eighth Solvay Conference, and their themes reflect the preoccupations of the participants.

The Meson Song

(Solvay Version)

The following song is an adaptation of a piece of classical American literature by Dr and Mrs H. C. Childs, Dr and Mrs R. E. Marshak, Dr and Mrs R. L. McCreary, Dr and Mrs J. B. Platt, Dr and Mrs S. N. Voorhis, all of the University of Rochester, and Georges E. Valley of the Massachusetts Institute of Technology.

[Acknowledgement by Edward Teller]

There are mesons pi, there are mesons mu
The former ones serve as nuclear glue
There are mesons tau, or so we suspect
And many more mesons which we can't yet detect

    Can't you see them at all?
    Well, hardly at all
    For their lifetimes are short
    And their ranges are small.

The mass may be small, the mass may be large,
We may find a positive or negative charge,
And some mesons never will show on a plate
For their charge is zero, though their mass is quite great

    What, no charge at all?
    No, no charge at all!
    Or, if Blackett is right
    It's exceedingly small

Some beautiful pictures are thrown on the screen,
Though the tracks of the mesons can hardly be seen,
Our desire for knowledge is most deeply stirred
When the statements of Serber can never be heard.

    What, not heard at all?
    No, not heard at all!
    Very dimly seen
    And not heard at all!

There are mesons lambda at the end of our list
Which are hard to be found but are easily missed,
In cosmic-ray showers they live and they die
But you can't get a picture, they are camera-shy.

> Well, do they exist?
> Or don't they exist?
> They are on our list
> But are easily missed.

From mesons all manner of forces you get,
The infinite part you simply forget,
The divergence is large, the divergence is small,
In the meson field quanta there is no sense at all.

> What, no sense at all?
> No, no sense at all!
> Or, if there is some sense
> It's exceedingly small.

EDWARD TELLER*

The Meson Song
(Original Version)

We have mesons pi and mesons mu,
And mesons that serve as nuclear glue.
We have mesons large and mesons small,
Plus charge or minus, or no charge at all.

Chorus: What? No charge at all?
       No! No charge at all.
       A very small rest mass,
       And no charge at all.

Vector or scalar or halfway between,
Sometimes the convergence can scarcely be seen.
Two hundred, four hundred, nine hundred mass,
All sorts of charges in every weight class.

Chorus

The forces exchange when at distances small;
There's the depth of the well and the height of the wall;
The quadripole moment a tensor demands
What very strange forces we have on our hands!

Chorus

Oh Bose! Oh Fermi! Perhaps Einstein, too,
Send a unified theory for both pi and mu;
With spin and statistics, adjustable range,
Nuclear structure is yours to arrange.

Chorus: What? No sense at all?
       No! No sense at all.
       A very small rest mass,
       And no sense at all.

Sommaire en forme de ballade

Les électrons et positons
Mésons et forces nucléaires
Deuterons, protons, photons
Particules permanentes ou très éphémères
Particules bien connues ou trouvées naguère
Ou bien particules pas encore observées
Tout cela a été notre affaire
Et c'est là le but du Conseil Solvay.

Les grands corps en rotation
Le magnétisme de la terre
Et des étoiles; les explosions,
Gerbes nées dans l'atmosphère
Et la doctrine complémentaire
Qui a clairement montré
Qu'il faut renoncer aux images vulgaires
Et c'est là le but du Conseil Solvay.

On a vu que le triton
N'est point un être légendaire.
Quant au problème de l'électron,
La situation théorique était claire
Quand Oppi a dit qu'il pourra se défaire
De toute divergence et de l'infinité,
Une fois qu'il sera devenu grand-père;
Et c'est là le but du Conseil Solvay.

Envoi.

Leprince Ringuet quoiqu'on ne sache guère
Si la τ-particule est une réalité
Nous voulons bien le croire, seulement pour vous plaire
Et c'est là le but du Conseil Solvay.

H. B. G. CASIMIR.

The Lion and the Unicorn

The Lion and the Unicorn
Were fighting for the crown
Of Nature's secrets, leading down
To matter's innermost recesses,
The depth of which they searched in vain.
The Unicorn with hanging mane
Peered down the bottomless abysses,
The Lion with his mighty paw
In vain did scratch his lofty brow.
For, like a meadow dug by moles,
The universe seemed full of holes:
Every electron in its course
Bored such a hole through space,
And so did each proton, of course;
But, oh what a disgrace,
Each infant meson newly born
Of the same mischief was suspected,
And this was still more unexpected.
'Let's try our hand', quoth Unicorn,
'At those holes we know best;
And later on cope with the rest.'
The Lion, sunk in deep concern,
Received the hint with scorn:
'Nature's great lesson we must learn',
He said unto the Unicorn;
'Half-hearted work Dame Nature shuns:
Whatever holes she has to fill,
She'll fill them all at once.'
This speech was but of little use:
The Unicorn he sat quite still
And did not change his views.
So on they fight for Nature's crown
A fight that brings them wide renown,
And praise of ages yet unborn
To Lion and the Unicorn.

L. ROSENFELD

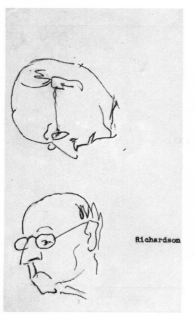
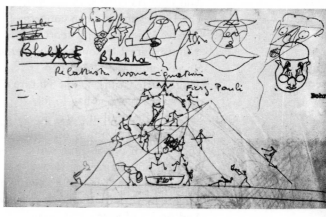

Drawings and Doodles by H. J. Bhabha during Lectures at the Eighth Solvay Conference (1948)

TWELFTH SOLVAY CONFERENCE 1961

S. MANDELSTAM   G. CHEW   M.L. GOLDBERGER   G.C. WICK   M. GELL-MANN   G. KALLEN   E.P. WIGNER   G. WENTZEL   J. SCHWINGER   M. CINI

A.S. WHIGHTMAN

I. PRIGOGINE   A. PAIS   A. SALAM   W. HEISENBERG   F.J. DYSON   R.P. FEYNMAN   L. ROSENFELD   P.A.M. DIRAC   L. VAN HOVE   O. KLEIN

S. TOMONAGA   W. HEITLER   Y. NAMBU   N. BOHR   F. PERRIN   J.R. OPPENHEIMER   Sir W. LAWRENCE BRAGG   C. MØLLER   C.J. GORTER   H. YUKAWA   R.E. PEIERLS   H.A. BETHE

11

Quantum Field Theory*

1. INTRODUCTION

The twelfth Solvay Conference on Physics took place in Brussels from 9 to 14 October 1961. The general theme of the conference was 'Quantum Field Theory', and it would seem that this was the last time that this subject stood at the centre of a theoretical description of particle phenomena. In the following years the old quantum field theory would lose its dominant role in elementary particle theory, and return only much later in a largely modified form.

The 1961 conference was also intended to celebrate the fiftieth anniversary of the first Solvay Conference in 1911, whose theme had been the theory of radiation and the quanta. The choice of quantum field theory as its theme was thus appropriate.

The Scientific Committee of the Conference consisted of: Sir Lawrence Bragg (London), President, E. Amaldi (Rome), C. J. Gorter (Leyden), C. Møller (Copenhagen), N. F. Mott (Cambridge), J. R. Oppenheimer (Princeton, N.J.), and F. Perrin (Paris). Amaldi and Mott were not able to attend.

The invited speakers included: N. Bohr (Copenhagen), R. P. Feynman (Pasadena, California), M. M. Gell-Mann (Pasadena, California), M. L. Goldberger (Princeton, N.J.), W. Heitler (Zurich), G. Källén (Lund), S. Mandelstam (Birmingham), A. Pais (Princeton, N.J.), and H. Yukawa (Kyoto, Japan).

Among the invited participants were: H. A. Bethe (Ithaca, N.Y.), G. Chew (Berkeley, California), M. Cini (Rome), P. A. M. Dirac (Cambridge), F. J. Dyson (Princeton, N.J.), O. Klein (Stockholm), Y. Nambu (Chicago, Illinois), R. E. Peierls (Birmingham), L. Rosenfeld (Copenhagen), A. Salam (London), S. Tomonaga (Tokyo, Japan), L. Van Hove (CERN, Geneva), G. Wentzel (Chicago, Illinois), G. C. Wick (Upton, N.Y.), and E. P. Wigner (Princeton, N.J.).

M. Born, N. Bogoliubov, L. D. Landau, T. D. Lee, F. E. Low, H. P. Lehmann, J. Schwinger, I. E. Tamm, and C. N. Yang, all of whom had been invited, could not attend. E. Schrödinger, who had accepted to attend, health permitting, had died.

For the first time, many of the invited speakers and participants came from the United States, indicating the important role high energy physics had played there during the preceding two decades. Among the invited participants, some had attended

---

* *La Théorie Quantique des Champs*, Rapports et Discussions du Douzième Conseil de Physique tenu à l'Université Libre de Bruxelles du 9 au 14 Octobre 1961, Interscience Publishers, New York, and R. Stoops, Brussels, 1962.

previous Solvay Conferences, but many belonged to the new generation of distinguished theoretical physicists.

The 1961 Solvay Conference essentially summarized the successes of quantum field theory in the previous thirty to thirty-five years. In this sense, this conference looked at the past and examined the current status, but the discussions were not 'fought out' debates. However, a group of theoreticians, who took part in the 1961 Conference, started to question the entire approach of quantum field theory. In the succeeding years they would consider quantum field-theoretical models, if at all, only as heuristic starting points in order to derive or to render plausible certain basic statements on which a future description of hadron phenomena should rest.

It seems remarkable, though not surprising, that this revolution, denying thirty years of theoretical development, was started by some of the most active and committed exponents of quantum field theory. Among these was L. D. Landau, who unfortunately did not attend the 1961 Conference. Since the early 1930s Landau had been struggling with the difficulties that had appeared in relativistic quantum theory.[1] In the mid-1950s Landau started a detailed examination of quantum electrodynamics, which brought him to the conclusion that there existed fundamental difficulties even in this most successful branch of quantum field theory. Because of this, Landau abandoned the shaky ground of quantum field theory and turned to a more general theory of elementary particles.[2]

At the 1961 Solvay Conference, the beginnings of the new directions were presented by Mandelstam and Goldberger. Most of the speakers, however, dealt with the various aspects of quantum field theory. Murray Gell-Mann, in his report, noted: 'So far all our serious attempts to describe elementary particle phenomena in mathematical terms really constitute a single body of theory, all of which I shall describe as "field theory". We assume some relativistic invariance, microcausality, positive energies, and positive probabilities totaling one. Then we try to construct a consistent theory that will explain as much as possible of the available information.

'Now there are various approaches to field theory, each of which has advantages in certain situations. There has been some controversy among the proponents of various methods, but in my view they all have something to teach us and should be considered together, so that ideas originating from one approach can be applied to the others.

'We may distinguish three main divisions:

'(1) *Lagrangian field theory*. Here the Lagrangian density L is specified as a function of certain "bare" fields and their gradients. We have one ψ for each "elementary" particle. The parameters in the Lagrangian are "bare" masses and coupling constants which do not appear explicitly in the final results. The equations of motion (together with the commutation relations) are solved by perturbation methods or some modification thereof. The S-matrix is calculated and expressed in terms of physical masses and coupling constants; the theory is usually considered acceptable only if at this stage there is no dependence on a cut-off (i.e., the theory is "renormalizable"). The method of Feynman diagrams expresses the results very neatly.

'Quantum electrodynamics was developed in this way and has been successful in describing the properties of electrons, muons, and photons to the high order of accuracy now attainable by experiment.

'(2) A second approach is that of Lehmann, Symanzik and Zimmermann, Wightman, and others.[3] For any particle to be discussed, elementary or not, free "fields" ψ_{in} and ψ_{out} are constructed.[4] For at least some of these, there are supposed to be local interpolating *renormalized* fields which reduce ψ_{in} and ψ_{out} at $t = \mp \infty$ respectively. The vacuum expectation of values of products of the interpolating fields are discussed, with the conditions of relativistic invariance, microcausality, and conservation of probability imposed. The S-matrix, which relates the ψ_{out} fields to the ψ_{in} fields, is calculated in terms of certain boundary conditions, the specification of which replaces the choice of elementary particles and of Lagrangian in the first approach. In (2) as in (1), we calculate scattering amplitudes not only for physical particles on the mass-shell, but also for virtual particles off the shell.

'(3) The third approach is that of dispersion theory.[5] As in (2), we impose directly the causality, relativity and unitarity conditions; however, we work only with amplitudes on the mass shell. Of course, we must pay the price of considering the never-never land in which $|\cos \theta| > 1$, certain momenta are imaginary, etc. The choice of theory corresponds in this case to the specification of the number and character of subtractions in the dispersion relations and the locations and strengths of "CDD poles". The concept of "field" is not made use of explicitly. Instead one works directly with the analytic properties of various amplitudes.'

The classification given by Gell-Mann is helpful in considering the reports themselves at the conference: The reports of Feynman, Pais, Heitler, and Yukawa, may be counted in class (1), Källen's report in approach (2), while Goldberger and Mandelstam discussed aspects of the approach (3). Gell-Mann's own report cut across all three approaches.

The conference opened with a report by Bohr on 'The Solvay Meetings and the Development of Quantum Physics', in which he discussed how the Solvay Conferences, which had started in 1911, 'have been unique occasions for physicists to discuss the fundamental problems which were at the centre of interest at the different periods, and have thereby in many ways stimulated modern development of physical science."*

2. QUANTUM ELECTRODYNAMICS

Feynman's report was devoted to 'The Present Status of Quantum Electrodynamics'. Since he discussed the results and problems rather than the general structure of the theory, we shall first present briefly the development of quantum electrodynamics since 1948, when it had been discussed at the eighth Solvay Conference.

* Bohr's account provided a thematic continuity by considering the earlier Solvay Conferences as reflecting the development of quantum physics. I have benefited from this account and, in certain respects, followed it.

2.1. DEVELOPMENT OF QUANTUM ELECTRODYNAMICS

Quantum electrodynamics in its present form originated at four different places, and was developed mainly by four scientists: E. C. G. Stückelberg in Switzerland, S. Tomanaga in Japan, and J. Schwinger and R. P. Feynman in the United States. All this work is rooted in the pioneering papers of Dirac, Heisenberg, Jordan, and Pauli. Dirac's work is the most frequently named point of departure.

The present quantum electrodynamics is a device to circumvent the divergence difficulties inherent in all relativistic local field theories with non-trivial interaction terms. The method is to absorb the divergent expressions into parameters by renormalization. In order to perform renormalization practically, two formal definitions have been most helpful. The first is the interaction or Dirac picture of states and operators which is defined by the equation,

$$|\psi(t)\rangle = e^{iH_0^s/\hbar} |\phi_s(t)\rangle, \tag{1}$$

where H_0^s is the free Hamiltonian in the Schrödinger representation and, $|\phi_s(t)\rangle$, the corresponding state vector. In the absence of the interaction, the state vector is constant in time and the time dependence of operators is given by the equation

$$i\hbar \frac{\partial Q(t)}{\partial t} = [Q(t), H_0]. \tag{2}$$

One may now define a time displacement operator, $U(t, t_0)$, which transforms a state $|\psi(t_1)\rangle$ to a state $|\psi(t_2)\rangle$,

$$U(t_2, t_1) = e^{iH_0 t_2} e^{-iH(t_2 - t_1)} e^{-iH_0 t_1},$$
$$H = H_0 + H_i(t). \tag{3}$$

If one takes the limits $t_2 \to +\infty$ and $t_1 \to -\infty$, it becomes the S-matrix as introduced by Heisenberg into particle theory in 1943. The S-matrix in the interaction (Dirac) picture is defined as,

$$S = \lim_{t_2 \to \infty} \lim_{t_1 \to -\infty} U(t_2, t_1) = \sum_{n=0}^{\infty} \left(-\frac{i}{\hbar}\right)^n \frac{1}{n!} \cdot \int_{-\infty}^{+\infty} dt_1 \dots \int_{-\infty}^{+\infty} dt_n \cdot P(H_i(t_1) \dots H_i(t_n)), \tag{4}$$

where P denotes a time ordering. Equation (4) is clearly related to the many-time formalism introduced by Dirac in 1932.

The interaction or Dirac picture proved to be particularly useful, since the field operators according to Equation (2) satisfy the free-field equations and the invariant commutation rules can be written down for all times in a straight-forward manner. This, in turn, leads to a relativistic generalization of the equation of motion of a Dirac state,

$$i\hbar \frac{\partial}{\partial t} |\psi(t)\rangle = H_i(t) |\psi(t)\rangle, \tag{5}$$

to its covariant form

$$i\hbar c \frac{\delta |\psi(\sigma)\rangle}{\delta\sigma(x)} = H_i(x)|\psi(\sigma)\rangle. \tag{6}$$

Here H_i is the interaction Hamiltonian density and σ is the spacelike surface. Equation (6) plays an important role in the approach of Tomonaga[7] and Schwinger.[8]

Independently, Feynman found his way to quantum electrodynamics. A remark of Dirac's relating the wave function at a time t to a later time $t+\varepsilon$ by a kernel, $K(x', x)$, brought him to consider the Ansatz,

$$\psi(x', t+\varepsilon) = \int dx\, A \exp\left[\frac{i\varepsilon}{\hbar} L\left(\frac{x'-x}{\varepsilon}, x\right)\right] \psi(x, t), \tag{7}$$

where A is an amplitude factor and L represents a Lagrangian. Composing a finite time of many infinitesimal displacements, ε, Feynman was led to the idea of the path integral which gave rise to a formulation of quantum mechanics different from matrix mechanics or wave mechanics. The advantage of the new theory was that it could immediately be written in a relativistically invariant manner. Feynman was thus led to the idea of space-time graphs by multiplying kernels; from this resulted the famous Feynman rules for calculating transition amplitudes in perturbation theory.[9] They allowed one to obtain the good and finite results of the old quantum electrodynamics. However, some characteristic graphs remained divergent, and had to be cured by renormalizations.

The first renormalization to be invented concerned the mass. In 1947 Kramers had suggested the principle of renormalization, and Bethe, stimulated by Kramers' ideas, interpreted the Lamb shift[10] as an effect of the self-energy of the electron. Relativistic calculations by Schwinger and Weisskopf,[11] and others, confirmed Bethe's non-relativistic result. In Feynman's technique, the infinite self-energy diagram can be isolated by using regulators. Further divergent graphs are the vacuum polarization and the vertex part.

Dyson not only proved the identity of the Feynman and Schwinger approaches but also greatly simplified their understanding.[12] Quantum electrodynamics is a finite theory, but for the renormalization constants z_1, z_2, and z_3. By Ward's identity one finds that the vertex and wave function renormalization are the same, i.e. $z_2 = z_3$.[13] Thus quantum electrodynamics is a finite theory if the electron mass and charge are subtracted properly.

Quantum electrodynamics was applied with great success to all phenomena which included only the interaction of leptons with the photon field.

2.2. THE PRESENT STATUS OF QUANTUM ELECTRODYNAMICS

(R. P. Feynman)

Feynman discussed the successes and problems of quantum electrodynamics, a quantum field theory of the first type in the classification of Gell-Mann. Such a quantum field theory has a more phenomenological than rigorous basis. Quantum electrodynamics is the theory of the electromagnetic field as well as the theory of the motion of electrons and muons. In his report at the first Solvay Conference in 1911, Lorentz had characterized the electromagnetic field as a set of harmonic oscillators. In 1927 Dirac quantized this set systematically in interaction with atoms, and a large variety of equivalent formalisms had been proposed until the work of Stückelberg, Tomonaga, Schwinger, Feynman and Dyson completed the theory.

The comparison with experiment is very favourable to the theory, and Feynman remarked at the outset: 'Considerable evidence for the general validity of Q.E.D. is, of course, provided by the enormous variety of ordinary phenomena which, under rough calculation, are seen to be consistent with it. The superfluidity of helium and the superconductivity of metals having recently been explained, there are to my knowledge no phenomena occurring under known conditions, where quantum electrodynamics should provide an explanation, and where at least a qualitative explanation in these terms has not been found. The search for discrepancies has turned from looking for gross deviations in complex situations to looking for large discrepancies at very high energies, or by looking for tiny deviations from the theory in very simple, but very accurately measured situations.'

Feynman then reported on the status of high energy experiments and considered, in particular, the elastic scattering of electrons from nucleons studied by Hofstadter and others.[14] These experiments show that the protons and neutrons have an extended structure, expressed by two form factors, F_1 and F_2, which are functions of the momentum transfer squared, q^2. They appear in the matrix element of the electric current, J_μ, between the nucleon states with momenta p_1 and p_2,

$$J_\mu = \gamma_\mu F_1(q^2) + \sigma_{\mu\nu} q_\nu F_2(q^2), \qquad q = p_1 - p_2. \tag{8}$$

For a charged particle, $F_1 = 1$, and $F_2 = 0$ for all q^2, but one found a sharp decrease of F_1 with increasing $|q^2|$. The Fourier transform of $F_1(q^2)$ may be interpreted as the charge structure of the nucleons.

The test of quantum electrodynamics in these electron-hadron scattering experiments consists in looking for a deviation from the photon propagator,

$$g_{\mu\nu}\left(\frac{1}{q^2}\right)\left(1 - \frac{q^2}{\Lambda^2}\right), \tag{9}$$

in terms of a 'cut-off' energy Λ. In 1961, Λ was found to exceed 500 MeV. Later on the limit was proved to be much higher, in the order of several GeV.

Precise low energy experiments are just as important as tests for the validity of quantum electrodynamics, the classic experiment being the detection of the Lamb-shift. Here the $^2S_{1/2}$ and $^2P_{1/2}$ energy separation in hydrogen, deuterium and ionized helium was determined. The effect was zero in the old 1928 Dirac theory of the electron. In quantum electrodynamics, several contributions add up: the emission and reabsorption of one virtual photon, which is the main effect, the higher order corrections to it, the fact that the potential of the nucleus becomes changed by the existence of virtual electron-positron pairs (vacuum polarization), as well as corrections for the finite nuclear mass and the extended structure of the nucleus. Most of these terms had been found to be infinite in the theory developed in the 1930s, but renormalization made them finite. The results of the experiment and the theoretical calculation agreed fully by 1958.[15] The error was 1 in 10^5, and the agreement showed that Coulomb's law is altered so little that, $\varepsilon < 10^{-10}$, in the Ansatz,

$$V_c = \frac{1}{r^{1+\varepsilon}}. \tag{10}$$

Other high precision experiments refer to the anomalous magnetic moments of electrons and muons. According to Dirac's theory of the electron, these particles have spin $\frac{1}{2}$ and a gyromagnetic ratio of 2. This value becomes altered by the emission of virtual photons and electron-positron pairs to,

$$g_{\text{el.}} = 2\left[1 + \frac{\alpha}{2\pi} - 0.328\left(\frac{\alpha}{\pi}\right)^2\right] = 2\,(1.001\,159\,61), \tag{11a}$$

and

$$g_{\mu} = 2\left[1 + \frac{\alpha}{2\pi} + 0.75\left(\frac{\alpha}{\pi}\right)^2\right] = 2\,(1.001\,165), \tag{11b}$$

which are in perfect agreement with the experiments performed in the 1950s for the electron and the CERN experiment [started in 1962] for the muon. The new experiments of the late 1960s enforced the evaluation of the sixth and eighth order perturbation calculations, and still the agreement increased.

A third characteristic test of quantum electrodynamics in low energy phenomena was obtained from the study of the hyperfine structure of spectral lines. Again virtual photons and electron-positron pairs lead to important corrections computed in terms of a perturbation series in powers of the fine structure constant α. These corrections are so subtle that they serve to determine the exact value of this coupling constant, and it was found that,

$$\alpha^{-1} = 137.0417 \pm 0.0025. \tag{12}$$

In the late 1960s a new method was found for determining α from the Josephson effect. The following research revealed errors in the hyperfine structure measurements, and again led to an even better proof of the accuracy of quantum electrodynamical calculations.

Finally Feynman reported on the calculation of the positronium atom consisting of an electron and a positron. Since the theory involves the relativistic description of two particles of equal mass, one should not attach too much significance to the agreement obtained between experiment and the Q.E.D. calculation.

Quantum electrodynamics indeed appeared to be the example of a perfect theory. No error in its predictions had been found until 1961 and deviations which became apparent during the following years disappeared on closer inspection, like the so-called violation observed in the production of muon pairs at 'large' angles of several degrees.

Perhaps the first problem occurred recently in the study of multi-hadron production from electron-positron pairs. This production should proceed, according to the rules of Q.E.D., mainly via the intermediate state of a photon; hence the cross-section depended on the energy like

$$\sigma_{\text{hadron production}} \propto (E)^{-2}. \tag{13}$$

The new data indicate that the production cross-section remains constant up to 25 GeV^2.

Despite the overwhelming successes of quantum electrodynamics, various problems remained in the theory. The renormalization of mass and charge of the electron seems to include infinities, e.g. in the perturbation calculation,

$$\Delta m_e = m_0 \frac{3\alpha}{2\pi} \ln \frac{\Lambda}{m_0}, \tag{14}$$

where Λ is a cut-off. The excuse that the electron's mass correction (14) is not observable in principle becomes shaky if one thinks of the mass differences in hadron multiplets, as between π^+ and π^0, where the 'electromagnetic' part of the mass becomes observable. Equally the renormalization of the charge becomes infinite.

The philosophy of renormalization, including the manipulation of infinities, prevents one from verifying an old dream, namely to calculate the fine structure constant and the electron-proton mass ratio. Hence the question arises whether one should not consider the present quantum electrodynamics as a phenomenological theory which is a very good approximation to a self-consistent scheme. But this is only speculation so far. A problem which Feynman did not discuss was the mathematical rigour on which this theory could be based.

On the other hand, quantum electrodynamics, although accurate, must be incomplete since charged hadrons exist and can be produced by high energy photons. Hence at energies sufficient to create a meson one enters the field of strong interactions. One cannot even calculate typical electromagnetic effects in hadron physics, such as mass splittings in isospin multiplets or the anomalous magnetic momenta of the nucleons.

The problems of the infinities in quantum electrodynamics arise from considering virtual processes in the calculation. One can try to avoid playing with unobservable quantities by using the approach of the dispersion theory. 'One serious question appears, however,' noted Feynman. 'Integrals over all energies are still required and at high energies the real processes involve all kinds of particles in considerable numbers. Thus the set of interconnected equations becomes enormously elaborate just when it becomes interesting. It is not clear how to get started grappling with this complexity.'

Quantum electrodynamics, Feynman concluded, should be viewed as a correct theory, which is incomplete, and one should look very hard for its failure.

In discussing Feynman's report, Heisenberg suggested that 'in Q.E.D. we might meet a situation very similar to the Lee model with local interaction.' There the situation is such that one cannot make it consistent without introducing 'ghost' states.[16] In Q.E.D., Landau's research seemed to propose a similar problem.[17] According to these studies the renormalized coupling constant in electromagnetic interactions becomes

$$e^2 = \frac{e_0^2}{1 + e_0^2 \dfrac{1}{3\pi} \ln \dfrac{\Lambda}{m_e^2}}, \tag{15}$$

i.e., negative in the limit of infinite cut-off or strict locality. Hence at energies larger than 10^{65} eV 'ghosts' would appear in quantum electrodynamics. Heisenberg had proposed a way out of this difficulty by using a special form of the indefinite metric which works in the case of the Lee model.[18]

3. WEAK INTERACTIONS

The second quantum field theory which came to be developed was the scheme describing nuclear beta-decay. This phenomenon had been known since the turn of the century when the rays appearing on Becquerel's photographic plates were analyzed closely. But only more than thirty years later did one arrive at some quantitative understanding. And it was twenty years later still that the theory was developed which described the decays more or less completely.

3.1. THE EARLY DAYS OF FERMI'S THEORY

Shortly after 1930, B. W. Sargent presented a compilation of the electron energy distribution of various beta-emitters, classifying these by observing groupings of radioactive nuclides in a plot of partial decay constants versus the upper energy limits of the electrons for various components of complex beta-spectra.[19] On the other hand, W. Pauli had already proposed in 1931 the hypothesis that the neutrino accounts for the continuous energy distribution of electrons in beta-decay.

Based on these preliminary ideas, Fermi constructed a theory of beta-decay, assuming that the beta-interaction was a new interaction-by-contact and that its interaction energy density was proportional to the product of the *four fermion wave fields* of the neutron, the proton, the electron, and the neutrino.[20] In other words,

$$\mathcal{L}_I = G(\bar{\psi}_p O \psi_n)(\bar{\psi}_e O \psi_v). \tag{16}$$

Here O indicates an operator in spin space, and it could be

$$O = 1, \gamma_5, \gamma_\mu \gamma_5, \sigma_{\mu\nu}, \gamma_\mu. \tag{17}$$

Fermi concluded from the large de Broglie wavelengths of the electrons and neutrinos that the nucleus would essentially act as a point-like source, and one could calculate the electron energy distribution of the 'allowed' transitions verifying the experimental spectra. The transition rate for the decays is obtained in Fermi's theory from the square of Fermi's coupling constant G multiplied by a known function, f, of the upper limit of the beta-spectrum. Hence the product

$$f\tau = \text{constant}, \qquad (17a)$$

where τ is the decay time, and Fermi could account for Sargent's empirical curves.

Fermi's theory was developed further by Gamow and Teller, who included the nuclear spin.[20a] The Gamow–Teller interaction described a new class of transitions forbidden in Fermi's original scheme, and one could classify,

$$O = 1, \gamma_\mu \qquad (18a)$$

as Fermi interactions, and

$$O = \gamma_\mu \gamma_5, \sigma_{\mu\nu} \qquad (18b)$$

as Gamow–Teller interactions, both leading to similar Sargent diagrams. The fifth type of Fermi interaction, the coupling $O = \gamma_5$, is suppressed in the nonrelativistic limit. The common coupling constant, G, is given by

$$G = 1.5 \times 10^{-49} \text{ erg cm}^3, \quad \text{or} \quad 10^{-23}\, m_e^{-2}, \quad \text{or} \quad 10^{-13}\, \lambda_\pi^{+2}, \qquad (19)$$

where λ_π is the Compton wavelength of the pion. From the beta-spectra it was concluded that either the scalar and tensor interactions, or pseudo-vector and vector interactions, were present. The conclusion was supported by a study of the angular correlations of the electron and neutrino. The situation remained like this until 1956.

3.2. PARITY VIOLATION AND V-A THEORY

The conjecture of parity violation in weak interactions was made, independently, by Salam, Landau, and Lee and Yang,[21] and confirmed by C. S. Wu[22] and collaborators subsequently. This led to a new study of beta-decays, since several classes of them could now be thought of: (i) parity-conserving weak interactions; (ii) parity-violating weak interactions; (iii) time reversal-violating weak interactions.

A study of the empirical data led Sudarshan and Marshak[23] to the following conclusions: either one had to choose the combination of vector and axial vector or the scalar and tensor interactions. The postulation of a maximal parity-violating V-A interaction was thus a deduction from experimental data, and some of the choice of data was later confirmed.

The discovery of the V-A interaction for nuclear beta-decay was very satisfactory from the point of view of universality. The universal Fermi interaction had been proposed by O. Klein[24] and G. Puppi[25] when they discovered marked similarities between the beta-decay, muon-decay, and muon-capture processes. After the discovery of parity violation one could also classify the pion-decay as a V-A interaction having the same coupling constant.

On closer inspection one found that the A-interaction was 1.2 times the V-interaction in nuclear beta-decay, but that they were exactly equal in muon decay. The former was explained as a renormalization effect due to strong interactions, and the equality was assumed as the original situation. This suggested that the chiral invariance plays an important role, a fact which had already been used by Feynman and Gell-Mann, as well as by Sakurai.[26]

3.3. Weak Interactions

(A. Pais)

Pais reported on the development of the V-A theory and its explanation of decay processes. The weak decay,

$$\mu^- \to e^- + v + \bar{v},$$ (20)

is described well by the effective interaction,

$$H = \frac{G}{\sqrt{2}} j_\lambda^{(e)}(x) j_\lambda^{(\mu)*}(x) + \text{hermitian conjugate},$$ (21)

including a coupling of the electron current, $j_\lambda^{(e)}$, and the muon current, $j_\lambda^{(\mu)}$,

$$j_\lambda^{(l)} = i\bar{l}\gamma_\mu(1 + \gamma_5) v, \qquad j_\lambda^{(l)*} = i\bar{v}\gamma_\lambda(1 + \gamma_5) l,$$ (22)

where $l, \bar{l},$ represent the wave functions of the leptons, electron and muon, and $v, \bar{v},$ those of the neutrinos. G is the universal coupling constant,

$$G \approx 10^{-5} m_N^{-2},$$ (23)

where m_N is the nucleon mass.

The interaction (21) is local, and the coupling constant G, Equation (23), has the dimension of a length squared. When one wants to compare the weak and electromagnetic interactions, a dimensionless 'constant',

$$g^2 = GL^{-2}pL,$$ (24)

with a characteristic length L, and p, a momentum characteristic of the specific problem, is constructed. The latter is of the order of MeV in the beta-decay of nuclei, and of the order of 100 MeV in μ- and K-decays. For all weak decays L can be chosen such that,

$$pL \ll 1,$$ (25)

and

$$g^2 \ll e^2.$$ (26)

Hence weak interactions are commonly said to be much weaker than electromagnetic ones, but Equation (24) suggests that this need not remain so in high energy regions, i.e. where p becomes large. Indeed, with the interaction (21), in the scattering process

$$e^- + v \to \mu^- + v,$$ (27)

the cross-section would rise indefinitely with the square of the c.m.s. momentum p, and reach the unitarity limit,

$$4G^2 p^2 / \pi = \pi / 2p^2$$ (28)

at

$$\bar{p} \approx \sqrt[4]{\left(\frac{\pi^2}{8G^2}\right)} \approx 300 \text{ GeV}. \tag{29}$$

Hence L^{-1} should be regarded as the effective momentum cut-off.

On the other hand, Pais argued, one can judge from the second order weak interaction effect, showing up in the $K_1^0 - K_2^0$ mass difference, that weak interactions perhaps do not grow indefinitely or even become strong with increasing energy. In addition, in strong interactions, parity-nonconserving contributions seem to be absent; thus one might follow

$$L \lesssim 30 \text{ GeV}^{-1}, \tag{30}$$

as a scale of energy where the present weak interaction theory breaks down. In beta-decays such a limit is far from being reached.

The parity violating current-current interaction describes three kinds of weak decays: (i) Leptonic, $\Delta S = 0$, processes like π, β, and μ decays; (ii) Leptonic, $\Delta S \neq 0$, processes like K_{l2}, K_{l3}, or Σ and Λ decays; (iii) Non-Leptonic, $|\Delta S| = 1$, processes like $K \rightarrow 2\pi$ or $\Sigma \rightarrow N + \pi$, etc.

In 1961, the purely leptonic and semi-leptonic, strangeness-conserving weak interactions were rather well-known; however, less was certain in the strangeness-changing semi-leptonic decays and only little information was available in the non-leptonic weak decays. All of the decays, however, showed a maximal violation of parity, P, and of charge conjugation, C, in 1961; hence CP seemed to be conserved throughout, as had already been suggested by Wick, Wightman, and Wigner some time previously.[27] Pais noted: 'It must be admitted, however, that so far there is no logical objection to the view that CP may hold for some but not all of the weak interactions. This is so because we may in practice neglect higher order interference effects between different weak interactions, as one has recently emphasized especially in connection with certain class (ii) reactions.'[28] Pais also noted that no 2π mode of the long lived K_2^0 had been recorded. T-invariance also seems to hold wherever it can be tested. Hence the CPT-symmetry is satisfied, thus far, in weak interactions, especially for the class (i) processes.

Important features of the weak interaction theory are the two-component theory of the neutrino, revived by Salam, Landau, and Lee and Yang in 1957, the local action of lepton currents, the lepton number conservation, and the muon-electron universality. The latter states that electrons and muons behave exactly in the same way and are only different in mass. This gives rise to the puzzle of how to understand that single difference.

The V-A interaction implies CP- and T-invariance, two-component neutrinos and lepton conservation. In neutron decay the hadronic current, J_λ, has the low energy structure,

$$J_\lambda^{\text{had.}} = \frac{1}{\sqrt{2}} \bar{p}(1 + \lambda\gamma_5) n, \tag{31}$$

with $\lambda = -G_A/G_V$, where G_V and G_A are Fermi and Gamow–Teller couplings respectively. One finds

$$G_V \approx G, \tag{32}$$

by a comparison of beta-decay with muon-decay; hence the meson cloud does not seem to renormalize the vector coupling. This observation led to the conserved-vector-current hypothesis.[29]

Little could have been learned in 1961 from the non-leptonic weak interactions which are connected with a transition $|\Delta S| = 1$ and $|\Delta I| = \frac{1}{2}$. In the strangeness-changing reactions, the rule, $\Delta S/\Delta Q = 1$, seemed to be satisfied. The general current entering the weak interactions may therefore be written as,

$$J_\lambda = -J_\lambda^{\text{had.}}(s=0) + J_\lambda^{\text{had.}}(s \neq 0) + j_\lambda^{(e)} + j_\lambda^{(\mu)}. \tag{33}$$

However, *deviating from unity*, the strangeness-changing current couples weaker than the other currents. In order to satisfy the isospin change-$\frac{1}{2}$ rule in non-leptonic decays one might introduce the hypothesis of *neutral hadronic currents*,

$$J_\lambda^{(0)\,\text{had.}} = J_\lambda^{(0)\,\text{had.}}(s=0) + J_\lambda^{(0)\,\text{had.}}(s \neq 0) \tag{34}$$

in the interaction like

$$H_{\text{int.}} = G\left[J_\lambda^* J_\lambda - \frac{1}{\sqrt{\lambda}} J_\lambda^{(0)*} J_\lambda^{(0)}\right]. \tag{35}$$

Finally, Pais dealt with two extensions of the scheme, the question of intermediate vector bosons and the question of two neutrinos. The intermediate vector boson had been proposed by Yukawa in 1935.[30] It was revived in 1960 by Lee and Yang. Such a boson should have spin 1 and be coupled with a dimensionless strength g to the $(V-A)$-current, Equation (33). Two charged $W^\pm$ would suffice in the case of absence of neutral currents; if they existed, as proposed in Equation (35), two further neutral W^0 would be needed. The intermediate bosons dissolve the pointlike four-fermion vertex. At low energies the couplings are related by the equation

$$G = \frac{g^2}{\mu^2}, \quad (\mu = \text{mass of } W). \tag{36}$$

This means that the vector mesons couple similar to the photons, and the usual decays are of second order in the constant g. If the mass μ is not too large one expects the production of W from the collision of neutrinos with nucleons.

From the fact that both the reactions

$$\mu \to e\gamma \quad \text{and} \quad \mu^- + N \to e^- + N \tag{37}$$

are forbidden, one expected further conservation laws in lepton physics, especially the existence of two kinds of neutrinos.[31] In addition to the electron neutrino, v_e, there should occur in all transitions involving muons, a muon neutrino, v_μ. This particle should have identical properties to the electron neutrino. Experiments on the collision of neutrinos arising from muon-decay with nuclei were in preparation in 1961. The results obtained a year later confirmed the two-neutrino assumption.[32]

Pais concluded by asking three questions which might play an important role in the future. First, he asked for the place of leptons in the world of elementary particles. Second, why do strong and electromagnetic interactions conserve parity? Third, what is the reason for the empirical rules like, $|\Delta S| = 1$, or $|\Delta I| = \frac{1}{2}$, in weak interactions?

3.4. FURTHER DEVELOPMENT OF THE V-A THEORY

After 1961 the experiments revealed many new facts about weak interactions, especially about the weak decays. First, the purely leptonic muon-decay was studied in detail. The differential spectrum of the electrons of energy $\varepsilon \cdot E_{max}$, $E_{max} = \frac{1}{2} m_\mu$, as a function of the angle θ between the muon spin and electron momentum, becomes

$$P(\varepsilon, \cos \theta)\, d\varepsilon\, d\Omega = \frac{m_\mu^5}{3\pi^4 \cdot 2^9} \cdot \varepsilon^2 \left\{ \left[1 + 4 \frac{m_e}{m_\mu} \eta \right]^{-1} \cdot \left[3(1 + \varepsilon) + 2\varrho\left(\tfrac{4}{3}\varepsilon - 1\right) + \right. \right.$$

$$\left. \left. + 6 \frac{m_e}{m_\mu} \frac{1 - \varepsilon}{\varepsilon} \eta \right] - \xi \cos \theta \cdot \left[1 - \varepsilon + 2\delta\left(\tfrac{4}{3}\varepsilon - 1\right) \right] \right\} d\varepsilon\, d\Omega . \qquad (38)$$

Here ξ is the asymmetry parameter, δ its dependence on the energy, ϱ the Michel parameter, and η, a low energy correction to the spectrum's shape. According to the V-A theory, the prediction was

$$\varrho = \delta = 0.750, \qquad \xi = 1.00, \qquad (39)$$

whereas experimentally one obtained,

$$\varrho = 0.747 \pm 0.005, \qquad \delta = 0.78 \pm 0.05, \qquad \xi = 0.978 \pm 0.030. \qquad (40)$$

The theoretical quantities (39) have to be corrected for electromagnetic effects which turn out to be finite and of the order α.

Important tests of the universality character were performed in cases where decays into muon and electron, respectively, had been measured. In all other cases, the strong interaction of the hadrons entered directly. For instance, in the neutron beta-decay the axial coupling,

$$G_A \sim -1.2 \, G_V ; \qquad (41)$$

this 20% deviation from the value -1 is caused by strong interactions, and was computed by Weissberger and Adler.[33]

Since the hadron currents in weak decays contain nucleon fields, the vector and axial-vector currents of beta-decay are both the charge-changing components of isotopic vectors. Adding the electric vector current as isotopic spin 3-component, one might expect that, in the vector beta-decay matrix element, the analogue of the anomalous magnetic moment exists. Experimentally such a component (weak magnetism term!) had been found.

On the other hand, the weak interactions of kaons and hyperons proceed by a factor of roughly 10 slower than the strangeness-conserving reactions. This had been expressed by Cabibbo's assumption of an angle θ_V and θ_A, such that the amplitudes of strangeness-violating processes are suppressed with respect to the strangeness-conserving ones by,

$$\frac{A_{(\Delta s \neq 0)}}{A_{(\Delta s = 0)}} = \tan \theta, \qquad \tan \theta_V = 0.22, \quad \tan \theta_A = 0.28. \tag{42}$$

Cabbibo's scheme is related to $SU(3)$ symmetry, but no consistent derivation of the angles θ has been proposed thus far.

The years since 1957 have confirmed the validity of the V-A theory in all weak decays. But one still lacks a feeling for weak interactions at higher energies. The four-fermion theory has become too perfect a description, and it does not allow for any extension into a wider domain. On the other hand, the V-A theory has stimulated an entire field which has been called the current algebra, and some of the important results of the 1960s have been derived from that scheme.

4. EXTENSION OF QUANTUM FIELD THEORY TO STRONG INTERACTIONS

At the 1961 Solvay Conference it was possible to report great progress on the field-theoretical models of electromagnetic and weak interactions. The understanding of strong forces, on the other hand, seemed to have come to a standstill. It was mainly about the problems of quantum field theory of strong interactions that Heitler talked about in his report on the 'Physical Aspects of Quantum Field Theory', and Yukawa in his report on the 'Extensions and Modifications of Quantum Field Theory'.

4.1. DIVERGENCES IN QUANTUM FIELD THEORY

(W. Heitler)

As Heisenberg remarked in the discussion, the problems faced by quantum field theory are of a two-fold nature. He said: '(i) In field theory the interactions are local while in quantum mechanics they are non-local. (ii) In field theory we have three boundary conditions while in quantum mechanics we have only two (at infinitely small and infinitely large distances of particles). The third boundary condition in field theory refers to an infinite number of particles.'

Heitler, in his report, addressed himself mainly to the first problem, and he asked the question whether one could not try to overcome it by constructing a theory with particles of finite size. Renormalization theory has taken the opposite direction; it removed the infinite rest mass of a point particle by subtraction, and did the same with the charge. Relativistic covariance was achieved by the fact that renormalization constants became infinite. 'The well-known, and very great, success of renormalized Q.E.D. rests on the occurrence of infinite quantities. Only as long as they are given by *infinite* expressions can they be removed in a Lorentz- and gauge-invariant manner.'

It has been argued that the infinity properties of renormalization constants depend on the perturbation theory. But it is hard to obtain reasonable results in Q.E.D.

except in a perturbation calculation. Secondly, some people believe that ghosts with negative norm eventually show up also in Q.E.D. Third, not all theories are renormalizable; e.g., the four-fermion scheme of weak decays is not. Moreover, a whole group of phenomena, namely the electromagnetic mass differences of hadrons, lies outside its range. Heitler noted that, 'This situation reinforces the necessity of achieving convergence as a *primary* demand to be made. We are then at once confronted with the difficulty of reconciling this convergence with Lorentz-invariance as well as with gauge invariance in the case of electrodynamics.' Heitler imagined a generalization of the Lorentz transformation which takes into account something like a 'smallest space and time region'. With that, small deviations, e.g. from Einstein's mass-velocity relation, might show up.

In the renormalization theory of Q.E.D., the infinities disappear, but certain ambiguities show up which cannot be settled on any mathematical grounds, but are treated by imposing the correct relativistic behaviour. In the case of the photon self-energy, not only Lorentz invariance but also gauge invariance has to be used in order to obtain the correct result.

Heitler argued that quantum electrodynamics must finally break down at high energies; but some descriptions, like that of cascade showers, will remain valid. On the other hand, the renormalizable pseudoscalar meson-nucleon coupling does not have any success similar to the one in quantum electrodynamics. Thus, Heitler concluded: 'Renormalization is not a basic principle of physics, and renormalizability is not a feature that particularly recommends a certain theory. Above all we must look for ways of making the theory finite.'

4.2. EXTENDED SOURCES AND ELECTROMAGNETIC MASS DIFFERENCES

We have earlier considered the partial success obtained in the 1950s in describing strong interaction phenomena by using nonrelativistic extended sources. Heitler now reviewed such a model in which the nucleon is considered as being extended in space, and its dimensions are given by the magnitude of the nucleon's Compton wavelength. One can then treat the low energy meson-nucleon scattering; one can explain the p-wave satisfactorily, and also predict the nucleon resonance with spin and isospin $\frac{3}{2}$, the only two parameters being the pseudo-vector coupling constant,

$$f_r^2 = 0.08 \tag{43}$$

and the cut-off constant,

$$k_0 \sim m_N c. \tag{44}$$

Heitler noted that a pseudo-scalar coupling gave a large and dominant s-wave in contradiction with experience.

If one explains the electromagnetic mass differences within isospin multiplets of hadrons, one needs, according to Heitler, hadrons of finite size. Thus one can obtain the right magnitude of the $\pi^- - \pi^0$ mass difference by using the same cut-off momentum (44). More problematic is the case of nucleons, since here the neutron is heavier than the charged particle. Hence Heitler turned to a mixed mesonic-electromagnetic interaction, which is of the second order both in the strong and the electromagnetic coupling. With pseudo-vector coupling, the result is,

$$(\delta M_p - \delta M_n)_{e^2 f^2} = C f^2 [-1 + 0.65\,\bar{\varrho}] \left(\frac{k_0}{Mc}\right)^3,\tag{45}$$

with $C \sim 10\, m_e$. Here $\bar{\varrho}$ describes a form factor with values between 0 and 1. Equation (45) gives a too small value if the 'renormalized' coupling constant (43) is used, but the sign is correct. However, further contributions can be expected. Similar results are obtained in the case of the kaons and the Σ-hyperons.

4.3. TOWARDS A CONVERGENT FIELD THEORY

Heitler proposed to extend the non-relativistic extended source model into the relativistic region by introducing a form factor F into the interaction which depends on differences between the space points at which the coupled fields are taken. This form factor is not relativistically invariant, nor strictly gauge invariant. As a specific example, Heitler chose the nonlocal Q.E.D. with the interaction term,

$$H_{\text{int}} = - ie \int d^4 (x' \ldots x''') \bar{\psi}(x') \gamma_\mu F(x - x', \ldots) \psi(x'') A_\mu(x'''),\tag{46}$$

and in particular the form factor,

$$F(x - x', x - x'', x - x''') = G(x - x', x - x'')\, \delta^4(x - x'''),\tag{47}$$

where $G(x - x', x - x'')$ has the Fourier transform

$$g(p, q) = \frac{\lambda^4}{\lambda^4 + p^2 q^2 - (pq)^2}.\tag{48}$$

λ is an invariant cut-off constant. The S-matrix thus becomes convergent in all orders of the power expansion.[34]

The function, $G(x-x', x-x''')$, with the Fourier transform (48), seems to extend over the whole space-time region. By averaging it over a small region x'' and x_0'' of the order of λ^2/m, $\bar{G}$ reduces to a smeared out δ^4-function; thus it is only different when $x-x'$ as well as $x_0 - x_0'$ are small.

Quantum electrodynamics with the form factor, Equations (47) and (48), satisfies all requirements except those of strict covariance and gauge invariance. Indeed the photon self-energy, which is finite, turns out to be smaller than zero. Also the magnetic moment of the electron turns out to be wrong. In order to restore agreement with experiment, Heitler added a term to the interaction energy operator that is bilinear in the electromagnetic field, A_μ,

$$H'_{\text{int}} = e^2 \int d^4 x' \cdot f_{\mu\nu}(x - x') \{A_\mu(x) A_\nu(x') - \langle 0 | A_\mu(x) \cdot A_\nu(x) | 0 \rangle\},\tag{49}$$

with a form factor $f_{\mu\nu}(x-x')$ such that the photon energy becomes zero, the anomalous magnetic moment of the electron becomes correct, and the Lamb shift takes on the empirical value.

The self-mass of the electron is no longer Lorentz invariant since

$$\delta_m = \delta_m (p = 0) \cdot \left[1 - \frac{1}{72} \frac{p^2}{m^2} + \cdots \right].$$ (50)

Such a deviation from covariance, Heitler noted, was still in agreement with the experimental results since $\delta m(0)$ is only a fraction of m_0.

Heitler concluded his report by noting that: 'So far no way is known of combining convergence with exact Lorentz and gauge invariance. Although no proof exists, it does look as if the two demands, convergence and exact invariance, were incompatible... . It is likely that all of the above mentioned concepts, including the Lorentz transformation, will have to undergo changes and generalizations before the problem of quantum field theory can be finally solved.'

In the discussion of Heitler's talk Heisenberg mentioned the fact that local interaction can be reconciled with finiteness if the indefinite metric is used. Since this particular question was treated extensively at the 1967 Solvay Conference, we shall omit its discussion here and take it up in its proper context. We shall now discuss Yukawa's report.

4.4. EXTENSION OF THE HILBERT SPACE

(H. Yukawa)

Quantum field theory, according to Yukawa, should not only be free from logical and structural defects, but also be able to give cogent reasons for the existence and interaction of various types of particles. Since the discovery of strange particles one might complain about the inability of the present schemes to predict the new particles and phenomena.

How should one alter the theory to obtain more satisfactory results? The first problem is whether the pointlike structure of particles assumed so far can be upheld. Particles with extended structure may be introduced.[35] Or, indefinite metric in Hilbert space may be used.[36] It is to this extension of quantum field theory that Yukawa addressed himself in his report. He noted: 'What we hope to achieve in such attempts is to construct a field theory which satisfies the following conditions, if it is ever possible: (1) Lorentz invariance, (2) Convergence, (3) Probabilistic interpretation, (4) Macroscopic causality.'

The extension of the properties of the usual Hilbert space is mainly contained in the new definitions of pseudo-Hermiticity and pseudo-unitarity of operators, notably,

$$H^* = \eta H \eta$$ (51)

and

$$U^{-1} = \eta U^* \eta,$$ (52)

where η is the metric operator of the pseudo-Hilbert space, and H^* and U^* denote the Hermitian conjugates of the operators H and U. In the simplest possible case of a two-dimensional pseudo-Hilbert space, η is, for instance,

$$\eta = \begin{pmatrix} 1 & 0 \\ 0 & -1 \end{pmatrix} \quad \text{or} \quad \begin{pmatrix} 0 & 1 \\ 1 & 0 \end{pmatrix}. \tag{53}$$

For pseudo-Hilbert spaces many inequivalent representations appear even in the case of finite dimensions. For instance, the two matrices in Equation (53) lead to such inequivalent representations, and as a result a pseudo-Hermitian operator has in the first case only real diagonal elements, whereas in the second case only complex conjugate ones.

The regularization of Feynman, Pauli and Villars can be viewed from a standpoint that the unitarity of the S-matrix is not an absolute restriction, but is applicable only to those processes in which the total proper energy of the system of colliding particles does not exceed the proper energy of the abnormal [negative metric] auxiliary particle with the smallest mass.[37] These auxiliary particles couple as strongly to the normal [positive norm] objects as the latter among each other, thus regularizing the infinities in the quantum field theory. Yukawa assumed that the auxiliary particles become so unstable that one can hardly call them particles any more, and that these broad resonances would determine the high mass spectrum.

Indefinite metric and non-locality are very closely related; thus the indefinite metric can be regarded as a result of an averaging process with respect to internal variables describing extended particles. The washing out of elementary particles leads to an abandoning of δ-functions on the light cone, and therefore regularizes the theory.

5. AXIOMATIC QUANTUM FIELD THEORY

In the 1950s there developed an interesting approach to quantum field theory. One hoped to put it on a mathematically more rigorous basis and looked for a set of simple assumptions or 'axioms' on which such a theory could be based. One of the hopes was also to avoid the use of perturbation theory and non-allowed but intuitive selection of limits in the renormalization scheme.

5.1. The Axiomatic Formulations of Q.F.T.

The aim of the partisans of axiomatic field theory has been to determine whether any local relativistic field theory exists at all. These investigations were led by Lehmann, Symanzik, and Zimmermann, and by Wightman.[38] They make the following assumptions about the quantum field theory:

(i) The usual postulates of quantum mechanics are valid; especially a Hilbert space (with positive metric) of state vectors exists, and the operators can be represented by self-adjoint operators in the Hilbert space.

(ii) The theory is invariant under inhomogeneous Lorentz transformations.

(iii) The energy-momentum spectrum is reasonable, containing an invariant vacuum state, the others being in the forward light cone of energy-momentum.

(iv) The theory is local, i.e. observables commute at spacelike distances. (The principle of micro-causality.)

The approaches of Wightman and Lehmann–Symanzik–Zimmermann (LSZ) differ a bit in so far as Wightman used, as the starting point, vacuum expectation values of field operators, whereas LSZ used vacuum expectation values of time-ordered or retarded products of field operators. Relativistic invariance, locality, and spectral conditions lead to linear relations between the functions, whereas unitarity yields nonlinear connections. The programmes themselves are very tricky, and involve a high degree of mathematical skill including, in Wightman's case, the theory of analytic functions of several variables. LSZ start from the fact that an S-matrix should be obtainable; thus they discuss ingoing and outgoing fields and in between, i.e. in the interaction region, interpolating fields. Within their scheme, the important results are the derivations of dispersion relations. Hence the axiomatic approach of the 1950s is connected with the third approach [in the classification given by Gell-Mann] to the quantum field theory discussed at the 1961 Solvay Conference.

5.2. SOME ASPECTS OF THE FORMALISM OF FIELD THEORY

(G. Källén)

In his report, Källén represented the stricter form of Q.F.T. He had worked on the consistency of the formalism of Q.E.D. earlier, and had performed numerous investigations on the analytic properties of products of field operators.

The canonical quantization of quantum fields leads to most satisfactory results like the description of particles and the spin-statistics theorem in the case of no interaction. In quantum electrodynamics one could extend the scheme formally to include the interaction between electrons and photons known from the classical theory. However, one did not succeed in finding the exact mathematical solutions. An approximation scheme had to be adopted which allows one to compute results that agree very well with the empirical data. But even in Q.E.D., a difficulty in principle appears. In the basic equation of Q.E.D.,

$$\left(\gamma^\mu \frac{\partial}{\partial x^\mu} + m \right) \psi(x) = ie\gamma^\mu A_\nu(x) \psi(x), \tag{54}$$

where ψ is the electron field, and A_ν, the photon field, a product of field operators occurs, leading to a product of delta-functions which does not always make sense. One has to make certain prescriptions, and these are condensed in the renormalization scheme which adds certain counter terms.

The problems become enhanced in pion-nucleon physics where one has to try various 'Ansätze' for the form of the interaction. But here no approximation scheme seems to be reasonable due to the large coupling. The difficulties with approximations might be understood from the Van Hove–Friedrich–Haag theorem which states that the expansion of physical states in terms of free states is very singular, i.e. the probability coefficients, $C_{nn'}$, in the equation

$$|n_{\text{phys.}}\rangle = \sum_{n'} C_{nn'} |n'_{\text{math.}}\rangle,\tag{55}$$

are practically zero.[39] The so-called axiomatic school has attempted to avoid the expansion technique and to extract as much information as possible from general symmetry properties and other physical assumptions about the theory.

The basic assumptions mainly derived from an extrapolation of our experience for macroscopic distances are: (i) Lorentz invariance; (ii) Reasonable mass spectrum (every physical state should lie inside the positive light cone); (iii) Local commutativity of the fields; (iv) Asymptotic condition, i.e. the fields should approach free fields in the distant past and in the distant future.

Frequently one sees some connection between the postulates (i), (ii) and (iii), and there are cases in which one may derive local commutativity from Lorentz invariance and the mass spectrum. For instance, one takes the vacuum expectation value of two scalar fields, and inserts a complete set of intermediate states, $|z\rangle$,

$$F^A(x_1, x_2) = \langle 0|A(x_1) A(x_2)|0\rangle = \sum_{|z\rangle} \langle 0|A(x_1)|z\rangle \cdot \langle z|A(x_2)|0\rangle.\tag{56}$$

Then one finds, by assuming translation invariance and making certain formal rearrangements, that

$$F^A(x_1, x_2) = i \int_0^\infty dx^2\, G(-x^2)\, \Delta^{(+)}(x_1 - x_2, x^2),\tag{57}$$

where

$$\Delta^{(+)}(x, x^2) = \frac{1}{i\pi} \lim_{\varepsilon \to 0} \int_0^\infty \frac{dx'^2\, \bar{\Delta}(x'^2, x^2)}{x'^2 + x^2 - (x_0 - i\varepsilon)^2},\tag{57a}$$

$\bar{\Delta}$ being a Bessel function. Hence the commutator is,

$$\langle 0| [A(x_1), A(x_2)] |0\rangle = i \int_0^\infty dx^2\, G(-x^2) \cdot \Delta(x_2 - x_1, x^2),\tag{58}$$

with

$$\Delta(x, x^2) = \frac{-i}{(2\pi)^3} \int dp \cdot e^{ipx} \delta(p^2 + x^2) \frac{p_0}{|p_0|}.\tag{58a}$$

The right hand side of Equation (58) disappears for space-like distances; thus the local commutativity is proved under the above manipulations for the vacuum expectation value.

In the above calculations certain analytic functions enter, such as $\Delta^{(+)}(x, x^2)$ in Equation (57). If one insists on their use, one can derive local commutativity once the commutativity is assumed for larger distances. Hence macroscopic causality plus analyticity immediately leads to microcausality.

From the asymptotic condition (iv), one can easily derive dispersion relations via the so-called reduction formula. This point was treated explicitly in the report of Goldberger, which we shall discuss in the following.

6. SCATTERING MATRIX AND DISPERSION RELATIONS

By 1961 the general view was that some minimal relations could be derived from quantum field theory without using debatable or even unallowed approximations. These results were condensed in the so-called dispersion relations, which Goldberger discussed at the 1961 Solvay Conference. The extensions of these single variable equations to several variable representations and equations was discussed by Mandelstam. The latter indicated more general properties of the scattering matrix whose study became a primary issue in particle theory of the 1960s.

6.1. THEORY AND APPLICATIONS OF SINGLE VARIABLE DISPERSION RELATIONS

(M. L. Goldberger)

Dispersion theory was invented [omitting its earlier use by Sommerfeld, *Phys. Z.* **8** (1907), *Ann. Phys.* (1914)] by Kramers and Kronig[40] in electrodynamics. It was reborn in 1946 when Kronig raised the question whether causality would restrict in any way the S-matrix of Wheeler and Heisenberg.[41] Shortly after 1950, Wheeler, Toll, Wigner, and Van Kampen studied some classical and quantum mechanical systems in this regard. The first attempt to study the problem within the framework of quantum field theory was made by Gell-Mann, Goldberger, and Thirring.[42] This initiated extensive studies, especially in the field of strong interactions, at a time when it was increasingly felt that the conventional theoretical approach had failed.

The original Kramers–Kronig relation connects the forward scattering amplitude, f, of light of frequency, ω, from atoms to an integral over the total cross-section for the removal of light from the beam,

$$
\begin{aligned}
\operatorname{Re} f(\omega) &= -\frac{e^2}{mc^2} + \frac{2\omega^2}{\pi} P \int_0^\infty d\omega' \cdot \frac{\omega'\,\operatorname{Im} f(\omega')}{\omega'^2 - \omega^2} \\
&= -\frac{e^2}{mc^2} + \frac{\omega^2}{2\pi^2 c} P \int d\omega' \frac{\sigma(\omega')}{\omega'^2 - \omega^2}.
\end{aligned}
\tag{59}
$$

In going from the first equation to the second in (59), the so-called optical theorem,

$$
\operatorname{Im} f(\omega) = \frac{\omega \sigma(\omega)}{4\pi c}
\tag{60}
$$

is used. In the usual language, Equation (59) represents a dispersion relation, with one subtraction (at frequency $\omega = 0$), which is expressed in terms of the mass m and charge e of the electron.

The theoretical basis of the Kramers–Kronig dispersion relations is the assumption that the measurements of the electromagnetic field carried out at points with spacelike separation should not interfere with each other. The mathematical expression for this assumption is normally local commutativity of the field operators at spacelike distances. Although one cannot exclude empirically the violation of this 'microcausality' condition at distances of the order of 10^{-13} cm, it seems difficult not to require the stronger assumption in quantum field theory.

The first important generalization of the Kramers–Kronig relations, Equation (59), was to treat the forward scattering of mesons by nucleons.[43] The important difference in the scattering of photons is the fact that the energy region between 0 and m_π is not accessible, hence assumptions must be made to obtain,

$$\text{Re } T^+ (\omega) - \text{Re } T^+ (m_\pi) = \frac{f^2}{m_N} \cdot \frac{k^2}{\omega^2 - (m_\pi^2/2m_N)^2}$$

$$+ \frac{k^2}{2\pi^2} P \int_{m_\pi}^{\infty} \frac{d\omega'\, \omega'}{2k'} \cdot \frac{\sigma_- (\omega') + \sigma_+ (\omega')}{\omega'^2 - \omega^2}, \qquad (61a)$$

and

$$\text{Re } T^- (\omega) = \frac{f^2}{m_N} \cdot \frac{\omega}{\omega^2 - (m_\pi^2/2m_N)^2} + \frac{\omega}{2\pi^2} P \int_{m_\pi}^{\infty} \frac{d\omega'\, k'}{2} \cdot \frac{\sigma_- (\omega') - \sigma_+ (\omega')}{\omega'^2 - \omega^2} \qquad (61b)$$

where $\omega^2 = m_\pi^2 + k^2$, with the momentum k of the incident positive $(+)$ and negative $(-)$ pions. The constant, f^2, describes the pion-nucleon coupling at the pole, $\omega = m_\pi/2m_N$, which is the only contribution of the unphysical region. The numerical value is, $f^2 = 0.08$. The sum of the cross-sections, σ_- and σ_+, increases more slowly than linearly in ω, whereas the difference goes faster to zero than $(\ln \omega)^{-1}$.

Goldberger noted: 'Verification of the theoretical predictions does not by any means prove that quantum field theory is correct, any more than that the Bohr atom predictions of the Balmer series proved that picture. On the other hand, if there should appear a discrepancy between theory and experiment the shock would be at least as profound as that caused by the non-conservation of parity, if not more so. As we shall see, the elements of field theory that go into the derivation are those which would be abandoned with only the greatest reluctance. From the standpoint of a pure dispersion theory form of a dynamical theory (to which we allude later) the effects would be fatal.'

If the scattering amplitudes are regarded as analytic functions of the energy or the momentum transfer, one may infer from observation certain poles at certain unphysical values.[44] Thus one may relate different processes by analytic continuation.

Dispersion relations may replace field theory to some extent if they can be generalized further, e.g. to the non-forward direction. Although this extension is troublesome, it has been achieved mathematically.[45] Other processes like the electromagnetic structure of nucleons and pions have also been considered in terms of dispersion relations, yet the result is not as spectacular as in the forward pion-nucleon dispersion relations. Another application of dispersion relations is the Pomeranchuk theorem; it states the fact that total cross-sections approach constant values at asymptotic energies, which are the same for particles and anti-particles.[46]

The derivation of dispersion relations for pion-nucleon scattering makes use only of the pseudo-scalar character of the pion, but not of any particular coupling to the nucleons. Goldberger concluded: 'If we could do serious and accurate computations in quantum field theory, the fact that the amplitudes had certain analytic properties would be an amusing but relatively uninteresting observation. The fact is that at the present we do not really know what we have theoretically, whether the axiomatic scheme is rich enough to contain all the embarrassingly large number of particles or even more modestly anything at all relevant to experience. In some ways the dispersion approach extracts the best of field theory and has begun to acquire a character of its own. The way this story unfolds, the next logical step after the single variable dispersion relations, and the crucial one, without which there is no hope of formulating a dynamical theory, is the subject of the next address [Mandelstam's report].'

In the discussion, several questions were raised concerning the application of dispersion relations to compute the 3-3 resonance explicitly, but the results were not satisfactory. The problem of deriving the dispersion relations for massive particles from axiomatic field theory was mentioned. Here Bogoliubov et al.[47] had made great progress, but Goldberger did not refer to it in his report.

Another question was whether the implied breakdown of local commutativity should be realized in dispersion relations. Precursors of the order of $a = 10^{-13}$ cm should still be admitted, although they change the form of the dispersion relations.

6.2. Two-Dimensional Representations of Scattering Amplitudes

(S. Mandelstam)

A further step towards a dynamical theory based on a minimum of field-theoretical assumptions is achieved by postulating two-dimensional representations of scattering amplitudes. In his report, Mandelstam presented these ideas, approximations, and speculations.

Using quantum field theory without a Lagrangian avoids many of the troubles connected with the latter approach. One can use the most important principles, unitarity and causality, in the S-matrix language. Dispersion relations have been established describing scattering amplitudes that are functions of two variables: the total energy squared, s, and the momentum transfer squared, t. But, essentially, only single integral relations have been applied. The conjecture of 'maximal analyticity' allows one to assume double dispersion relations of the form:

$$A(s, t) = \frac{1}{\pi^2} \int ds' \, dt' \, \frac{\varrho(s', t')}{(s' - s)(t' - t)} + \text{crossed terms}, \qquad (62)$$

where ϱ is the double spectral function.

In order to compute in practical cases, one must make approximations, and it is not even sure whether double dispersion relations always follow from a kind of maximal analyticity. Moreover, maximal analyticity is not a consequence of local quantum field theory either. But maximal analyticity is the essential concept for giving a basis to S-matrix theory, independent of the quantum field theory.

In using double dispersion relations two problems have to be solved: first, the function ϱ must be determined; second, the low lying angular momentum states have to be obtained. The greatest success was achieved in dealing with the electromagnetic form factor of the nucleons and predicting the ϱ-resonance.[48] It was later found, not exactly at the energy computed, but 'close enough' to it.

7. SYMMETRY ASPECTS OF Q.F.T.

As we have noted, at the 1961 Solvay Conference three different approaches to the phenomena of particle physics had been discussed, all based on some kind of quantum field theory. Not related to any specific approach was the report given by Gell-Mann.

7.1. SYMMETRY PROPERTIES OF FIELDS

(M. Gell-Mann)

Gell-Mann divided his talk into two sections on exact and approximate symmetries, respectively, counting among the exact symmetries those which come under Poincaré transformations and CPT-invariance as well as 'superselection' rules. The latter state the fact that the superpositions of states having a definite charge, baryonic number, etc., are not admitted in particle theory. The remaining laws include the conservation of charge, baryonic and leptonic numbers, which are perhaps all connected with gauge symmetries.

Approximate symmetries are symmetries which are valid in strong interactions, but maybe not in weak or electromagnetic processes, as for instance, the isospin symmetry. Consequently the conservation laws are broken only by small perturbations. The strongest interaction allows for the largest symmetry group. This includes the $SU(3)$ group which Gell-Mann discussed briefly. 'The development of physics has been characterized thus far by the fact that no theory has been exact,' Chew confessed in the discussion and proposed that one should deal first with the strong interactions. But then one would lose the wealth of information coming, for instance, from weak interactions.

In the years immediately following 1961, the symmetry approach to particle physics became an important field. Gell-Mann and Ne'eman proposed a scheme, the 'eight-fold way', which dealt rather successfully with the pattern of particles, on the one hand, and their properties, on the other. As a consequence, an organization of the resonances found since 1961 was accomplished, while none of the above-mentioned problems of quantum field theory could be solved.

7.2. GENERAL DISCUSSION

The general discussion at the 1961 Solvay Conference covered all aspects of the reports and suggested hints for the interpretation of various details. It also dealt with the relation between the different approaches to quantum field theory.

One question that came up concerned the features of quantum electrodynamics that cannot be reproduced by the axiomatic approach. This has to do with the space-time description in axiomatic and S-matrix theory. The systems described by the usual quantum field theory develop in time, and perturbation theory, with the equations of motion, presents the only method of handling the problem, as Källén pointed out. The deeper point is that there must be a connection with a classical description, and this implies the use of quantum field theory. Nevertheless it is possible to start with the S-matrix theory and derive, say, the Lamb shift. The question is how far does one have to specify the Lagrangian beforehand. One must always put something in, which tells if one is talking about electrons, muons, pions, and what interactions they do have.

Chew pointed out that in the case of a non-relativistic potential approach to strong interactions, 'one can compute the *complete* solution' from the S-matrix theory, 'not just the asymptotic behaviour.' Also, as Feynman remarked, 'there is a feature of the study of S-matrices that may warrant further analysis. Given the S-matrix for $A+B$ scattering, for $B+C$ and for $C+A$, what limitations etc., can we deduce on the scattering for A, B and C together? We try to understand physics in complicated situations by combining knowledge in simpler systems.'

The discussion reminded the participants of the reasons why one had to abandon the Lagrangian theory. The knowledge of the divergence difficulties had brought Heisenberg in the 1940s to introduce the S-matrix. But, said Chew, 'the enthusiasm died down because people were not bold enough, then, to assume analyticity in all momentum variables and so found the theory lacking in dynamical content.' To which, Heisenberg replied: 'When I had worked on the S-matrix for a while in the years 1943 to 1948 I came away from the attempt of construction of a pure S-matrix theory for the following reason: when one constructs a unitary S-matrix from simple assumptions (like a hermitian η-matrix by assuming $S=e^{i\eta}$), such S-matrices always become non-analytical at places where they ought to be analytic. But I found it very difficult to construct analytical S-matrices. The only simple way of getting (or guessing) the correct analytical behaviour seemed to be a deduction from a Hamiltonian in the old-fashioned manner.'

Heisenberg continued: 'It should be possible to define the S-matrix by postulating some underlying groups as basis of the theory, adding the postulate of unitarity and analyticity and calculating the masses, etc., from some condition of consistency, without any use of the metric in Hilbert space. My criticism comes only from a practical point of view. I cannot see how one could overcome the enormous complications of such a programme.'

Feynman pointed out that the great difficulty in all the approaches remains the lack of success in computing the phenomena of strong interactions. In favour of the S-matrix approach, Gell-Mann said: 'A good feature of the dispersion theory approach is that it works with quantities that are observable or nearly so. While the S-matrix theory is being built, we are learning to understand a great deal about the experiments. We could mention as examples the use of forward scattering and form factor dispersion relations, polology and the current study of high energy diffraction.

These applications have not only helped to interpret data, but have stimulated a great deal of work.'

As to the content of the S-matrix theory, Heisenberg proposed the following classification: 'The S-matrix theory has the widest axiomatic frame: one only postulates the existence of a unitary S-matrix with reasonable (causal) analytic properties, representing the group structure given by the experiments.

'If one adds, in order to get a more narrow frame, the postulate of the existence of a local field operator commuting (or anti-commuting) at space-like distances, one comes to a field theory with a (possibly) indefinite metric, which would allow the description of a local interaction.

'If one further adds the axioms that the asymptotic operators should be sufficient to construct the complete Hilbert space, and that its metric shall be definite, one comes to the orthodox field theory.'

About the usefulness of the various approaches, Wentzel remarked: 'Abandoning field theory altogether in favour of a dispersion-theoretic scheme, seems to me similar in spirit to abandoning statistical mechanics in favour of phenomenological thermodynamics. We believe in statistical mechanics as a comprehensive theory although only in simple cases can one actually calculate a partition function. If this calculation is technically too difficult, we may content ourselves with applying thermodynamics at the cost of feeding in more experimental data. Still nobody would want to do without statistical mechanics as a higher ranking discipline. Similarly, I would hope that also field theory, in one form or another, will retain its place as a superior discipline.'

REFERENCES

1. L. D. Landau and R. E. Peierls, *Z. Phys.* **62**, 188 (1930).
2. L. D. Landau, *Nucl. Phys.* **13**, 181 (1959).
3. H. Lehmann, K. Symanzik, and W. Zimmermann, *Nuovo Cim.* **1**, 1425 (1955); A. S. Wightman, *Phys. Rev.* **101**, 860 (1956).
4. C. N. Yang and D. Feldman, *Phys. Rev.* **79**, 972 (1950); G. Källén, *Ark. Fys.* **2**, 371 (1950).
5. For the programme of using dispersion and unitary relations as a basis for calculation in field theory, see, for example, M. Gell-Mann, *Proc. of the Sixth Annual Rochester Conference* (1956); L. D. Landau, *Proc. of the Kiev Conference* (1959); S. Mandelstam, *Phys. Rev.* **115**, 1741 (1959); G. F. Chew, *Lectures at the Summer School of Theoretical Physics*, Les Houches (1960).
6. L. Castillejo, R. H. Dalitz, and F. J. Dyson, *Phys. Rev.* **101**, 453 (1956).
7. S. Tomonaga, *Prog. Theor. Phys.* **1**, 27 (1926).
8. J. Schwinger, *Phys. Rev.* **74**, 1439 (1948).
9. R. P. Feynman, *Phys. Rev.* **76**, 749, 769 (1949).
10. W. E. Lamb and R. C. Retherford, *Phys. Rev.* **72**, 241 (1947).
11. J. Schwinger and V. F. Weisskopf, *Phys. Rev.* **73**, 1272 A (1948).
12. F. J. Dyson, *Phys. Rev.* **75**, 486, 1736 (1949).
13. J. C. Ward, *Phys. Rev.* **77**, 293L (1950); **78**, 182L (1950).
14. R. Hofstadter *et al.*, *Phys. Rev. Lett.* **6**, 293 (1961).
15. See, e.g., A. Peterman, *Fortschr. Phys.* **6**, 505 (1958).
16. T. D. Lee, *Phys. Rev.* **95**, 1329 (1954); W. Pauli and G. Källén, *Kgl. Danske Vid. Selsk. Mat-fys. Medd.* **30**, No. 7 (1955).
17. See Landau's contribution in *Niels Bohr and the Development of Physics*, Pergamon Press, London (1955), p. 52.

18. W. Heisenberg, *Nucl. Phys.* **4**, 532 (1957).
19. B. W. Sargent, *Proc. Camb. Phil. Soc.* **28**, 538 (1932); *Proc. Roy. Soc.* **A 139**, 659 (1933).
20. E. Fermi, *Z. Phys.* **88**, 161 (1934).
20a. G. Gamow and E. Teller, *Phys. Rev.* **49**, 895 (1936).
21. A. Salam, *Nuovo Cim.* **5**, No. 1, 299 (1957);
    L. D. Landau, *J. E. T. P.* **32**, 405 and 407 (1957);
    T. D. Lee and C. N. Yang, *Phys. Rev.* **104**, 254 (1956), **105**, 1671 (1957).
22. C. S. Wu, *et al. Phys. Rev.* **105**, 1413 (1957).
23. E. C. G. Sudarshan and R. E. Marshak, *Padua-Venice Conference on Elementary Particles* (157); *Phys. Rev.*, **109**, 193 (1958).
24. O. Klein, *Nature*, **161**, 897 (1948).
25. G. Puppi, *Nuovo Cim.* **5**, 587 (1948).
26. R. P. Feynman and M. Gell-Mann, *Phys. Rev.* **109**, 193 (1958);
    J. J. Sakurai, *Nuovo Cim.* **7**, 649 (1958).
27. G. C. Wick, A. Wightman, and E. P. Wigner, *Phys. Rev.* **88**, 101 (1952).
28. R. Sachs and S. Treiman, *Phys. Rev. Lett.* (1961).
29. R. P. Feynman and M. Gell-Mann, *Phys. Rev.* **109**, 193 (1958).
30. H. Yukawa, *Proc. Phys. Math. Soc. (Japan)*, **17**, 48 (1935).
31. B. Pontecorvo, *JETP*, **10**, 1236 (1960).
32. G. Danby, *et al.*, *Phys. Rev. Lett.* **9**, 36 (1962).
33. W. I. Weissberger, *Phys. Rev. Lett.* **14**, 1047 (1965);
    S. L. Adler, *Phys. Rev. Lett.* **14**, 1051 (1965).
34. E. Arnous, W. Heitler, and Y. Takahashi, *Nuovo Cim.* **16**, 671 (1960);
    E. Arnous and W. Heitler, *Nuovo Cim.* **11**, 443 (1959);
    E. Arnous, *et al.*, *Nuovo Cim.* **16**, 785 (1960);
    L. O'Raiferteigh, *Helv. Phys. Acta*, **33**, 783 (1960).
35. H. Yukawa, *Phys. Rev.* **77**, 219 (1950); **80**, 1047 (1950); **91**, 415 (1953); *Science*, **121**, 405 (1955).
36. D. Bohm, *et al.*, *Progr. Theor. Phys.* **24**, 761 (1960).
37. R. P. Feynman, *Phys. Rev.* **74**, 1439 (1948);
    W. Pauli and F. Villars, *Rev. Mod. Phys.* **21**, 434 (1949);
    E. C. G. Stückelberg and D. Rivier, *Phys. Rev.* **74**, 218, 986 (1948).
38. H. Lehmann, K. Symanzik, W. Zimmermann, *Nuovo Cim.* **1**, 1425 (1955); **6**, 319 (1957);
    A. S. Wightman, *Phys. Rev.* **101**, 860 (1956).
39. L. Van Hove, *Physica*, **18**, 145 (1952);
    K. O. Friedrich, *Comm. Appl. Math.* **5**, 349 (1952); **4**, 161 (1951);
    R. Haag, *Dansk. Vid. Selsk., Mat., Fys. Medd.*, **29**, No. 12 (1955).
40. H. A. Kramers, *Congresso Int. di Fisici, Como* (1927), and R. Kronig, *J. Opt. Soc. America*, **12**, 547 (1926).
41. R. Kronig, *Physica*, **12**, 543 (1946).
42. M. Gell-Mann, M. Goldberger, and W. Thirring, *Phys. Rev.* **95**, 1612 (1954); **97**, 508 (1955).
43. M. L. Goldberger, *Phys. Rev.* **99**, 979 (1955);
    M. L. Goldberger, H. Miyazawa and R. Oehme, *Phys. Rev.* **99**, 986 (1955).
44. G. F. Chew and F. Low, *Phys. Rev.* **113**, 1640 (1959).
45. G. F. Chew, M. L. Goldberger, F. E. Low, and Y. Nambu, *Phys. Rev.* **106**, 1337, 1345 (1957);
    K. Symanzik, *Phys. Rev.* **105**, 743 (1957);
    H. Bremmerman, *et al.*, *Phys. Rev.* **109**, 2178 (1958);
    N. N. Bogoliubov, *Fortschr. Phys.* **6**, 169 (1958).
46. Ia. Pomeranchuk, *Z. Exp. Teor. Fiz.* **34** (1958).
47. N. N. Bogoliubov, *Fortschr. Phys.* **6** (1958).
48. W. R. Frazer and R. J. Fulco, *Phys. Rev. Lett.* **2**, 365 (1959).

FOURTEENTH SOLVAY CONFERENCE 1967

F. CERULUS -   L. MICHEL   G.F. CHEW -   J. HAMILTON   D. SPEISER   A.S. WIGHTMAN   P. CASTOLDI

Cl. GEORGE - I.B. OKUN - S. MANDELSTAM - D. RUELLE - S.L. ADLER -   C.F. v. WEIZSACKER - H. UMEZAWA -   R. HERMAN -   R. HOFSTADTER - E.C.G. SUDARSHAN - F. HENIN -   F. MAYNE

S.A. WOUTHUIJSEN - L.A. RADICATI - M. LEVY - R. OMNES - M. FROISSART - S. FUBINI - B. FERRETTI - F.E. LOW - A. TAVKELIDZE - L. ROSENFELD - S. SAKATA - N. CABIBBO - R. HAAG - C. SCHOMBLOND - M. EVRARD-MUSETTE - J. BIJTEBIER

I. PRIGOGIGNE - R.E. MARSHAK - E.P. WIGNER - J. SCHWINGER - W. HEISENBERG - C. MØLLER - E. AMALDI - F. PERRIN - J. GEHENIAU

Absents : S. WEINBERG et M. GELL-MANN

12

Fundamental Problems in Elementary Particle Physics*

The fourteenth Solvay Conference took place in Brussels in October 1967. Its general theme was the 'Fundamental Problems of Elementary Physics'.

The Scientific Committee consisted of C. Møller (Copenhagen), President, E. Amaldi (Rome), Sir Lawrence Bragg (London), C. J. Gorter (Leyden), W. Heisenberg (Munich), F. Perrin (Paris), I. Tamm (Moscow), S. Tomonaga (Tokyo, Japan), and J. Géhéniau (Brussels), Secretary. Bragg, Tamm and Tomonaga were not able to attend. J. R. Oppenheimer, who was the President of the Scientific Committee, had died on 18 February 1967; his place was taken by Møller, who rendered a homage to the memory of Oppenheimer.

Among the invited speakers at the 1967 Solvay Conference were: G. F. Chew (Berkeley, California), H. P. Dürr (Munich), M. Gell-Mann (Pasadena, California), R. Haag (Hamburg), W. Heisenberg (Munich), G. Källén (Lund), E. C. G. Sudarshan (Syracuse, New York), and A. Tavkhelidze (Dubna, U.S.S.R.).

The large number of invited participants included: S. L. Adler (Princeton, N.J.) N. Cabibbo (CERN, Geneva), B. Ferretti (Bologna), M. Froissart (Saclay), S. Fubini (Turin), J. Hamilton (Copenhagen), R. Hofstadter (Stanford, California), M. Lévy (Paris), F. E. Low (Cambridge, Mass.), S. Mandelstam (Berkeley, California), R. E. Marshak (Rochester, N.Y.), L. Michel (Bures-sur-Yvette), L. B. Okun (Moscow), R. Omnes (Strasbourg), L. A. Radicati (Pisa), T. Regge (Turin), L. Rosenfeld (Copenhagen), D. Ruelle (Bures-sur-Yvette), S. Sakata (Nagoya, Japan), J. Schwinger (Cambridge, Mass.), H. Umezawa (Tokyo, Japan), C. F. von Weizsäcker (Hamburg), S. Weinberg (Cambridge, Mass.), A. S. Wightman (Princeton, N.J.), E. P. Wigner (Princeton, N.J.), and S. A. Wouthuijsen (Amsterdam).

1. NEW ASPECTS OF PARTICLE PHYSICS IN THE 1960s

Shortly before 1960 the two large accelerators at Brookhaven and CERN had been completed. Research at these big laboratories started immediately and soon the first results were available. The reactions leading to only a few final states were studied in particular, and knowledge of these objects and their interactions grew systematically. Apart from this growth of hadron physics a few more or less isolated facts were discovered, which were connected with the field of weak interactions.

---

* *Fundamental Problems in Elementary Particle Physics*, Proceedings of the Fourteenth Conference on Physics at the University of Brussels, October 1967, Interscience Publishers, New York, 1968.

1.1. The Realm of Resonances

If one looks at a table of elementary particles of the late 1950s one finds about thirty objects listed: the photon, the leptons (neutrino, electron, muon), the pions, the kaons, the nucleons (proton, neutron), and the hyperons (lambda, sigma, and xi). A few years later this number had increased very much to include several hundreds of new objects, and there was a period in which almost every week a new particle was discovered and subsequently announced in the *Physical Review Letters* or *Physics Letters*. Tables of particles and resonances appeared twice a year in which the new candidates were listed and the question marks after old ones were removed.

All of the new particles shared one property. They were observed not as 'stable' objects, but in enhancements of two or more mesons resulting from the inelastic scattering of two hadrons. The first example was the ω-meson, discovered by the Alvarez group at Berkeley in the analysis of the reaction

$$\bar{p} + p \to \pi^+ + \pi^- + \pi^- + \pi^0. \tag{1}$$

Many pictures were taken in the 72-inch liquid hydrogen bubble chamber, and the study of the track geometry revealed the interesting fact that when the invariant mass of three pions, M_3^2,

$$M_3^2 = (p_1 + p_2 + p_3)^2, \tag{2}$$

with the four momenta, p_i, was plotted versus the number of events, a clear peak showed up in the $\pi^+ \pi^- \pi^0$ distribution at $M_3 = 790$ MeV.[1] The width of the peak, Γ, was later determined to be 9.4 MeV, indicating a lifetime for the new resonance of

$$\tau = \frac{\hbar}{\Gamma} \sim 10^{-22} \text{ sec}. \tag{3}$$

Hence the ω-meson is similar to the pion-nucleon resonance N^* or Δ discovered by Fermi's group in 1952, which has a lifetime of 5×10^{-24} sec and a width of over 100 MeV.[2]

Further studies had to be made in order to determine the properties of these fast decaying resonances. Since the ω appeared only in the neutral $\pi^+ \pi^- \pi^0$ enhancement its isospin was determined to be zero. The spin is obtained from observing the distribution in the so-called Dalitz plot, and for the ω one found $J^P = 1^-$. Thus the ω-meson represented the first example of a vector meson. Soon afterwards a second member of the same family was seen, the ϱ-meson.[3] The ϱ-meson has a mass of the order of 760 MeV and is broader than 100 MeV. It appears in all three charged states, i.e. ϱ^+, ϱ^0, and ϱ^-, and decays essentially into two pions.

Around the same time another meson was discovered, whose spin and parity assigned it to the family of pions and kaons, since $J^P = 0^-$. Finally an object became known by analyzing the $(K\pi)$ final state at mass 890 MeV. It is the 'strange' member of the vector meson family.

The study of final states was extended to include baryons, and already in 1960 the Alvarez group saw an enhancement in the $(K\pi)$ distribution at 1385 MeV. Its quantum numbers were found to be,

$$\text{isospin spin parity} = I J^p = 1 \tfrac{3}{2}^+, \tag{4}$$

hence one put it into the family of the nucleon isobar $\Delta (I P^J = \tfrac{3}{2} \tfrac{3}{2}^+)$. The best known member of the family is perhaps the Ω^- which was predicted by the so-called $SU(3)$ scheme and discovered shortly afterwards.

1.2. HIGHER SYMMETRIES AND PARTICLE PATTERN

The discovery of the strange particles had led people to speculate about new symmetries beyond the isospin scheme. The Gell-Mann–Nishijima scheme had introduced the concept of a hypercharge, with the properties

$$Q = I_3 + I_2 \tag{5a}$$

and

$$Y = S + B, \tag{5b}$$

saying that it contributes, besides the third component of the isospin, to the electric charge Q, and that it is composed of the baryonic number of the particle with its strangeness.

The grouping of the resonances into families of same spin and parity indicated that a higher symmetry might act in strong interaction physics. The most successful proposal was made independently by Gell-Mann[4] and Ne'eman,[5] the 'eightfold way'.[6] This group has eight generators, among them the components of the isospin I_+, I_-, and I_3, and the hypercharge. Its rank is two, since two of the generators commute. The representations of this group should be related to the pattern of particles, similar to the grouping of particles in isospin multiplets. The lowest representation, containing three members, does not seem to be used in nature. Out of it were constructed the octet and the singlet, and one could immediately fill two eight-dimensional representations with the members of the pseudo-scalar meson and the members of the vector-mesons. Another octet was formed by the nucleons and the hyperons (Λ, Σ, Ξ). When looking for a place for the baryon resonances $(N^*, \Sigma^* = Y^*, \Xi^*)$, Gell-Mann found the decouplet representation adequate, and predicted another yet unknown object, the Ω^-.[7] It was subsequently discovered by V. E. Barnes, et al.[8]

In predicting the existence of Ω^-, one had to go beyond the pure $SU(3)$ scheme and assume symmetry-breaking. The reason is that if the symmetry is exact, all members of a multiplet should have equal masses. In nature this is not satisfied already in the isospin multiplets, but one relates the mass difference between proton and neutron to the existence of electromagnetic interaction which breaks isospin symmetry. Gell-Mann similarly assumed that, in strong interaction dynamics, two parts are to be separated in the Hamiltonian; the very strong being $SU(3)$ invariant, and the medium strong breaks the symmetry. Hence the interaction hierarchy now reads,

$$H = H_{VS} + H_{MS} + H_{EM} + H_W, \tag{6}$$

where H_{EM} and H_W refer to the electromagnetic and weak interactions, respectively. By assuming the simplest possible form for H_{MS}, breaking $SU(3)$, that is

$$H_{MS} = a + b U_3, \tag{7}$$

with U_3 as the third component of the so-called U-spin, one found the following mass-splitting formula,

$$M = a + bY + c\left[I(I+1) - \tfrac{1}{4}Y^2\right], \tag{8}$$

with the constants a, b, and c for each multiplet.[9]

Applying Equation (8) to the baryon resonances, and neglecting isospin mass-splitting, one obtains equal spacing between the masses,

$$M(N^*) - M(Y^*) = M(Y^*) - M(\Xi^*) = M(\Xi^*) - M(\Omega). \tag{9}$$

Hence, Ω^- should have a mass around 1676 MeV, which was indeed found.

The mass formula (9) represented the greatest triumph of the $SU(3)$ scheme, and yet it pointed to the main problem: the symmetry-breaking terms are not small, and the perturbation procedure used in making the prediction cannot be justified on theoretical grounds. A similar situation occurred in dealing with the other properties. The $SU(3)$ scheme made predictions about couplings, decay times etc., but though many of them are roughly satisfied one hardly has much feeling about the deviations.

Finally the $SU(3)$ scheme was extended to include the spin, and still higher $SU(6)$ multiplets were constructed, with similar successes and failures.[10] In the $SU(6)$ scheme the problem of the relativistic nature of spin could not be accounted for. The particle pattern, however, is described satisfactorily, especially when one bases the representations on fundamental triplets composed of three 'quarks'.[11]

1.3. New Developments in Weak Interactions

The $SU(3)$ scheme was also applied to the properties of weak interactions. Cabibbo made the additional assumption, by introducing the angle, θ, that the strangeness-conserving hadron current is multiplied by its cosine, while the strangeness-violating current is multiplied by its sine. By this Ansatz, which proved to be very successful, the $SU(3)$ symmetry is not broken.

The high energy neutrino beams at the Brookhaven and CERN machines allowed one to detect the second neutrino, connected with the muon.[12]

Another important experiment studied the decay of neutral kaons. A prominent decay mode is the decay into two pairs, $\pi^0\pi^0$ or $\pi^+\pi^-$. Since these modes are eigenstates of the combined transformation CP, namely belonging to $CP = +1$, one may ask about the $CP = -1$ mode. Further K^0 and $\bar{K}^0$ are transformed into one another by the CP-transformation, and one can form the combinations

$$|K_1\rangle = \frac{1}{\sqrt{2}} (|K^0\rangle + |\bar{K}^0\rangle) \tag{10a}$$

and

$$|K_2\rangle = \frac{1}{\sqrt{2}} (|K^0\rangle - |\bar{K}^0\rangle). \tag{10b}$$

Indeed these two artificial combinations exist in nature since one observes two neutral kaons, the K_1 decaying with a half time of $\tau = 0.9 \times 10^{-10}$ sec into two pions, and the K_2 decaying in roughly 10^{-8} sec into mainly three pions. This had been predicted by Gell-Mann and Pais, and confirmed by experiment.[13] However, a closer inspection showed that the long-lived K_2 also decays occasionally into two pions violating CP-invariance.[14] One found a branching ratio,

$$\frac{\Gamma (K_2 \to \pi^+ \pi^-)}{\Gamma (K_2 \to \text{all charged modes})} = 2 \times 10^{-3}. \tag{11}$$

Hence a 'fifth interaction' seems to exist in nature besides the strong, electromagnetic, weak and gravitational ones. The 'superweak' interaction violates CP-symmetry and is of the order of 10^{-3} of the weak interaction. Various speculations have been made about, and various consequences drawn from, the existence of this effect in the neutral K-system. Proposals to relate it to electromagnetic and gravitational interaction had to be refused and the best description which occurred, isolated only in the neutral kaon-decay, was provided by a theory of Wolfenstein.[15]

2. S-MATRIX THEORY WITH REGGE POLES

(G. F. Chew)

The detailed study of quasi two-body reactions, or reactions with few final particles, called for specific models to describe the strong processes. Since the quantum field-theoretic attempts had not been very successful, one tried systematically the road of the S-matrix theory. Chew's report to the 1967 Solvay Conference was devoted to this subject, and he could discuss considerable progress since the previous conference on quantum field theory in 1961.

As Chew noted, 'The idea of a nuclear theory based on the S-matrix goes back to the 1943 papers of Heisenberg. The motivation then, as now, was to avoid unobservable local space-time concepts which became troublesome when quantum principles are combined with relativity. Although macroscopic space-time is an essential component of S-matrix theory, there is no way to construct on-mass-shell wave packets whose spatial localization (in the particle rest frame) is sharper than the particle's Compton wavelength. The divergences plaguing conventional local field theory correspondingly have difficulties finding their way into S-matrix theory.

'Heisenberg clearly identified two key S-matrix properties, Lorentz invariance and unitarity, and partially appreciated a third, analyticity in momentum variables. The special aspect of the latter that interested him was the correspondence between poles and particles, an essential idea pinpointed by Kramers. S-matrix theorists of the 1940s, however, did not appreciate the concept now called "maximal analyticity", and correspondingly they came to believe that interparticle forces are necessarily ambiguous without appeal to local space-time concepts.'

2.1. DEVELOPMENT OF S-MATRIX THEORY

S-matrix theory was not treated extensively in the 1950s, when the hope was fostered by the success of Q.E.D. that perhaps the strong interactions also could be handled by the renormalization scheme. But it became increasingly evident that rather insurmountable difficulties had piled up. At that time one began to look for a replacement of the field-theoretic perturbation approach. Single variable dispersion relations were developed by Goldberger and others, and the dynamical content of analyticity was found. Mandelstam proposed the dispersion relations. A further principle, that of crossing, was discovered by Gell-Mann and Goldberger.[16] By this principle one understands the relation between processes obtained from each other by exchanging states in the incoming and outgoing channels. In quantum field theory the crossing property appears as an algebraic relation satisfied by the Lagrangian, whereas in the S-matrix view it is ascribed to the analyticity properties of scattering amplitudes. From this, one may conclude that the analyticity of the scattering matrix refers both to angle and energy variables, or in relativistic language to the variables t and s.

Another important step forward was connected with the fact that in S-matrix theory the difference between elementary and composite particles seems to disappear. In developing a relativistic generalization of the concept of strong forces, Chew and Mandelstam found that all strongly interacting particles are on the same composite level, i.e. there exists a fundamental nuclear democracy.[17]

Chew commented: 'Elementary particles were no longer essential to the dynamics, but they were not excluded. On the other hand, it continues today surprisingly often to be forgotten that, once a composite hadron exists, general principles require it to generate forces of a strength and range determined by its mass and partial widths. These forces cannot be "turned off" at the convenience of the theorists in favour of conjectured forces due to the exchange of elementary particles.'

Some distinction between different types of S-matrix poles had arisen in connection with the work of Castillejo, Dalitz and Dyson, but no general reason for this so-called CDD poles is found.[18]

S-matrix theory received great help from Regge's idea that bound states lie on certain trajectories in the plane of complex angular momentum J.[19] Chew and Frautschi then proposed that all hadrons should lie on these so-called Regge trajectories.[20] Chew remarked: 'If it were found that certain hadrons did not exhibit the characteristic Regge recurrence at a succession of different spin values, these particles should be classed as elementary. Our suggestion was coupled with the conjecture, still tenable today, that there might exist a unique hadronic S-matrix which not only was Lorentz-invariant, unitary, and analytic, but which contained only Regge poles. It was proposed in other words that the combination of unitarity, Lorentz invariance and maximal analyticity might suffice to define a "complete bootstrap" theory of strong interactions, without the need of input parameters or "master equation of motion".'

Chew continued: 'Far from fearing that Lorentz invariance, unitarity and maximal analyticity are insufficient to define a complete dynamical theory, I worry that these requirements may be too much for *any* S-matrix."

2.2. MAXIMAL ANALYTICITY

One key principle of S-matrix theory, already implicit in Mandelstam's double dispersion relations, states that only the momentum singularities occur at the analytic continuation of the kinematic constraints corresponding to physical multiple processes. By a 'multiple process' is meant a reaction that proceeds via a succession of two or more *macroscopically* separated collisions. What one obtains from unitarity and analyticity are the so-called Landau singularities.[21]

Another principle connected with analyticity is the factorizability of amplitudes. Although it arises naturally in a Feynman graph picture, it should not be viewed, according to Chew, as derived from field theory. Finally one includes crossing and 'hermitian analyticity'. The latter property ensures real masses and coupling constants for stable particles.

So far the analyticity structure determines all singularities, once the pole structure is given, but the latter remains completely arbitrary. To create a self-consistent bootstrap by which the poles are determined one has to add other analytic properties. An important step is the assumption that all hadrons lie on Regge trajectories.

The conjecture of Gribov,[22] and Chew and Frautschi, was that Regge poles control the high energy behaviour of scattering amplitudes at fixed momentum transfers. This assumption gained much evidence through the experiments in the 1960s. One can relate integrals over the low energy resonance region (in the so-called s-channel) to Regge parameters (in the t-channel) that control high energy.[23] Hence one can try to make a bootstrap calculation by using crossing-symmetry according to this 'duality' assumption which seems to be well-satisfied in nature.

2.3. REGGE TRAJECTORIES

Experimental observation puts particles, which are separated only by a spin difference of 2 and the mass, on linear trajectories in the plane made up of the real part of angular momentum and mass squared. Whereas on meson trajectories one has not seen more than two partners, baryon trajectories are occasionally fixed by four or five candidates. No convincing theoretical model describing this fact exists.

The trajectories normally have a slope of roughly 1 GeV^2 per spin unit characterizing the strong interaction. The trajectory determined best by exchange reaction is the one for the ϱ-meson. It plays the dominant role in the charge exchange reaction,

$$\pi^- p \to \pi^0 n, \tag{12}$$

and its slope can be obtained from the differential cross-section given by the formula,

$$\frac{d\sigma}{dt} \propto f(t) \left(\frac{s}{s_0}\right)^{2\alpha(t)-2} \tag{13}$$

$$\alpha(t) \approx \alpha(0) + \alpha'(t), \quad (s_0 \sim 1 \text{ GeV}^2).$$

In Equation (13), $\alpha'(t)$ is the slope and $\alpha(0)$, the intersection with the real J-axis at roughly $J = \frac{1}{2}$. On the other hand, the trajectory has to pass, for positive t, through spin 1 at $t = m_\varrho^2$. The linearity, as well as this fixed point, have been confirmed by experiment.

Although the case of the ϱ-meson is the only clear situation in which the idea of the Regge trajectory seems to describe the experimental situation rather well, one can often fit the differential cross-sections roughly by a certain number of trajectories belonging to the particles that are admitted for exchange. According to the Regge theory, two other predictions about the exchange reaction are made. First, the diffraction peak shrinks with increasing energy. Second, the total cross-sections of reactions depend on the intercept, $\alpha(0)$, like

$$\sigma_{\text{tot.}} \propto \left(\frac{s}{s_0}\right)^{\alpha(0)-1}, \quad (s_0 \sim 1 \text{ GeV}^2). \tag{14}$$

Therefore it decreases to zero for asymptotic energies, if $\alpha(0)$ is less than 1; which is satisfied in all known cases. Experimentally, cross-sections that were either constant or decreasing had indeed been found by 1967.

A constant cross-section was explained by the exchange of the vacuum trajectory which is assumed to pass through $J = 1$ at precisely zero energy (Gribov, Chew and Frautschi, 1961). The same vacuum or Pomeranchuk trajectory also has the following consequences: (i) all forward elastic amplitudes become purely imaginary at very high energies; (ii) for a common target particle and anti-particle total cross-sections approach the same limit; (iii) the total cross-sections for all members of a multiplet (of $SU(3)$ or isospin!) approach the same limit. Before the 70 GeV accelerator in Serpukhov went into operation, the preceding predictions seemed to have been satisfied. The rising cross-sections for nearly all reactions except for $\bar{p}p$-scattering seemed to contradict important consequences of the Regge theory.

The Pomeranchuk 'trajectory' presents further problems. If it rises with a slope, $\alpha'(t)$, comparable to the ϱ-trajectory, it causes shrinkage of the forward peak as the energy increases, but this has not been confirmed in most cases. A conceivable alternative to the vacuum trajectory is a fixed pole at $J = 1$, but it has other disadvantages. To summarize, the Pomeranchuk exchange presents various difficulties to the understanding; however, one may hope to find a dynamical explanation within the S-matrix scheme which is satisfactory.

The exchange of Regge trajectories leads to a semi-quantitative description of scattering processes in which particles with arbitrary spin are involved as providing the forces. On the other hand, the exchange of a pion pole within the framework of quantum field theory proved to be rather successful. The pion is the least massive hadron, and an exchanged pion pole comes closer to the physical region than any other singularity. The question is whether this object really lies on a trajectory, and if so, can one explain its mass? The success of one-pion-exchange models seems to give a negative answer to these important questions.

2.4. Multi-Regge-Exchange, Regge Daughters, and Conspiracy

The Regge exchange model has been extended to include more complicated processes than quasi two-body reactions. These involve the exchange of several trajectories and the assumption of factorization. A typical example is multi-particle production.

One can derive from symmetry arguments the existence of certain relations between residues of different Regge poles (conspiracy). The existence of daughter trajectories is motivated, especially, by Lorentz invariance. Both extensions of the original Regge scheme helped to overcome some of the difficulties which had hindered progress in the description of hadron reactions.

Chew concluded his report rather optimistically. S-matrix theory of hadrons, he said, may require no further physical ingredients. 'Properties already identified, with solid experimental backing, are more than theorists can presently handle. The problem seems not to be the discovery of detailed basic principles, but rather the understanding of the mechanism by which recognized properties manage to be mutually compatible.' For example, it could well be that the only poles consistent with unitarity and normal analyticity are Regge poles, then one would not need the extra assumption of all particles lying on Regge trajectories.

3. SYMMETRIES AND INTERACTIONS

The application of general symmetry schemes to particle properties was a dominant feature of high energy theory of the 1960s. At the 1967 Solvay Conference, three reports dealt directly with these aspects: H. P. Dürr on 'Goldstone Theorem and Possible Applications to Elementary Particle Physics'; the report of M. Gell-Mann (not included in the proceedings of the conference); and A. Tavkhelidze's report on the 'Simplest Dynamic Models of Composite Particles'. We shall start our analysis with the last report, although it presented the most difficult problems and offered only model solutions.

3.1. The Quark Model of Elementary Particles

The quark hypothesis of Gell-Mann and Zweig viewed the hadrons as being composed of the most elementary representations of the $SU(3)$ group, of objects having fractional charges, $\frac{1}{3}e$ and $\frac{2}{3}e$, called the proton-quark, neutron-quark, and λ-quark. Mesons are made essentially out of a quark and an anti-quark, whereas baryons contain three quarks. The fact that fractionally charged particles have not been observed so far may be explained by assuming a large mass, exceeding, say, 5 GeV for these objects.

The results of the Gell-Mann–Ne'eman scheme are very easy to derive by means of the quark hypothesis; hence one may consider the quark content of hadrons as a useful shorthand. The advantage of the quark model is that it can account similarly for some breaking of the SU (3)-scheme in a fundamental and simple way, if one gives the different quarks slightly different properties like mass, magnetic moment etc. Then the composite hadrons will reflect these symmetry-breaking terms due to its formation. Not only the static properties, like magnetic moments, will be naturally 'explained' in this way, but also the magnitude of total cross-sections. The important assumption which goes into these consequences of the quark model is the *additivity* of properties of the constituent quarks.

The understanding of the additivity is not so easy. If one thinks of the quarks as very heavy objects then the binding energy comes out as very large and may spoil the additivity altogether. In addition, there is the great difficulty of understanding the special quark forces that prefer bound states of two quarks and three quarks. Finally, one may doubt that using the heavier constituents of the hitherto known elementary particles does not solve the very problem of relativistic interaction, which appears again as soon as the kinetic energy approaches the rest mass of the quarks. Tavkhelidze spoke about work on the problem of extending the non-relativistic quark model, which had been performed by Bogoliubov and collaborators.[24] He assumed the quarks to be described as Fermi fields by a Dirac equation, with a spherically symmetric potential, in a semi-relativistic model. The baryons transform like a 56-plet of the SU (6) group and one requires the s-state to have the lowest energy. Then the magnetic moment of quarks in baryons, postulating minimal electromagnetic interaction of the free quarks, is given by

$$\mu_q = \frac{e_q}{2E_0} (1 - \delta), \tag{15}$$

where e_q is the charge of the quark, E_0 is the bound state energy ($E_0 = \frac{1}{3} m_N$ for the nucleons), and δ the mean value of the third component of the orbital momentum which is always positive. Consequently one finds,

$$\mu_p = 3 (1 - \delta) \text{ nuclear magnetons}, \tag{16a}$$

and,

$$\mu_n = - 2 (1 - \delta) \text{ nuclear magnetons}. \tag{16b}$$

The same parameter δ also appears in the expression for the weak interaction of the nucleons since the ratio of the axial over the vector coupling constants there is,

$$\frac{G_A}{G_V} = \tfrac{5}{3}(1 - 2\delta) \tag{17}$$

which is $\tfrac{5}{3}$ in the static limit $\delta = 0$. In a scalar potential well $\delta = 0.17$, hence $\mu_p = 2.49$ (experimental, $\mu_p = 2.79$), and $G_A/G_V = 1.1$ (experimental, 1.2).

Tavkhelidze and Logunov had attempted to extend this promising model relativistically. For this purpose, the potential was written in a relativistic form and the
intrinsic motion of the quarks was displayed relativistically. For solving the problem,
a Bethe–Salpeter approximation was tried, yielding the result

$$\delta = \frac{\langle \mathbf{q}^2 \rangle}{6m^2} + \cdots, \tag{18}$$

with $\mathbf{q}$, the internal momentum of the quarks, in the potential.

The quark model used above assumed totally symmetric spin-unitary wave functions
for the quarks; hence these objects are treated with Bose statistics. Another possibility
is to ascribe para-statistics to quarks. In any case, a closer study of the quark model
leads to difficulties out of which one has to figure out a clever way.

3.2. ALGEBRAS AND INTERACTIONS

Electromagnetic and weak interactions are described by currents. The electromagnetic
four-current density,

$$J_{EM}^\lambda(x) = (\varrho(x), \mathbf{J}(x)) \tag{19}$$

has been known since the work of Lorentz; the weak currents were established in
1958.[25] The discovery of approximate $SU(3)$ invariance of strong interactions suggested the use of this symmetry for a unified description of both weak and electromagnetic interactions.[26] Hence the weak and electromagnetic interactions are represented by an octet of vector and axial vector currents,

$$F_g^\lambda(x) \qquad \text{and} \qquad F_j^{5\lambda}(x), \tag{20}$$

where j gives the unitary spin index ($j = 1, \ldots, 8$) and λ, the spatial index ($\lambda = 0, 1, 2, 3$).

The first three unitary spin components are the isotopic spin current densities,
while the eighth component is proportional to the hypercharge current density. The
components, $F_4^\lambda, \ldots, F_7^\lambda$, change strangeness. In the absence of $SU(3)$-breaking, we have

$$\frac{\mathrm{d}}{\mathrm{d}x^\lambda} F_k^\lambda(x) = 0, \tag{21}$$

which represents the octet generalization of the *conserved vector current* hypothesis
of Gell-Mann and Feynman.[27] In the case of symmetry-breaking, the components
$k = 4, 5, 6, 7$ are proportional to the $SU(3)$-breaking parameter.

The axial vector current is only partially conserved due to the small mass of the
pion.[28] One finds the extension,

$$\left\{ \begin{aligned} \frac{\partial}{\partial x^\lambda} (F_1^{5\lambda} + iF_2^{5\lambda}) &= c\phi_{\pi^+}^+, \end{aligned} \right. \tag{22a}$$

$$\left\{ \begin{aligned} \frac{\partial}{\partial x^\lambda} (F_1^{5\lambda} - iF_2^{5\lambda}) &= c\phi_{\pi^-}^+, \end{aligned} \right. \tag{22b}$$

$$\frac{\partial}{\partial x^\lambda} F_3^{5\lambda} = c\phi_{\pi^0}^+, \tag{22c}$$

$$\frac{\partial}{\partial x^\lambda} (F_4^{5\lambda} \pm iF_5^{5\lambda}) = c'\phi_{K^\pm}^+, \tag{22d}$$

$$\frac{\partial}{\partial x^\lambda} (F_6^{5\lambda} + iF_7^{5\lambda}) = c'\phi_{K^0}^+, \tag{22e}$$

$$\frac{\partial}{\partial x^\lambda} (F_6^{5\lambda} - iF_7^{5\lambda}) = c'\phi_{\bar{K}^0}^+, \tag{22f}$$

and

$$\frac{\partial}{\partial x^\lambda} F_8^{5\lambda} = c''\phi_\eta^+, \tag{22g}$$

where, for instance,

$$c = f_\pi = \frac{\sqrt{2}\, m_N m_\pi^2 g_A}{g_r(0)} \tag{22h}$$

in the pion case, with the off-mass-shell pion-nucleon coupling constant, $g_r(0)$, and the weak axial coupling constant, g_A. These relations state that the divergences of the axial currents are proportional to the pseudo-scalar meson fields ϕ_π, etc. Due to the small mass of the pion, the isovector components, $F_1^{5\lambda}, \dots, F_3^{5\lambda}$, are nearly conserved, whereas the other components are much less so.

Now the electromagnetic current becomes

$$J_{EM}^\lambda = e(F_3^\lambda + \tfrac{1}{3} F_8^\lambda), \tag{23}$$

whereas in the weak interaction other currents F_j^λ and $F_j^{5\lambda}$ enter in the hadronic current. Cabibbo puts,

$$J_{had}^\lambda = (F_1^\lambda + iF_2^\lambda - F_1^{5\lambda} - iF_2^{5\lambda}) \cos\theta_c + (F_4^\lambda + iF_5^\lambda - F_4^{5\lambda} - iF_5^{5\lambda}) \sin\theta_c. \tag{24}$$

The *current algebra hypothesis* of Gell-Mann states that the time components (charge densities) of the physical vector and axial-vector octets (i.e. those measured in the weak and electromagnetic interactions of hadrons as shown in Equations (23) and (24)) satisfy the equal time commutation relations of the (free) quark model,

$$[F_k^0(x), F_l^0(y)]_{x^0 = y^0} = i\delta(\mathbf{x} - \mathbf{y}) f_{klm} F_m^0(x) \tag{25a}$$

$$[F_k^0(x), F_l^{50}(y)]_{x^0 = y^0} = i\delta(\mathbf{x} - \mathbf{y}) f_{klm} F_m^{50}(x) \tag{25b}$$

$$[F_k^{50}(x), F_l^{50}(y)]_{x^0 = y^0} = i\delta(\mathbf{x} - \mathbf{y}) f_{klm} F_m^0(x), \tag{25c}$$

where f_{klm} are the structure constants of the $SU(3)$ group.

The motivation of Equations (25) is to specify the sense in which the $SU(3)$ symmetry is exact in hadron physics. The algebraic relations (25), or the weaker integrated form of the charges, $\int d^3x\, F_j^0(x)$ and $\int d^3x\, F_j^{50}(x)$, should fix the symmetry content despite the fact that the symmetry is broken in the interaction. In addition, the current algebra postulate fixes the scale of the weak currents. The leptonic charges $W_\pm$,

$$W_+^{\text{lepton}} = \tfrac{1}{2} \int d^3x \left[v_\mu^+ (1 - \gamma_5) \mu + v_e^+ (1 - \gamma_5) e \right], \qquad (26)$$

$$W_-^{\text{lepton}} = W_+^{+\,\text{lepton}},$$

satisfy, at equal times, the relation

$$\left[W_+^{\text{lepton}}, W_-^{\text{lepton}} \right] = 2W_3 = \tfrac{1}{2} \int d^3x \left[v_\mu^+ (1 - \gamma_5)_{v_\mu} - \mu^+ (1 - \gamma_5) \mu \right.$$
$$\left. + v_e^+ (1 - \gamma_5) v_e - e^+ (1 - \gamma_5) e \right]. \qquad (27)$$

Gell-Mann also required the hadronic charges to satisfy the $SU(2)$ commutation relations,

$$\left[W_+^{\text{had}}, W_-^{\text{had}} \right] = 2W_3^{\text{had}}, \qquad (28\text{a})$$

and,

$$\left[W_3^{\text{had}}, W_\pm^{\text{had}} \right] = \pm\, W_\pm^{\text{had}}. \qquad (28\text{b})$$

Now, the weak current Ansatz, Equation (24), and the current algebra, imply the *universality* of weak interactions as expressed by Equation (28).

Gell-Mann also extended the algebra of current densities to the space components of the currents. The difficulty arises in commutators between current densities where Schwinger terms may show up. Fortunately many results do not depend on their existence or detailed structure. There is some evidence for all of the integrated (over the three-dimensional space) current commutation relations of Gell-Mann.

Current algebra leads to many applications. The most fruitful ones perhaps are the low energy theorems for soft pions expressing the matrix element for the emission of a pion of very small four-momentum in terms of the corresponding matrix element in the absence of the 'soft' pion and certain equal time commutators of currents.[29] Thus one can relate weak decay modes of, say, the kaon which differ only by one extra pion. Another application of current algebra is to derive sum rules which take a particularly simple form when the infinite momentum frame is used.[30]

3.3. BROKEN SYMMETRIES

The application of the higher symmetry schemes, especially the groups $SU(3)$ or $SU(3) \times SU(3)$ (as in weak interactions), to particle physics had been very fruitful. It gave rise to some very interesting results which could be confirmed by experiment. As one has known from the very beginning, $SU(3)$ is not an exact symmetry, not even in strong interactions. Now there exists an important theorem, usually called the 'Goldstone theorem', which states that under certain conditions a dynamical theory which is invariant under a particular symmetry group, and where this symmetry group is broken by the ground state, i.e. the vacuum state, in a relativistic field theory, there must exist particles of mass zero.

In nature, a number of massless particles are present, such as the photon, the neutrinos, and possibly the graviton. It is tempting to correlate these particles with particular breakdowns of symmetries, like the photon with the breaking of isospin in electromagnetic interactions, the neutrinos with the breakdown of parity in weak interactions, etc. [The isospin-breaking was first proposed by Heisenberg, *et al.*, *Z. Naturf.* **14a**, 441 (1959).]

Unfortunately, the Goldstone theorem requires certain properties of the mass zero objects to break a certain symmetry and in none of the cases mentioned are these satisfied.

H. P. Dürr, who discussed the symmetry-breaking at the 1967 Solvay Conference, looked for a generalization of the Goldstone theorem.[31] In the derivation of the Goldstone theorem several assumptions enter: first, there must exist a conserved current to begin with; second, there must be no long-range forces involved. Models of Higgs and Kibble show that under certain conditions the Goldstone particle can become massive.[32] But the original symmetry remains meaningful only if one mass zero particle survives.[33] The Higgs–Kibble mechanism may thus offer a possibility to understand the breaking of isospin and parity as suggested in the non-linear spinor theory of Heisenberg and collaborators.

4. QUANTUM FIELD THEORY: PROGRESS IN FORMALISM

Although quantum field theory had been proclaimed to be dead by many of those working in high energy physics, work in this field continued to be carried on. A number of theoretical physicists continued to remain interested in constructing a mathematically clean theory of particles, the axiomatic theory, constantly looking for new tools for dealing with the problems of quantum field theory. R. Haag gave a report on these new mathematical aspects at the 1967 Solvay Conference.

4.1. MATHEMATICAL ASPECTS OF QUANTUM FIELD THEORY

(R. Haag)

Haag started by emphasizing the difficulties that arise when one deals with the description of particle phenomena. One has to treat functions with infinitely many variables; or, in a more economical way, one wants to know infinitely many amplitudes each of which is a function of several momentum variables. These amplitudes are related to each other by a (non-linear) unitarity equation. Further constraints (Lorentz invariance, locality, positive metric, etc.) are imposed via linear equations. Only a few non-interesting cases have been soluble so far. Approximation methods can rarely be justified and progress in calculating experimentally interesting quantities is rather slow.

One might, therefore, first study the methods of functional analysis which are not concerned with numerical computational techniques but with the classification of structures. Haag discussed the basic mathematical quantities of the theory, such as the self-adjoint element A of an algebra $\mathscr{R}$, which might represent observables if they satisfy locality and causality. Symmetry transformations act as mappings on the elements A. One may hope to extend the scheme so as to include the situations occurring in particle physics. Many of the results have been obtained in the framework of the so-called C^*-algebras.

4.2. Different Approaches to Field Theory, Especially Q.E.D.

(G. Källén)

For quantized field theories, various techniques of approach have been developed over the years, and G. Källén tried to characterize the typical examples in his report at the 1967 Solvay Conference.

The Lagrangian approach to quantum field theory contains, as an essential step, the expansion of physical states in terms of fictitious mathematical states which are present when there is no interaction. This expansion is questionable from the very beginning since infinities occur in the interaction Hamiltonian if treated in the canonical scheme. Moreover, even in cases where the Hamiltonian should present no troubles the expansion is singular if no cut-off is applied. Therefore, one should not introduce mathematical states at all when dealing with quantum field theory.

In the 1940s it had been learned that the expansion of physical states in terms of the states of the free Hamiltonian can be avoided. One passes from the Schrödinger picture to the Heisenberg picture (or to the Dirac picture, as is usually done), in which the state vectors are constants. In Q.E.D., one has then to solve the equations of motion for the electromagnetic field, $A_\mu(x)$, and the electron field, $\psi(x)$, with the potentials $j_\mu(x)$ and $f(x)$,

$$\Box A_\mu(x) = -j_\mu(x), \tag{29a}$$

$$\left(\gamma^\mu \frac{\partial}{\partial x^\mu} + m\right)\psi(x) = f(x). \tag{29b}$$

Classically, one uses the Ansatz,

$$A_\mu(x) = A_\mu^{(\mathrm{in})}(x) + \int D_R(x - x')\,j_\mu(x')\,\mathrm{d}x', \tag{30a}$$

$$\psi(x) = \psi^{(\mathrm{in})}(x) - \int S_R(x - x')\,f(x')\,\mathrm{d}x', \tag{30b}$$

where D_R and S_R are the retarded Green's functions of the scalar and the spinor fields, and $A_\mu^{(\mathrm{in})}(x)$ and $\psi_\mu^{(\mathrm{in})}(x)$ correspond to those fields present long before the sources had the opportunity to influence the fields. In quantum theory, j_μ and f are operators; hence one does not know the solution of Equations (30). Since, however, the sources $j_\mu(x)$ and $f(x)$ are multiplied at least with one power of e, which is a small constant, one may start an iteration procedure where the fields enter the sources in a linear approximation. The essential result which follows from the corresponding calculations is that the complete Hamiltonian, expressed in terms of the Heisenberg fields as an operator, is identical with the free particle Hamiltonian expressed as a function of the incoming field. Therefore, the same states which diagonalize the incoming free particle Hamiltonian, also diagonalize the complete interacting Hamiltonian and, therefore, are the physical states. Roughly it follows that diagonalizing the free Hamiltonian becomes equivalent to solving the eigenvalue problem of the full Hamiltonian. This statement can also be verified independently of the perturbation theory. Unfortunately an assumption goes into the proof, which had been derived so far only in the perturbation theory of an arbitrary order. Moreover, one should note that the procedure sketched thus far is different from the usual Q.E.D.; however, like the latter, it involves renormalization.

In the years following the formulation of Q.E.D., one tried to draw some more general conclusions from the basic postulates of quantum field theory. As an example, Källén derived, essentially from Lorentz invariance, the vacuum expectation value of a product of two conserved vector fields like, for instance, the currents in Q.E.D.,

$$\langle 0 | j_\mu(x) j_\nu(x') | 0 \rangle = \left(\frac{1}{2\pi} \right)^3 \int d^4 p \, e^{ip(x-x')} \Pi(p^2) \cdot [p_\mu p_\nu - \delta_{\mu\nu} p^2] \, \theta(p), \qquad (31)$$

where $\Pi(p^2)$ is a weight function which also appears in the photon self energy integral. Hence it should be zero. By this argument the so-called Schwinger terms appearing in the commutator,

$$\langle 0 | [j_\mu(x), j_k(x')] | 0 \rangle |_{x_0 = x'_0} = \frac{\partial}{\partial x_k} \delta(\bar{x} - \bar{x}') \int_0^\infty da \, \Pi(-a) \qquad (32)$$

should vanish.

The technique using spectral functions, of which the above is a typical assumption, is not restricted to vacuum expectation values. Unfortunately the extension leads to considering functions of several complex variables which are technically very difficult to handle.

4.3. Q.F.T. with Decaying Particles

(I. Prigogine)

The formalism of quantum field theory, as opposed to the S-matrix theory, has one disadvantage since it does not include unstable particles as primary objects. Prigogine devoted his report on 'Dissipative Processes, Quantum States, and Field Theory' precisely to the study of quantum states and quantum levels corresponding to a finite lifetime. This study involved the use of concepts from statistical mechanics, such as the density matrix and non-equilibrium methods.

Certain types of problems which exist in field theory and in statistical mechanics are very similar, since in both cases one studies systems with essentially infinitely many degrees of freedom. In the time description of systems of interacting particles, great progress has been achieved by Prigogine and his collaborators, and one entertains the possibility of generalizing these results to Q.F.T. The important point is that unstable particles cannot be discussed in an S-matrix approach. But the latter may be extended and one discusses the transition probabilities rather than the particle itself at $t = \pm \infty$.

One proceeds from the von Neumann equation for the density matrix ϱ,

$$i \frac{\partial \varrho}{\partial t} = [H, \varrho], \tag{33}$$

H being the Hamiltonian. One then derives equations for the expectation values, ϱ, and finds, for instance, for the 'vacuum of correlation' element

$$i \frac{\partial \varrho_0(t)}{\partial t} = \Omega \psi(0) \varrho_0(t), \tag{34}$$

where $\psi(z)$ is the Laplace transform of the Green's function of the problem and the operator, Ω, becomes

$$\Omega = 1 + \psi^1(0) + \tfrac{1}{2}\psi''(0)\,\psi(0) + [\psi^1(0)]^2 + \cdots. \tag{35}$$

Any correlations, $\varrho_v(t)$, can be split into two parts; one disappears asymptotically, whereas the other follows an equation corresponding to the creation of correlations from ϱ_0. Hence the entire situation resembles the Boltzmann equation with a collision term. An application of this procedure to the interaction between radiation and matter yields Planck's equation.

The transformation theory of the dissipative systems involves non-unitary transformations, since $\Omega \psi$ is not hermitian. One may introduce a dressing operator, χ, by which the equation for the new $\tilde{\varrho}$ becomes

$$i \frac{\partial \tilde{\varrho}}{\partial t} = \phi \tilde{\varrho}, \quad \text{with} \quad \tilde{\varrho} = \chi^{-1} \varrho_0, \tag{36}$$

where $i\phi$ is a Hermitian operator. For non-dissipative examples, dressing leads to the usual renormalization. But no renormalized wave function will be defined. One obtains a consistent dynamical description of unstable physical particles, but the states corresponding to a finite lifetime are defined outside the Hilbert space. As Prigogine pointed out, 'The dressing process in dissipative systems is an irreversible process involving an increase of entropy. It is precisely because it can take into account this "thermodynamic" aspect of renormalization in dissipative systems, such as coupled fields, that our theory goes beyond the usual thermodynamical approach.'

Thus Prigogine came to the following conclusion: either one uses the bare fields, then one has to give up the particle description; or, the concept of physical particles is used, but then the field as a dynamical notion is eliminated.

5. NEW QUANTUM FIELDS

Quantum field theory had led to a disappointment in the description of strong interaction processes. Numerous distinguished physicists declared it to be completely dead and wrong beyond any doubt. And yet, quantum field theory got another chance to reappear in a new form, less connected with the Hamiltonian structure which it had assumed from the correspondence principle. The question could be asked in the following way: Is there any correspondence to a classical scheme when strong nuclear forces are involved? The answer is not altogether no; and this had some empirical basis, as is demonstrated in the case of the so-called vector dominance. This concept led to a new consideration of elementary fields and interactions, about which E. C. G. Sudarshan reported at the 1967 Solvay Conference. He also discussed the non-local character of fields describing the strongly interacting objects. Finally, W. Heisenberg presented a brief outline of the ideas and results of a quantum field theory, based on an indefinite metric, which sought to unify all interactions and to be finite.

5.1. VECTOR DOMINANCE

When studying processes involving the photon, on the one hand, and the ϱ-meson at the same time, on the other, one found a great similarity in the corresponding amplitudes F. Explicitly, one obtains roughly

$$F\left(\gamma_{\text{isovector}} + A \to B\right) = \left(\frac{e}{f_\varrho}\right) F\left(\varrho^0_{\text{transverse}} + A \to B\right). \tag{37}$$

As an example one may compare the photo-production reaction,

$$\gamma + n \to \pi^- + p \tag{38a}$$

and the ϱ-production reaction,

$$\pi^- p \to \varrho^0 + n. \tag{38b}$$

This simply amounts to a ϱ-photon analogy substitution rule,

$$eA^{(\text{isovector})}_\mu \leftrightarrow f_\varrho \varrho^0_\mu, \tag{39}$$

where f_ϱ denotes a coupling constant of the order of 2.5. Such an analogy was proposed in 1960 by J. J. Sakurai,[34] at a time when the ϱ had not even been identified in a production process. Later on, the nonet of vector mesons ϱ, ω, ϕ, and k^* was detected, and it was claimed that they are coupled uniquely with a unifying coupling constant f_ϱ (apart from $SU(3)$ factors) to the photon. Indeed, the current-field identity

$$j^{(\text{isovector})}_\mu = \left(\frac{m^2_\varrho}{f_\varrho}\right) \varrho^0_\mu \tag{40}$$

was proclaimed between the isovector part of the electric current and the (neutral) vector meson field ϱ^0_μ.[35]

This relation, called vector dominance, clearly brought people to think about analogies even where the photon couples to hadrons. As one remembers from the electromagnetic form factors of the nucleons, these could be described rather well by a 'dipole representation' in the space of momentum transfer, q,

$$G(q) \propto \frac{1}{\left(1 + \dfrac{q^2}{0.71}\right)^{-2}}, \tag{41}$$

and one might think of fitting it by an average vector meson. Unfortunately the situation whether the vector dominance model explains the form factor (41) is not so clear even now, but it tends in the right direction.

5.2. PRIMARY INTERACTIONS OF ELEMENTARY PARTICLES

The important role of vector mesons in the electromagnetic interaction of hadrons suggested that perhaps a unified description of strong and electromagnetic interactions could be achieved in which the coupling through the vector mesons is the primary interaction. Sudarshan devoted his report to these aspects and their generalization. He proposed the postulate that the weak and electromagnetic interactions of the nucleon are not the primary interactions, but are consequences of the weak and electromagnetic interactions of the vector and axial vector meson field. 'The primary interactions are the direct couplings of the vector and axial vector meson fields with leptons and the photon.'

The primary interactions include electromagnetism, where electrons and muons are coupled directly to the photon field A_μ, but the nucleons are coupled via the neutral vector meson fields like

$$- e \left(\frac{m_\varrho^2}{f_\varrho} \varrho_\mu + \frac{m_\omega^2}{f_\omega} \omega_\mu \right), \tag{42}$$

where $f_\varrho = f_\omega$.

Leptons do not have any strong interactions. The hadrons, however, couple strongly via vector and axial vector fields, where the vector fields are divergence free but the axial vector fields have the form

$$A_\mu = B_\mu + \left(\frac{\xi}{m} \right) \partial_\mu \phi, \tag{43}$$

with $\delta^\mu B_\mu = 0$. B_μ describes the pseudo-vector meson and ϕ, a pseudo-scalar meson of mass m; ξ is a characteristic constant.

The purely leptonic weak interactions have the V-A structure. Nucleons again couple to leptons only via the vector and axial vector mesons.

The scheme of primary interactions allows one to explain the electromagnetic form factors of the nucleons at least qualitatively. Similarly the so-called 'weak magnetism' term is explained as well as the absence of neutral lepton currents. Using some symmetry considerations, Sudarshan also found a satisfactory explanation of the relation of the renormalized axial vector to vector coupling constants. Strangeness-changing transitions are described in the scheme of primary interactions by introducing strange vector mesons, V'_μ, the pseudo-vector mesons, B'_μ, and the pseudo-scalar mesons, ϕ'. Due to the higher mass of the latter (the K-meson), the ratio of the decay rates of the kaon to that of the pion can be explained without introducing an extra parameter, the Cabibbo angle.

Similarly, the strong interactions might be viewed as being mediated through vector and axial vector fields. For example the primary pion-nucleon interaction is given by,

$$\left(\frac{f'}{m_\pi}\right) \bar{N}\gamma^\mu\gamma_5 \frac{\partial}{\partial x^\mu} (\tau \cdot \phi) N,\tag{44}$$

with $f'=\frac{1}{2} g \cdot g_A$, ($g_A=1.2$). Here N describes the nucleon field, and τ is the 'three-vector' of Pauli's (iso)spin matrices.

The scheme of primary interaction has certain attractive features. For instance, the conservation of vector currents may be understood by the absence of scalar mesons. Sudarshan noted, however, that the problem of divergences still remained in the theory.

5.3. Indefinite Metric and Nonlocal Field Theories

The above mentioned extension of the old quantum field theory may be regarded as being on a rather phenomenological level. It organizes various interactions and unifies them into a scheme which fits, roughly, the knowledge one obtains from experiments. It, however, does not solve the main problem of interacting fields, the divergence problem. This defect can be traced back to the local coupling of the fields. From a rigorous mathematical point of view, the local coupling is not allowed in the form in which it stands; since fields are thought of as distributions, the local product of such quantities is not, in general, defined. The renormalization procedure is not satisfactory since it involves infinite quantities. Moreover, it cannot treat the fundamental V-A theory of weak interactions.

The regularization method of removing infinities, however, recalls attention to a possible generalization of the mathematical framework first discussed by Dirac.[36] This possibility of using a linear vector space with an indefinite form for the inner product as the space of state vectors was discussed by Sudarshan in his second report.

In quantum theory, one talks about the states, $|A\rangle$, and introduces an inner product which is Hermitian,

$$\langle A \mid A \rangle = \langle A \mid A \rangle^*.\tag{45}$$

The quantity (45) is larger than zero in the usual quantum theory. Relaxing this condition to include negative norms also brings certain problems. In particular, the probability interpretation breaks down unless some restriction is placed on which kind of vectors correspond to physical states. The simplest possibility is to restrict the physical states to form a non-negative norm subspace of the larger indefinite metric space. One wants to deal with the large space because it is in this space that one has the local relativistic fields. The locality is destroyed when attention is restricted to the small space.

The divergence difficulties of Q.F.T. can be cured by nonlocal interactions. However, it is very difficult to prove practically the relativistic invariance of nonlocal Q.F.T. The alternative is to use propagators which fall off faster than the canonical ones used in Feynman's theory. These propagators must be explained in terms of subtractions, e.g.

$$\frac{1}{q^2(q^2-m^2)} = \frac{1}{q^2} - \frac{1}{q^2-M^2},\tag{46}$$

i.e. the propagator indicates that two fields are employed at the same time, the first with a positive norm, the second with a negative norm. If the energy of collisions becomes large enough to produce the state with the negative norm and mass M, one has to apply a selection rule restricting the physical space to the one with positive or semipositive norm.[37]

This introduction of negative norm states into the interaction eliminates the δ- and δ'-singularities on the light cone. Hence the theory becomes equivalent to a non-local one. The use of the indefinite metric thus removes the divergence difficulties of field theories.

The questions in the discussion referred to some crucial points of the theory. First the causality problem was mentioned. It was shown later on that the usual cluster decomposition and, therefore, also the usual statements of causality may break down in physical theories with the indefinite metric. However, macrocausality will be upheld. The requirement that no 'ghosts' appear in the final states can be ensured by abandoning the analytic structure of scattering amplitudes to a certain extent. This leads to the above mentioned relaxed causality.

5.4. NONLINEAR SPINOR THEORY OF ELEMENTARY PARTICLES

(W. Heisenberg)

An example of a quantum field theory using the indefinite metric was discussed by W. Heisenberg, who reported on the current situation of the unified field theory of elementary particles proposed by his group.[38]

He started from the Heisenberg–Pauli equation, which, in the form without spin, reads

$$i\sigma_\nu \frac{\partial \chi}{\partial x_\nu} + \sigma_\nu : \chi(\chi^* \sigma^\nu \chi): \qquad (47)$$

Here χ represents a fundamental spinor ('Urfeld'), the σ_ν are the Pauli matrices, and the double dots indicate a generalized Wick-ordering of the local interaction term.

The propagator of this 'Urfeld' falls off faster in momentum space than the canonical one. For their calculations, Heisenberg's group had used two subtractions of the kind (46), leading to a structure with $(q^2)^3$ in the denominator, where q is the momentum. This involves the introduction of a 'ghost' and a 'dipole ghost' in the spectral representation of the fundamental spinor.

The calculations are performed in the lowest order Tamm–Dancoff method. The latter starts from the assumption that, in any realistic problem, the number of particles contributing to a specific process is limited to a good approximation. The results of this scheme include the prediction [in 1959] of the η-meson with isospin zero at approximately 500 MeV, the calculation of the pion-nucleon constant $(g^2/4\pi = 15)$, and the prediction of nearly as large a coupling of the η-meson to the nucleons. Exploiting the group structure, in particular the idea of broken symmetries [which had been first explicitly mentioned in their 1959 paper two years before Goldstone], Heisenberg and collaborators were able to derive the electromagnetic coupling constant. They found α to be roughly 10^{-2}. If one makes an explicit Ansatz for the breaking of the isospin symmetry, an $SU(3)$ type spectrum for the pseudoscalar mesons and spin $\frac{1}{2}$ baryons follows, having approximately the right mass splitting.

6. THE SITUATION IN PARTICLE THEORY

In the general discussion some other material was added to the reports. For example, S. L. Adler expanded on the experimental tests of the local current algebra. He mentioned the Adler–Weissberger sum-rule explicitly, which allows one to determine the ratio of the renormalized axial weak coupling constant to the vector coupling constant by inserting the pion-proton cross-sections, and the Cabibbo–Radicati sum rule which involves the electromagnetic static properties of the nucleons and an integral over the difference of the electron-nucleon cross-sections. A sharp exchange took place on whether one can hope to proceed further with local fields and some forms of the sum rules or with the S-matrix theory in strong interaction. But this question still remains open today [1975].

Several years have elapsed since the last [1967] Solvay Conference on elementary particle physics. Meanwhile, more experimental data have become available. The superhigh energies reached in collisions of particles, produced by the 400 GeV NAL accelerator or by the ISR colliding beam machine, allowed one to study in detail the phenomena of multiparticle production. A host of old and new models have been applied to interpret the results. One of the main outcomes is that the total cross-sections of most collision processes (except for the antiproton-proton collision) rises for laboratory energies higher than 50 GeV, thus violating the accepted versions of the Pomeranchuk theorem. Moreover, the cross-sections for higher multiplicities of the secondaries seem to grow faster than expected with increasing energy.

None of the 'more fundamental' particles like quarks, weak bosons, etc., have been found so far, and theoretical speculations have been proposed to explain this absence. No spectacular breakthrough in the understanding of particle phenomena has come about, and one is faced with the task of putting little stones on top of each other in order to build the edifice of an adequate description of high energy processes.

REFERENCES

1. B. C. Maglic, *et al.*, *Phys. Rev. Lett.* **7**, 178 (1961).
2. H. L. Anderson, *et al.*, *Phys. Rev.* **91**, 155 (1955).
3. A. R. Erwin, *et al.*, *Phys. Rev. Lett.* **6**, 628 (1961).
4. M. Gell-Mann, Caltech Report CSTL-ZO (1961); *Phys. Rev.* **125**, 1067 (1962).
5. Y. Ne'eman, *Nucl. Phys.* **26**, 222 (1961).
6. A precursor of the $SU(3)$ scheme was the model of strongly interacting particles of Sakata in which all particles consist of proton, neutron and Λ. S. Sakata, *Prog. Theor. Phys.* **16**, 686 (1956).
7. M. Gell-Mann, *Proc. Int. Conf. High Energy Phys.*, *Geneva* (1962), p. 805.
8. V. E. Barnes, *et al.*, *Phys. Rev. Lett.* **12**, 204 (1964).
9. S. Okubo, *Prog. Theor. Phys.* **27**, 949 (1962).
10. F. Gürsey and L. Radicati, *Phys. Rev. Lett.* **13**, 173 (1964);
    B. Sakita, *Phys. Rev.* **136**, B, 1756 (1964).

11. M. Gell-Mann, *Phys. Lett.* **8**, 214 (1964);
    G. Zweig, CERN Reports No. 8182/TH 401; 8419/TH 412 (1964), unpublished.
    See also S. Sakata, *Prog. Theor. Phys.* **16**, 686 (1956).
12. G. Danby *et al.*, *Phys. Rev. Lett.* **9**, 36 (1962).
13. M. Gell-Mann and A. Pais, *Phys. Rev.* **97**, 1378 (1955).
14. J. H. Christenson, *et al.*, *Phys. Rev. Lett.* **13**, 138 (1964).
15. L. Wolfenstein, *Phys. Rev. Lett.* **13**, 562 (1964).
16. M. Gell-Mann and M. L. Goldberger, Proc. of the 1954 Rochester Conference, Talk by M. L. Goldberger.
17. G. F. Chew and S. Mandelstam, *Phys. Rev.* **119**, 467 (1960).
18. L. Castillejo, R. Dalitz, and F. Dyson, *Phys. Rev.* **101**, 453 (1956).
19. T. Regge, *Nuovo Cim.* **14**, 951 (1959).
20. G. F. Chew and S. Frautschi, *Phys. Rev. Lett.* **7**, 394 (1961).
21. D. Olive, *Phys. Rev.* **135**, B745 (1964).
22. V. N. Gribov, *JETP*, **41**, 677, 1962 (1961).
23. D. Horn and C. Schmid, *Phys. Rev. Lett.* **19**, 402 (1968).
24. A. Logunov and A. Tavkhelidze, *Nuovo Cim.* **29** (1963).
25. M. Goldhaber, *Proc. Int. Conf. on High Energy Physics at CERN*, (1958), p. 233.
26. N. Cabibbo, *Phys. Rev. Lett.* **10**, 531 (1963).
27. M. Gell-Mann and R. P. Feynman, *Phys. Rev.* **109**, 193 (1958), and also S. S. Gershtein and I. B. Zeldovitch, *Soviet Phys. JETP*, **2**, 576 (1956).
28. PCAC hypothesis of M. Gell-Mann and M. Lévy, *Nuovo Cim.* **16**, 705 (1960).
29. Y. Nambu and D. Lurie, *Phys. Rev.* **125**, 1429 (1962).
30. S. Fubini and G. Furlan, *Physics* **1**, 229 (1965).
31. J. Goldstone, *Nuovo Cim.* **19**, 154 (1961).
32. P. W. Higgs, *Phys. Lett.* **12**, 132 (1964).
33. P. W. Higgs, *Phys. Rev. Lett.* **13**, 508 (1964).
34. J. J. Sakurai, *Annals Phys. (N.Y.)*, **11**, 1 (1960).
35. N. M. Kroll, T. D. Lee, and B. Zumino, *Phys. Rev.* **157**, 1376 (1969).
36. P. A. M. Dirac, *Proc. Roy. Soc.* **A 180**, 1 (1942).
37. E. C. G. Sudarshan, *Phys. Rev.* **123**, 2183 (1961).
38. H. P. Dürr, W. Heisenberg, H. Mitter, S. Schlieder, and K. Yamazaki, *Z. Naturf.* **14a**, 441 (1959).

FIFTEENTH SOLVAY CONFERENCE 1970

G. REIDEMEISTER   J. DEENEN   R. CEULENEER   J. BLOMQVIST   P. CAMIZ   C. VAN DER LEUN   G. E. BROWN   D.R. INGLIS   D.M. BRINK   P. MACQ   M. DEMEUR   M. RAYET   R. ARVIEU   M. BOUTEN

DO TAN-SI   C. LECLERCQ-WILLAIN   J.L. VERHAEGHE   P. KRAMER   P. VAN LEUVEN   C. BLOCH   K. T. HECHT   B.R. MOTTELSON   G. RIPKA   J.P. ELLIOTT   M. LIBERT-HEINEMANN   M.C. BOUTEN DENYS

G. BERTSCH   B. R. JUDD   M. VENERONI

J. FLORES   M. FULD-ROUSEREZ   C. QUESNE   J.B. FRENCH   J. MEHRA   M. MOSHINSKY   L. C. BIEDENHARN   I. TALMI   H. FRAUENFELDER   H. J. LIPKIN   L. RADICATI   C. GILLET   J. HUMBLET

A. BOHR   J. GEHENIAU   W. HEISENBERG   E. AMALDI   C. MØLLER   E. P. WIGNER   F. PERRIN   D.H. WILKINSON

13

Symmetry Properties of Nuclei*

1. INTRODUCTION

The fifteenth Solvay Conference on Physics was held in Brussels from 28 September to 3 October 1970. The general theme of the Conference was the 'Symmetry Properties of Nuclei'.

The Scientific Committee of the Conference consisted of: E. Amaldi (Rome), President, A. Abragam (Saclay, France), L. Artsimovich (Moscow), Sir Lawrence Bragg (London), C. J. Gorter (Leyden), W. Heisenberg (Munich), C. Møller (Copenhagen), F. Perrin (Paris), A. B. Pippard (Cambridge), and J. Géhéniau (Brussels), Secretary. Artsimovich, Bragg, and Pippard were not able to attend.

Among the invited speakers and participants at the Conference were: A. Bohr (Copenhagen), D. M. Brink (Oxford), J. P. Elliott (Sussex, England), K. T. Hecht (Ann Arbor, Michigan), P. Kramer (Tübingen), M. Moshinsky (Mexico City), B. R. Mottelson (Copenhagen), L. Radicati (Pisa), D. H. Wilkinson (Oxford), L. Rosenfeld (Copenhagen), A. Sandage (Pasadena, California), I. Tamm (Moscow), S. Tomonaga (Tokyo), and E. P. Wigner (Princeton, N.J.). Sandage, Tamm, and Tomonaga were unable to attend. Among the other invited members, who also gave reports, were L. Biedenharn (Durham, North Carolina), J. B. French (Rochester, N.Y.), H. J. Lipkin (Rehovot, Israel), B. R. Judd (Baltimore, Maryland), and H. Frauenfelder (Urbana, Illinois).

The last Solvay Conference on nuclear structure was held in 1933. The neutron had been discovered in 1932, and Heisenberg had used the Pauli spin matrix formalism to describe neutrons and protons as the different states of one and the same entity, the nucleon. The main properties of the nuclei could be understood on the basis of these nucleons. The discovery of the positron in the same year, 1932, helped to establish the neutron-proton symmetry. Shortly after the 1933 Solvay Conference Fermi proposed a theory of beta-decay.

In the succeeding years, attempts were made to understand nuclear structure and nuclear reactions completely on the basis of the quantum theory of protons and neutrons. In order to understand the forces between nucleons two different empirical methods could be used: the study of the interactions of two nucleons and the study of the properties of complex nuclei. The latter research yielded the first significant results, and success was achieved from a consideration of the binding forces. It was found that the average binding energy of a nucleon within a nucleus is roughly con-

* *Symmetry Properties of Nuclei*, Proceedings of the Fifteenth Conference on Physics at the University of Brussels, 28 September to 3 October 1970, Gordon and Breach, Publishers, New York, 1974.

stant, but only for the very light nuclei. A semi-empirical formula was devised by Bethe and Weizsäcker, which could be understood if the nuclei were considered as liquid drops of an incompressible fluid of very high density.[1] It gave,

$$\text{Binding Energy} = a_v A - 4a_c \frac{Z(Z-1)}{A^{1/3}} - a_s A^{2/3} - a_\tau \frac{(A-2Z)^2}{A} + E_\delta, \quad (1)$$

where A and Z are the mass number and atomic number respectively. The first term may be regarded as a volume energy, the second as the total Coulomb energy between the protons, and the third term represents the surface energy. The fourth term expresses the fact that an excess of neutrons in the nucleus, expressed by the number $A-2Z$, tends to lower its binding energy. Finally the last term expresses the fact that nuclei with both odd number of protons and neutrons are the least stable objects, whereas those with both numbers even are the most stable ones, such that

$$E_\delta = \frac{\delta}{2A}, \qquad \text{if } Z \text{ and } A \text{ are even}, \tag{2a}$$

$$= 0, \qquad \text{if } A \text{ is odd}, \tag{2b}$$

$$= -\frac{\delta}{2A}, \qquad \text{if } Z \text{ and } A - Z \text{ are odd}. \tag{2c}$$

Empirically, it is found that,

$$
\begin{aligned}
a_v &= 14.0 \text{ MeV}, \\
a_c &= 0.146 \text{ MeV}, \\
a_\tau &= 19.4 \text{ MeV}, \\
a_s &= 13.1 \text{ MeV, and} \\
\delta &= 270 \text{ MeV}.
\end{aligned}
\tag{3}
$$

The first successful model of nuclear fission, that of Bohr and Wheeler, also used the liquid drop model.[2] Although its basis is highly empirical, and it became very difficult to derive it from a completely theoretical scheme of many interacting nucleons, the liquid drop model constituted the greatest success of nuclear theory in the years before the Second World War.

The second step towards an understanding of nuclear structure again rests on a semi-empirical theory. The basic observation is that especially stable nuclei result when either the proton number, Z, or the number of neutrons, $N = A - Z$, is equal to one of the following numbers

$$2, \quad 8, \quad 20, \quad 50, \quad 82, \quad 126. \tag{4}$$

These numbers are called *magic numbers*, and they have been interpreted as indicating a shell structure of protons and neutrons in a nucleus. For example, the nuclei He^4 and O^{16} consisting of two protons and eight protons and the same number of neutrons, respectively, turn out to be particularly stable. The empirical data was analyzed by M. Goeppert-Mayer, and by O. Haxel, J. H. D. Jensen and H. E. Suess, and led to the establishment of the nuclear shell model.[3]

The shell model had been explained by assuming that each nucleon moves within the nucleus independent of all other nucleons, and the orbits are determined by a potential, $V(r)$, which takes into account the average effect of all interactions with the other nucleons. The form of this potential turns out to be, empirically, something between a square-well potential and an oscillator potential. The Pauli exclusion principle applied to protons and neutrons selects the magic numbers 2, 8, 20, 40, 70, 112, 168, the last four of which contradict observation. However, a strong spin-orbit coupling changes them into the magic numbers, given by Equation (4). Thus symmetry considerations play an important role in nuclear shell theory.

The potential is assumed to be spherically symmetric, hence the nuclei in this theory have a spherical form. However, it was found that some nuclei having atomic numbers, $150 < A < 190$, and, $A > 220$, show rotational spectra. The theory of these ellipsoidal nuclei was worked out by A. Bohr and B. Mottelson. In non-spherical nuclei, the projection of the total angular momentum on the symmetry axis, Ω, plays a fundamental role, and the rotational energy is given by

$$E_{\text{rot}} = \frac{\hbar^2}{2I} J(J+1) + (-1)^{J-1/2} b(J + \tfrac{1}{2}), \tag{5}$$

with $J = \Omega, \Omega + 1, \ldots$.

In Equation (5), I represents the moment of inertia of the nucleus with respect to an axis perpendicular to its symmetry axis, and b is a constant. The second term corresponds to the rotation of the nucleus without the last nucleon.

The symmetry principles have played an important role in the theory of nuclear structure, and already as early as 1936 Wigner had proposed a systematic symmetry approach in his lecture at the Harvard Bicentennial Conference.[4] At that time, many facts about nuclear forces and weak interactions were not known, and results obtained in the 1950s put Wigner's theory on a secure basis. Radicati gave a report on the progress of Wigner's 'supermultiplet theory' at the 1970 Solvay Conference.

Before going into a discussion of this topic, however, we shall deal with the other reports at the Solvay Conference that dealt with subjects using only a part of the symmetries, like the isotopic symmetry which became established on the basis of research on elementary particles. Indeed, the considerable progress obtained in the understanding of nuclear structure came about, to a large extent, from the successful application of concepts employed in elementary particle phenomena. They provided a basis for the relation between symmetries and interactions. The same kind of interactions, which are responsible for the reactions between elementary particles, also determine nuclear structure. Whereas from the complicated many-particle systems of complex nuclei only indirect information about the properties of the various forces could be drawn, specific simple processes between elementary particles could be used to find out the characteristic conservation laws obeyed by the strong, weak and electromagnetic forces. Special attention must be paid to the symmetry under rotations in the isotopic spin space, which is exact in strong interactions but broken by electromagnetic and weak forces. On the other hand, parity breaks down for weak decay processes as became apparent in 1956–57. Finally, one relation proved to be true in all cases, that between spin and statistics: even spin objects are described by Bose statistics, whereas odd ones are described by Fermi statistics (the CPT-theorem).

After the rules of the fundamental theory of nuclear structure had been established, one could proceed to put the semi-empirical models of the 1930s and 1940s on a more secure basis by developing the approximation schemes. This work was greatly supported by the fact that symmetry approaches became fashionable in particle physics. Thus the old ideas of Wigner and Weyl were studied again and extended, and research in nuclear physics soon followed the fashionable path, perhaps even with greater success than in the field of elementary particles.

The 1970 Solvay Conference demonstrated the progress which had been made by the symmetry approach to nuclear structure in the following reports: D. M. Brink on 'Symmetry of Cluster Structures of Nuclei'; B. R. Mottelson on 'Vibrational Motion in Nuclei'; B. R. Judd on 'Lie Groups in Atomic Spectroscopy'; J. B. French on 'Symmetry and Statistics'; L. B. Biedenharn on 'SL3R Symmetry and Nuclear Rotational Structure'; H. J. Lipkin on 'SU (3) Symmetry in Hypernuclear Physics'; L. A. Radicati on 'Wigner's Supermultiplet Theory'; H. Frauenfelder on 'Parity and Time Reversal in Nuclear Physics'; D. H. Wilkinson on 'Isobaric Analogue Symmetry'; A. Bohr on 'Rotational Motion'; P. Kramer on 'Permutation Group in Light Nuclei'; M. Moshinsky and C. Quesne on 'Oscillator Systems'; K. T. Hecht on 'The Nuclear Shell Model – Doublets and Pseudo SU (3) Coupling Schemes', and J. P. Elliott on 'Shell-model Symmetries'.

We shall first consider the exact symmetry conservation and violation laws of the interactions that play a role in nuclear physics. Then we shall discuss Radicati's report on Wigner's supermultiplet theory and related topics before dealing with the reports on the shell model and more refined theories of the nucleus.

2. EXACT AND LESS EXACT SYMMETRIES IN NUCLEAR PHYSICS

Research in fundamental nuclear physics since 1930 has brought about a deeper knowledge of the properties of the interactions involved. These properties can be expressed by symmetry laws. The first such law was that of the conservation of isotopic spin in strong nuclear interactions. It had first been put forward explicitly by G. Young in 1935. Soon afterwards charge independence, except for Coulomb interactions, became established.

2.1. ISOBARIC ANALOGUE SYMMETRY

(D. H. Wilkinson)

Wilkinson reviewed the status of isotopic spin conservation in nuclear physics. He first discussed the topics of charge symmetry and charge independence. The former says that in corresponding states the *nn* and *pp* nuclear forces are the same; while the latter says that, in addition, the np force equals the other two. These symmetries remain approximately true for the total forces between nucleons because the electromagnetic interactions are so much weaker than the strong ones. The low energy test makes use of the effective range formula comparing the scattering lengths and the effective ranges. If applied to the pp, pn, and nn systems respectively, one finds that charge symmetry is satisfied to within 1%, and charge independence up to 2% in nuclear forces.[5]

Mass differences between mirror nuclei like ^{13}N and ^{13}C, and between even mass nuclei such as ^{10}Be and ^{10}B, can be accounted for by differing Coulomb energies.[6] The charge distribution in nuclei became known through electron scattering and muonic X-ray work; hence a better calculation should be possible, but the complicated structure has thus far [1970] prevented quantitative success.

The multiplet mass equation is

$$M(I_3) = a + bI_3 + cI_3^2, \tag{6}$$

where b and c indicate the charge dependence of the mass due to electromagnetic interaction, and I_3 represents the third component of the isospin. In practical cases, $b \approx Ac$, where A is the mass of the nuclei belonging to an isospin multiplet.

The charge independence can also be tested in reactions like,

$$A + p \rightarrow B + {}^3\text{H} \quad \text{or} \quad B' + {}^3\text{He}, \tag{7}$$

where B and B' are members of an isospin multiplet, and where the ratio of cross-sections should only depend on the isospin. In addition, the total gamma-decay widths of members of an isospin multiplet follow a formula similar to Equation (6).

Whereas most of the nuclei are in rather pure isospin states, in some cases considerable mixing is possible, as in $I=1$ of ^{8}Be, which does not exist but appears as two states that behave rather like ^{7}Li$+p$ and ^{7}Be$+n$.

The allowed Fermi transitions take place only between members of an isospin multiplet if isospin is a perfect quantum number. Adopting the conserved vector current hypothesis of Gell-Mann and Feynman (1958) one concludes that all nuclear Fermi matrix elements should be determined solely by the appropriate isospin Clebsch–Gordan coefficients if no great isospin mixing occurs in the states in question. This can be approximately demonstrated. Other situations in beta-decay offer a wide field of study of isotopic spin symmetry and strong interactions as well as the details of weak interactions like second class currents.

2.2. PARITY AND TIME REVERSAL IN NUCLEAR PHYSICS

Whereas Wilkinson discussed the improved evidence for isotopic spin symmetry in strong nuclear interactions, Frauenfelder turned to a subject which did not arise until the late 1950s or even the mid 1960s. Moreover, the questions came about mainly from experimental findings and were not suggested by bold theoretical assumptions.

In 1956, Lee and Yang suggested that weak interactions do not conserve parity, and subsequent experimental research of Wu and collaborators verified a 100% parity violation in a particular weak decay.[7] The successful V-A theory of weak interactions stemmed from these investigations. Parity violation in nuclear physics is found everywhere, whereas the other effect called the time reversal violation has been confirmed so far only in the neutral kaon system.[8]

2.3. SU (3) SYMMETRY IN HYPERNUCLEAR PHYSICS

(H. J. Lipkin)

The violations of symmetry mentioned earlier are found in systems having weak interactions. In the early 1960s elementary particle physicists invented a higher symmetry, the SU (3) symmetry, which is so refined that even strong nuclear forces do not conserve it [M. Gell-Mann, Y. Ne'eman, 1961]. In the SU (3) scheme, isotopic spin and hypercharge Y, defined as

$$Y = \tfrac{1}{2}(B + S), \tag{8}$$

where B is the baryon number and S, the strangeness, are combined in one larger group. It might be applied in all cases where particles with strangeness, like hyperons etc., enter.

Lipkin discussed hypernuclear physics, dealing with the nuclei which contain a Λ-hyperon among their constituents. Considerable knowledge has been obtained about the ground states of hypernuclei, since no Pauli principle prevents the Λ from going down into the lowest $1s$ orbit. More information might be available from studying the excited states of hypernuclei [D. H. Davis and J. Sacton, 1969]. The possible situations arising are such that either the Λ or a nucleon is excited from a s-shell to the p-shell, and in general the two states are mixed.

The description imitates the one in isotopic spin formalism. One defines 'strangeness analogue states' whose members differ only by the fact that nucleons might be changed into Λ-hyperons via the U- or V-spin operator of the SU (3) scheme. Other hyperons will not play an important role because their masses are too high, and, say, a Σ-hyperon will decay 'immediately' into a Λ-hyperon. Hence the SU (3) scheme relevant in nuclear physics will resemble the Sakata model, in which the $np\Lambda$-triplet is the essential concept rather than the modern Gell-Mann–Ne'eman model based on the quark idea.[9]

Hypernuclei can be described by a Hamiltonian in which there is one term completely symmetric in Λ, p and n, and another one that breaks the symmetry between Λ and the nucleons. This symmetry-breaking term is small in general, since in a normal hypernucleus there are many nucleons but only one Λ. Hence by considering the strangeness analogue state one can easily calculate the corresponding hypernuclei properties. Lipkin presented a few rough estimates of the excitation energy of hypernuclei. The agreement with known data is satisfactory.

2.4. Lie Groups in Atomic Spectroscopy

(B. R. Judd)

B. R. Judd discussed, in a general way, the role of Lie groups in atomic spectroscopy. Pioneering work in this field of applying a rather mathematical point of view to complicated atomic situations was performed by Racah.[10] Racah had attempted to calculate the matrices of the inter-electronic Coulomb interaction for the f-shell. He organized the groups in question in a hierarchical order. The seven orbital functions of the f-electrons span the space for the unitary group U_7 or the rotation group R_7; both these large groups contain the ordinary rotation group, R_3, in three space dimensions. Racah inserted the group G_2 of Cartan between R_7 and R_3. Although the Coulomb interaction is only a scalar in R_3, he succeeded in writing it in terms of the operators that belong to R_7 and G_2.

Racah's procedure opened a way of dealing with even more general problems, the important task being to find the intermediate group that replaces G_2 in the example of f-electrons. One may try in some cases to treat the spin as a good quantum number, etc. Special use is made of quasiparticle operators. The progress obtained by the introduction of symmetry arguments in the calculation of atomic spectra over the last ten years has been remarkable. It may be hoped that the application of this rather mathematical approach to nuclear physics problems turns out to be successful in a similar way.

3. SYMMETRIES AND THE SUPERMULTIPLET THEORY

The central topic of the 1970 Solvay Conference on symmetries in nuclear physics was perhaps the application of ideas developed by Wigner in the 1930s. They revealed their full glory only much later when symmetry considerations became fashionable in particle physics and, as a result, many physicists paid attention to them in nuclear physics also. We shall first describe the essential ideas of Wigner's supermultiplet theory, and then discuss Radicati's report on the progress achieved recently in this field and turn to special applications afterwards.

3.1. Wigner's Supermultiplet Theory

In the understanding of nuclear structure, that is the ground state as well as the excited state, two methods have been applied from the very beginning. First, models were constructed that explained various situations dynamically, e.g. what the ground state looks like and what the first excited state is. Second, the symmetry properties of the system under consideration were investigated, thus clarifying the energy levels, etc. The second method was advocated primarily by Wigner who had applied it already in 1931 to atomic spectra.[11] In this case the dynamical laws were well known, namely the electromagnetic interaction between the atomic nucleus and the electrons, but the main difficulty was presented by the complex situation involving the interaction of many charged particles. The investigation of the group of operators under which the Hamiltonian is invariant led to a classification of the energy levels of the electrons, and to a description of the fine structure of spectral lines. However, the position of the levels could not be predicted.

The classification of the energy levels proceeds in three steps.[12] First, one makes use only of the symmetry properties of space-time and arrives at the total (inner) quantum number and parity. The next step involves assumptions concerning the nature of nuclear forces. Two assumptions can be made: first, that the forces are independent of the orientation of the spins of the nucleons, and, second, that they are also independent of the nature of the nucleons (isotopic spin independence); if one drops the first assumption one arrives at a better description of the actual situation (second approximation). None of these assumptions is strictly valid, although isotopic spin independence is satisfied by the strong nuclear forces [Heisenberg, 1932; Breit, Condon and Present, 1936, and numerous others]. Certainly the existence of Coulomb forces gives a natural limitation. Similarly the assumption of spin independence of forces between nucleons is violated, e.g. by the existence of a quadrupole moment in deuterium nuclei.

The second approximation leads to an energy splitting in an isospin multiplet, labelled by $I_s = \frac{1}{2}(n_n - n_p)$, where n_p and n_n are the numbers of protons and neutrons, which is given by the mass differences of proton and neutron and the electrostatic forces.[13] The resulting equation is rather well satisfied.

The first approximation, which includes the spin-independence of nuclear forces, leads to an even higher degeneracy of the states having the same total spin and isospin. The breaking by spin-orbit coupling, however, destroys this symmetry to a large extent.

In the simplest case of a two-nucleon system one obtains a six-fold degeneracy of states in the first approximation. Spin-dependent forces destroy that degeneracy and lead to a splitting of the spin levels. In general the spin-dependent forces are less important for many-particle nuclei. Thus it makes sense to divide the Hamiltonian into three parts,

$$H = H_0 + H_1 + H_2, \tag{9}$$

where H_0 refers to the interaction that is independent of both ordinary and isotopic spin, H_1 is that part of the interaction which depends on ordinary spin but not isotopic spin, while H_2 is the electrostatic interaction which depends also on the isotopic spin coordinates. The study of the magnetic moments of nuclei seems to support the first approximation (only H_0) nearly in all cases.

The weak interactions bring in many complications into the supermultiplet scheme and the above-mentioned approximations, but they do not destroy the usefulness of the concepts introduced by Wigner. [See also Feenberg and Wigner, 1941, Ref. 4.]

3.2. 34TH ANNIVERSARY OF WIGNER'S SUPERMULTIPLET THEORY

(L. A. Radicati)

In many respects Wigner's theory of 1937 came too early. Little was known about the symmetries of interactions at that time. Later on the isospin became a good quantum number in strong interactions and, therefore, Wigner's second approximation is well-satisfied there and may serve as a natural starting-point for dealing with all kinds of forces between nucleons. The structure of weak interactions became well-known, and the approximate $(V-A)$ interaction leads in a natural way to extending the $U(2)$ symmetry, generalized by the vector current, to a larger group $SU(4)$, generated by both the vector and the axial vector current. The symmetry, however, is rather badly broken, hence the transition from $U(2)$ to $SU(4)$ corresponds to a much less accurate description of nuclei.

Wigner's supermultiplet theory of 1937 appears in the static limit as a natural generalization of isospin invariance expressed by the groups $SU(2)$ or $U(2)$. Nowadays one writes the charged hadronic currents, coupled to the leptonic field in weak current-current interaction, as

$$ h_\mu^\pm (x) = (j_\mu^1 \pm i j_\mu^2) + (g_\mu^1 \pm i g_\mu^2), \tag{10} $$

where $j_\mu^{1,2}$ are the isospin components of the vector current, and $g_\mu^{1,2}$ are the isospin components of the axial vector current. The third vector current component is related to the electric current, J_μ, and charge, ϱ, by the equations

$$ J_k = \int d^3x\, j_k^3(x), \quad \text{and} \quad \varrho = \int d^3x\, j_0^3(x), \tag{11} $$

where $k=1, 2, 3$. With the currents j_μ^i and g_μ^i, $(i=+, -, 3)$, one can now construct a current algebra $SU(2) \times SU(2)$. A weaker term of this scheme arises when only the integrated quantities like J_k and ϱ, Equation (11), i.e. the $SU(4)$ algebra, are used.

The $SU(4)$ scheme has been applied by Franzini and Radicati, and the prediction that the splitting within multiplets is smaller than the separation between different multiplets has been confirmed.[14] Also Wigner's result[15], that all allowed weak decays occur only between states belonging to the same multiplet, has been confirmed. Violations of that rule are due to the mixing of $SU(4)$ supermultiplets.

Radicati stated: 'It thus seems fair to conclude that: (i) The position of the supermultiplets is in good agreement with the prediction of Wigner's mass formula. (ii) The amplitude for super-allowed transitions is given by the matrix elements of the space integrals of the weak currents, i.e. by the operators $SU(4)$. The group $U(4)$ can therefore be considered as an approximate symmetry for nuclear systems.'

Another example of testing the prediction of Wigner's $SU(4)$ supermultiplet theory is μ-capture.[16] Here one had been able to estimate even the $SU(4)$-breaking effects.

Radicati summarized the new results as follows: 'The main point I wanted to make in this report was to illustrate the deep connection which in Wigner's theory exists between the symmetry of strong interactions and the form of the hadron-lepton coupling. This connection could not be fully understood at the time of Wigner's writing, and his theory was thus essentially an approximate description of nuclear systems. Today the symmetry between spin and isotopic spin inherent in the supermultiplet theory appears as a consequence of the almost symmetrical role played, in the static limit, by the vector and axial vector currents. Of course, the two currents are not entirely equivalent and we thus cannot expect to deduce from their analogy an exact symmetry for nuclei. However, the validity of the mass formula, the *ft*-values for the SU (4)-forbidden transitions and the good agreement obtained in the prediction of the μ-capture rate, are, in my opinion, a convincing proof that SU (4) is on its 34th anniversary even more relevant to the description of nuclei and their transitions than it was at its birth.'

In the discussion of Radicati's report the question was raised as to how critically the results claimed depended on SU (4) symmetry. It is very hard to answer this question. Another point was whether one should approach the symmetry by Wigner's original argument of spin independence or by the new current algebra.

3.3. PERMUTATION GROUP IN LIGHT NUCLEI

The supermultiplet scheme also plays a role in the phenomena which are caused by the permutation properties existing in nucleons. According to Pauli's principle, nuclear states have very simple transformation properties under permutation. One can organize the consequences by combining orbital and spin-isospin states with antisymmetric states described by three permutation parameters p, p', and p'', which occur already in Wigner's old theory. A particularly successful scheme includes cluster configurations.[17] It may also be extended to describe nuclear reactions.

3.4. SYMMETRY AND STATISTICS

The problem which might spoil the usefulness of the symmetry approach is the fact that the symmetry method is hardly able to deal with non-trivial cases of badly admixed symmetries. By generalizing the methods, however, in a way that takes into account statistical concepts introduced by Wigner in the 1950s one may bypass this difficulty.[18]

The idea is not to treat the properties of single states, or transitions between pairs of them, but to deal with distributions of the energy levels and transition strengths, and with certain refinements of these quantities. The distribution method assumes that there is a statistical simplicity in the behaviour of many-particle systems caused by the microscopic structure. The distributions of the levels and related quantities tend, as particle number increases, to characteristic, close to Gaussian, forms which are defined only by a few parameters, and these parameters depend on the microscopic symmetries involved. The calculations provide a good description of even ground state properties. One should expect that the statistical method works better for the higher lying states.

4. THE SHELL MODEL AND SYMMETRIES

In the reports and discussions at the 1970 Solvay Conference the shell model of Jensen and Goeppert-Mayer took its appropriate place. In the general theme of the conference the emphasis was on the application of symmetries, and many connections with the supermultiplet theory were revealed in the reports of J. P. Elliott, K. Hecht and others.

4.1. Shell Model Symmetries

(J. P. Elliot)

Rotational symmetry is perhaps the only true symmetry in nuclei. It leads to the conservation of angular momentum, J, the familiar selection rules for multiple operators, and to the $(2J+1)$-dimensional multiplet degeneracies. All the other symmetries, like parity, time-reversal, and isotopic symmetry, are broken or might be broken to a more or less known extent. A still weaker symmetry is connected with the separation of spin and orbital angular momentum as proposed first by Wigner in his supermultiplet theory. J. P. Elliott discussed the meaning of these weaker symmetries in the framework of the shell model of nuclei.

In the shell model, which considers the nucleons within a nucleus to lie on shells, such that the inner ones are completely occupied, leaving N nucleons for the last shell, one can try various symmetries. They certainly must contain as sub-groups the rotation group, R_3, and the isospin group $SU(2)$. An example which is sometimes useful is the $SU(3)$ group. It is exact when the nucleons in the outer shell move in a three-dimensional harmonic oscillator potential. In nuclear physics this $SU(3)$ group provides a superstructure above the $SU(4)$ group of Wigner and can explain certain configuration mixing of Wigner's supermultiplets. Further possibilities include the use of group structures that do not conserve particle number as weak symmetries. Thus there are many supergroups which may help to classify states when there is mixing of configurations. Particularly important are, however, those which serve to mix adjacent oscillator shells.

4.2. The Nuclear Shell Model in Terms of Pseudo Spin-Orbit Doublets and Pseudo $SU(3)$ Coupling Schemes

(K. T. Hecht)

The nuclear shell model rests on the assumption that strong spin-orbit coupling exists for the nucleons. This leads to considerable difficulties when discussing complex nuclei unless one is justified in considering only a few nucleons even in complex nuclei. To these one may apply coupling schemes, of which K. T. Hecht presented a review at the 1970 Solvay Conference.

In the coupling schemes in question one ascribes a pseudo spin and a pseudo-orbital momentum to a set of nucleons. There are many examples in which the model of a pseudo spin-orbit doublet fits the data rather well, like the spectra of ^{140}Ce. A more refined scheme includes the assumptions of a pseudo-$SU(3)$ symmetry into which groups of nucleons can be classified. This coupling scheme arises if one adds to a series of pseudo-doublets a singlet state of opposite parity and given angular momentum.

4.3. OSCILLATOR SYSTEMS

(M. Moshinsky and C. Quesne)

The shell model can be derived on the assumption that the nucleons of a nucleus move in a certain oscillator potential. The starting point of Moshinsky and Quesne's discussion was the fact that the potentials are invariant under a symmetry group of canonical transformations. Hence one should search for these groups and their unitary representations.

One finds that the group of linear symplectic transformations, $Sp(2)$, in two dimensions is the symmetry group of the one-dimensional harmonic oscillator. The result can be extended to the N-dimensional harmonic oscillator where the symplectic group $Sp(2N)$ and its unitary group $U(N)$ replace $Sp(2)$. When applied to n particles in three-dimensional oscillator potentials the irreducible representations of chains of subgroups of $U(3n)$ become important.

These groups can be used in approximations (Hartree-Fock, etc.) to realistic problems like levels of nuclei and electromagnetic form factors. Moshinsky proposed to explore further the consequences arising from the study of oscillator systems and to apply them to all questions of nuclear physics. Thus far, however, it seems that things are still on a rather formal level, and G. Brown excused himself for his comments by saying that, 'I should apologize for trying to inject physics into the discussion.'

5. VARIOUS TOPICS OF NUCLEAR PHYSICS AND SYMMETRY

The remaining topics reported at the 1970 Solvay Conference cannot be organized too specifically, but they all had to do with some application of symmetries.

5.1. SYMMETRY OF CLUSTER STRUCTURES OF NUCLEI

(D. M. Brink)

D. M. Brink reported about the connection between cluster structures, that seem to be apparent in nuclei, and symmetry principles. The fact that certain light nuclei are very unstable with respect to decay into α-particles, such as

$$
\begin{aligned}
^8\text{Be} &\to 2\alpha \\
^{16}\text{O} &\to {}^{12}\text{C} + \alpha,
\end{aligned}
\tag{12}
$$

suggests that nucleons tend to form clusters of α-particles.[19] It is the task of modern theory to construct many-nucleon wave functions which satisfy the requirements of the Pauli principle and certain clustering effects. Nuclei having four nucleons outside a shell particularly demonstrate clustering.

One can show how clustering effects are computed in Hartree–Fock calculations, e.g. in ^{20}Ne and ^{12}C. Similarly some simple oscillator shell-model wave functions can be related to the clustering wave function, and finally clustering wave functions have been used successfully in variational calculations. In the α-cluster model of the nucleus ^{20}Ne, the rotational and vibrational spectra have been obtained.

5.2. VIBRATIONAL MOTION IN NUCLEI

(B. R. Mottelson)

B. R. Mottelson introduced his report on the vibrational motion in nuclei by a short summary of the development of nuclear physics since 1932. This development may be thought of as culminating into the shell model. However, from there arises the problem of collective motions. Collective distortions in the average nuclear density generate a field exerted on one particle. This field causes vibrational motions of the particle.

In nuclear physics, the simplest collective modes are the shape oscillations and the neutron-proton oscillations (that may cause the giant dipole resonance in the nuclear photo-effect). One finds modes from a classical calculation and, in addition, typical quantum modes which involve charge exchange or spin flip of the nucleons. As examples, the low frequency quadrupole oscillations have been recognized since 1955, and a little later the octopole shape oscillations were found. The dipole resonance observed in the nuclear photo-effect is the main example of a collective isovector oscillation in the nucleus. Spin-dependent modes became apparent only in 1971.

5.3. ROTATIONAL MOTION

The existence of rotational motion in molecules was recognized already very early, but in nuclear physics it did not seem to play a role for a long time. When it was discovered that many nuclei have equilibrium shapes that deviate from spherical symmetry, it became evident that collective rotational degrees of freedom must be excited. And these motions cause Coriolis and centrifugal forces which may perturb the structure of the rotating object. At the 1970 Solvay Conference, A. Bohr discussed rotational motion, while Biedenharn added a report on the 'SL3R Symmetry and Nuclear Rotational Structure'.

The occurrence of a rotational spectrum corresponds to the possibility of obtaining an approximate separation of the motion represented by a total wave function of the product form:

$$\Psi = \varphi_{in}(q)\, \Phi_{rot}(\omega), \tag{13}$$

where the angular variables, ω, specify the orientation of the system, while the coordinates, q, characterize the intrinsic motion with respect to the body- (fixed) frame with orientation, ω. The rotations that leave the deformations of the nucleus invariant form a sub-group of all possible rotations. These would not lead to observable effects.

One can calculate the rotational spectra, and one finds a band structure which fits the data provided certain phenomenological parameters are adjusted. To derive these parameters from a consistent theory one uses a semi-microscopic theory which has been fairly successful, even when large quantum numbers become involved.

Biedenharn applied the SL3R model to nuclei, since it possesses as irreducible representations the band structures. The group can be derived from the consideration of a generalized symmetric top. Since SL3R involves only a quadrupole transition operator, its applications are limited to nuclei in which the electric quadrupole transitions are of major significance. Biedenharn was able to classify transitions of the nuclei ^{154}Gd and ^{168}Er.[20] The difficulty of the model is that it is hard to generalize. If other transitions become important a new group has to be invented.

6. CONCLUDING REMARKS

At the 1970 Solvay Conference the application of symmetries to various problems in nuclear physics was discussed very thoroughly. Two important points emerged from the reports and discussions. First, although the stimulus for discussing one symmetry or the other was frequently provided by particle physics, the result is not the same. In other words, the $SU(3)$ used in nuclear physics has only little to do with the Gell-Mann–Ne'eman scheme in particle physics. Second, in nuclear physics, approximate symmetries seem to play a very large role. Frequently these symmetries were badly broken from the very beginning. There thus arose considerable discussion between the particle and nuclear physicists.

The leading personality in the symmetry approach, E. P. Wigner, was asked to present the concluding remarks at the conference. He first criticized the collective model for the reason that it uses the classical picture too extensively. For instance, the statistical fluctuations in the shape cannot be neglected, and the question arises as to what the shape of the nucleus is.

The group theory that was presented fell into two classes. First, one had considered group theory as a tool for calculation, typically in the reports of Judd, Moshinsky and Hecht. The groups used do not necessarily have to do with realistic properties of nuclei. Second, group theory was applied in the old-fashioned sense via studying the transformation properties, exact or approximate, of a Hamiltonian. However, Wigner remarked that, 'most of the confirmation (of consequences from a symmetry) is derived by comparison with other calculations, and only to a small degree by comparison with experimental data. This is a little bothersome. The agreement between two calculations is not as significant as the agreement with experiment and we are only too often reminded of this in the course of our discussions. It is, of course, much easier to calculate something on the basis of a more or less confirmed theory than to persuade an experimental physicist to measure something which may be very difficult to measure. Nevertheless there is no real substitute for experimental confirmation.'

REFERENCES

1. H. A. Bethe, *Rev. Mod. Phys.* **8** (1936).
2. N. Bohr and J. A. Wheeler, *Phys. Rev.* **56**, 426 (1939).
3. M. Goeppert-Meyer, *Phys. Rev.* **74**, 235 (1948);
   O. Haxel, J. H. D. Jensen, and H. E. Suess, *Phys. Rev.* **75**, 1766 (1949).

4. E. P. Wigner, *Phys. Rev.* **51**, 106 (1937);
   E. Feenberg and E. P. Wigner, *Rep. Prog. Phys.* **8**, 274 (1941).
5. E. M. Henley in *Isospin in Nuclear Physics*, ed. D. H. Wilkinson, Amsterdam (1969).
6. Feenberg and Wigner, 1941, see Ref.4.
7. T. D. Lee and C. N. Yang, *Phys. Rev.* **104**, 254 (1956);
   C. S. Wu, *et al.*, *Phys. Rev.* **105** 1413 (1957).
8. J. H. Christenson, *et al.*, *Phys. Rev. Lett.* **13**, 243 (1964).
9. S. Sakata, *Prog. Theor. Phys.* **16**, 686 (1956).
10. G. Racah, *Phys. Rev.* **76**, 1352 (1949).
11. E. P. Wigner, *Gruppentheorie und ihre Anwendung auf die Quantenmechanik der Atomspektren*, Brunswick (1931).
12. E. P. Wigner, *Phys. Rev.* **51**, 106 (1937);
    F. Hund, *Z. Phys.* **105**, 202 (1937).
13. E. P. Wigner, 1937, Ref, 12.
14. P. Franzini and L. A. Radicati, *Phys. Rev. Lett*, **6**, 322 (1963).
15. E. P. Wigner, *Phys. Rev.* **56**, 519 (1939).
16. L. L. Foldy and J. D. Walecka, *Nuovo Cim.* **34**, 1026 (1964).
17. K. Wildermuth and W. McClure, *Cluster Representations of Nuclei*, New York (1966).
18. J. B. French, in *Nuclear Structure*, ed. by A. Hossain *et al.*, Amsterdam (1967).
19. W. Wefelmeier, *Naturwiss.* **25**, 525 (1937);
    D. Dennison, *Phys. Rev.* **57**, 544 (1940); **96**, 376 (1954).
20. L. Weaver and L. Biedenharn, *Phys. Lett.* **32B**, 326 (1970).

NINTH SOLVAY CONFERENCE 1951

R. GASPART W.M. LOMER A.H. COTTRELL G.A. HOMES H. CURIEN E. HENRIOT

C.W. RATHENAU W. KÖSTER E. RUDBERG L. FLAMACHE O. GOCHE E. OROWAN W.O. BURGERS W. SHOCKLEY A. G'JINIER C.S. SMITH U. DEHLINGER J. LAVAL F.C. FRANK

C. CRUSSARO N.P. ALLEN Y. CAUCHOIS C.O.G. BORELIUS Sir W.L BRAGG C. MOLLER F. SEITZ J. H. HOLLOMON

14

Solid State Physics*

1. INTRODUCTION

The first eight Solvay Conferences on Physics had dealt with the general development of quantum physics and the structure of matter. Since the creation of quantum mechanics and its application to a wide variety of phenomena, new fields of physics had been growing up. The subject of solid state physics had emerged into prominence and many fundamental problems occupied physicists working in it in different countries. The ninth and tenth Solvay Conferences were devoted to a discussion of these problems in solid state physics. The questions of crystal growth, grains and dislocations in 'The Solid State' were discussed at the ninth Conference. The theme of the tenth Conference was 'The Electrons in Metals'. The electrical conductivity of metals had been discussed at the fourth Solvay Conference in 1924, but since quantum mechanics was then not available the solution of these problems lay in the future. The variety of fields and problems discussed at the ninth and tenth Solvay Conferences is evident from the following accounts of the reports presented in them.

The ninth Solvay Conference on Physics, dealing with 'The Solid State', took place in Brussels from 25 to 29 October 1951. The Scientific Committee of the Conference consisted of: Sir Lawrence Bragg (Cambridge), President, H. A. Kramers (Leyden), C. Møller (Copenhagen), N. F. Mott (Bristol), W. Pauli (Zurich), F. Perrin (Paris), J. R. Oppenheimer (Princeton), and F. H. van den Dungen (Brussels), Secretary. Perrin and Oppenheimer were not able to attend the Conference.

Among the principal invited lecturers were: W. G. Burgers (Delft), F. C. Frank (Bristol), A. Guinier (Paris), W. Köster (Stuttgart), N. F. Mott (Bristol), F. Seitz (Urbana, Illinois, U.S.A.), W. Shockley (Murray Hill, N.J., U.S.A.), C. S. Smith (Chicago, U.S.A.), and E. Rudberg (Stockholm). Others who were invited to present reports were: C. O. G. Borelius (Stockholm), A. H. Cottrell (Birmingham), C. Crussard (Paris), U. Dehlinger (Stuttgart), J. Laval (Paris), E. Orowan (Cambridge, Mass., U.S.A.) and G. W. Rathenau (Eindhoven).

1.1. INTERFACE BETWEEN CRYSTALS

(C. S. Smith)

Smith reported that though the interface between fluids has been experimentally

---

* *L'État Solide*, Rapports et Discussions du Neuvième Conseil de Physique tenu à l'Université libre de Bruxelles du 25 au 29 Septembre 1951, R. Stoops, Brussels, 1952.

Les Électrons dans les Métaux, Rapports et Discussions du Dixième Conseil de Physique tenu à l'Université libre de Bruxelles du 13 au 17 Septembre 1954, R. Stoops, Brussels, 1955.

studied for three centuries and satisfactory theories have existed for nearly one, crystal boundaries have until recently been ignored by the theorist.

Different assumptions have been made about the interface. Either one considers that a layer of certain thickness is in a kind of amorphous state, or that the layer is very thin, even monatomical, and the interface is only a two-dimensional surface where the rearrangement takes place.

One could distinguish between two phenomena which take place at the interface. One is a topological disorder arising from the interpenetration of the two lattices; second, there are some strains on the boundaries which may have an influence on the whole system.

Smith discussed the variation of the grain boundary energy with orientation, and noted that, 'An examination of the microstructure of an annealed metal or single phase alloy reveals a high proportion of 120° angles between smoothly curved boundaries, which suggest that the energy of a boundary is independent both of the orientation of the crystals that it separates and of the direction of the boundary in relation to the crystal axes.'

Smith gave the results of a theory due to Shockley and Read, based on the simple dislocation model. Their treatment for a boundary between two crystals, which have a principal crystallographic direction in common, gives the following relation for an isotropic cubic metal,

$$E = \frac{Ga\,(\cos\varphi + \sin\varphi)}{4\pi\,(1-\sigma)}\,\theta\,(A - \ln\theta), \tag{1}$$

where E is the total grain boundary energy; G is the shear modulus; σ, the Poisson ratio; a is the slip vector of the dislocation; A a constant, representing the energy associated with each dislocation; θ is the orientation difference between the lattices; and φ represents the inclination of the boundary from the position of exact symmetry.

Smith discussed the detailed technical aspects of the work of Shockley and Read, and pointed out that this theory was in very good agreement with the experimental results.

He explained the methods by which absolute measurements of grain boundary energy can be performed. The principal method is to measure the free energy of liquid surfaces or interfaces by standard capillary techniques and to equilibrate such interfaces with solid-liquid interfaces, whose relative energies are then obtained by measuring the equilibrium angles and using Equation (1). He gave the results obtained with this technique by different authors.

Smith then described the changes in composition at single phase crystal interfaces. He showed, using the standard Gibbs–Guggenheim treatment, that large effects may be expected at the interfaces due to the presence of impurities. He treated the problem of the interfaces between different crystalline solids and showed that in almost all cases the interphase-interface has a lower energy than does the grain boundary in either phase separately.

Smith also reported the results on the interface between a crystal and a liquid. He discussed the melting of grain boundaries and gave some well-known results about alloys. He showed the importance of the grain size for the physical properties of the system under study, and gave explanations of the shapes of the grains.

1.2. GRAIN GROWTH OBSERVED BY ELECTRON OPTICAL MEANS

(G. W. Rathenau)

Rathenau pointed out that whereas the profound influence of interfacial tensions on the microstructure of solids has been well proved, comparatively little work has been done with the aim of determining how local equilibrium at grain boundaries is actually established.

Though electron emission microscopy furnishes only qualitative results, it allows direct observation of grain boundary movement and straightforward information regarding the factors by which it is influenced.

Rathenau discussed how, by the evaporation of an electropositive metal, such as Ba and Cs, it was possible to follow, with an emission microscope, the grain structure of a metal surface at different temperatures. He considered the grain growth in different cases and pointed out that when the grain growth in an array of small grains of very similar orientation is in contact with only one large grain of deviating orientation, this large grain is generally concave towards the grains of the matrix it is going to absorb. He noted that the grain movement at an interface between two nearly identically oriented crystals is very restricted, and discussed the grain growth accompanying the phase transition of alloys.

1.3. RECRYSTALLIZATION AND GRAIN GROWTH IN SOLID METALS

(W. G. Burgers)

Burgers discussed the structural changes taking place in metals during heat treatment, in so far as they have an influence on the number, size, shape, state or orientation of the constituent crystallites. The structural changes are due to allotropic transition or, in general, due to transgression of a boundary line in the phase diagram of a system; Burgers either did not consider them or only in so far as the considerations applied to recrystallization phenomena occurring within the range of thermodynamic stability of a given phase. Thus he did not discuss the changes in actual crystal structure, but only the changes in the grain structure and texture of crystalline solids.

Burgers discussed the questions of atomic processes that take place during annealing. He noted that annealing is a process by which imperfections are removed from a crystal, generally by heat treatment. The general imperfections thus removed are of Schottky and Frenkel types, as well as lines or screw dislocations, sometimes called the Taylor and Burgers dislocations.

He discussed the following processes concerning different types of dislocation motions: (a) mutual elimination ('neutralization') of two suitably 'oriented' dislocations of opposite signs (edge dislocations), or opposite sense (screw dislocations); (b) rearrangement of a set of dislocations inside a coherent lattice region, for example an accumulation into definite planes of dislocations originally scattered throughout the domain considered; (c) collective migration of an array of 'parallel' dislocations of equal sign or of two non-parallel arrays under the influence of over-all stresses. All these movements take place in a sense leading to a decrease in free energy. Burgers noted that grain structure changes were more likely in (b) and (c), but not in (a).

Burgers defined the nomenclature of recrystallization phenomena. As long as one observes the growth of new crystals at the cost of the deformed matrix until the stage that they impinge upon each other, one speaks of *primary recrystallization*. If the size of the primary crystallites is sufficiently small, for example with diameters of the order of 0.01 to 0.1 mm, then, on prolonged heating, a gradual coarsening of the grain structure may occur, due to the growth of some crystallites at the cost of the neighbouring ones. This phenomenon is called 'grain growth' or 'coalescence'. It may also happen that only one or two crystals develop at the expense of the primary matrix grains; this is called *secondary recrystallization*.

Burgers discussed the course of recovery as a function of time and temperature of heating. The two phenomena that occur are called *meta-* and ortho-recovery. The kinetics of these two processes are different, in the sense that for a given temperature ortho-recovery appears to continue until complete restoration, whereas meta-recovery practically ceases short of complete recovery, the 'end value' being closer to the completely annealed state for higher temperature of annealing.

Burgers then discussed the various theories of recovery, including the recovery of sheared crystals, and the subject of polygonization. Polygonization is caused by an accumulation of dislocations along more or less well-defined planes forming the boundaries of lattice regions (polygons), with generally slightly different orientations, upto a few tenths of a degree. He noted that the sub-boundaries are far more mobile than ordinary grain boundaries and that annealing produces a coarsening of the sub-structure.

Burgers briefly discussed polygonization in relation to the types of deformation. The problem which had been posed but not solved was whether there exists a relation between the shape of the polygons and the stresses which produce them. This is important for the study of nucleation.

Discussing the kinetics of nucleation, Burgers explained that a test piece with small grains will form new crystals of spherical shape when submitted to a deformation of a few percent (and then heated). Their number will grow with time as a function of the temperature.

As for the form of nuclei, according to almost all authors, they are very small crystals around which crystallization takes place. Burgers reported several experiments confirming this view. According to Van Arkel and Van Liempt, the potential nucleus has to be activated in order to become transformed into a central growth nucleus. This point of view is confirmed by the possibility of stimulating growth by the presence of a new crystal.

In Cahn's conception of the nucleation process, growth nuclei are actually formed in the most distorted parts of the lattice, and the process which transforms potential nuclei into growth nuclei is essentially the process of 'polygonization'. Experimental evidence exists for this polygonized state.

Burgers discussed the question of boundary migration, which occurs when two lattice blocks try to have a boundary for which the tension is minimal. The boundary surfaces have the tendency to arrange themselves in such a way that they include definite angles, corresponding to a state of equilibrium between the surface tensions acting on the boundary planes. This point of view was particularly adopted by C. S. Smith in 1948. Burgers gave examples of such an effect.

In order to understand other features observed during boundary migration, it is important to realize that the excess energy in the boundary depends on the 'misfit' between the two adjacent lattices, i.e. both on the difference in relative orientation of the lattices and on the position of the boundary plane.

Read and Shockley (1950) calculated the dependence of boundary energy on the 'misfit' based on dislocation models in a simple cubic lattice. For the case that the two lattices are rotated with regard to each other about the common Z-axis over a small angle θ, the resulting expression is

$$E = E_0 [A - \ln \theta], \tag{2}$$

where E_0 and A are constants depending to some extent on the orientation of the grain boundary and the macroscopic elastic constants. Burgers noted that the complete dependence of energy on 'misfit' is very complicated, but the theory and experiment are in good agreement.

Continuing the context of boundary migration (crystal growth), Burgers discussed the following subjects: the connection between boundary displacement and orientation difference of the adjoining lattice regions; the influence of a difference in internal state of adjoining lattice regions on boundary displacement; the growth of large secondary crystals in a primary recrystallized fine-grained matrix; and growth selectivity.

Burgers then discussed the origin of recrystallization structures. A direct consequence of the concept of nucleation is the assumption that the orientations of the crystals formed after recrystallization correspond to those of definite lattice regions already present in the deformed matrix, independently of whether the crystals are formed by primary or secondary crystallization. In those cases in which the recrystallization texture closely resembles the deformation texture this condition presents no difficulties. Moreover, in several instances of clearly different textures, it has been found that the recrystallization texture can be traced, be it weakly, in the texture of the cold-worked test piece.

In order to explain the difference between recrystallization and deformation texture, two assumptions must be made. (1) Those lattice domains in the cold-worked matrix which correspond in orientation to the crystallization texture, are more favourable for 'nucleation' than the lattice domains occupying the main orientations present in the deformed matrix, i.e. nucleation is 'oriented'; (2) the polygonized nuclei can actually grow.

Burgers treated the problem of the influence of growth selectivity on the formation of recrystallization structures. It is clear that a definite selectivity is inherent to the process of boundary displacement. However, according to one theory the definite recrystallization texture in the recrystallization matrix comes from an oriented growth, while in the other theory, it is due to 'oriented nucleation'. Both theories seemed to have some experimental support.

Finally, Burgers discussed the formation of annealing twins. There are two possible causes of this phenomenon: (1) The mechanical nucleation hypothesis, which suggests that the preceding cold-work has given rise to small mechanical twins, which serve as 'nuclei' for the annealing twins formed on subsequent recrystallization; and (2) The growth fault hypothesis, which assumes that twins are formed in the course of the annealing process itself, when the advancing grain boundary encounters some kind of discontinuity, which will induce a twinning accident.

Burgers mentioned that several arguments had been brought forward in favour of the second hypothesis. The twin band enclosed within a grain does not increase during grain growth. Burgers concluded his report with a discussion of the influence of impurities on recrystallization phenomena.

1.4. RECENT WORK ON SOLID STATE TRANSFORMATIONS IN SWEDEN

(E. Rudberg)

Rudberg gave a review of recent research work in Sweden dealing with the solid state, in particular regarding transformations. Most of the pioneer work of crystal structure determinations by X-ray methods in Sweden had been accomplished or inspired by Westgren at Stockholm University, beginning early in the 1920s. Westgren and his pupils investigated inorganic compounds in the crystalline state, making systematic studies of existing phases, limits of solubility and type of solution in a number of alloy systems, and solving for atomic arrangement in reasonably simple cases. The school founded by Westgren at Stockholm continued to engage in problems of structures of increasing complexity under his successor, Ölander. Hagg, one of Westgren's first associates, built up at Uppsala University a laboratory for structure studies, which soon became an active research centre in this field.

An important aspect of the work at Hagg's laboratory was the structure and properties of borides of transition elements, investigated by Kiessling. The structure is characterized by chains of borons making 120° angles, or rows of linked hexagons, etc.

The possibility and importance of chemical reactions involving solid substances, where reactions in a gaseous or liquid phase are not responsible or essential for the changes observed, were emphasized by Hedvall at Gothenburg. Hedvall and his co-workers investigated a large number of reactions in which solid substances take part. A large group of these are cases which exhibit a marked change in reaction rate as a function of temperature at a transformation point for the solid substance involved. Rudberg gave as examples the oxidation of several metals and alloys, and commented on the reactivity of the various substances involved. In all these reactions in solids, the process exhibiting a change in reaction rate occurs at the boundary of a solid phase towards a gas, a liquid, or a second solid. The altered reactivity of that boundary region could well be associated with definite changes in the structure of the main body of the solid.

Rudberg reported on the systematic studies, in Stockholm under Borelius, on order-disorder transformations in binary alloys. The main tools for the experimental side of this work were resistivity measurements, fine structure X-ray determinations and calorimetric methods of several kinds. Borelius had used a thermodynamic approach for interpreting his results, and Rudberg gave an account of Borelius' work on a binary alloy system.

Rudberg also reported on Linde's determination, in Stockholm, of the resolved critical shear stress for the principal planes and directions of easy shear, from measurements of the yield points of single crystals of known orientation in simple tension.

Axel Hultgren, also in Stockholm, conducted extensive research on the different reactions which take place in austerite when held at different, in each case constant, temperatures. The material used in these investigations comprised a number of steel alloys, with a carbon content usually near 50% and a second alloying element of one to a few percent. Thin samples were austenized, usually at 1300°C, then held at the chosen reaction temperature in a lead or lead alloy bath for a series of increasing time intervals, terminated by quenching in water or oil. Metal microscopy of polished and suitably etched sections served to determine the nature and relative amount of the different phases formed, as a function of time at a given temperature, and also yielded information as regards orientation relationships, particularly between neighbouring parts of the same phase. Rudberg also described Hultgren's work on ortho- and para-reactions in various alloy steels.

1.5. PERIODICITY DEFECTS IN THE LATTICES OF SOLID SOLUTIONS

(A. Guinier)

Guinier pointed out that a solid solution is not a crystal, because the perfect periodicity of atoms does not exist. He outlined the theories of order-disorder phenomena of Yvon and Fournet, and commented on the possible better approximation to experimental results from these theories compared to the earlier theories of Bragg and Williams, and Bethe and Cowley.

Guinier discussed the diverse applications of X-rays to the study of disorder in structures and noted that, in spite of considerable progress, there remained fundamental difficulties in the description of the precise constitution of a solid solution without introducing hypotheses external to the results of X-ray studies.

The mechanical or electrical properties of a mixed crystal are very sensitive to lattice perturbations. Inversely, the question arises, whether from the latter one can deduce its mechanical or electrical properties. In the first case, the theory has still not advanced very much, and all experimental measurement is the result of just a few factors for which one can isolate the effect of the perturbation due to foreign atoms. Among the electrical properties, for instance, one knew how to calculate the resistivity due to the thermal vibrations of atoms, but the problem of the propagation of electronic waves in a mixed crystal, knowing the real distribution of the atoms, had still not been treated. In the case of electrical measurements, a favourable element is that one can, by following the variation of the resistance with the temperature, isolate the part due to thermal vibrations from the resistance of the alloy in the absence of the vibrations. Qualitatively, these measurements of resistivity serve to characterize the order of a solid solution, or to follow the steps in the precipitation of a supersaturated solution. All efforts to calculate the real structure involve the introduction of numerous hypotheses.

In spite of their imperfection, it seems that the methods based on X-ray analysis have the greatest possibility of leading to the understanding of the real structure of solid solutions. The results obtained by X-rays might thus serve in establishing the theory of the effects of periodicity defects on the other properties of solids, thereby providing the means of controlling other conceived structures.

1.6. New Studies of Hardening

(W. Köster)

Köster gave an account of the reasons for supposing that the phenomena of 'cold' and 'hot' age hardening have different physical origins, and proceed independently in the crystal. He pointed out that the 'cold' hardening process, which reveals itself through an altered resistivity, is attended by no change of the thermoelectric force in the Al-Ag case, but involves some change in the case of the Al-Cu alloy. The warm hardening is produced by a slow nucleation of a new phase, which starts concurrently and independently, and which finally drains the regions which produced the cold hardening and produces separate crystals of the new phase.

1.7. The Elasticity of a Crystalline Medium

(J. Laval)

The first theory of elasticity of a crystalline medium was due to Cauchy[1], who defined the 21 elastic coefficients deemed sufficient to determine the tensions and forces necessary for the production of all the deformations in conformity with Hooke's law. Cauchy had considered a lattice of identical particles, attracting each other along the line joining them; the attraction between any two of them being determined completely by the distance between them, decreasing rapidly with the distance. Later on Voigt[2] developed Cauchy's theory systematically. Voigt cast it in the language of tensors, and verified a number of conclusions on the basis of experimental results. Voigt, as against Cauchy, did not consider the reticular structure. He applied the laws of classical elasticity to a crystalline medium, considering this medium as being homogeneous throughout.

It was Born[3] who re-established crystalline elasticity on the basis of atomic structure. He verified all the conclusions of Voigt, but in order to do so he had to appeal to central forces: he assumed that two atoms interact with each other along the line joining them with a force proportional to the distance between them. The atoms in crystals were regarded as rigid impenetrable spheres; they were also endowed with a uniform electronic density, possessing spherical symmetry.

Now these atoms are not bounded by indeformable surfaces and separated by empty spaces. The crystalline medium is continuous; nowhere does the electronic density vanish. Nevertheless, one can distinguish in a crystal the positive ions, which are approximately spherical and rigid, and which are immersed in a kind of fluid medium consisting of the weakly bound valence or conducting electrons. When the crystal undergoes a deformation the positive ions approximately conserve their shape and their electronic density; the surrounding fluid is, however, necessarily deformed. The electronic eigenfunctions are modified, and the electronic density changes.

The positive ions form almost the entire mass of the crystal. Their mutual interaction gives rise to restoring forces which can be analyzed into two parts: the first arises from the relative displacement of the two ions; it would be the only contribution if the fluid medium were suppressed and the crystal were reduced to an assembly of positive ions in vacuum. Since the positive ions retain an approximately spherical electronic density, this first contribution is approximately a central force. But two interacting ions also act upon one another through their surrounding electronic medium. The associated restoring force is no longer central. While the calculations of the elastic constants have recently been the object of important studies[4], the deformation of the electronic medium surrounding the positive ions has never been systematically considered. Yet if the electrons do not conserve the same wave function, they also do not conserve the same total energy. Their correlation energy changes, as does their exchange energy, with the electrons which form part of the positive ions. Most important is the change of energy of each electron considered individually in the force field of the positive ion. The Fermi energy itself is modified.

The effect of this second non-central contribution to the restoring force is weakest in ionic crystals such as rock salt, and strongest in metallic crystals. Consequently, for ionic crystals the ratios between the elastic constants obey the Cauchy relations, which is not the case for metallic crystals.

Thus the restoring forces acting upon the atoms of the crystalline medium may not be treated as central forces, except for certain ionic crystals, and then only as an approximation restricted to within a small temperature range. One must therefore reconsider the atomic theory of elasticity in crystals and exclude the hypothesis of central forces.

Laval had devoted himself to such a study. He restricted himself to deformations which produce restoring forces proportional to the displacement. Purely elastic deformations, strictly satisfying Hooke's law, are an idealization. While Hooke's law is an approximation, deviation from it is infinitesimal if the deformation is weak.

Laval developed the mathematical formalism of the elasticity theory of a crystalline medium based on non-central forces, and, as a special case, applied his formalism to Born's considerations derived from the assumption of central forces. He showed that under these conditions his coefficients can be identified with those of Voigt.

Laval had introduced forty-five elastic coefficients. This led Lawrence Bragg to remark in the discussion: 'Mr Laval's paper makes considerable demands on our courage! The theory of crystalline elasticity seemed formidable enough when we believed there were 21 coefficients in the general case. We know now that we must envisage 45. Nevertheless he deals with a very practical point which must be taken into consideration when we consider the stresses and strains of a distorted crystal.'

REFERENCES

1. A. M. Cauchy, *Mémoires de l'Académie des Sciences*, **9**, 114; **10**, 293; **18**, 153.
2. W. Voigt, *Lehrbuch der Kristallphysik*, Teubner, Berlin (1910).
3. M. Born, *Dynamik der Kristallgitter*, Teubner, Berlin (1915); *Atomtheorie des Festen Zustandes*, Teubner, Berlin, pp. 536 and 548 (1923).
4. J. M. Bardeen, *J. Chem. Phys.* **6**, 367 (1938).

1.8. Crystal Growth and Dislocations

(F. C. Frank)

The history of the theory of crystal growth is divided principally into two parts. One is the theory of ideally perfect crystals, starting with J. Willard Gibbs[1] (1878), and then developed between 1920 and 1948 by various workers, notably Volmer[2] (1920 onwards), Kossel[3], Stranski[4], Becker and Döring[5], Frenkel[6], and Burton and Cabrera[6a]. The second is the theory of the growth of imperfect crystals, commencing with Burton, Cabrera and Frank (1949)[7]. The second theory makes use of the first: basic to both is some understanding of the atomic nature of a crystal surface in equilibrium. In the early work, this point was only tacitly assumed, and its theoretical study was undertaken in 1949 by Burton and Cabrera.

Frank discussed the equilibrium state of a crystal surface. It has usually been assumed that the faces of low index of a perfect crystal, when in equilibrium with its vapour (or solution), are atomically smooth, with the exception of small numbers of absorbed molecules, surface vacancies, and successively smaller numbers of pairs, triads, etc., of these. Provided these numbers are small, it is not difficult to make some calculation of their magnitude. In particular the proportion of surface sites occupied by single adsorbed molecules (or surface vacancies) is expressed approximately by

$$n_s = \exp(- W_s/kT), \tag{3}$$

where W_s is about half the total evaporation energy W. To derive this result, the interactions between the surface singularities are neglected.

If the interactions between singular points is taken into account, as Burton and Cabrera had done, it is seen that for every rational face of a crystal there is a critical temperature, at which its surface roughness on the atomic scale increases rapidly with the increase of temperature. This has been called a 'surface melting temperature'. The 'melting', however, does not appear at the same temperature for different faces.

At all temperatures below which there exists a set of 'unmelted' faces, the structure is characterized by 'steps' and 'kinks' which make a sideways displacement. According to Frenkel, Burton and Cabrera, if n_+ and n_- are the numbers of kinks of opposite sign per atomic spacing measured along the nearest principal direction, then the mean kink density is given by

$$n_+ + n_- = [4 \exp(-2w/kT) + \tan^2 \theta]^{1/2}, \tag{4a}$$

where,

$$\tan \theta = n_+ - n_- ; \quad \text{and} \quad n_+ n_- = \exp(-2w/kT), \tag{4b}$$

and w is of the order of $\frac{1}{12}$ of the evaporation energy for close-packed crystals with homopolar binding. This is usually quite a high density. The lowest temperature at which one is likely to grow a crystal from its vapour is that at which its vapour pressure is 10^{-10} atmospheres; w/kT is then about 2. However, even in the closest packed direction of a step, kinks occur every three or four atoms apart.

An important consideration about the state of the surface, first emphasized by Volmer, is the ease of surface diffusion. He had shown that if x_s is the mean distance diffused by a molecule between arrival from the vapour and re-evaporation, then

$$x_s = a \exp[(W'_s - U_s)/2kT], \tag{5}$$

where W'_s is the work of desorption, U_s, the activation energy for surface migration, and a is the interatomic spacing. One concludes from this formula that the rate of direct arrival of molecules from the vapour at any particular point on a crystal surface is generally small compared with the rate of indirect arrival by way of surface migration.

Frank discussed the theory of growth of perfect crystals. As Gibbs had first noted, this is essentially a nucleation process, like the formation of droplets from vapour. He described the different possible situations in nucleation processes as well as the growth of imperfect crystals. X-ray diffraction studies had shown that there occur various types of dislocations in crystals. These dislocations, when far enough from one another, serve as points of nucleation. The theory of growth kinetics for such crystals was developed by Burton, Cabrera, and Frank.[8] The dislocations seem to arise essentially from impurities and variations of composition.

REFERENCES

1. J. W. Gibbs, 1878, *Collected Works*, p. 325 (1928).
2. M. Volmer, *Kinetik der Phasenbildung*, Dresden and Leipzig, Steinkopf (1939).
3. W. Kossel, *Nach. Ges. Wiss. (Göttingen)*, p. 135 (1927).
4. I. N. Stranski, *Z. Phys. Chem.* **136**, 259 (1928).
5. R. Becker and W. Döring, *Ann. Physik*, **24**, 719 (1935).
6. J. Frenkel, *Kinetic Theory of Liquids*, Oxford, Clarendon Press)1946).
6a. W. K. Burton and N. Cabrera, *Disc. Faraday Soc.* **5**, 33, 40 (1949).
7. W. K. Burton, N. Cabrera and F. C. Frank, *Nature*, **163**, 398 (1949).
8. W. K. Burton, N. Cabrera and F. C. Frank, *Phil. Trans. Roy. Soc.* A **243**, 299 (1951).

1.9. Interference of Thermal Agitation Waves in Crystals

(C. Crussard)

Crussard analyzed the interferences of thermal agitation waves in a crystal and showed how they could produce local slippings (or distortions) in small 'embryonic volumes' containing up to 1000 atoms. Such cooperative movements can explain the germination of many transformations in the solid state, as well as the formation of dislocations.

Thermal agitation occurs in the form of a wave train of exceptional amplitude. Once the transformation has started locally, the flux of the remaining wave train should contribute to extending its domain. Crussard emphasized that this point was still not clear in the theory. He showed that a dislocation can occur by thermal agitation alone inside a perfect crystal only under tensions of the order of 10 to 100 kg/mm^2, which are far higher than the elastic limit of metallic crystals. On the other hand, it seems that a dislocation can also occur on the surface.

1.10. On the Generation of Vacancies by Moving Dislocations

(F. Seitz)

Seitz pointed out that the experiments of Molenaar and Aarts, Blewitt and others, had confirmed his view that vacant lattice sites, and possibly interstitial atoms, are generated during plastic flow in ductile crystals, most particularly in metals. The average temperatures near a moving dislocation are probably not sufficiently high to evaporate vacant lattice sites or interstitial atoms as a result of thermal effects alone. Instead one must apparently conclude that imperfections are generated either by purely geometrical means during the looping of dislocations about appropriate obstacles, as the result of dynamical instability in the motion of a dislocation, possibly near a jog, or in the very high thermal pulses or 'spikes' which are generated either in the zone where two dislocations of opposite sign annihilate one another or near impediments where dislocations are strongly curved.

Seitz thought that a pair of vacancies is probably stable near room temperature and may diffuse more rapidly than a single vacancy, and that vacancies retained during the quenching of Al–Cu alloys and those generated by cold work play an important role in the precipitation process.

He discussed the origin of work hardening in single crystals, and presented several alternative interpretations, which involve the impediment of Frank–Read generators either directly or indirectly as a consequence of the generation of vacancies.

Seitz emphasized the importance of prismatic dislocations formed by condensation of vacancies. He discussed the role that vacancies formed by cold work may play in determining the stored energy, and decrease in density in affecting processes such as creep and hardening of latent slip planes. He proposed a number of experiments which could prove decisive in isolating the influence of vacancies.

1.11. CALORIMETRIC STUDIES OF ISOTHERMAL RECOVERY

(G. Borelius)

In a very brief report, Borelius gave the results obtained in the physical laboratory of the Technical University of Stockholm. They had, by the method of isothermal calorimetry, made measurements on the heat evolved during recovery at raised temperatures after 'cold-working' at room temperature. The results were still preliminary.

Al and Cu rolled at 23 °C, and immediately after rolling, measured at 60 °C, gave a power-time (P.t) curve, the first part of which was given by

$$P = \frac{a}{t} - b, \tag{6}$$

where a and b are constants.

For samples of Al and Cu, which after rolling were aged for some time at room temperature, and then measured at 60 °C, the curves take on the form

$$P = \frac{a}{t + t_0} - b. \tag{7}$$

From experimental curves, t_0 is found to be 0.25 hours. Borelius concluded that the state of recovery reached after 24 hours at 24 °C is reached after 15 minutes at 60 °C.

1.12. DISLOCATION MODELS OF GRAIN BOUNDARIES

(W. Shockley)

The suggestion that an array of dislocations constitutes the interface between two crystal grains was discussed by W. L. Bragg (1940) and J. M. Burgers (1940). Shockley thought that for boundaries between crystals of very small orientation difference, such a model for the boundary seems as natural as do dislocations themselves as one form of imperfection in nearly perfect crystals.

In the late 1940s, a number of workers considered the energetic consequences of this model. The most striking conclusion is that if other quantities are held constant then E, the energy per unit grain boundary, should vary with θ, the orientation difference, according to the law,

$$E = E_0 \theta [A - \ln \theta], \tag{8}$$

where E_0 and A are independent of θ. This formula was derived independently at several places, including Cambridge and Bristol, and was first published by Read and Shockley (1949).

The formula (8) occupies a unique position in the field of dislocation theory. The reason is that only in the case of a grain boundary and in the case of a crystal growing by the Frank mechanism (see F. C. Frank, Section 1.8) is there reason to believe that one knows what the dislocation pattern is in detail. Of these two cases, only the grain boundary has properties determined directly by the energies of the dislocations. The crystal growth case is concerned with geometry rather than energy. Thus the grain boundary case is the one in which direct comparisons between theoretical and experimental energies may best be made.

These comparisons are aided by the fact that E_0 and A depend on different aspects of the energy of the dislocations. The quantity E_0 is independent of the 'core energy' of the dislocation which arises in part from broken interatomic bonds. Shockley showed how it may be calculated in terms of the geometry of the dislocation array and elasticity theory. He discussed that it may thus be possible, from experimental measurements of E_0, to establish that a grain boundary consists of some particular array of dislocations. Beyond this it may be possible to determine just what part of A is due to local distortions at the core of the dislocation. Shockley discussed the way in which theory and experiment could be compared for core energies.

In addition to being static energy systems, grain boundaries are dynamic and will move under applied stresses. Shockley discussed the possible mechanisms of motion for a small angle boundary, and extended the considerations to large angle boundaries. The treatment of large angle boundaries, as layers of liquid metal, suggests a model of a molten metal as a solid with a high density of dislocations, and Shockley pursued this point to a limited extent.

Shockley pointed out how well the formula (8) has been found to fit experimental data over a large range of angles. He disposed of the criticisms that this agreement might be accidental or spurious in discussions.

1.13. THE YIELD POINT IN SINGLE CRYSTAL AND POLYCRYSTALLINE METALS

(A. H. Cottrell)

Cottrell reviewed experimental and theoretical work on the yield phenomenon. He described experiments which show that single crystals of various metals possess yield points when they contain certain impurities in solution, mainly carbon and nitrogen. The yield phenomenon of iron is more pronounced in polycrystals than in single crystals. Cottrell reviewed the dislocation theory which explains the yield points in terms of the segregation of solute atoms to dislocations. He proposed that the observed yield point is not the theoretical stress required to release a dislocation from its atmosphere of solute atoms. Dislocations begin to break away in regions of high stress concentration while the applied load is still small, and the observed yield point is the macroscopic stress at which yielding failure can spread from these centres into the rest of the specimen. Cottrell then described experiments that had been made on the kinetics of the strain ageing process in order to test certain predictions of the dislocation theory.

Definitions

Certain metals, if they contain small quantities of certain impurities, show a peculiar feature in the stress-strain curve, known as the *yield phenomenon*. It has three main features: (1) *The yield point* – plastic flow begins abruptly at a critical stress (upper yield stress) and can then be continued, either at the same stress or, more usually, at a lower stress (lower yield stress); (2) *Overstraining* – once a specimen has yielded it will not show another yield point, but a smooth stress-strain curve, if it is immediately retested; (3) *Strain ageing* – the specimen regains its ability to show a yield point if it is rested or annealed after overstraining.

1.14. DIFFUSION, WORK-HARDENING, RECOVERY AND CREEP

(N. F. Mott)

Mott undertook to review the theoretical ideas on work-hardening, recovery and creep. Within crystals there exist sources of dislocation rings, of the type postulated by Frank and Read.[1] For each source there is a critical stress which has to be applied for it to act. When this stress is applied to the crystal, the source generates a series of rings. These give rise to slip bands. In cubic crystals rings generated on parallel planes join up to give a single ring; this is the origin of the cross slip, as Mott had described in his Guthrie lecture.[2] In cubic crystals, the origin of hardening is normally to be found in a phenomenon described alternatively as deformation or kink bands: dislocations coming from opposite directions interact with each other, and by so doing build up a planar obstacle which opposes both their own motion and that of further dislocations from the same and other sources.

Mott suggested that:

(a) The reason why a pair of edge dislocations approaching each other can stick, or jam, and form an obstacle is connected with cross slip which has occurred before they meet; in hexagonal crystals, if there is no cross slip, this does not occur. For arguments given on this point, Mott referred to his Guthrie lecture.[2]

(b) The mechanism for stabilizing deformation bands, so that they do not disappear when the stress is removed, always involves slip on one of the other sets of planes, not the ones on which the primary slip bands are formed. This was first suggested by Lomer[3], and represents a viewpoint differing from Mott's[2].

(c) Deformation bands will only form if, at a given point early in the deformation, the density of moving dislocations is high enough for the proposed jamming to occur. This is most likely to be the case if the stress applied varies from point to point. If the stress is uniform, one would expect only a few Frank–Read sources to operate, namely those for which the critical shear stress is lowest. One thus expects that the onset of work hardening for single crystals by this mechanism would be delayed and would depend rather critically on the experimental conditions, such as the method of applying the stress and the state of the surface. Experiments by Kochendörfer and Röhm[4], and by Andrade[5], suggest that this may be so.

(d) It is possible that another slower mechanism of hardening exists also, which takes over when deformation bands are not formed, as in single crystals of hexagonal metals and in cubic crystals under specially controlled conditions. Mott made some speculations about this point.

Turning to recovery and creep, Mott made the hypothesis that this is connected with self-diffusion, which he assumed to take place through the movement of vacancies. An edge dislocation is capable of moving out of its slip plane by giving off or absorbing vacancies. The activation energy for such a process is normally rather greater than that of self-diffusion, but in regions of very high stress within a deformation band it may be somewhat less. Mott assumed this to be the cause of the drop in the activation energy for recovery and creep that, according to some authorities, accompanies increased cold-work or increased stress. At low temperatures, however, a certain amount of transient creep can occur without the intervention of moving vacancies. In this case, the relation between strain and time is logarithmic.

Mott discussed the problems concerned with work-hardening, in particular (i) The effect of the surface on the observed form of the slip bands; (ii) The stability of the deformation bands; (iii) The temperature-dependence of work-hardening. He then developed some of the considerations on recovery and creep which he had treated earlier.[2]

REFERENCES

1. F. C. Frank and W. T. Read, *Phys. Rev.* **79**, 722 (1950).
2. N. F. Mott, *Proc. Phys. Soc.*, B **64**, 729 (1951).
3. W. M. Lomer, *Phil. Mag.* **42**, 1327 (1951).
4. F. Röhm and A. Kochendörfer, *Z. Naturforschung*, **3a**, 648 (1948).
5. E. N. da C. Andrade and C. Henderson, *Phil. Trans. Roy. Soc.* **244**, 177 (1951).

1.15. THE DYNAMICS OF SLIP

(E. Orowan)

The main questions of the dynamics of slip are: (1) Why does slip occur at stresses so much below the usual theoretical estimate for perfect crystals? (2) What determines the observed value of the yield stress, and what is the mechanism of strain hardening? (3) What is the cause of the formation of slip zones (slip bands)?

Orowan presented a brief critical summary of the views concerning these questions. He noted the following points:

(1) There is a curious lack of experimental evidence for the reality of the high theoretical estimate of the yield stress of crystals, and new experiments for clearing up this point would be in order.

(2) If the reality of the discrepancy between calculated and observed yield stress is accepted, the only possible way of accounting for it seems to be the assumption that every microscopic crystal or polycrystalline metal contains at least a few dislocations, and that a mechanism for the multiplication of dislocations exists.

(3) The Frank–Read 'dislocation mill' alone cannot act as a multiplication mechanism spreading slip to other planes, as is observed in kinking; nor can the presence of a sufficient number of dislocation mills in an undeformed crystal be assumed. Consequently, there is a need for a multiplication mechanism for dislocations capable of spreading slip to other planes, and a mechanism for producing dislocation mills.

Both mechanisms could be replaced by one producing mills in slip planes not yet in operation.

(4) Cross slip, as observed microscopically, is bound to produce dislocation mills in the two main slip planes, unless slip in the main and subsidiary planes progress exactly synchronously. If, however, cross slip arises from two independent slip processes in different planes, bridged by slip in an oblique subsidiary plane, mills can arise only in slip planes that are already in operation.

(5) If dislocations can acquire velocities of the order of magnitude of the Rayleigh velocity, they must have at these velocities a tendency to swing into oblique slip planes. This would give both a multiplication mechanism of dislocations, and a mechanism for creating cross slip and thus dislocation mills in not yet operative slip planes.

The deflection effect is completely analogous to a similar effect occurring in the propagation of tensile cracks which is responsible for the characteristic features of brittle fracture (specular feature).

(6) The primitive yield stress may be the stress at which dislocations are accelerated to the critical velocity needed for the deflection mechanism of multiplication to act.

(7) Strain hardening in single crystals seems to include two different effects. First, an increase of the operating stress of the dislocation mills in the slip zones, due to the increase of the number of mills with the slip, and the consequent decrease of the spacing of the obstacles between which the dislocations must be squeezed through by the applied stress. Second, the increase of the stress required for opening up new slip zones which is inversely proportional to the largest slip zone spacing present, and thus increases with the existing number of slip zones.

(8) This picture of strain hardening, with the simplest assumptions, leads to stress-strain curves of simple parabolic type. It explains the decrease of the spacing of slip zones for a given strain with decreasing temperature and/or increasing rate of deformation. It also explains why strain hardening is much less rapid in hexagonal metals in which the available subsidiary slip planes have a higher yield stress than the basal plane, so that the number of cross-slips, and thus the number of dislocation mills, produced per unit shear in slip zones is less than in cubic metals.

TENTH SOLVAY CONFERENCE 1954

C.J. GORTER   C. KITTEL   B. MATTHIAS   L. ONSAGER   J. SMIT   H. JONES   P.O. LÖWDIN   A. SEEGER   P. KIPFER   C. BALASSE

I. PRIGOGINE   A.P. PIPPARD   F. FUMI   J.H. van VLECK   O. GOCHE   J. GÉHÉNIAU

C. MOLLER   W. PAULI   Sir W.L. BRAGG   N.F. MOTT   L. NÉEL   W. MEISSNER   D.K.C. MACDONALD   C.C. SHULL   J. FRIEDEL

K.A. MENDELSSOHN   H. FRÖHLICH   D. PINES

2. THE ELECTRONS IN METALS

The tenth Solvay Conference on Physics took place in Brussels from 13 to 17 September 1954. The general theme of the Conference was 'The Electrons in Metals'.

The Scientific Committee of the Conference consisted of: Sir Lawrence Bragg, President, J. R. Oppenheimer, W. Pauli, N. F. Mott, C. Møller, F. Perrin, and A. P. Pippard, Secretary. Oppenheimer was not able to attend.

Among the principal invited lecturers were: D. Pines (Illinois), D. K. C. MacDonald (Ottawa, Canada), A. P. Pippard (Cambridge), C. Kittel (Berkeley, California), J. Friedel (Paris), L. Néel (Grenoble), and H. Fröhlich (Liverpool).

Among the other invited speakers and participants were: J. H. Van Vleck (Cambridge, Mass.), L. Onsager (New Haven, Conn.), B. Matthias (Murray Hill, N.J.), C. G. Shull (Oak Ridge, Tennessee), K. A. Mendelssohn (Oxford), H. Jones (London), A. P. Pippard (Cambridge), P. Aigrin (Paris), C. J. Gorter (Leyden), J. Smit (Eindhoven), A. Seeger (Stuttgart), W. Meissner (Munich), P. O. Löwdin (Uppsala), and F. Fumi (Milan).

We shall now give a résumé of the reports presented at this conference.

2.1. The Collective Description of Electron Interaction in Metals

(D. Pines)

Pines stressed the problems arising in the treatment of electron interaction on the basis of the conventional Hartree–Fock theory. The usual treatment of electron motion in metals is based on an independent electron model, in which the motion of a given electron is assumed, in the first approximation, to be independent of the motion of all the other electrons. The effect of the other electrons on this electron is then represented by a smeared-out potential, which may be determined by using a self-consistent field method. In its simplest version, the Hartree approximation, in which all correlations in the position and energy of the electrons due to their mutual interaction are neglected, this model is quite successful qualitatively and, in many cases, quantitatively. Examples of its success include the specific heat, X-ray band width and magnetic behaviour of the alkali metals.

The success of the simple Hartree approximation is surprising for two reasons. First, on physical grounds, one might expect electronic correlations to play an important role in metals. Since the Coulomb interaction is a long range interaction, and many hundreds of particles interact simultaneously, one anticipates a high degree of long-range correlation in their motion. Second, one might expect that any improvement in the Hartree theory which takes into account electronic correlations would lead to even better agreement between theory and experiment. Such is not the case. In fact, if one uses the Hartree–Fock approximation, in which correlations in the motion of electrons of parallel spin are introduced via the exclusion principle, one finds that agreement between theory and experiment for the above-mentioned metallic properties is completely destroyed.

In the description of a metal it is necessary to take into account the electronic correlations. The Hartree approximation fails badly in the calculation of the cohesive energy of metals, the difficulties being related to the fact that correlation energy terms lead to a significant reduction of the density of states near the Fermi level. One way of getting out of this difficulty is to treat the correlation of anti-parallel spin electrons by second order perturbation theory, but this term diverges badly because of the very long range order of Coulomb interactions. A way out of this dilemma was proposed by Wigner.[1] He calculated the correlation energy[2] by using determinantal wave functions in which the wave function of an electron of given spin is assumed to depend on the positions of all the electrons of opposite spin. However, this method, which represents a departure from the independent electron model, could be applied only to an electron gas of very high density.

The viewpoint of collective description for the electron interaction is diametrically opposed to the one-electron model. The correlated electron motion can be understood from a study of the classical analogue of the electrons in a metal, i.e. the plasma in which screening, collective oscillations and fluid character are important. In the treatment of a classical electron gas, Pines showed that the behaviour of the Fourier components of the density can be split into two parts, one of which characterizes pure collective behaviour (for wavelengths greater than the Debye length) and the other represents individual particle behaviour. The individual particles component represents a collection of individual electrons surrounded by co-moving clouds of charges which act to screen out the field of any given electron within a Debye length. Thus the electron gas is found to be capable of both collective and individual particle behaviour.

A quantum mechanical generalization of the density fluctuation method is straightforward. Pines considered the physical basis for the collective description and sketched the mathematical derivation of the Hamiltonian which is used as a starting point for discussing electron interaction in metals. The basic Hamiltonian for the free electron gas model is given by,

$$H = \sum_i \frac{p_i^2}{2m} + 2\pi e^2 \sum_{ijk} \frac{e^{i\mathbf{k}\,(\mathbf{x}_i - \mathbf{x}_j)}}{k^2} - 2\pi n e^2 \sum_k \frac{1}{k}. \tag{9}$$

The first term corresponds to the electrons' kinetic energy, the second to their Coulomb interaction and the third to a subtraction of their self energy. One seeks to analyze the correlated behaviour of the electrons and then to investigate the character of the electronic motion once the collective aspects are isolated. A field energy term,

$$H_{\text{field}} = \sum_{k < k_c} \frac{p_k^* p_k}{2}, \tag{10}$$

is added to the Hamiltonian for the description of the collective behaviour of electrons. The summation is limited to a value k_c, since only long wavelengths are responsible for the collective behaviour.

By means of a canonical transformation the Hamiltonian can be split into a set of terms describing the plasma oscillation and the free electron energy, and another describing the short range part of the Coulomb interactions. Two other terms remain, one giving a weak electron-plasma coupling and another which can be ruled out by random phase approximation.

The Hamiltonian for the electron interaction in metals turns out to be,

$$H = \sum_i \frac{p_i^2}{2m}\left(1 - \frac{\beta^3}{6}\right) + \sum_{k<k_c}\left\{\frac{p_k^* p_k + \omega^2 q_k^* q_k}{2} - \frac{2\pi n e^2}{k^2}\right\} + 2\pi e^2 \sum_{k>k_c} \frac{e^{i\mathbf{k}\cdot(\mathbf{x}_i-\mathbf{x}_j)}}{k^2} + H_{\text{r.p.}},$$

(11)

with $\beta = k_c/k_0$, and the plasma oscillation frequency is given by,

$$\omega \cong \omega_p\left(1 + \frac{3}{10}\frac{k^2 v_0^2}{\omega_p^2} + \frac{k^4}{8m^2\omega_p^2}\right),$$

(12)

where v_0 is the velocity of the electron at the top of the Fermi distribution. This Hamiltonian leads to the following view of the system: (1) a collective motion of the electrons for a frequency $\omega(k)$ much larger than the frequency of an electron at the top of the Fermi distribution; (2) a collection of electrons interacting via a short-range interaction. Since the plasma oscillations will not be excited at ordinary temperatures, the independent electron model would provide a reasonable approximation for the motion of the electrons in metals.

Pines discussed in some detail the correlation energy and considered the influence of correlation and exchange on the state density, magnetic susceptibility and band width of electrons in metals, confining attention to the alkali metals for which a free electron model is particularly appropriate. He also discussed the excitation of collective oscillations by a fast charged particle and the damping of these oscillations by interaction with the core electrons.

Finally, Pines discussed the role of electron-electron interactions in determining the electron-phonon interaction in metals. The importance of electron-electron interactions in this problem arises from the general tendency of the electrons to screen out the field of ions. A given conduction electron first of all interacts with the ions, whose motion may be described in terms of the phonon field. For long wavelength phonons, however, this interaction is radically altered by the field due to the other electrons, which in the course of responding to the ionic motion move in just such a way as to produce a field which very nearly cancels the ionic field. Bardeen[3] had established this by treating the electrons in the Hartree approximation, and thereby obtaining a self-consistent solution for the coupled motion of the ions and electrons.

The phonon frequency is found to be greatly influenced by the electronic response to the ionic motion. Staver and Bohm[4] had discussed this problem, treating the ions and electrons as a set of coupled plasmas. Working with the equations of motion, they obtained the dispersion relation for the phonon field as modified by the electron-electron interactions.

The effect of electron-electron interactions is to bring about short-range collisions between the electrons. This may be regarded as a relatively small perturbation on the motion of a given electron. However, in superconducting elements, the electron-phonon interaction is a violent one, so much so that the electrons involved cannot be treated by perturbation-theoretic methods. One expects that compared to such a strong interaction, the comparatively weak electron-electron interaction may properly be neglected. Pines noted that it may play a small role in determining the energetics of the transition, but it should not materially influence it in the first approximation. He then discussed, in a speculative fashion, the extent to which the model of electron-electron interactions could be modified so as to be applicable to all metals.

REFERENCES

1. E. P. Wigner, *Phys. Rev.* **46**, 1002 (1934); *Trans. Faraday Society*, **34**, 678 (1938).
2. The correlation energy is defined here as the difference between the energy calculated in the Hartree–Fock approximation and that calculated by using any better approximation.
3. J. Bardeen, *Phys. Rev.* **52**, 688 (1937).
4. D. Bohm and T. Staver, *Phys. Rev.* **84**, 836 (1952).

2.2. AN EXTENSION OF THE HARTREE–FOCK METHOD TO INCLUDE CORRELATION EFFECTS

(P. O. Löwdin)

Löwdin reported on an attempt to include correlation effects in the independent electron model by a direct extension of the conventional Hartree–Fock scheme, which is different from the plasma model developed by Bohm and Pines.

The basic idea of the 'independent-particle model' is that, in a first approximation, one can neglect the mutual interaction between the N-electrons in constructing the total wave function and assume that the electrons are moving independently of each other in a set of N-spin-orbitals, $\psi_k(X)$. The total wave function is then a simple product. However, between the electrons i and j there is in reality a repulsive Coulomb potential $H_{ij} = e^2/r_{ij}$, which, for small distances, $r_{ij} \approx 0$, becomes tremendously large. This repulsive potential tries to keep the electrons apart, and, since this 'correlation' between the movements of the particles is entirely neglected in forming the product wave function, the corresponding total energy is affected by an error usually called the 'correlation energy'.

The main problem is to take the correlation between electrons having different spins into proper account. In this regard, the Hartree–Fock scheme in its conventional form does not take full advantage of the degrees of freedom for the electronic motion provided by the independent-particle model. For instance, in constructing the singlet ground state of an electronic system, one starts usually from the assumption of the existence of a certain number of *orbitals* in ordinary space, which are then assumed to be occupied each by two electrons having opposite spins; the corresponding determinant then represents really a singlet state. In treating the atoms of the periodic system, this assumption was natural for historical reasons as a consequence of the Pauli-principle in its original formulation, but from the point of view of correlation it is energetically rather unfavourable since it compels electrons having opposite spins to be in the same orbital, i.e. in the same places in ordinary space leading to a high Coulomb repulsion.

The conventional scheme can be improved by avoiding the idea of 'doubly filled orbitals', and to let the best *spin-orbitals* be determined by the variational principle. The possibility of having different orbitals for different spins was first mentioned by Hartree and Hartree[1], but it was never used by them. The importance of this possibility for the description of magnetic properties was stressed by Slater[2], and self-consistent-field calculations for atoms on this basis have been carried out. Mulliken[3] pointed out the importance of finding a generalization of the self-consistent-field procedure from 'closed shell' to 'open shell' configurations with the number of orbitals doubled.

The theory can be developed without any additional assumptions, and one has only to let the N undetermined spin-orbitals, $\psi_k(x)$, be occupied each by *one* electron. The Slater determinant then usually does not represent a pure spin state, but the total wave function may simply be formed by taking a *projection* of this determinant giving the spin state of required multiplicity, and the spin degeneracy problem can be solved.

Löwdin considered the electrons in metals. By investigating the cohesive energy, one can get an idea of the validity of the extension of the Hartree–Fock scheme for treating the Coulomb repulsion, since the energy calculations are very sensitive to correlation effects. It is characteristic for a 'naive' metal theory, in which the total wave-function is approximated by a *single* Slater determinant, built up of Bloch orbitals doubly filled within a particular Fermi surface, that the curve for the cohesive energy as a function of the lattice parameter shows an entirely wrong asymptotic behaviour for separated atoms. The model permits electrons of opposite spins to accumulate on the same atom and give rise to negative and positive ions, having higher energy together than the dissociation products occurring in nature, where the Coulomb repulsion prevents the excessive formation of ions. By giving up the notion of doubly-filled orbitals, one may determine the new spin-orbitals by applying the variational principle. One can get a connection with the conventional theory by developing the new spin-orbitals in the system of ordinary and virtual Bloch spin-orbitals, and the expansion coefficients may be determined by a self-consistent-field procedure.

The resulting wave function leaves the electrons distributed on both sides of the conventional Fermi surface with some electrons excited above the surface by the influence of the Coulomb repulsion. This method seems to be able to take so much of the correlation into account that the energy curve gets at least a correct asymptotic behaviour.

This approach seemed to be promising, but it was still too early to decide whether the 'independent particle model' could be saved along these lines and whether one could discard the Hartree–Fock method in treating metallic properties.

REFERENCES

1. D. R. Hartree and W. Hartree, *Proc. Roy. Soc.* **A 154**, 588 (1936).
2. J. C. Slater. *Phys. Rev.* **82**, 538 (1951).
3. R. S. Mulliken, *Proc. Nat. Acad. Sci.* **38**, 160 (1952).

2.3. RESISTANCE OF METALS AT LOW TEMPERATURES

(D. K. C. MacDonald)

The application of quantum theory, together with the concept of Fermi statistics of conduction electrons, by such people as Houston, Bloch, Sommerfeld, Wilson, Mott, Fröhlich and others, to electron transport in metals constituted a great advance in the understanding of these processes. Thus in the case of resistance due to thermal vibrations it was found that the so-called Grüneisen–Bloch law,

$$R \propto \frac{T^5}{\Theta^6} \int_0^{\Theta/T} \frac{x^5 \, dx}{(e^x - 1)(1 - e^{-x})},$$

(13)

appeared to describe rather well the experiments for a number of metals over a fair range of temperature. This law was originally derived analytically by Bloch as the low temperature limit only $(T/\Theta \to 0)$, but Sondheimer and Rhodes have maintained that this law should not depart from the true expression by more than 10% to 15% in the region $T/\Theta \sim 1$.

Apart from the specific 'scaling' parameter, Θ, the Bloch–Grüneisen equation predicts for all metals a universal temperature-dependence of the electrical resistance due to thermal vibrations. However, departures are expected to arise from the following factors, each of which has been investigated by numerous workers: (1) Varying ion-electron scattering properties from metal to metal; (2) Asphericity of Fermi surface; (3) Non-uniform relaxation time; (4) Departure of frequency spectrum from the Debye model; (5) The influence of anharmonicity of lattice vibrations; (6) The immediate influence of thermal expansion (on Θ and the electron-ion interaction energy); (7) Resistance due to electron-electron scattering.

MacDonald discussed in detail the resistance component due to static imperfections. After Bloch and Houston had shown that electrons would not be scattered in a perfectly periodic lattice (including the influence of zero point energy) it was evident that the 'residual' resistance, as $T \to 0$, must be ascribed to static imperfections (physical and/or chemical) in the lattice. This had been familiar experimentally since the statement of Matthiessen's rule (1860) of the approximate additivity of 'residual' and thermal components of resistance.[1] With the wider availability of very low temperatures, many workers have made detailed studies of the resistivity of alloys since 1930. Nordheim and Mott contributed to a general understanding of the data, especially the influence of homovalent and heterovalent impurity scattering.

A particularly puzzling feature is that of the *minimum of electrical resistance* at low temperatures. It was first observed systematically in gold by de Haas in 1933, and it has since been found definitely in a few other metals, although not in the alkali family. However, it appears to be generally agreed that the resistance minimum would *not* occur in the ideally pure metal. There is also some evidence linking the mechanical properties of the metal with the resistive minimum, and it has also been suggested that grain boundaries alone are sufficient to produce the minimum. This phenomenon also seems to be linked with very marked anomalies in the thermoelectric behaviour at low temperatures.

In the purest available metals, the electron mean free path increases by a factor of 10^3 to 10^4 as the temperature is lowered from room temperature to that of liquid helium. Thus, for example, in the purest sodium a mean free path of the order of 0.1 mm can be attained. Consequently, the geometrical boundaries of the specimen may readily contribute to the scattering.

In order to understand the anomalous electron transport under a thermal gradient it is necessary to focus attention on the interpretation of thermoelectric power and more generally on the influence of a thermal gradient on electron diffusion and conduction phenomena. Peierls was the first to question the adequacy at low temperatures of Bloch's assumption that during electron transport (electrical conductivity) the *lattice* remains essentially in thermal equilibrium. This leads to the consideration of lattice relaxation processes, and Peierls introduced the 'Umklapp-prozess' for maintaining a momentum balance. In the course of this process, a conduction electron of selective de Broglie wavelength imparts its momentum *directly* to the rigid lattice, suffering an internal Bragg reflection, instead of dissipating its momentum through the more 'usual' intermediary of thermally excited lattice waves ('phonons'). Such Umklapp-processes in the case of monovalent metals seem rather improbable at low temperatures, although one does not have sure knowledge of the shape of the Fermi surface and its relation to the Brillouin zone structure. Houston and Peierls have suggested that multiple electron processes ought also to be considered.

When a temperature gradient is applied, the lattice equilibrium is to some extent directly disturbed, but the conventional theory of electron-transport assumes that this perturbation may be neglected in relation to the electronic effects. If this is so, then the absolute thermo-electric power of a metal, S, arising solely from electron diffusion is given by,

$$S = \frac{\pi^2 k^2 T}{3} \left(\frac{d \ln \sigma(E)}{dE} \right)_{E=\zeta} \tag{14}$$

where E=conduction-electron energy, ζ=Fermi level, and σ=electrical conductivity. The agreement with experiment is, however, not very good. Various theoretical approaches were made to clear up this problem, taking into account the lattice effect, and showing that the absolute thermo-electric power due to phonon-electron interaction will be given by

$$S \sim \frac{C}{N_e \cdot e} (\alpha) \dots, \tag{15}$$

where C=specific heat of the solid, $N_e e$=density of mobile charges, and α=a numerical factor between 0 and 1 governed by the relative importance of phonon-electron scattering to other phonon-scattering processes. For $T > \Theta$, and in a monovalent metal, one has

$$S \sim \frac{3k}{e} \cdot \alpha, \tag{16}$$

which, assuming free electrons as carriers, gives

$$S \simeq -250 \, \alpha \, \mu V/°K. \tag{17}$$

The observed absolute thermo-electric power of sodium around room temperature is ~ 5–$6 \, \mu V/°K$, and one must assume that phonon-electron scattering is of little importance at higher temperatures, in agreement with the usual conception that at high temperatures lattice anharmonicity forms the major limitation to lattice heat flow. At low temperatures, it is assumed that in metals and dilute alloys phonon-electron scattering becomes the dominant factor.

There arise a number of problems in phonon-electron and other phonon interaction processes. Thus, one is led to ask whether there is justification in speaking about a thermodynamic electron temperature. In the statistical theory, due originally to Lorentz, it is assumed that even under a thermal gradient one can assign at each point, x, of the conductor a unique temperature to the electron distribution function so that one may write,

$$\left(\frac{\delta f}{\delta x}\right)_{px} \approx \left(\frac{df_0}{dT}\right) \cdot \left(\frac{dT}{dx}\right), \tag{18}$$

where f_0 is the thermodynamic unperturbed distribution. On the other hand, following Lorentz, electron-electron collisions are specifically assumed to be negligible whereas one might have surmised that a sufficiently *short* electron-electron relaxation time would be necessary to ensure the validity of using an electron temperature. There are also questions involved in the statement of the Boltzmann equation for electrons in a metal under a thermal gradient which might give cause for concern. The contribution to metallic resistance which could arise from electron-electron collisions and the role of electron-electron interaction in relation to dielectric behaviour have been investigated in detail. MacDonald noted that it would be valuable to undertake a detailed assessment of: (a) Electron-electron relaxation; (b) Electron-lattice (including phonon-electron) relaxation; and (c) Lattice relaxation (phonon-phonon, etc.).

The latter process is the direct source of thermal resistivity in insulators, and has been treated by Peierls[2] and Klemens[3] in terms of a phonon-statistical model. Since, however, thermal resistivity due to lattice anharmonicity increases with temperature, one might expect that a rather direct *classical* analysis should be possible giving a quantitative result at higher temperatures, linked directly, in particular, with the thermal expansion of the solid. MacDonald discussed the various aspects of active research related to these problems.

REFERENCES

1. A. Matthiessen, *Ann. Phys. Chemi, Poggendorff*, **110**, 190 (1860).
2. R. E. Peierls, *Ann. Phys.* **3**, 1055 (1929).
3. P. G. Klemens, *Proc. Roy. Soc.* A **208**, 108 (1951).

2.4. Some Experiments on Thermal Conductivity of Metals

(K. Mendelssohn)

Mendelssohn reported on the programme of systematic investigations on transport phenomena in metals that was started at Oxford after the Second World War. Work on the electrical conductivities at low temperatures of the alkali metals and the alkaline earths showed that sodium represents a good approximation to an 'ideal' metal with quasi-free electrons while the other metals, including lithium, exhibited a more complex behaviour. These experiments were later on supplemented by detailed and comprehensive measurements of the thermal conductivities, covering the whole periodic system, establishing the part played by the metal electrons in transport effects.

There appears to exist a minimum, at a few degrees above absolute zero, in the electrical resistance of certain specimens of gold, magnesium and copper. The question arises whether there exists a corresponding effect in the thermal conductivity which at the lowest temperatures seems to obey in general the Wiedemann–Franz law quite well. Routine measurements on magnesium provided some indication of an anomaly, since the heat conduction at $\sim 4\,°K$ did not extrapolate to the value zero at absolute zero. Since the effect is small and structure dependent, the only convincing evidence can be provided by simultaneous determination, on the same specimen, of the electrical *and* the thermal conductivities. The result of this experiment shows that in the region of the resistance minimum the thermal conductivity, K, indeed appears to be slightly increased. One also finds that the electrical conductivity does not rise above the normal residual value at the lowest temperatures.

Mendelssohn mentioned two other experiments which, while different in character, give information on the lattice conduction K_g. A comparison of these two experiments then allows an estimate of the scattering of phonons by the free electrons. The first type of measurement is the determination of the heat conductivity of metals in high magnetic fields. Some metals show a marked decrease in thermal conductivity with increasing magnetic field, and one assumes that there is a decrease in the electronic heat conduction K_e. The heat conduction by phonons is uninfluenced by the magnetic field, and sets a limit to the extent to which the total heat conductivity can be decreased. Actually, the contribution of the phonon conduction is found to be negligible, amounting to less than one percent of the total heat conductivity.

A quite different possibility of determining K_g is provided by observations on superconductors. In a pure metal the heat conductivity in the superconductive state is at any temperature smaller than that in the normal state. This can be understood from the fact that the conduction electrons which pass into the superconductive state cease to contribute to the entropy. It is known, from a comparison of the measured electronic entropy in the normal state and the entropy defect when the metal becomes superconductive, that with decreasing temperature the number of these 'superconductive electrons' increases and that at absolute zero all conduction electrons have passed into the superconductive state. For this reason the heat conductivity of a superconductive metal near absolute zero is of special interest since one may expect the electronic heat conduction to become small in comparison with that of the lattice.

The result of such a measurement on a lead single crystal demonstrates that, in the temperature region of 1/20 of the transition temperature, the heat conductivity of the metal is proportional to T^3. This is the temperature dependence which has been postulated by theory for a dielectric crystal. This, together with the absolute value of heat conduction, indicates that below 1 °K in the lead single crystal heat is carried by the phonons only and that electronic conduction becomes negligibly small. Accordingly, at low enough temperatures, a superconductor behaves like an insulator as far as its heat conductivity is concerned.

The two values of K_g, obtained from superconductivity and the upper limit derived from the measurements in high magnetic fields, differ by a factor of 500 to 1000; that derived from the superconductivity results being the larger. The reason for the enormous discrepancy between the two K_g-values is due to the different parts played by the conduction electrons in both cases. In the heat conduction in high magnetic fields, the ability of the electrons to transport heat is greatly diminished but the electrons retain their function as scattering centres for the phonons. In the superconductor also, the electrons carry no heat but they have also ceased to scatter lattice vibrations. Under the conditions studied in these experiments the reduction of K_g by scattering of phonons on electrons had roughly the same value as K_e in the normal metal. One is thus faced with the surprising result that, at least at low temperatures, the conduction electrons in a metal lower, through scattering, the lattice conduction by about the same amount as they themselves contribute to the heat conductivity.

Finally, Mendelssohn reported on another series of measurements which promised to give new information on the behaviour of electrons in metals. These are heat conductivity measurements on semiconductors, especially germanium. The samples vary from extremely pure single crystals with a negligible number of current carriers to specimens showing definitely metallic behaviour. As the number of conduction electrons gradually increases from sample to sample, it is possible to trace step by step their contribution to the heat conduction as well as their scattering properties. The great advantage of this work is that it can be carried out on the same substance, which moreover permits comparison with such an ideal dielectric as the diamond with which it shares the same crystal structure. The first results, obtained on very pure germanium and silicon, strongly resembled the observations on diamond, showing the form to be expected for a pure dielectric crystal. Quantitative analysis revealed that the mean free path of the phonons is determined by the size of the crystal, which in the case of the germanium specimen was identical with its geometric dimensions.

2.5. METHODS FOR DETERMINING THE FERMI SURFACE

(A. B. Pippard)

Although in elementary treatments of the theory of metals the assumption of a free, or quasi-free, electron gas is made, it soon becomes clear when comparison is attempted of the results of experiment with the theoretical predictions that the model is at best very approximate, except perhaps for alkali metals and the noble metals of group 1. In order to achieve better agreement, the influence of the periodic field of the lattice on the electronic motion must be considered, but in trying to do so serious difficulties of calculation arise. There are vast gaps in knowledge between the delineation of the Brillouin zone structure and the quantitative explanation of any single property of a real metal.

Perhaps the main experimental obstacle to an understanding of the true behaviour of a metal is that most properties which are strongly influenced by the detailed electronic structure depend on so many factors that it is virtually impossible to sort them out from the experimental data. Thus, to take as an example a phenomenon whose very existence depends on departures from the free electron model, a complete explanation of the magneto-resistance effect involves knowledge of the form of the energy surfaces in the neighbourhood of the Fermi energy and of the relaxation time at all points of the Fermi surface. In other words, one needs to know the polar distribution of electron trajectories as well as the Fermi velocity, effective mass tensor and relaxation time for all directions of motion. These parameters are so intimately mingled in the general theory of the magneto-resistance effect that they could not be extricated from the experimental results.

There are two phenomena, however, the Anomalous Skin Effect and the de Haas–van Alphen Effect, which seem to offer an attack on the problem, since certain aspects of the effects are interpretable in terms simply of the geometrical form of the Fermi surface, uncomplicated by other factors. Moreover, the connection between the geometry and the observed behaviour is such that a systematic survey of the anisotropy of the effects in single crystals promises to be sufficient to determine the Fermi surface completely, at least in the simpler metals.

Two basic assumptions are common to the theories of both effects. The first is that interactions between the electrons may be neglected, so that the electron assembly may be treated as a perfect gas obeying Fermi statistics, but having an arbitrary dependence of energy E on wave number k. The second assumption is that the electrons may be treated semi-classically, quantization being introduced where necessary by means of the Bohr–Sommerfeld–Wilson rule.

Anomalous Skin Effect was first observed by H. London[1], and was investigated experimentally by Pippard[2], Chambers[3] and Fawcett[4]. The quantitative theory for an isotropic metal was given by Reuter and Sondheimer[5].

Pippard considered the differences between the normal and the anomalous skin effects. At low temperatures the mean free path is much larger than the skin depth, which rules out the usual theory. In this case, only those electrons which move nearly parallel to the surface can execute a substantial fraction of a free path within the surface layer which contains the field. Pippard introduced the 'ineffectiveness concept', which leads to a correct resistance at low temperatures, and derived the relation between the Fermi surface and the resistance for a two-dimensional system. The curvature, S, of the Fermi surface is found to be directly related to the resistance, in a band around a fixed axis defined by the surface. The study of crystals of different shapes, i.e. for which the surface makes different angles with respect to the symmetry axis of the crystal, allows a full description of the Fermi surface.

Pippard then considered the de Haas–van Alphen effect. It was found by de Haas and van Alphen that at very low temperatures the magnetic susceptibility of bismuth in high fields (~ 10 kG) showed a periodic variation with field strength. At first regarded as a peculiarity of bismuth, the effect has since been demonstrated in many other metals. An explanation of the effect was first given by Peierls[7] for a free electron model, and the theory has been subsequently refined by many workers. In his report, Pippard considered only those aspects of this effect which directly concern the study of the Fermi surface, drawing largely on the ideas expounded by Onsager.[8]

It is convenient to consider the theory of the effect in a two-dimensional free-electron metal, in which at $0°K$ the electrons are uniformly distributed within a circle in p-space, having radius corresponding to a Fermi energy, $E_0 = p_0^2/2m$. The orbits of the electrons under the influence of the magnetic field are quantized, and the only possible energies are $(2n+1)\beta H$, where n is an integer, β the Bohr magneton and H the magnetic field. The generalization of the theory to three dimensions, along the lines pointed out by Onsager, involves no new principles. If an electron is represented in p-space at a point on the closed curve $E(p) = $ constant, its velocity in real space is $\mathrm{grad}_p E$ and is directed along the normal to the curve. Thus in the presence of a magnetic field the Lorentz force, $e\mathbf{H} \times \mathbf{v}$, lies along the curve, and the resulting motion of the electron is such as to make its representative point trace out the curve of constant energy. Combining this result with the preceding quantization rule, one finds that the permitted trajectories in momentum space are those for which the energy curve encloses an area A such that,

$$A = (n + \tfrac{1}{2})\, heH. \tag{19}$$

The magnetization curve has thus a constant period, he/A, when M is plotted against $1/H$, and from such a curve A is deducible. It is this general result which gives the de Haas–van Alphen effect its potential importance as a tool for determining the form of the Fermi surface. This result can be generalized to a three-dimensional metal.

Pippard finally described the experimental techniques used in observing the two effects mentioned above. The main difficulties are related to the fact that the sample must be very pure. Moreover, the magnetic field has to be large and the detection very precise, because even in a field as high as 10^5 G the spacing of the oscillations is only 17 G.

In the case of tin the anomalous skin effect has been studied, but the interpretation is difficult due to the great complexity of the Brillouin zone. In the de Haas–van Alphen effect, the periods of oscillation are much larger than expected, and in metals such as tin, zinc, and mercury, there are inner zones which are almost filled.

REFERENCES

1. H. London, *Proc. Roy. Soc.* **A 176**, 522 (1940).
2. A. B. Pippard, *Proc. Roy. Soc.* **A 191**, 385 (1947).

3. R. G. Chambers, *Proc. Roy. Soc.* **A 215**, 418 (1952).
4. E. Fawcett, *Proc. Phys. Soc.* **A 66**, 1071 (1953).
5. G. E. H. Reuter and E. H. Sondheimer, *Proc. Roy. Soc.* **A 195**, 336 (1948).
6. W. J. de Haas and P. M. van Alphen, *Comm. Phys. Lab. Univ. Leiden*, 212a (1930); 220d (1932).
7. R. E. Peierls, *Phys.* **81**, 186 (1933).
8. L. Onsager, *Phil. Mag.* **43**, 1006 (1952).

2.6. Resonance Experiments and Wave Functions of Electrons in Metals

(C. Kittel)

Kittel reviewed three principal types of resonance experiments which give information bearing on the wave functions in solids. These experiments deal with: (a) Cyclotron resonance; (b) Nuclear spin resonance; (c) Electron spin resonance.

Cyclotron resonance absorption of radiofrequency power occurs when the orbital rotation frequency of a conduction electron moving in an applied constant magnetic field is approximately equal to the frequency of the *rf*-field. The first successful experiments were performed in germanium crystals. The experiments have been successful in determining accurately the effective masses, symmetries and degeneracies of the band edges of the conduction and valence bands in germanium and silicon.

In principle, cyclotron resonance is ideally suited to the determination of the actual energy surfaces in crystalline solids. In practice, it appears that the successful application of the method is probably limited to semi-conductors and insulators. The experimental attempts to observe cyclotron resonance in metals and superconductors have not been successful.

The nuclear spin resonance experiments on pure metals, on the other hand, give information relating to conduction electron wave functions in a number of ways, including: (a) Knight shift; (b) Nuclear spin-lattice relaxation times; (c) Nuclear exchange coupling via conduction electrons.

The Knight[1] shift was observed in 1949 in the nuclear magnetic resonance frequencies in metals from frequencies observed for the same nucleons in chemical compounds in the same constant magnetic field. Townes[2] suggested that the Knight shift was caused by the hyperfine interaction of the nuclear magnetic moment with the magnetic moment of the conduction electrons. The shift produced by this interaction is related to the Pauli spin susceptibility, and consequently to the density of states on the Fermi surface. Values of the shift as large as 2 percent of the resonance frequency have been reported for the heavy elements; in Na^{23} the shift is 0.10 percent. Kittel gave a theoretical derivation of the Knight shift under the assumption that the electron charge distribution has cubic symmetry about the nucleus, and neglecting the electron spin-orbit interaction.

Another type of interaction is the nuclear spin-lattice relaxation time. Frequently in metals the dominant mechanism of energy exchange between the nuclear spin system and the lattice is by means of the hyperfine coupling of the nuclear magnetic moments with the conduction electrons. This process was discussed first by Heitler and Teller[3]; the detailed theory was given by Korringa[4].

Ruderman and Kittel[5] have shown that there exists in metals a coupling between the magnetic moments of two nuclei via their hyperfine interaction with conduction electrons. It is an indirect exchange interaction, provided that the contact part of the hyperfine interaction is dominant.

Finally, the electron spin resonance experiments on conduction electrons in metals provide information about the state of the conduction electrons in several ways, including: (1) Intensity of the resonance signal; (2) Shift in the g-value; (3) Overhauser shift accompanying saturation; (4) Spin-lattice relaxation time; (5) Electron diffusion effects on the line shape.

Overhauser[6] discovered the remarkable result that under appropriate conditions the population distribution of nuclear spins in a metal among the nuclear magnetic sublevels is determined essentially by the magnitude of the electronic magnetic moment μ_B, rather than by the nuclear moment μ_I. His conditions are that the electron spin resonance of the conduction electrons should be saturated, and that the principal spin-lattice relaxation mechanism of the nuclear spins should be the $\mathbf{I} \cdot \mathbf{S}$ hyperfine coupling with the conduction electrons. The predicted enhancement of nuclear polarization on saturating the electron resonance has been detected experimentally in metallic lithium by Carver and Slichter[7].

It was further suggested by Overhauser that there should be, especially in metals, an important shift in the position of the electron spin resonance line as the line is saturated. The shift is the result of the hyperfine interaction with the nuclear moments. At room temperatures over 1 °K the nuclear moments under equilibrium conditions cause a shift in the position of the electron spin resonance in metals by less than 1 part in 10^5. However, if the nuclear polarization is enhanced by a factor of the order of 10^3 by the Overhauser effect, the electron shift may be of the order of 1 in 100, which should be observable.

In sodium, potassium, and cesium the electron spin resonance line width increases as the temperature increases. In lithium and beryllium the width is independent of temperature, but (at least in lithium) varies from sample to sample and is known to be sharper in the purer samples. The principal causes of electron spin-lattice relaxation include: (a) The Overhauser[8] mechanism interaction between the magnetic moment of one electron and the field caused by the translational motion of a second electron; (b) The Elliott[9] mechanism electron spin flipping caused by spin-orbit effects in the collision of a conduction electron with thermal phonons and with impurities and imperfections.

Finally, Kittel discussed electron diffusion effects on the line shape. It was noticed by Griswold[10] that conduction electron resonance could be observed at microwave frequencies in thick plates of alkali metals. It had been thought previously that the rapid diffusion ($\sim 10^{-9}$ sec) of electrons out of the skin depth would broaden the resonance line proportional to the reciprocal of the diffusion time, so that it was not expected that a line could be observed in thick plates at high frequencies. However, the line observed by Griswold has a normal width but an unusual shape, and the intensity is quite weak. These results were explained by Dyson[11], who obtained an exact solution to the line shape problem in the normal skin effect region. His results have been extended to the anomalous skin effect region which obtains in the measurements at low temperatures. Physically, it may be said that the reason why a sharp line is observed is that the electrons diffusing out of the skin depth may have a very good chance of diffusing back in a number of times before the spin relaxes.

REFERENCES

1. W. D. Knight, *Phys. Rev.* **76**, 1259 (1949).
2. C. H. Townes, C. Herring and W. D. Knight, *Phys. Rev.* **77**, 852 (1950).
3. W. Heitler and E. Teller, *Proc. Roy. Soc.* A **155**, 637 (1936).
4. J. Korringa, *Physica*, **16**, 601 (1950).
5. M. A. Ruderman and C. Kittel, *Phys. Rev.* **96**, 99 (1954).
6. A. W. Overhauser, *Phys. Rev.* **92**, 411 (1953).
7. T. R. Carver and C. P. Slichter, *Phys. Rev.* **92**, 212 (1953).
8. A. W. Overhauser, *Phys. Rev.* **89**, 689 (1953).
9. R. J. Elliott, *Phys. Rev.* **94**, 564 (1954).
10. T. W. Griswold, Berkeley Thesis (1953).
11. F. J. Dyson, *Phys. Rev.* **98**, 349 (1955).

2.7. PRIMARY SOLID SOLUTIONS IN METALS

(J. Friedel)

Friedel reported on the theoretical study of metallic solid solutions. The experimental analysis of phase boundaries led Hume-Rothery[1] to distinguish essentially between three factors the action of which seemed to be additive. These are: the *size* of the constituents, defined, for instance, by the atomic sphere radius of the pure elements; the place they occupy in the periodic table defined by their *valence*, and the *period* to which they belong.

In a rough model the matrix and the atoms dissolved in the system are treated as a continuous and homogeneous medium. In this model, there occurs first of all an increase in the free energy term due to elastic constants, a short range interaction between the nearest neighbour dissolved atoms and a volume change of the whole system, if finite, due to the change of volume taking place when atoms are introduced in the system; and finally a change in the energy and entropy terms.

As for the 'period', the effect is small, in particular for substitutional alloys of heavy elements. These effects have been studied by Mott[2] and Huang.[3]

In the discussion of interstitial or substitutional solutions, in which the solute and the solvent have different valences, the factors of 'size' and 'period' are either neglected or treated separately. The model is an approximation of the generalized Thomas–Fermi model, using molecular orbitals. It is particularly useful for the study of the electronic density in the vicinity of the Fermi surface. All the valence electrons are assumed to move in the same potential, which is the sum of the periodic potential of the pure solvent and a perturbation, treated as being locally constant. With this assumption, the model yields the same band structure as the solvent, but translated by a term equal to the mean perturbation.

Friedel discussed the effect of the perturbation V_p on the Fermi level, and showed that the displacement of this level is very small at low concentrations and becomes significant only when the interactions between the atoms dissolved in the system is important. The concentration for which the change ΔE_m in the Fermi level begins to appear can be determined by the Knight shift. For tin, for instance, it is 3 percent. From the curve of ΔE_m against the concentration it can be deduced that the curvature of the curve, $\varepsilon(c)$, for the formation energy of alloys, is proportional to the charge, Z, of the foreign atom. Friedel showed how this conclusion fits with the experimental data.

The example of ferromagnetic alloys shows that the 'rigid band' model leads to the same conclusions as 'localized screening'. If several bands overlap at the Fermi level, the rigid band model shows that supplementary electrons introduced with each atom of the solute fill the different bands in proportion to their densities at the Fermi level. The magnetic moments of ferromagnetic alloys can be explained in this way, and Friedel gave examples of metals in which this situation exists. The overlapping of bands leads to a screening of an atom of the solution. The change in the charge density is then proportional to the states at the Fermi level, yielding a change in the magnetic moment of this band. Finally, Friedel discussed how the rough model he had employed could be improved.

REFERENCES

1. W. Hume-Rothery, *The Metallic State* (Oxford 1931).
2. N. F. Mott and H. Jones, *Metals and Alloys* (Oxford 1936).
3. K. Huang, *Proc. Phys. Soc.* **60**, 161 (1948).

2.8. THE CREATION AND MOTION OF VACANCIES IN METALS

(F. Fumi)

The experimental study of the Kirkendall effect has shown that for a number of metals the transport of matter within the solid does not occur by an exchange process either of the traditional two atom type or of the Zener ring type: lattice defects are involved in an essential way in the diffusion process in these metals. From experiments it is not clear whether the defects involved are single vacancies, interstitials, or perhaps more extended regions of disorder in the lattice. Theoretically, the only quantitative calculations on the activation energies of creation and motion of simple defects in metals are those of Huntington and Seitz[1], which show fairly convincingly that, for Cu, vacancies are the important defect of thermal origin. In order to understand the process of matter transport within a metal it is useful to calculate these energies for a number of metals, and for this one constructs simple models for the processes of creation and motion of defects in metals.

The method followed by Fumi was to calculate the energy required to create a vacancy, based on a formula of Friedel[2] which relates the phase shifts of the conduction electrons at the Fermi level to the excess of defect charge of the impurity they screen in a given matrix. The energy change connected with the expansion of a monovalent metal, for which the number of atoms in the metal sphere equals the number of conduction electrons, is found to be

$$\Delta E_{el} = \tfrac{4}{15} E_F. \tag{20}$$

This value provides an upper limit for the experimental energy of creation of a vacancy.

REFERENCES

1. H. B. Huntington and F. Seitz, *Phys. Rev.* **61**, 315 (1943); *Phys. Rev.* **76**, 1728 (1949); H. B. Huntington, *Phys. Rev.* **91**, 1092 (1953).
2. J. Friedel, *Phil. Mag.* **43**, 153 (1952), Equation (2).

2.9. Neutron Diffraction Studies of Transition Elements and Their Alloys

(C. G. Shull)

Shull reported on the neutron scattering characteristics of transition elements and certain of their alloy systems. Since the neutron possesses a small but definite magnetic moment (only about 10^{-3} that of an electron), it is capable of interacting with any atomic magnetic moments present within a scattering sample; and this interaction shows itself as a scattering contribution to the diffraction pattern. The information obtained from the scattering process helps in understanding the electronic structure present in these elements and alloys, and thus offers some guidance in the formulation of theories of metallic structure in general.

Shell reviewed three kinds of studies: (1) various transition elements at normal or low temperatures; (2) the ferromagnetic iron and nickel at high temperatures up through their Curie transition and, (3) some typical alloys of the elements.

Neutron diffraction studies of W, Mo, Nb, V, Cr, Mn, Fe and Co, i.e. the transition elements, were made. W, Mo, Nb and V did not reveal any superstructure diffraction lines indicative of an antiferromagnetic lattice. The appearance of the diffuse scattering also ruled out the presence of paramagnetically-coupled magnetic moments of strength indicated by the electronic distribution in the free atoms. From the observed intensities it was possible to set an upper limit of a few tenths of a Bohr magneton as the maximum atomic magnetic moment seen in neutron scattering in either ordered or disordered orientation. The upper limit is very much smaller than the free atom electronic structure would predict, for instance $3\mu_B$ for vanadium or $4\mu_B$ for niobium.

Cr and Mn exhibit superstructure line intensity at reflection positions not permitted for nuclear scattering. Accordingly they are considered to be anti-ferromagnetic.

As for Fe and Co, earlier studies on these ferromagnetic elements had shown the presence of magnetic scattering which for a simple ferromagnetic lattice is to be found superimposed on the normal nuclear reflections. A study of the magnetic form factor in angular dependence allows one to determine the magnetic scattering component and the ferromagnetic moment, and the results are in agreement with those obtained by magnetization saturation for Fe and Co.

Shull reported on studies of the ferromagnetic elements in the high temperature region above the Curie temperature. The paramagnetic susceptibility data for Fe and Ni have long been considered anomalous with respect to the ferromagnetic magnetization so that it is of interest to compare the results of paramagnetic neutron scattering with that obtained below the Curie temperature. There also exists a fundamental uncertainty as to whether paramagnetic scattering is observable in such transition metals. It could be suspected that electron spin relaxation effects might influence or eliminate the paramagnetic scattering. For these reasons, nickel and iron were studied at temperatures up to about 1000 °C, well above their Curie temperatures of 348 °C and 770 °C respectively. The diffraction patterns for iron at room temperature show the body-centered-cubic reflections; and at 971 °C the face-centered-cubic reflections of γ-iron are obtained. No magnetic superstructure lines are found in the γ-iron pattern showing that the face-centered-cubic structure is definitely not anti-ferromagnetic. In the high temperature patterns, there appears the diffuse scattering which should contain ferromagnetic disorder scattering and paramagnetic scattering. A paramagnetic scattering is indeed seen to exist for iron at high temperatures and the absolute intensity is thought to be consistent with the ferromagnetic moment. Studies on nickel have also shown the presence of paramagnetic scattering above T_c, but with an intensity somewhat less than that calculated with the ferromagnetic moment.

Neutron diffraction data for a series of ferromagnetic binary alloys of transition elements provide information about the individual atomic moments existing in the alloy mixture. In studies of the alloys of Fe-Cr series, Ni-Fe series, Co-Cr series, and ordered Ni_3Mn, it has been demonstrated that there exist within the alloys differences in the magnetic moments among the various atomic species; moreover, these atomic magnetic moments differ from the values known to exist in the pure element form. In all cases the magnetic scattering which is observed for the alloy systems, whether of disordered or of superstructure form, exhibits a magnetic form factor dependence on the scattering angle, which is satisfactorily matched with that characteristic of pure iron.

2.10. Anti-ferromagnetism and Meta-magnetism

(L. Néel)

Néel sought to examine the extent to which it would be possible to understand the properties of meta-magnetic substances by considering them as anti-ferromagnetics. Meta-magnetic substances, so named by J. Becquerel[1] in 1939, such as $CoCl_2$, $NiCl_2$, $FeCl_2$, obey a Curie–Weiss law in the paramagnetic region, with a positive paramagnetic Curie point. Also, much below this point, these substances do not become ferromagnetic as would be expected, and in the high field of 30000 Oe no saturation is observed.

A possible explanation of such effects by means of an arrangement of anti-parallel atomic moments was proposed by Landau[2]. Starr[3] also proposed a theory based on an anti-parallel grouping. However, the simplest way of treating such problems seems to be by means of the molecular mean field approximation. Néel made use of the usual division of the lattice into two sub-lattices, and treated the interaction within and between them with the help of the mean field approximation.

Néel studied the susceptibility in a direction perpendicular to the anti-ferromagnetic direction. He considered the case of a uniaxial substance and showed that for a field H, perpendicular to this axis, the resulting magnetization J is given by,

$$J = \frac{H}{n + 2K/M^2},\tag{21}$$

where K is a constant arising from the definition of the magneto-crystalline energy, n is another constant, and $M/2$ is the spontaneous magnetization of each sub-lattice. If K is smaller than nM^2, and can be neglected, then J becomes a constant independent of the field, H, and the temperature, T. This result is confirmed experimentally.

In the case of the susceptibility, parallel to the anti-ferromagnetic direction, MH is sufficiently small and J is dependent both on the temperature and the field, having a relation of the type

$$J = \frac{H}{n - \dfrac{n + n'}{2} \cdot \dfrac{M}{T} \dfrac{dT}{dM}},\tag{22}$$

where n' is a constant and the derivative, dT/dM, is taken along the thermal variation curve of the spontaneous magnetization M.

As the strength of the field is increased, the phenomena become more complicated because of the changes in orientation of the spontaneous magnetization of the two sub-systems. In this case, at low temperatures, the system has three different equilibrium configurations depending upon the relative values of H and M.

Néel then treated the problem of paramagnetic hysteresis, showing that the existence of three regions of equilibrium leads to a possible hysteresis cycle. However, the hysteresis should be small since the ratio of the magneto-crystalline energy to the exchange energy is small. In the case of ferromagnetics, however, the coercive field should theoretically be much larger than the one observed, because of the existence of a very small region which remains oriented in the direction opposite to the field.

Néel discussed the variation with the field of the susceptibility of antiferromagnetic poly-crystalline substances. As the temperature approaches the transition temperature, T_n, the susceptibility increases and tends to $1/n$. Nagamiya and Yoshida[4] had calculated the corrections for the temperature dependence of the susceptibility. Néel noted that the symmetries of the system lead to other possible antiferromagnetic directions. Comparing with experiments the values obtained for isotropic systems and systems having a cubic symmetry, he concluded that there must exist a mechanism which prevents the badly oriented domains to orient themselves in the correct direction. Néel discussed the cases of polycrystalline systems which are situated between parallel and perpendicular fields respectively, relative to the axis of the crystal, and applied his results to meta-magnetic systems.

Néel then discussed the variation of the interactions as a function of the atomic number. But for $CuCl_2$, all the chlorides of the type MCl_2 have a structure formed of parallel layers. The M ions form a hexagonal lattice in two dimensions, and these planes are separated by two layers of chloride atoms. The antiferromagnetic structure is thus formed very probably by successive layers of atoms magnetized in reverse directions. With the help of a table for the coefficients n, n' and K, Néel showed that the molecular field of the coefficient n is transmitted through the chloride atoms. It is thus a kind of 'superexchange'. This magnetization decreases when one passes along the elements Ni, Co, Fe, Mn, and Cr, in this order.

The main point is therefore the existence of important orientation actions between magnetic atoms exerted through two intermediate atoms. Néel gave examples of substances for which the antiferromagnetic character can be explained with his theory. He noted that the classical theory of antiferromagnetism gives a satisfactory account of most of the magnetic properties of substances which had hitherto been considered in the meta-magnetic category.

Néel considered the thickness and energy of the wall in ferromagnetic substances, including the magnetic moments of the wall, and obtained his results by comparison and analogy with ferromagnetic domains. The main difference arises from the small-ness of the domains, smallness which is due to the lack of the unifying effect of the field. The wall energies are much smaller than for the ferromagnetics, but still lead to the existence of a small permanent magnetic moment.

A possible theoretical mechanism for the understanding of the behaviour of the ferrites of rare earths is also provided by the mean field approximation. Néel showed how the ferrites could be treated in analogy with the case of Fe_2O_3. He assumed the existence of two sub-lattices coupled by a negative molecular field in ferrites. The first sublattice is formed of Fe^{+++} ions and has properties similar to Fe_2O_3; the second is formed of M^{+++} ions, weakly coupled, having nearly pure paramagnetic character-istics. Assuming an interaction between these two lattices, which can be described by a molecular field, Néel obtained a theoretical explanation of the observed phenomena.

REFERENCES

1. J. Becquerel, *J. Physique*, **10**, 10 (1939).
2. L. Landau, *Sow. Phys.* **4**, 675 (1933).
3. C. Starr, *Phys. Rev.* **58**, 984 (1940).
4. T. Nagamiya, *Prog. Theor. Phys.* **6**, 342 (1951);
   K. Yoshida, *Prog. Theor. Phys.* **6**, 691 (1951).

2.11. SUPERCONDUCTIVITY

(H. Fröhlich)

The modern theory of superconductivity is due to Bardeen, Cooper, and Schrieffer. Their collaborative effort began in 1955 and was published in 1957.[1] During the previous decade Fröhlich had contributed important insights to the understanding of superconductivity. His brief report on this subject at the tenth Solvay Conference in 1954 thus came on the eve of a major conceptual development and is of historical interest.

In order to explain the properties of the superconductive state, the model of a metal in which the electrons are treated as free has to be refined by introduction of an inter-action between the conduction electrons. It used to be generally accepted that this interaction does not involve the ionic lattice, but Fröhlich[2] showed in 1950 that the interaction of the electrons with the lattice vibrations, first introduced by Bloch[3] in 1928 in order to account for ordinary conductivity, necessarily leads to an interaction between the electrons carried by the field of lattice displacements. Thus without intro-ducing a new hypothesis it had been shown that an interaction between electrons exists, which hitherto had been unjustifiably neglected. It could be shown that the dynamic part of this interaction may strongly influence the properties of electrons near the Fermi surface, provided a certain interaction parameter, F, satisfies $F > 1$. The value of this F could be estimated from the magnitude of ordinary conductivity at high temperatures, and it was found that for most superconductors F is larger than for normal metals, a property closely connected with the empirical fact that at high temperatures most superconductors are poor normal conductors. The conjecture was, therefore, advanced that the dynamic part of the interaction of electrons through the field of lattice displacements is responsible for superconductivity. This conjecture was soon afterwards strongly supported by the discovery of the isotope effect.

Attempts to give a satisfactory treatment of this interaction in the case, $F > 1$, led to serious mathematical difficulties[4]. The case, $F < 1$, was treated in a satisfactory way by Buckingham and Schafroth[5]; it leads essentially to an increase in electronic level density at low temperatures as compared with that at high temperatures. As a result the electronic specific heat at low temperatures should be of the form, $C_v = \gamma T$, with a factor, γ, which, however, would be obtained by using numbers of free electrons deduced from high temperature measurements (e.g. from Hall effect).

In view of the above-mentioned difficulties it was impossible to decide whether the interaction mentioned actually yields the main properties of superconductors. It might be argued that although the isotope effect proves the importance of the lattice displace-ments, the whole model normally used in the electron theory of metals is inadequate. However, this theory is based on a hypothesis rather than on a derivation from first principles.

It seemed important, therefore, to find a limiting case of the usual model which permits a simple treatment of the electron-lattice interaction and still shows features of a superconductor. This was carried out by Fröhlich in 1954[6].

The one-dimensional case has many unrealistic features. Fröhlich emphasized, however, that it does not involve any new assumptions but represents a limiting case of the ordinary model of free electrons, interacting with the lattice displacements. The use of a self-consistent method which converges in this limit suppresses the isotope effect because such a method does not take account of dynamical properties. Fröhlich thought that it was from an analysis of this feature that further progress would be made.

Perturbation theory gives a correct order of magnitude of the energy of the ground state, but leads to very bad wave functions. The self-consistent field (Hartree) method suppresses dynamic features of the interaction, but yields some detailed properties in a satisfactory way. This method, however, cannot be applied in a straightforward way to three dimensions. It is desirable to find a new method which would form a link between the two methods by introducing dynamical features into the self-consistent field method. A similar problem arises in the much simpler case of a single electron in an ionic crystal, which Fröhlich treated by applying the two methods.[7]

Finally, Fröhlich suggested certain experiments which he thought would be helpful in connection with these questions. As mentioned earlier, the result of Buckingham and Schafroth showed that for normal metals the electronic specific heat is of the form, $C_v = \gamma T$ (where the factor γ is larger than γ_0), a value which would be obtained by using the number of free electrons measured at high temperature (e.g., from the Hall coefficient). Furthermore, γ increases with the increasing interaction constant F, until F reaches a critical value, $F = 1$, when the metal becomes a superconductor and a qualitative change in C_v takes place. Fröhlich therefore thought that it would be of interest to investigate the electronic specific heat together with the Hall coefficient at high temperatures for alloys whose composition can be changed gradually, leading from normal to superconductive materials. It should be expected then that F should increase as the composition is approached in which the material becomes superconductive. On the other hand γ/γ_0 is a direct measure for F. Alloys with the required composition have been found by B. Matthias.[8]

REFERENCES

1. J. Bardeen, L. N. Cooper, and J. R. Schrieffer, *Phys. Rev.* **108**, 1175 (1957).
2. H. Fröhlich, *Phys. Rev.* **79**, 845 (1950).
3. F. Bloch, *Z. Phys.* **52**, 555 (1928).
4. J. Bardeen, *Rev. Mod. Phys.* **23**, 261 (1951);
   H. Fröhlich, *Physica* **19**, 755 (1953).
5. M. Buckingham and R. Schafroth, *Proc. Phys. Soc.* A **67**, 828 (1954).
6. H. Fröhlich, *Proc. Roy. Soc.* A **223**, 296 (1954).
7. H. Fröhlich, *Advances in Chemical Physics* (1954).
8. B. Matthias, *Phys. Rev.* **92**, 874 (1953).

2.12. THE EMPIRICAL RELATION BETWEEN SUPERCONDUCTIVITY AND THE NUMBER OF VALENCE ELECTRONS PER ATOM

(B. T. Matthias)

Matthias reported on the empirical findings that superconductors with highest transition temperatures are those which have an average valence electron per atom ratio near 5 or 7. Furthermore, the superconducting transition temperatures in the horizontal rows of the periodic system seem to be more or less symmetrical with respect to the sixth column. Matthias discussed how the qualitative dependence of the transition temperature depends on the valence electron per atom ratio. He gave examples of how this curve was traced experimentally: (a) Superconducting compounds with the beta-W structure; (b) The rhodium-selenium and rhodium-tellurium systems; (c) The molybdenum and tungsten alloys; (d) Solutions of rhodium in zirconium.

Whereas all alloys indicate a close connection between superconductivity and the number of valence electrons per atom, it appears that the Zr-Rh system is the most interesting. Here the electron density alone seems to determine the transition temperature, and one could seek to obtain a relation between the electronic term of the specific heat and the transition temperature by a continuous variation of the electron density, as suggested by Fröhlich.

2.13. HALL EFFECT IN FERROMAGNETICS

(J. Smit)

The Hall effect in ferromagnetics is not directly proportional to the applied magnetic field, but an extra term connected with the spontaneous magnetization has to be added. If the sample is saturated in the Z-direction, then

$$\frac{E_y}{i_x} = R_0 B + R_1 M_s, \tag{23}$$

where R_0 is the normal Hall coefficient, R_1 the extraordinary Hall coefficient, and B is the mean field inside the specimen. B is used instead of H because the magnetic dipoles are not dumb-bell dipoles, but have to be regarded as being caused by circular currents. If we say that the mean field B acts on a conduction electron, then we have to allow for the penetration of this electron into the interior of the circular current of the dipole. This has peculiar consequences if one recognizes that these magnetic moments are mainly those of the spins of the $3d$-electrons. The $4s$ conduction electron has then to penetrate the $3d$-electrons. These difficulties are formally explained in the Dirac theory of the electron, and one can sail around them by bearing in mind that the electrons are smeared out, and with them their spin currents, which then have dimensions of the electronic orbits.

The normal Hall coefficient R_0 has the same order of magnitude as the one for non-ferromagnetic metals, and permits the application of free electron theory for finding the number of free electrons. It is found to be of the order of one electron per atom, or somewhat less. Iron shows a positive Hall coefficient (electron holes).

The most striking property of the spontaneous Hall effect is its strong temperature dependence, as Smith found for nickel in 1910, and Jan and Gijsman[1] at low temperatures in 1952. In these samples R_1/R_0 was about 120 at room temperature, but at the lowest temperatures it decreased to 20, while it increased to about 1000 near the Curie temperature. The change in R_1 is especially remarkable below the room temperature, since the magnetization does not vary appreciably. The only property of the material whose magnitude increases in a comparable way is its electrical resistance. If a correlation does exist, one should expect a much smaller variation of R_1 for an alloy below room temperature than for a pure metal.

These considerations were the starting point of the measurements by Volger and Smit[2] on the Hall effect, together with the resistivity of several Ni alloys and of some Ni specimens of different purity, at temperatures of liquid hydrogen and nitrogen and at room temperature. It was found that the purest Ni sample has no extraordinary Hall effect R_1 near $T=0$ °K, whereas R_1 for alloys was still finite there. In general R_1 varied approximately in the same manner as the resistivity ϱ; in some cases a power law, $R_1 \sim \varrho^n$ with $n \approx 1.5$, is satisfied.

R_1 and R_0 appear to be extremely sensitive to the chemical composition; several Ni samples with different impurity content indicated that. Ni has negative R_0 and R_1. Addition of Co or Fe lowers $|R_1|$. For 30 percent Co, or 16 percent Fe, R_1 is positive, just as for pure iron. R_1 is extremely large for Ni alloys with non-magnetic metals, like Si, Al, or Sn. At the same time these additions increase the resistivity, so that it may be assumed that R_1 is closely related to the resistivity of the material.

REFERENCES

1. J. P. Jan and H. M. Gijsman, *Physica*, **5**, 277 (1952).
2. J. Smit and J. Volger, *Phys. Rev.* **92**, 1576 (1953).

2.14. THE ELECTRON THEORY OF TRANSITION METALS

(A. Seeger)

Seeger reported on the electronic bands and crystal structure of transition metals. He proposed a model which sub-divides the entire d-band according to crystal symmetry. In this way, one is led to the view that certain sub-groups of d-electrons give rise to predominantly van der Waals binding, and others to homopolar binding. Thus one can draw conclusions about the occurrence of three simple metallic structures (most dense cubic and hexagonal packings, and cubic space-centred lattice) from a certain configuration of the outer electrons, which are essentially confirmed experimentally. Also the conclusions concerning the behaviour of the electronic specific heat and magnetic susceptibility in the antiferromagnetic transition elements agree with experiments. Seeger discussed the significance of the results of certain new experiments (neutron interferences) and calculations (plasma method and configuration exchange interaction) for the electronic theory of ferromagnetism, especially in relation to the properties of the ferromagnetic transition metals. He discussed such a calculation (the four electron problem with configuration exchange interaction) in detail.

15

Astrophysics, Gravitation, and the Structure of the Universe*

1. INTRODUCTION

The science of astronomy was the precursor of physics. Astronomy developed before any physical problems were considered, and it provided remarkably accurate numbers before any other science even described things qualitatively. It was again astronomy which introduced the new age of science in the sixteenth century. The laws of planetary motion displayed their great power in the new mechanics of Newton. Great progress was achieved in physics and chemistry through patient laboratory work, while astronomy became a study of the motions of celestial bodies, dealing with detailed explanations of orbits. With the improvement of optical instruments in the nineteenth century, dark (Fraunhofer) lines were discovered in the spectra of luminous stars. Their relationship to phenomena occurring in flames in a laboratory was established by Kirchhoff and Bunsen, and the science of stars became part of the new and modern field of the study of radiation; astronomy became a part of physics.

The revolutionary ideas of physics of the twentieth century, the quantum and relativity theories, became extremely fruitful in astrophysics. Quantum theory opened the way for the understanding, first, of the atmospheres of stars and then of the nuclear energy processes in their interiors. Modern physics also brought about the understanding of the variety of classes of stars and their history from red giants to white dwarfs, the neutron stars or the black holes. The latter concept goes back to K. Schwarzschild (1916) who studied Einstein's equations of general relativity and looked for their spherically symmetric solution. The theories of special and general relativity not only provided an understanding of specific phenomena which show up in fast moving objects, but also led to the discussion of the entire history of the universe. The structure of space and time, and the physics of gravitation, became directly entangled, and the macroscopic and microscopic phenomena could not be separated any more.

Technological progress during the Second World War opened new horizons, since more penetrating radiation of micrometer wavelength became observable. The realm of radio stars and very distant radio galaxies, as well as quasars which show extremely high red shift, was now opened up for investigation. Rockets and satellites made it

* *La Structure et L'Évolution de l'Univers*, Rapports et Discussions du Onzième Conseil de Physique tenu à l'Université libre de Bruxelles du 9 au 13 Juin 1958, R. Stoops, Brussels, 1958.

The Structure and Evolution of Galaxies, Proceedings of the Thirteenth Conference on Physics at the University of Brussels, September 1964, Interscience Publishers, New York, 1965.

Astrophysics and Gravitation, Proceedings of the Sixteenth Solvay Conference on Physics at the University of Brussels, 24 to 28 September 1973, Editions de l'Université de Bruxelles, Brussels, 1974.

ELEVENTH SOLVAY CONFERENCE 1958

O.S. KLEIN

W.H. McCREA   W.W. MORGAN   B.V. KUKASKIN

J.N. OORT   G. LEMAITRE   V.A. AMBARZUMIAN

F. HOYLE   H.C. van de HULST   A.R. SANDAGE   H. ZANSTRA   L. LEDOUX

M. FIERZ   W. BAADE   J.A. WHEELER   H. BONDI   T. GOLD   L. ROSENFELD   A.C.B. LOVELL   J. GÉNÉNIAU

C.J. GORTER   E. SCHATZMAN

W. PAULI   Sir W.L. BRAGG   J.R. OPPENHEIMER   C. MOLLER   H. SHAPLEY   O. HECKMANN possible to register X-rays coming from distant parts of the universe. A triumph of observation was achieved when one discovered the 3 °K background radiation. The new phenomena served to unite the physics of the earth and the heavens, and made astrophysics one of the most exciting frontiers of present day science. One learned that effects occur at scales hitherto unknown, such as gravitational collapse, when a large amount of inertia is enclosed in a comparatively small volume.

Astrophysical topics had occasionally appeared in the reports and discussions of Solvay Conferences, e.g. in E. Teller's contribution to the 1948 meeting on elementary particles. The Solvay Conference which was devoted, for the first time, entirely to astrophysics was the eleventh, the conference on 'The Structure and Evolution of the Universe', which took place in Brussels from 9 to 13 June, 1958. The thirteenth Solvay Conference in September, 1964, dealt with a similar but more restricted topic, 'The Structure and Evolution of Galaxies', while considerable progress in the understanding of 'Astrophysics and Gravitation' was reported at the sixteenth Solvay Conference in September, 1973.

2. THE STRUCTURE AND EVOLUTION OF THE UNIVERSE

Einstein's theory of general relativity (1915) not only predicted new gravitational effects, such as the bending of light, but also gave new impetus to the efforts to describe the structure of the universe. If gravitation due to matter was responsible for causing the geometry of space, then the new laws of gravitation should impose restrictions on the nature of the cosmos. Einstein himself entered the field of cosmology after 1916 and constructed the first static, spherically symmetric models. Later Friedmann produced a solution of the Einstein–Hilbert equations of gravitation which corresponded to an expanding universe. This fitted nicely into the systematic observations of Hubble, leading to the conclusion that cosmic systems seem to run away from each other with velocities which are greater the farther the systems are separated.

This situation was well-established already in the 1930s, yet the time was apparently not ripe enough to devote a Solvay Conference to astrophysics and the consequences of general relativity. Nuclear physics had made great progress during those years, and it was suggested that thermonuclear processes provide the energy of the stars. Teller discussed some of the results at the 1948 Solvay Conference on elementary particles.

General relativity, which for long had been the field of research of a few devoted theoreticians such as Einstein, Schrödinger and some others, gained a new interest in the 1950s. The old papers were studied again, and new or improved solutions of Einstein's equations were proposed. Altogether this entire field had acquired a new enthusiasm when the eleventh Solvay Conference on 'The Structure and Evolution of the Universe', dealing with the union of relativity and astronomy, took place in 1958. Some of the most distinguished astronomers and experts in general relativity were invited to present reports.

The Scientific Committee of the Conference consisted of: Sir Lawrence Bragg, President, C. J. Gorter, C. Møller, J. R. Oppenheimer, W. Pauli, F. Perrin, and

F. H. van den Dungen (Brussels), Secretary. N. Mott, another member of the Committee, could not attend. Among the invited speakers were: V. A. Ambartsumian (U.S.S.R.), W. Baade (Mt. Wilson and Palomar), F. Hoyle (Cambridge), O. Klein (Stockholm), G. Lemaitre (Louvain), A. C. B. Lovell (Jodrell Bank), J. H. Oort (Leyden), A. T. Sandage (Mt. Wilson and Palomar), and H. C. Van de Hulst (Leyden). T. Gold (Cambridge, Mass.), W. Morgan (Yerkes), E. Schücking and O. Heckmann (Hamburg), and J. A. Wheeler (Princeton, N.J.) also presented papers.

2.1. GENERAL STATEMENTS ABOUT COSMOLOGICAL THEORY

General relativity theory has serious implications for the structure of the universe. Wheeler presented a note on this subject, and summarized five conceivable observational tests of Einstein's general relativity. This theory does not only mean the set of Einstein–Hilbert equations*, but according to Einstein [in the fifth edition of *The Meaning of Relativity* (Princeton, 1955)] the additional statements: (i) the universe is closed, and (ii) no 'cosmological' term is to be added to the field equations. Hence it becomes possible to deduce the equations of motion of concentrations of mass-energy from field theory.

Assuming a spherical isotropic uniform universe, the kinematics can be shown to be completely determined by the maximum radius and by the equation of state of the medium that fills the space. As observations are extended to greater distances the curvature of space should show up. Likewise the total energy density as a function of time becomes completely determined. The limits set by Einstein's theory are far from being reached. The assumption of a uniform and isotropic curvature of space is by no means necessary, as hydrodynamic-like instabilities might occur in principle. Finally, Wheeler *et al.* considered the possibility of a gravitational collapse which might occur when too much matter is concentrated in a small volume. Although no definite answer as to what happens at the 'untamed frontier between elementary particle physics and general relativity' was available in 1958, Wheeler stressed the importance of this subject for the theory of the structure of the universe and it would become a field of major interest ten years later.

Heckmann and Schücking discussed world models which can be classified according to whether the shear and rotation of incoherent matter vanishes everywhere or not. In the first case only homogeneous and isotropic world models consistent with energy-momentum conservation arise. The knowledge of anisotropic world models is still very incomplete.

In his report on 'The Primeval Atom Hypothesis and the Problem of the Cluster of Galaxies', Lemaitre drew more detailed consequences from general relativity. Quantum theory, he said, must have a bearing on cosmology, and the uncertainty principle opens up essentially new possibilities for it. Quantum theory as well as thermodynamics demand evolution. The cosmos started from a 'primeval atom'.

---

* The field equations of gravitation in general relativity had been discovered independently by David Hilbert at the same time as Albert Einstein. For details, see J. Mehra, *Einstein, Hilbert, and the Theory of Gravitation*, D. Reidel, Dordrecht and Boston (1974).

Although these conclusions are far from being rigorous, the implications of a strict quantum logic or consequent quantum mechanical models (including the modern hadron bootstrap of Hagedorn and Omnes) on the theories of evolution of the universe seem to be very strong. In other words, quantum theory might be consistent only with 'big bang' models.

From the expansion of the primeval atom, Lemaitre tried to explain the phenomena of cosmic rays, formation of gas clouds and galaxies. Friedmann's equation with a cosmological term allows reasonable solutions of these problems. Lemaitre asserted that each evolutionary step has been incompletely fulfilled, and primitive matter still exists in cosmic rays; as much gaseous matter is spread between the galaxies as forms the galaxies, and the cosmological term is directly related to the Hubble constant.

In the discussion of Lemaitre's report, Wheeler contradicted this view since the cosmological term would mean something like a mass of the graviton; moreover, the cosmological term had been introduced by Einstein to obtain a universe of unchanging size, and it loses its justification when the universe is not constant in size. Oscar Klein argued against the cosmological constant from a different point of view. If one introduces a minimal length of 10^{-32} cm (derived from the gravitational constant and Planck's constant) and expresses the cosmological term in this unit, then one obtains a coefficient of the order of magnitude of 10^{-118}. Such a coefficient seems to be very strange in a fundamental theory, and one might as well put it equal to zero.

A less speculative point of view was taken up by Klein in his report on 'Some Considerations Regarding the Earlier Development of the Systems of Galaxies'. He studied the evolution of a hypothetical gas cloud in the framework of general relativity, from which many universes might follow.

F. Hoyle discussed his 'steady state theory' and its implications. This theory assumes that the present situation concerning matter density, etc., is kept unchanged even when the universe is expanding. In other words, new matter is constantly produced as the frontiers of space move out. This effect need not violate energy conservation, since as soon as matter is created energy is lost by its coupling to a gravitational potential which is negative. Steady state theory would also allow one to give a deeper meaning to ratios of the fine structure constant and gravitational constant, etc., because they wouldn't change as time goes on. The problems which seem to arise in the formation of irregularities and of magnetic fields can be overcome. On the other hand, the spatial density of galactic material and the energy spectrum of cosmic rays are well represented in this theory. Steady state theory predicts that galaxies are being formed all the time.

In a short note addressed to the Solvay Conference, Gold related the direction of time to the expansion of the universe. Outgoing radiation is then always absorbed by the expanding matter of constant density of steady state theory, and the Feynman–Wheeler considerations apply, according to which the total absorption of radiation determines its (unique) propagation. The physical laws still remain symmetric with respect to time, and the arrow of time comes in because of the large scale behaviour of the universe.

2.2. Experimental Data on the Universe

After the theoretical reports on what the structure of the universe should or may look like, two eminent astronomers presented observational data. First J. H. Oort talked about the 'Distribution of Galaxies and the Density in the Universe', then A. C. B. Lovell discussed 'Radio-Astronomical Observations which may give Information on the Structure of the Universe'.

The universe is not homogeneous, and matter is concentrated in galaxies to the extent of 10^7 to 10^{12} solar masses per galaxy. The distribution of mass in galaxies and the angular momentum may reflect some reminder of the original situation of the early universe. The shape and structure of double galaxies hint at the fact that the dimensions of the rotational elements of the primeval universe were of the size of galaxies, the density being of the order of 10^{-25} g cm^{-3}. Galaxies tend to form clusters, and the latter might even agglomerate in groups. It appears that galaxies and groups of clusters do not expand or increase their size in general. Radio observations more or less confirm this picture, although peculiar objects occur whose radio emission is particularly large and might be ascribed to collision phenomena.

The experimentalists had concluded that the situation in 1958 did not allow one to draw specific conclusions about the structure of the universe. In other words, many cosmological models could fit the data. On the other hand, the powerful means of radio astronomy had not yet been developed to be able to identify very distant stars, but it was hoped that this situation would improve very soon.

2.3. The Evolution of Galaxies and Stars

Whereas the evolution of the large aggregations of mass as well as the formation and history of stars are more specific problems of astrophysics, the distribution of matter in the universe is directly related to the theory of gravitation. The progress made in astrophysics was presented from various points of view at the 1958 Solvay Conference.

First, W. Baade discussed the observations on galaxies and stars, in particular the aspects connected with the formation of stars. V. A. Ambartsumian had analyzed the data and drawn certain conclusions concerning the evolution of galaxies. Since the study of multiple stars and stellar clusters sheds light on the problem of the origin and evolution of stars, the observation of multiple galaxies and groups and clusters of galaxies deserve special attention. In almost all cases, the members of a group are formed at the same time or have a close relation to each other. In addition, the investigation of radio galaxies shows traces of non-stability in multiple systems. Since the period of revolution of clusters of galaxies is of the order of their lifetimes, it is difficult to determine whether the systems disintegrate unless velocity dispersion is particularly large. Special examples of separation have been observed in radio astronomy (Perseus A and Cygnus A). The blue coloured ejections of some elliptical galaxies represent very young galaxies. New galaxies and spiral arms are formed at the expense of matter in the nuclei of galaxies; hence large masses of prestellar matter exist in nuclei. Ambartsumian noted that stars and gas are produced together from protostars.

A typical large late-type galaxy is our own system, which had been explored by means of radioastronomy to some extent during the years since 1945. H. C. van de Hulst summarized the results and discussed both the structure of the disc, in which neutral hydrogen gas is concentrated and gives rise to the 21 cm wavelength, and of the halo in which electrons seem to emit synchroton radiation in a magnetic field of 10^{-5} G.

W. W. Morgan presented a report on 'Some Spectroscopic Phenomena associated with the Stellar Population of Galaxies'. He distinguished between two types of galaxies: on the one hand, systems containing mainly violet spectra of stars of classes B or A and, on the other, systems whose large inner parts have violet spectra of late G and K (especially large ellipticals). However, one cannot decide as to which systems are actually the younger ones since, as Morgan pointed out, the spectral class of stars does not have anything to do with their age, but rather with the initial conditions under which they were formed.

In 1911 Hertzsprung had plotted the luminosity of stars in galactic clusters versus their colour index and found two lines in the diagram: the 'main sequence' and the 'giant branch'. A similar diagram containing the stars in the vicinity of the sun was given by Russell in 1913. According to the early theories of stellar evolution one assumed that stars started as giants and eventually ended on the main sequence. F. Hoyle reported on the new views about stellar evolution in his talk on the 'Origin of Elements in Stars'. Stars, he noted, are formed from the interstellar gas (or from proto-stars, as Ambartsumian had assumed) at all epochs within our galaxy. Stars also disintegrate to a certain extent, especially from giants. All elements, except hydrogen, are created within the stars by nuclear reactions especially in the hot centre. Energy is released in these processes at the same time, and there exists an adjusted balance. The higher the temperatures, the larger are, in general, the atomic numbers of the elements produced, and one can fairly well explain the abundance of various nuclei. With the help of nuclear reactions and their consequences the phenomena of supernova can also be explained; when at very high temperatures the nuclear reactions become endo-thermic, the hydrostatic balance within the star is finally lost.

2.4. GENERAL DISCUSSION

In the general discussion at the eleventh Solvay Conference, all the main points of theory and observation were considered. H. Bondi asked for crucial tests which could decide between the evolutionary and the steady state universe. It was argued that perhaps one finds a change with time in the ratio of various galaxies. But the classi-fication of galaxies becomes uncertain as the distance grows beyond 10^9 light years. On the other hand, Baade noted that all galaxies are of the same age; hence the evolu-tionary picture seems to be favoured. The distribution of galaxies in the universe also follows naturally from the expansion of the universe. The need was stressed for more information about radio 'stars', especially objects different from normal galaxies. For cosmological considerations, the determination of the distance independent of the red shift would be highly desirable. Research along these lines was carried out until the astrophysicists met again to review the situation at the 1964 Solvay Conference.

THIRTEENTH SOLVAY CONFERENCE 1964

P. SWINGS   L. SPITZER   F.D. KAHN   L. BIERMANN   K.H. PRENDERGAST   J.C. BOLTON   W.A. FOWLER   F. HOYLE   G. BURBIDGE   A. SANDAGE   M. SCHMIDT   R. MINKOWSKI

E. SPIEGEL   P. BOURGEOIS   H. ALFVÉN   B. ROSSI   J. VAN MIEGHEM   B. LINDBLAD   J.H. OORT   Sir BERNARD LOVELL   G. LEMAITRE   L. WOLTJER   E. SCHATZMANN   V. AMBARTSUMIAN   P. LEDOUX   E.E. SALPETER   K. CUYPERS

B. STRÖMGREN   I. PRIGOGINE   C.J. GORTER   E. AMALDI   M. BURBIDGE   R. OPPENHEIMER   W. HEISENBERG   C. MØLLER   F. PERRIN   J. GEHÉNIAU   R. COUTREZ

3. GALAXIES

The thirteenth Solvay Conference on astrophysics, dealing with the 'Structure and Evolution of Galaxies', was held in Brussels in September 1964. It was devoted to similar topics as the 1958 meeting. It was, however, less ambitious than the previous one, as it dealt with galaxies rather than the universe as a whole. Since the 1958 meeting, considerable progress had been achieved in the specific problems concerning the structure and evolution of stars, and a part of the 1964 conference was devoted to these. Moreover, in accordance with the hope expressed at the 1958 meeting, radio sources outside our own galaxy had been studied and could be discussed.

The Scientific Committee of the Conference consisted of: J. R. Oppenheimer, President, E. Amaldi, Sir Lawrence Bragg, C. J. Gorter, W. Heisenberg, C. Møller, N. F. Mott, F. Perrin, I. Tamm, S. Tomonaga, and J. Géhéniau, Secretary. Bragg, Mott, Tamm and Tomonaga were not able to attend.

Among the invited speakers and participants were: V. Ambartsumian (U.S.S.R.), J. G. Bolton (Sydney, Australia), G. and M. Burbidge (San Diego, California), F. Hoyle (Cambridge, England), R. Minkowski (Berkeley, California), J. H. Oort (Leyden), E. E. Salpeter (Ithaca, N.Y.), M. Schmidt (Pasadena, California), L. Spitzer (Princeton, N.J.), L. Woltjer (New York), H. Alfvén (Stockholm), L. Biermann (Munich), W. A. Fowler (Pasadena, California), F. D. Kahn (Manchester), Sir Bernard Lovell (Manchester), B. Rossi (Cambridge, Mass.), A. Sandage (Pasadena, California), E. Schatzmann (Paris), and B. Strömgren (Princeton, N.J.).

3.1. THE STRUCTURE AND EVOLUTION OF GALAXIES

The 1964 Conference opened with a report by V. Ambartsumian 'On the Nuclei of Galaxies and Their Activity'. In his report at the 1958 meeting, Ambartsumian had already discussed the important role that the nuclei play in forming new galaxies and in giving them their observed structure and shape. His hypothesis at that time, that very active radio galaxies do not necessarily result from collisions of galaxies, but arise from the giant explosions taking place in the nuclei and resulting in large clouds of relativistic electrons, contradicted the general opinion. New observational data now supported Ambartsumian's ideas, revealing an outflow of matter from the core of galaxies, sometimes as explosions, at others as jets. On a minor scale these phenomena show up in our galaxy, but they happen much more dramatically, for instance, in the galaxy M 82. One might therefore expect that most of the processes connected with the formation of new galaxies and their structure (spiral arms, subsystems) start from the nuclei. Galaxies can have small or extended nuclei which, in turn, are either quiet or active as in Seyfert galaxies. The Seyfert type nuclei contain gaseous matter in violent motion. Finally there exist quasi-stellar radio objects, some kind of super-massive bodies of non-stellar nature.

These facts suggest the following interpretation of the evolution of galaxies. The giant galaxies begin their life as elliptical systems in which the nuclei do not contain many stars. The more active are the nuclei, the brighter they look (galaxies of types Sa,

Sb, SBa, SBb). The galaxies Sc, SBc and the irregular ones (containing the Magellanic Clouds) seem to be the oldest systems. Nuclei of high luminosity are rarely encountered in galaxies of the Sc type, while no nuclei are to be seen in SBc's and irregulars. This point of view contradicts the usual opinion which assumes that objects of the Magellanic Cloud type are young. Ambartsumian expressed his conviction that the relativistic theory of gravitation might explain the phenomena occurring in the nuclei of galaxies.

The view that it is the nuclei that determine the structure of galaxies, and not galaxies the nuclei, was partly contradicted by F. Hoyle and others in the discussion of Ambartsumian's report, especially the assumption that the age of the galaxy M 82 is only about 10^8 years. No definite conclusion was reached, however. In any case, Ambartsumian's report reflected the great progress that had been achieved on the basis of the radio-astronomical observations of galaxies.

J. H. Oort's report on 'Some Topics Concerning the Structure and Evolution of Galaxies' depended heavily on the new observational techniques. He described the latest knowledge concerning our own galaxy. The Milky Way is composed of three ingredients: stars, interstellar gas and cosmic rays, the bulk of the mass being represented by the stars, whereas the gas represents 5 to 10 per cent of the total mass. Cosmic rays are contained by the magnetic fields. Most of the gas is concentrated in the disc, while the halo is made up by stellar objects of the so-called Population II. Heavier elements seem to appear in much smaller amounts than in Population I, to which essentially the stars of the disc belong.

The conventional view about the formation of galaxies assumes that they originate from irregularities in the expanding universe: from the primeval gas, first the stars of Population II emerge, then the gas contracts to build the rotating disc and stars of Population I, in which the elements heavier than helium must be more abundant. The time of the collapse might be of the order of 2×10^8 years, or about 2 percent of the age of the universe. The spiral structure of most galaxies is probably caused by the dynamics of gaseous matter; the nuclear disc of 800 parsec radius contains particularly dense atomic hydrogen in rapid rotation around the centre, forming expanding spiral arms whose origin might have been a gigantic explosion similar to that observed in some radio galaxies today.

In the discussion B. Lindblad, of the Stockholm Observatory, tried to reconcile the apparently contradictory phenomena that have to do with the existence of spiral arms. According to Lindblad's assumption, rings are formed in the plane of rotation and they break into spirals. These spirals attract and assimilate matter from their vicinity; spiral arms might exchange matter, and its momentum gives the 'receiving' arm an outward deviation from the circular motion. F. D. Kahn proposed to modify the first evolutionary step of the formation of a galaxy by assuming that, first, matter contracts and forms a massive object, which evolves rapidly and explodes, driving out the remaining material of the proto-galaxy and compressing it in a shell.

The discussion of the structure and evolution of galaxies was concluded by L. Woltjer's report on 'Galactic Magnetic Fields'. Non-thermal radio emission, interstellar polarization and Faraday effect in radio sources provide direct evidence of galactic magnetic fields of the order of 10^{-5} G. The most natural assumption is that the magnetic flux connected with it is primeval, and was frozen in when pre-galactic matter condensed to form a galaxy. (The magnetic field of the intergalactic space is lower by a few orders of magnitude.) In some quasi-stellar objects the magnetic field even seems to be much larger (up to 10^{-3} G). In our galaxy the magnetic field of the halo serves to confine the cosmic radiation of the highest energy. Magnetic fields might help to explain the structure of galaxies, but Woltjer did not propose any definite model.

In the discussion of Woltjer's report, H. Alfvén mentioned that the magnetic fields of planets and stars should be caused by hydromagnetic processes in their interior. The same mechanism might also create the magnetic field in a galaxy, hence no primeval field need be introduced.

3.2. The Structure and Evolution of Stars

Relativity and quantum theory, as well as atomic and nuclear physics, made it possible to attempt a study of the structure and evolution of stars. First the stars had been considered as spheres formed by layers of gas [Emden]. One applied the conditions of hydrostatic equilibrium between the pressure acting towards the periphery and gravitation. Since energy is created in stars and radiation is transported, one has to include it in the stability considerations. The energy transport to the surface of the star occurs through radiation and convection. With these ingredients an equation of state for a star can be derived. Nuclear processes were then considered as the sources of energy in stars, a typical example of which was discovered in the carbon cycle [Bethe and von Weizsäcker]. Knowledge about the structure of, and the processes in, stars was greatly improved in the years after 1945. The knowledge of nuclear reactions and the possibility of using large computers helped in this progress, and a fair picture of normal stars had emerged by 1957. The development after the 1958 Solvay Conference was especially rapid. Objects like supernovae were studied and important new features were discovered. At the 1964 conference, therefore, a considerably more detailed picture of stellar objects could be presented.

The first report in this field was presented by L. Spitzer who discussed 'Physical Processes in Star Formation'. He noted that young stars have been observed to be formed within the gas clouds in galactic spiral arms, and theoretical studies have clarified the types of processes that can occur. The first step in star formation consists in the formation of interstellar clouds by shock waves which might be caused by expanding H II regions (of ionized hydrogen) or supernovae shells. As a second step the interstellar material contracts under its gravitational self-attraction, which becomes effective once the density of the gas reaches a critical value, which is about 10^6 times the average density in the present universe. Such conditions are observed to exist in real clouds. In a third step the condensed gas might fragment into smaller and smaller masses [Hoyle]; these fragments contract further, losing energy, angular momentum and magnetic flux.[1] Spitzer concluded that there was no overwhelming theoretical reason why stars of the early type should not form from the observed clouds of interstellar matter. Ambartsumian stressed the point, however, that perhaps the processes mentioned above take too much time, and certain observations indicate the existence of a less conservative procedure which he had outlined in his report.

E. E. Salpeter developed the account a little beyond where Spitzer had left it. The important work of Hayashi (1962) showed that the gravitational contraction of stars takes place in times much shorter than the age of the universe (10^4 to 10^9 years).[2] Given certain initial conditions, e.g. the mass of the star, one can compute its history till it reaches the main sequence. For instance, a star with mass below 0.08 $M_\odot$ will never heat up to burn hydrogen and cool off towards an invisible black dwarf. The larger the initial mass, the faster the evolution proceeds. In the later stages the stars leave the main sequence branch when the hydrogen in the core is burnt out. The temperature goes up and helium burning, and later carbon burning, sets in. One may use this knowledge about the evolution of stars to construct models which explain the statistical data observed today in the Milky Way [M. Schmidt, 1959].[3] And, finally, one may extend the study to external galaxies.

W. A. Fowler raised some doubts whether the 'Hayashi effect', i.e. the fact that evolution towards the main sequence is mainly due to convection, and takes place in a period much shorter than the age of the universe, is not spoilt by rotation and magnetic fields. However, L. Biermann presented the results of calculations about the later evolution which predicted the position of variable stars reasonably well.

The last two reports in the session on the evolution of stars dealt with a particularly spectacular type of stars, the so-called supernovae. These objects show in their (observed) history a sudden explosive increase of luminosity after which they essentially disappear. The energy radiated is of the order of 10^{49} ergs, and a considerable part of the star's mass is blown away, as R. Minkowski mentioned in his report on the available data concerning the supernovae. F. Hoyle sought to connect these observations with the understanding of stellar evolution. Since more than 10^{50} ergs may be contained in the expanding envelopes, blown off the original star, it is generally agreed that supernovae should be related to more advanced phases of stellar evolution. According to the original state of the star, i.e. whether the explosion occurs in non-degenerate or degenerate material, the two observed types of supernovae may be explained. The energy set free in the process is obtainable from burning one solar mass of hydrogen or helium, but it may be that the gravitational energy is more important. Hence two types of mechanisms seem to be possible in the phenomena of supernovae: a catastrophic triggering of nuclear fuels in comparatively extended configurations, or gravitational effects in highly compact configurations of the order of 10^6 cm. The former mechanism does not occur in hydrogen burning, but the process

$$C^{13} + \alpha \rightarrow O^{16} + n, \tag{1}$$

might become explosive if the core is degenerate. Another possibility of instability might be found when endo-ergic reactions cause implosions, such as O^{16} burning catastrophically at 2.5×10^9 °K. [Fowler and Hoyle, 1964].[4] A nuclear explosion in the envelope of the star is then triggered, leading to a Type II supernova in which several solar masses expand rapidly outwards. The remnants of supernovae could be black holes or neutron stars. The latter possibility has been confirmed by observations of X-rays from the Crab Nebula.

REFERENCES

1.   F. Hoyle, *Astrophys. J.* **118**, 513 (1953).
2.   C. Hayashi *et al.*, *Progr. Theor. Phys. Suppl.* 22 (1962); *Progr. Theor. Phys.* **30**, 460 (1963).
3.   M. Schmidt, *Appl. Phys.* **129**, 243 (1959).
4.   W. A. Fowler and F. Hoyle, *J. Appl. Phys. Suppl.* **9**, 201 (1964).

3.3. EXTRAGALACTIC RADIO SOURCES

The last part of the 1964 Solvay Conference was devoted to 'peculiar' radio sources. It had been anticipated in the general discussion at the 1958 conference that one expected to learn much about cosmology from a detailed study of the structure and nature of these sources. J. G. Bolton presented a survey of the findings about them.

The first discrete source was discovered in 1948 when the Crab Nebula was identified as one. In 1951 the most intense extragalactic source, Cygnus-A, was recognized as a distant galaxy, and later on galaxies with red shifts of the order of 5 were discovered through radio observation. Finally, around 1960, the first quasi-stellar system, a distant galaxy of unprecedented luminosity was discovered. The progress achieved was mainly due to the tremendous improvement in resolving the positions of radio astronomical objects to a few tenths of an arc.

The uniform distribution of radio sources suggests that most of them are extra-galactic, a substantial fraction having distances over 10^9 parsec. Only fewer than 100 sources could be resolved in some greater detail, and frequently they showed a double structure, thus hinting at the existence of two close or interacting galaxies. The radio radiation of most of the sources shows some degree of polarization which should occur when the synchrotron mechanism is responsible. The most spectacular objects are the so-called radio galaxies, emitting up to 10^6 times the (long wavelength) radiation of normal galaxies, being the quasi-stellar systems which are the brightest known objects in the universe.

In the discussion of Bolton's contribution, R. Minkowski pointed out the importance of far distant radio sources for the testing of cosmological models, provided the time variation of the strength of the sources is known. Thus for an evolutionary universe one finds that in the early stages bright radio sources occurred more frequently. B. Rossi reported on the observations and nature of X-ray sources. Until 1964, essentially four localized sources had been discovered, and the main assumption was that the X-rays are generated by fast electrons (in Bremsstrahlung, inverse Compton effect, and synchrotron radiation).

An important step in the identification of extragalactic and other special discrete radio sources consists in the search for optical evidence of these objects. M. Schmidt of Pasadena discussed the major successes in his report on 'Spectroscopic Observations of Extragalactic Radio Sources'. The first source to be identified as a (faint) optical galaxy was Cygnus-A in 1954 [Baade and Minkowski].[1] Since then many radio objects could be related, in most cases, to weak optical phenomena. A new type of radio object was first noticed in 1963 when it was realized that some of the star-like objects that had been identified with radio sources were very distant, compact, super-luminous extragalactic objects. In 1964, observations on nine quasi-stellar sources were available whose optical lines revealed red shifts that were 5 to 10 times larger than the red shifts found in radio galaxies of equal apparent brightness. The assumption that these objects are far outside our galaxy can only be avoided if the red shifts are generated by gravitation. But even then they will not be in our galaxy.

Greenstein and Schmidt had discussed the different interpretations of the red shifts in detail and concluded that the gravitational red shift hypothesis most likely does not apply to the quasi-stellar objects.[2] The surviving hypothesis, i.e. that the red shifts are cosmological, is adopted, leading immediately to the distances of four quasi-stellars with known red shifts. The corresponding luminosities are 10^{45} to 10^{46} ergs/sec, up to two powers of ten larger than those of other galaxies. The study of the spectra leads to the following conclusions: the quasi-stellars are composed of a small body (less than one light-year), producing most of the (variable) optical continuum; they are surrounded by a tenuous, filamentary gas (about 10 to 30 parsecs radius), producing the emission lines; which, in turn, is surrounded by a volume of about 500 parsecs radius producing the radio emission.

The quasi-stellar objects, or quasars as they are called now, represented perhaps the most exciting discovery presented at the 1964 Solvay Conference. The report by G. B. Burbidge and E. M. Burbidge on 'Theories of the Origin of Radio Sources' attempted to summarize the understanding that had been achieved on the mechanisms creating the long waves. Since many radio sources within the galaxy and all external sources are non-thermal, one assumes the radiation to be mainly synchrotron emission. The simplest explanation of the phenomena may be obtained when one assumes that the violent releases of radio energy take place only over a short period.

The Burbidges listed four categories of energy production: (1) The energy is released by the interaction of a galaxy with material that was previously unconnected with it; (2) the energy is taken from the internal energy of a galaxy (gravitational potential energy, rotational energy, turbulent energy, and magnetic field energy released by some catastrophic process); (3) the energy is released in the evolution of stars (gravitational energy, nuclear energy, rest mass energy); (4) gravitational or nuclear energy is released in the collapse of a massive object.

Although the first proposal seemed to be most convincing in the beginning [Baade and Minkowski, 1954], it can be excluded in all cases that collision between galaxies generates the energy. The second hypothesis has obtained new interest in the explanation of quasars. G. Field had proposed that they might be spherical galaxies in the process of formation. Since supernovae are to be explained by release of nuclear energy and lead to synchrotron radiation (Crab Nebula), a similar mechanism seems to be possibly at work in quasars. However, explosions of 10^5 to 10^{11} solar masses were needed to describe the emission. For instance, a chain reaction of the supernovae was proposed.

Hoyle and Fowler had discussed hypothesis (4): a massive object might collapse to a radius close to its Schwarzschild limit.[3] The difficulty then is to obtain the outgoing energy. Hoyle and J. V. Narlikar had assumed that a hypothetical C-field of negative energy would stop the collapse, and the object would start to oscillate near its Schwarzschild radius. On the other hand, Hoyle and Fowler assumed two different phenomena to occur in quasars: first, a collapse, which occurred 10^5 years ago or more, causing the large radio luminosity; second, another object which is responsible for the weak optical luminosity. In any case, the Burbidges concluded, the problem of understanding the mechanism of gravitational collapse had developed into one of the most important puzzles of contemporary physics.

REFERENCES

1.   W. Baade and R. Minkowski, *Astrophys. J.* **119**, 206 (1954).
2.   J. L. Greenstein and M. Schmidt, *Astrophys. J.* **140**, 1 (1964).
3.   F. Hoyle and W. A. Fowler, *Monthly Notices Roy. Astron. Soc.* **125**, 169 (1963a); *Nature*, **197**, 533 (1963b).

3.4. GENERAL DISCUSSION

The 1964 Solvay Conference was dominated by the recent discovery of observational phenomena in our galaxy and, presumably, in remote ones. Most of these phenomena are not properly understood. In the galaxies, one has no reasonable and quantitative explanation of the very high energies, the very sharp jets and the very marked shock fronts. This might involve a detailed theory of hydromagnetic phenomena. Second, the problems of energy conversion from one form into another are entirely unsolved. Perhaps one might speculate about the annihilation of matter and antimatter, as H. Alfvén proposed in the discussion.

Alfvén started from O. Klein's cosmological model, which assumes that the initial state of the universe is a homogeneous mixture of matter and antimatter. Under the action of gravitation, a contraction of the meta-galaxy starts, but is stopped and changed into expansion by a great release of energy due to annihilation. The important point is whether the separation of the primeval 'ambiplasma' into matter and anti-matter is possible on a large scale. The theory of Alfvén and Klein predicts that either every second galaxy consists of antimatter or that thin sheets of matter and antimatter exist in one galaxy. The annihilation process should give rise to neutrinos and gamma rays which, however, cannot be detected as yet, but the radio waves emitted from fast electrons are observable. Alfvén concluded that maybe some radio stars originate from matter-antimatter annihilation.

Another mechanism of energy conversion, proposed by Zeldovitch, was discussed at the Solvay Conference by Ambartsumian. If stars pass each other they might be trapped in a spiral orbit which approaches the gravitational radius asymptotically. A collapse might occur eventually. I. D. Novikov had considered a Friedmann expanding cosmos and explained the quasars as collisions of expanding matter, with matter falling from the outside, assuming that the expansion of the universe does not occur in a moment but in several outbursts.

SIXTEENTH SOLVAY CONFERENCE 1973

H. ARP - M. BOVIJN - B. CARTER - R. OMNES - S.R. de GROOT - J. EHLERS - C.J. PETHIK - A.G.W. CAMERON - E.P.J. VAN DEN HEUVEL - J. HEISE - F. PACINI

V. CANUTO - L. CELNIKIER - R. COUTREZ - P. LEDOUX - P.M. MATHEWS - R. PENROSE - I.M. KHALATNIKOV - C. RYTER - M. SCHMIDT - R. RUFFINI - R. DEBEVER

J. LEROY - G. COCCONI - P. SWINGS - G. HENSBERGE - J. DEMARET - M. BURGER - M.J. REES - R. HOFSTADTER - E.C.G. SUDARSHAN - J. MEHRA - Y. NE'EMAN - J.A. WHEELER - A. TRAUTMAN - A. SCHILD

L. HALPERN - J. WILLIAMS - G. DERIDDER - H. HENSBERG - M.A. RUDERMAN - J. SHAHAM - G. BAYM - R.A. ALPHER - L. WOLTJER - F.G. SMITH

G. BURBIDGE - A. HEWISH - R. GIACCONI - E. SCHATZMAN - D. PINES - R. HERMAN - H. SATO - V.R. PANDHARIPANDE - J. BAHCALL

J. SENGIER - A. ABRAGAM - L. ROSENFELD - M. BURBIDGE - B AMALDI - C. MØLLER - F. PERRIN - C.J. GORTER - J. GEHENIAU - I. PRIGOGINE

4. ASTROPHYSICS AND GRAVITATION

4.1. INTRODUCTION

The sixteenth Solvay Conference on physics was held in Brussels from 24 to 28 September 1973. Its general theme was 'Astrophysics and Gravitation'.

Among the members of the Scientific Committee who attended the Conference were E. Amaldi, President, A. Abragam, C. J. Gorter, C. Møller, F. Perrin, L. Rosenfeld, and J. Géhéniau, Secretary. W. Heisenberg, A. B. Pippard, A. Sandage, S. Tomonaga, L. Van Hove, and E. P. Wigner were not able to attend. The invited speakers were: J. and N. Bahcall (Princeton, N.J.), A. G. W. Cameron (Cambridge, Mass.), V. Canuto (New York), R. Giacconi (Cambridge, Mass.), F. Pacini (Rome), V. R. Pandharipande (Urbana, Illinois), D. Pines (Urbana, Illinois), M. J. Rees (Cambridge, England), M. Schmidt (Pasadena, California), J. A. Wheeler (Princeton, N.J.), and L. Woltjer (New York). I. D. Novikov (Moscow) was not able to attend the Conference.

A large group of invited participants attended the conference, including numerous well-known physicists and astrophysicists, and quite a few among them took part in the discussions.

The development of astrophysics during the past fifty years has been quite steady. The present situation in astrophysics could perhaps be compared with the years between 1911 and 1950 which saw a steady development of atomic and particle physics. The reports at the Solvay Conferences of 1958, 1964, and 1973 summed up the status of the field at the respective periods and indicated its development. Problems that were barely hinted at in 1958 could be answered reasonably well by 1964. New problems were found to exist in 1964, and through hard and successful work a partial solution was achieved by 1973.

The 1964 Solvay Conference had already included a report on the X-ray sources in the universe. Since that time, with the increased availability of rocket and satellite measurements, it had been possible to make a careful study and localization of the sources of short wave radiation. From the theoretical point of view the ideas on the evolution of stars had been developed so far as to include the study of supernovae within the framework of nuclear physics. The suggestion was made that the remnants of supernovae emitting X-rays were neutron stars. Pulsars came under observation and turned out to be immediately useful in the theory. On the one hand, cosmological consequences concerning the structure of space and the model of evolution were drawn; on the other, the serious problems connected with the aggregation of a large mass in a comparatively small space were studied. Many new theoretical results about black holes and similar singularities were obtained. As a result astrophysics became one of the most exciting fields in modern physics, a field in which even the particle theorists sought to test their ideas about quarks, monopoles, etc. Recently the idea of neutral currents has made its way into the description of supernovae phenomena (J. Wilson, 1974).

4.2. Cosmic X-ray Sources and Neutron Stars

R. Giacconi gave a report on 'Observational Results on Compact Galactic X-ray Sources'. X-rays from cosmic sources can only be observed from positions outside the earth's atmosphere. Whereas rocket shots had revealed the existence of the first X-ray source, the new era of astronomical discoveries started with the advent of orbiting X-ray observatories. A catalogue has been compiled which includes some 160 sources ranging in apparent luminosity from Sco X-1 (the first observed X-ray star and the most intense in the 1 to 10 keV energy region) to 10^{-4} times Sco X-1. Of these sources, about 60 appear to be extragalactic and about 100 galactic.

The most important aspects of galactic sources are: (1) the large variations in intensity is the rule (often the luminosity changes by a factor of 100 within a few days, by a factor of 2 in months to minutes; some sources emit pulses in time scales of 100 milliseconds or less); (2) individual stars can emit luminosities of the order of 10^{38} ergs/sec; (3) several galactic sources occur in the binaries; (4) the existence of fast time variations in several X-ray sources and the measured parameters of the components of binaries strongly indicate that galactic X-ray sources correspond to stars near the end point of their evolution (white dwarfs, neutron stars). All galaxies emit X-rays, mainly as the integrated effect over all individual stellar sources (10^{39} to 10^{46} ergs/sec.) Some special galaxies (Seyfert, N-type, quasars, radio galaxies) have particularly high luminosity (10^{42} to 10^{46} ergs/sec). Clusters of galaxies may emit bremsstrahlung X-rays. Some galaxies, showing no visible or microwave emission, have also been detected. The best known stars are Hercules X-1 and Cygnus X-1, Cygnus X-3; all of them are in binaries, X-ray sources, and are probably very compact objects.

The details of binary X-ray sources were discussed by J. N. Bahcall and N. A. Bahcall in their report on 'Optical Properties of Binary X-ray Sources', by A. Cameron in a short contribution, and by E. P. J. van den Heuvel in his discussion remarks on 'The Evolutionary Origin of Massive X-ray Binaries and the Total Number of Massive Stars with a Collapsed Close Companion in the Galaxy'. From optical observations one can derive important parameters like masses; one finds the X-ray emitters to have masses slightly above 1 $M_{\odot}$. Cameron tried to explain the X-ray emitting part of a binary as a white dwarf rather than as a neutron star. Since a neutron star might have great difficulties in forming a close binary system, he made the assumption that the X-rays are created by shocks in the atmosphere of the white dwarfs.

Van den Heuvel also dealt with the question of the evolution of massive X-ray binaries. The systems consist in general of a massive primary ($\gtrsim 15\ M_{\odot}$) and a less massive secondary ($\sim 1\ M_{\odot}$). He interpreted the latter as neutron stars or black holes, and outlined the most likely evolutionary story (Kippenhahn and Weigert, 1967).[1] After the end of the core burning, the helium core of the massive primary star contracts until helium burning begins; during this stage hydrogen, in its shell, still burns and the hydrogen-rich massive envelope is transferred to the secondary. In a rather short time the remnant of the first star reaches the supernova state and explodes. Finally a neutron star or black hole would be formed which rotates very fast (pulsar). The new massive secondary would continue its evolution very slowly. The important consequence is that there exist examples for all stages of this history, the periods of the binary system changing continuously.

M. J. Rees reported on other aspects of the physics of X-ray sources. The X-ray binaries involve all the ill understood problems connected with ordinary binaries, together with a whole range of new phenomena connected with the compact object. One generally assumes that the companion star fills its Roche lobe, and that material flows across the Lagrangian point. This matter will form a differentially rotating disc which is mainly accreted by the compact object. As it falls on the star, energy is released mainly through a bremsstrahlung mechanism or thermal radiation.

In a series of reports presented at the 1973 Solvay Conference, it was assumed that the X-ray emitting stars are neutron stars of very condensed matter. F. K. Lamb and C. J. Pethick discussed specific aspects of 'Accretion on to Magnetic Neutron Stars' and the interplay between the position of the so-called Alfvén surface and changes in the X-ray pulsation frequency. D. Pines talked about 'Observing Neutron Stars: Information on Stellar Structure from Pulsars and Compact X-ray Sources'. Although the suggestion that the neutron stars are remnants of a supernova explosion had been made by Baade and Zwicky[2] already in 1934, observations confirmed it only in the late 1960s. The phenomena occurring in pulsars can only be explained if one assumes that they are formed of very dense neutron matter. There exist phenomenological, and less phenomenological, descriptions of changes in the pulsation period using the idea of superfluid neutrons etc. Starquakes may occur and what not!

The principal report on 'Neutron Stars' was given by A. G. W. Cameron and V. Canuto. Although Landau[3] had already commented on the existence of highly condensed stable objects in 1932, and Baade and Zwicky had speculated about their connection with supernova explosions a little later, interest in these questions was not renewed until the 1958 Solvay Conference when Harrison, Wakano and Wheeler discussed neutron stars as the final state reached in the collapse of a massive astrophysical body. The equation of state of a neutron star had been discussed by Wheeler, and later on by Cameron who pointed out the possible existence of hyperons in neutron stars.

When pulsars were discovered, literature in this field exploded. The successful ideas, which had been developed in nuclear physics, solid state physics and condensation phenomena, quickly helped to establish the basis of a theoretical description of pulsars. Detailed models of layers in neutron stars, with the density increasing up to 10^{16} gm cm^{-3} towards the centre, have been developed. For this purpose, the equation of state of matter having densities higher than 10^6 gm cm^{-3} must be considered. Beyond that density, the electrons get squeezed out of atomic matter and bare electrons survive, forming a crystalline structure. Then 'neutronization' sets in, since the electrons are captured according to the process

$$e^- + p \rightarrow n + \nu. \tag{2}$$

The neutrons are finally in a superfluid state, but at the same time hyperons come into the picture. At a density higher than 10^{16} gm cm^{-3} this scheme fails to provide an adequate description. Starting with an equation of state, and taking into account the magnetic field of the order of 10^{12} G, one can try to construct models of neutron stars.

In the second part of his report Cameron discussed briefly the evolution of a star, ending as a neutron star. Only the star starting with a mass higher than 15 solar masses will undergo nuclear fusion processes, turning its core eventually into iron. The outer parts, however, become unstable and are blown off. Some of it falls back, but when the neutron star becomes a pulsar a strong pressure is exerted on the infalling material which re-accelerates it outwards. The theory of pulsars is still in a preliminary state, though many possible mechanisms have been applied. Supernovae explosions are thought to be predominant sites for cosmic ray acceleration, since strong electromagnetic waves might be radiated away during the slowing down mechanism of the pulsar.

Physics of high density nuclear matter was further discussed in some detail by V. R. Pandharipande. Following his report a discussion on the time equation of state developed between Ruffini, Canuto, Ne'eman and others.

The discussion of neutron stars was rounded up by the progress report on 'Pulsars' by F. Pacini. The first pulsar was discovered in 1968, and by the end of the same year two others had been found at the sites of supernovae explosions. It was also discovered that the Crab pulsar is losing rotational energy at the rate of 10^{38} ergs sec^{-1}, which is about the amount needed to keep the nebular content of relativistic particles and magnetic fields in an approximately steady state; hence the pulsar seems to be the basic energy source of the overall nebular activity.

On the basis of these observations a fairly detailed understanding of the conversion of rotational energy of the pulsar into relativistic particles, and the mechanism of the sharp pulses and the highly coherent radio-emission, could be provided. While collapsing, the star will obtain both a high magnetic field and a fast rotation because of the conservation of angular momentum. The space outside the pulsar is divided into two different regions: in the first (co-rotating magnetosphere) the field lines close within the speed of light cylinder and the charge co-rotates with the star, while in the open magnetosphere a continuous outflow of ions and protons occurs. The evolution of the magnetic field, which takes on large values of say 100 G in the beginning and then drops to 10^{-3} G, in the extended nebula of today, is a particularly important feature of a pulsar which is considered to be the remnant of a supernova.

REFERENCES

1. R. Kippenhahn and A. Weigert, *Z. Astrophys.* **65**, 251 (1967).
2. W. Baade and F. Zwicky, *Phys. Rev.* **45**, 138 (1934).
3. L. Landau, *Phys. Z. Sovjetunion*, **1**, 284 (1932).

4.3. BLACK HOLES

At the 1964 Solvay Conference several participants had mentioned the problem of gravitational singularities and pointed to the task of trying to understand the possible phenomena connected with it. At the 1958 Solvay Conference, J. A. Wheeler had reported on the implications of Einstein's theory of general relativity for the structure of the universe, and had mentioned the problem of Schwarzschild singularities. Now, at the 1974 conference, Wheeler gave a survey of the physics of 'The Black Hole'. He divided his report into 'three times three parts': the three levels of gravitational collapse, the three handles required to specify the properties of a black hole, and the time signals from a black hole. The three levels of gravitational collapse begin with the universe: big bang at the start and recontraction at the end; the second level is the collapse of a single star to a black hole or a similar collapse of a collection of stars in compact galactic nuclei; the third level is the quantum fluctuations at the Planck scale of distances,

$$L^* = \sqrt{(\hbar G/c^3)} = 1.6 \times 10^{-33} \text{ cm}. \tag{3}$$

The mass, M, the electric charge, Q, and the angular momentum, J, determine the properties of a black hole. The mass of a black hole changes according to the changes in the 'irreducible' mass, M_{irr}, Q, and J, the irreducible mass increasing with the collisions.

Three signals come from a black hole: a pulse of gravitational radiation at the moment of formation or mass augmentation; X-rays accreting into the black hole and subsequently heated to 10^{10} to 10^{11} °K by gravitational compression before crossing the horizon of the black hole; and 'activity', that is, interaction of a rotating hole with the matter surrounding the horizon.

A sufficiently dilute but extended cloud of dust will eventually collapse to within the Schwarzschild radius,

$$r = 2\frac{G}{c^2}M. \tag{4}$$

An outside observer will never realize the approach to r, while a falling one will find that eventually the radius zero is obtained by the mass. The 'horizon', however, can be determined by light rays.

The physics of black holes was intensively studied after 1968 [Hawking, Penrose, and others].[1] According to Einstein's ideas the universe has a closed surface; but then the universe itself might collapse at a later era. Indications are that the actual time of the universe is smaller than the 'Hubble time', or time linearly extrapolated back to the start of the expansion; hence the expansion has slowed down already. Also, the universe, which is visible, seems to contain less mass than necessary for closure; hence it might exist in the form of dark matter. Moreover, quantum fluctuations of the order,

$$\Delta g \sim L^*/L, \tag{5}$$

should exist, leading to a foam-like structure of charges

$$q \sim \sqrt{(\hbar c)} \sim 10 \ e. \tag{6}$$

A black hole has no structure (Newman–Kerr geometry), and lepton or baryon number loses all meaning. A black hole can move on an orbit, rotate, and attract

particles. Energy can escape as gravitational or electromagnetic waves, and by elementary processes, if the hole is 'live'. Black holes might emit particles as if they were
bodies with temperature (Hawking 1974),

$$T = \frac{\hbar g}{2\pi} \approx 10^{-6} \frac{M_\odot}{M} \,^\circ K. \tag{7}$$

The laws of black hole physics have been established by various authors, giving a mass
change law, an entropy law, etc. Important questions arise, such as whether there
are 'white holes' vomiting all their mass as particles and radiation. Black holes, or
nearly black holes, emit radiation (gravitation, X-rays), and they seem to be a much
better source of the energy of the Crab nebula than other objects.

   Wheeler's colourful report was followed by a report of I. D. Novikov, read by
Ruffini, on the 'Search for Observational Evidence for Black Holes'. Since black holes
were predicted by Oppenheimer and Snyder[2] in 1939, nearly 35 years had passed
before specialists thought of their existence in binary X-ray sources. Black holes may
be thought to exist in globular clusters, in nuclei of galaxies and quasars. All the massive stars should turn into black holes if they do not lose mass to below $3 \, M_\odot$; hence
stellar black holes should be present in the galaxy in a considerable number. Many
other candidates have been proposed. They should be observable with a weak luminosity, but more chance exists of finding them in binary systems emitting X-rays.

   Novikov noted that the discovery of the black hole, together with the cosmological
red shift, would be one of the greatest discoveries of astrophysics in the 20th century.
The detection of star death would require 'Weber detectors' having a sensitivity at
least 10^4 times higher than those claimed to be available at present. Cyg X-1 is supposed to be the best candidate for a black hole object, but they might be more frequent.
Estimates are that several thousands of normal main sequence OB stars in the galaxy
might have a close black hole companion, some 20 may be within 1 kpc distance.
Other events may take place in the nucleus of a galaxy. Steady contraction of the
central star cluster may lead to several supernovae; the rest coalesces to give a single
massive object which eventually collapses.

REFERENCES

1. See S. W. Hawking, 'Black Holes in General Relativity', *Commun. Math. Phys.* **25**, 152 (1972);
    R. Penrose, 'Gravitational Collapse: The Role of General Relativity', *Nuovo Cimento* **1**, special
    number, 252 (1969).
2. J. R. Oppenheimer and H. Snyder, *Phys. Rev.* **56**, 455 (1939).

4.4. THE PROBLEMS OF QUASARS

The final topic of the 1973 Solvay Conference was the development in the problems
posed by quasi-stellar objects. E. M. Burbidge discussed the 'Recent Observations
of Quasi-Stellar Objects'. She first discussed the results on absorption spectra which
sometimes had a different red shift from the emission lines $(Z_a < Z_{em.})$, frequently
yielding individually different Z_a per line system. Moreover, a system of two 'close'

quasars, having very different red shifts of $Z=0.436$ and 1.901, was found in 1973. These observations add support to the non-cosmological red-shift hypothesis. According to Burbidge, sufficient knowledge does not yet exist either to construct or to rule out theories of gravitational red shifts. Perhaps the quasars do not form a uniform sample. In any case, cosmology would become definitely more complicated if all quasars have cosmological red shifts. H. Arp, in a report entitled 'Distances of the Quasars and Evidence for Non-velocity Redshifts', supported the view that at least some quasars are associated with nearby galaxies. He discussed how all quasars might originate from galaxies with much lower red shift.

L. Woltjer discussed the 'Theories of Quasars'. In the literature, red shifts of 198 quasi-stellar radio sources and of 61 quasi-stellar radio objects, without noticeable radio emission, had been reported; their red shifts were $0.1 \leqslant Z \leqslant 3.53$, eight of them having $Z > 2.2$. Variation in luminosity and polarization of radio emission had been observed. Outbursts of 10^{54} to 10^{55} ergs occur, provided the red shifts are of cosmological origin. Gravitational, kinematic, and cosmological red shifts have been considered. Woltjer concluded that gravitational local kinematical explanations seem rather unlikely, since either too high masses or too high velocities in our galaxy are needed. On the other hand, cosmological explanation is favoured by the facts that 'larger' objects are closer and quasars of a subclass are the fainter, the larger their red shifts are. Finally, clusters of quasars having similar red shifts do exist. Cosmological theory faces the problem of how to explain the energy supply. While large numbers of pulsars could contribute some part, and also a massive rotating object which collapses seems to be a possible source of energy, massive stars and multiple supernovae events lead to problems.

Assuming that the redshifts of quasars are of cosmological origin, M. Schmidt discussed the 'Distribution of Quasars in the Universe'. His aim was to find the number of quasars per unit co-moving volume of space of optical luminosity, F_{opt}, and radio luminosity, F_{rad}, at red shift Z. The result is that the quasars do not seem to have a uniform distribution in space. Quasars with very large red shifts apparently occur at a considerably lower rate.

Schmidt's observations, if confirmed by new data, would have important cosmological consequences, in so far as the outer frontier of the universe might become visible through the quasar counts. This would close the circle on the 1958 Solvay Conference dealing with astrophysics, at which the cosmological considerations had played an important role. At the 1964 conference these speculations were not discussed, but for certain specific objects and processes. Has the time now arrived to decide about the cosmological model?

Too many questions remain. For instance, the problem of energy production and of mass distribution, to which G. B. Burbidge devoted his report on 'The Masses of the Galaxies and the Mass-Energy in the Universe'. Galaxies have masses of the order of 10^{10} to 10^{11} solar masses, and the counts of galaxies in the universe yields a density of 10^{-31} gm cm^{-3} contributed by visible objects. The popular Friedmann cosmological model relates the mean density to other cosmological parameters like the Hubble

404 CHAPTER FIFTEEN

constant, H_0, and $q_0 = \frac{1}{2}$, yielding

$$\varrho_c = \frac{3H_0^2 q_0}{8\pi G} = 4.7 \times 10^{-30} \text{ gm cm}^{-3}. \tag{8}$$

The discrepancy of a factor of 30 to 60 has given rise to much speculation. Now, if we do not live in a steady state universe with $q_0 = \frac{1}{2}$, but in a Friedmann universe with $0 \leqslant q_0 \leqslant \frac{1}{2}$, the difference might be less relevant. In addition, other energy sources might be found in neutrinos (which are not easily detectable), relativistic particles (which should not contribute much), gravitational radiation, electromagnetic radiation (which may yield as much as 10^{-31} gm cm^{-3}), diffuse gas (which is less than 10 times the mass concentrated in galaxies), and in discrete dark objects. Burbidge concluded that one cannot exclude the possibility that the bulk of the mass-energy is present in such exotic forms as neutrinos and black holes. Other possibilities include the existence of gammas which might be observed in bursts [R. Hofstadter].

With the knowledge obtained from the new objects, the door has been opened again to the study of the structure and evolution of the universe. The next Solvay Conference on astrophysics should reveal definite steps in this direction, based on the pursuit of the present lines of research.

Index of Names

Italic figures at the end of a reference refer to plates

Aarts 350
Abragam, A. 323, 397; *396*
Abraham, Max xxi
Adler, S. L. 282, 296, 299, 320; *298*
Aigrin, P. 357
Albert, King of the Belgians xv, xviii, xxiv, xxv, xxvi, xxvii, xxxii, 10, 11
Alfvén, H. 389, 391, 395, 399; *388*
Allen, N. P. *338*
Alphen, P. M. van 367, 368, 369
Alpher, R. A. 260; *396*
Alvarez, L. W. 300, 301
Amaldi, Edoardo 269, 299, 323, 389, 397; *298, 322, 388, 396*
Ambartsumian, V. A. 384, 386, 387, 389, 390, 391; *382, 388*
Anderson, C. D. viii, xviii, xx, 210, 216, 217, 218, 232, 235, 242, 260
Anderson, H. L. 320
Andrade, E. N. da C. 353, 354
Arkel, van 342
Arnous, E. 296
Arp, H. 403; *396*
Artsimovich, L. 323
Arvieu, R. *322*
Aston, F. W. 102
Aubel, E. van 95, 115, 133; *94, 114*
Auger, P. xix, 140, 142, 239, 246, 247; *238*
Avogadro, A. 40, 42

Baade, W. 384, 386, 387, 393, 394, 395, 399, 400; *382*
Babcock, H. W. 240
Bahcall, J. N. 397. 398; *396*
Bahcall, N. A. 397, 398
Balasse, G. *238, 356*
Bancelin, M. 46, 51
Bardeen, J. M. 348, 359, 360, 376, 378
Barkla, C. G. 77, 81, 95, 100, 103
Barlow, W. 75, 85, 86, 88; *74, 94*
Barnes, V. E. 301, 320
Barnett, S. J. 109, 202, 203
Bartlett, J. H. 223
Bauer, E. 48, 51, 183, 194, 201, 203, 207; *208*
Bauer, H. E. G. 115, 124, 127; *114, 182*
Baym, G. *396*
Becker, R. 215, 218, 348, 349

Becquerel, J. 199, 277, 374, 376
Benedicks, C. 88, 92, 121, 122
Bertsch, G. *322*
Besso, Michele xiv, xx, xxxi, 10
Bethe, H. A. 242, 256, 260, 261, 269, 273, 309, 324, 336, 345, 391; *268*
Bhabha, H. J. xx, 232, 239, 250, 251; *238, 266*
Biedenharn, L. 323, 326, 335, 336, 337; *322*
Bieler 213
Biermann, L. 389, 392; *388*
Bijtebier, J. *298*
Blackett, P. M. S. viii, xviii, xix, 207, 210, 211, 216; *208, 238, 266*
Blewitt 350
Bloch, C. *322*
Bloch, F. xix, 184, 197, 199, 235, 239, 255, 261, 361, 362, 363, 377, 378; *238*
Blomqvist, J. *322*
Bogoliubov, N. N. 269, 292, 296, 308
Bohm, D. 296, 359, 360
Bohr, A. 323, 325, 326, 335; *322*
Bohr, Niels vii, viii, xi, xvi, xvii, xviii, xix, xxiii, xxiv, xxx, xxxi, xxxii, 39, 42, 76, 77, 78, 81, 95, 96, 97, 99, 101, 104, 106, 107, 108, 109, 110, 111, 112, 117, 120, 122, 123, 128, 133, 134, 135, 136, 139, 140, 141, 142, 146, 149, 150, 151, 152, 153–79, 181, 183, 184, 185, 186, 190, 193, 198, 199, 205, 207, 213, 214, 220, 222, 226, 230, 235, 239, 246, 260, 269, 271, 291, 295, 324, 336, 367, 368, 373; *132, 182, 208, 238, 266, 268*
Bolton, J. G. 389, 393; *388*
Boltwood, B. B. xxxi, 50, 51
Boltzmann, Ludwig xiii, 4, 6, 13, 15, 20, 21, 23, 24, 25, 26, 31, 37, 38, 42, 60, 69, 70, 72, 83, 135, 153, 315, 364
Bondi, H. 387; *382*
Bopp, F. 261
Borel, Emile xiv
Borelius, C. O. G. 121, 122, 339, 345, 351; *338*
Born, Max vii, xi, xvi, xxi, xxiv, 70, 72, 89, 91, 92, 133, 134, 135, 138, 141, 144, 145, 146, 147, 148, 149, 150, 151, 152, 157, 181, 228, 235, 261, 347, 348; *132*
Bose, S. N. 135, 149, 250, 251, 263, 309, 326
Bothe, W. xx, 141, 142, 207, 215, 218; *208*
Bourgeois, P. *388*

Bouten, M. *322*
Bouten Denys, M. C. *322*
Bovijn, M. *396*
Bradt, H. 261
Bragg, Sir William H. 75, 76, 82, 83, 84, 85, 86,
    95, 108, 112, 115, 122, 127, 129, 136, 139,
    142; *74, 114*
Bragg, Sir W. Lawrence xxvii, 76, 82, 83, 84,
    95, 133, 135, 136, 137, 138, 141, 181, 239,
    269, 299, 323, 339, 345, 348, 351, 357, 363,
    389; *94, 132, 238, 268, 338, 356, 382*
Bravais, A. 85
Breit, G. 330
Bremmerman, H. 296
Bretscher, E. 232
Bridgman, P. W. 115, 120, 121, 122, 124, 128,
    130; *114*
Brillouin, L. 46, 95, 99, 115, 122, 124, 133, 142,
    146, 152, 181, 183, 199, 205, 363, 367, 368;
    94, 114, 132, 182
Brillouin, M. xiii, xv, 8, 9, 13, 18, 59, 75, 86,
    87, 95, 115; *12, 74, 94, 114*
Brink, D. M. 323, 326, 334; *322*
Brode, W. 245
Broek, A. van den 100, 104
Broglie, Louis de xvi, xviii, xix, xxx, xxxi,
    xxxii, 133, 134, 135, 142, 143, 144, 145, 146,
    151, 152, 156, 181, 207, 239, 251, 261; *132, 208*
Broglie, Maurice de xi, xxii, xxx, xxxii, 11, 13,
    48, 75, 83, 95, 96, 104, 112, 207, 213, 277,
    363; *12, 74, 94, 208*
Broniewski, W. 115, 126; *114*
Brown, G. 334; *322*
Brown, R. H. 23, 244, 261
Buckingham, M. 377, 378
Bunsen, R. W. 38
Burbidge, E. M. 389, 394, 395, 402, 403; *388,
    396*
Burbidge, G. B. 389, 394, 395, 403, 404; *388,
    396*
Burger, M. *396*
Burgers, J. M. 351
Burgers, W. G. 339, 341, 342, 343, 344; *338*
Burton, W. K. 348, 349
Butler, C. C. 244, 261

Cabibbo, N. 283, 299, 302, 310, 317, 320, 321;
    298
Cabrera, B. xix, 183, 184, 191, 192, 193, 194,
    207, 348, 349; *182, 208*
Cahn 342
Calvin, J. xx
Cameron, A. G. W. 397, 398, 399, 400; *396*
Camiz, P. *322*
Canuto, V. 397, 399, 400; *396*
Carnot, S. 38, 39, 44, 92, 119
Cartan, E. 329

Carter, B. *396*
Carver, T. R. 370, 371
Casimir, H. B. G. 251, 264; *238*
Castillejo, L. 295, 304, 321
Castoldi, P. *298*
Catálan, M. A. 96
Cauchy, A. M. 346, 347, 348
Cauchois, Y. *338*
Celnikier, L. *396*
Cerulus, F. *298*
Ceuleneer, R. *322*
Chadwick, J. viii, xviii, 100, 101, 104, 207, 210,
    211, 213, 214, 215, 216, 223, 226; *208*
Chambers, R. G. 367, 369
Chandrasekhar, S. 260
Cherwell, Viscount *see* Lindemann, F. A.
Chevenard, P. 201
Chew, G. F. xxix, 269, 293, 294, 295, 296, 299,
    303, 304, 305, 306, 307, 321; *268, 298*
Chissick, S. S. xxxi
Christenson, J. H. 321, 337
Churchill, Winston xxii
Cini, M. 269; *268*
Clark, R. W. xxxi, xxxii
Clausius, R. E. 6, 42, 43, 44, 87, 119
Clebsch, R. F. A. 327
Cocconi, G. *396*
Cockcroft, J. D. 207, 210, 211, 212, 215, 226;
    208, 238
Compton, A. H. vii, xvi, xx, xxx, 99, 101,
    104, 133, 134, 135, 136, 137, 138, 139, 140,
    141, 142, 152, 156, 158, 162, 181, 230, 258,
    259, 278, 284, 303, 393; *132*
Condon, E. U. 222, 226, 330
Conversi, M. 242, 260
Cooper, L. N. 376, 378
Cosyns, M. xx, 207; *208*
Cotton, A. 183, 184, 199, 203, 204, 205; *182*
Cottrell, A. H. 339, 352; *338*
Coulomb, C. A. xvi, 78, 80, 110, 111, 213, 251,
    275
Coutrez, R. *388, 396*
Cowley 345
Crookes, W. 49
Crowther, J. A. 77, 81, 100, 103
Crussard, C. 339, 350; *338*
Curie, Irène Joliot- xiv, xviii, xx, 207, 210,
    215, 216, 217, 223, 231; *208*
Curie, Jacques 42, 126
Curie, Marie xiii, xiv, xv, xvi, xx, xxii, xxx,
    13, 20, 39, 42, 50, 51, 53, 59, 60, 61, 75, 77,
    80, 81, 84, 95, 103, 105, 106, 107, 115, 122,
    127, 128, 133, 142, 183, 184, 192, 193, 194,
    199, 200, 201, 203, 204, 205, 207, 213, 215,
    217, 222, 373, 374; *12, 74, 94, 114, 132, 182,
    208*
Curie, Pierre 42, 126, 199

Curien, H. *338*
Cuypers, K. *388*

Dalitz, R. H. 295, 300, 304, 321
Dalton, J. 42
Danby, G. 296, 321
Dancoff, S. M. 319
Darwin, C. G. 82, 84, 100, 103, 136, 137, 139, 183, 190, 194, 199; *182*
Davis, D. H. 138, 328
Davisson, C. J. 134, 142, 145
Debever, R. *396*
Debierne, A. 81
Debye, P. xix, xxx, 28, 40, 70, 71, 72, 80, 83, 84, 89, 90, 91, 92, 115, 120, 122, 123, 124, 127, 129, 133, 135, 136, 138, 139, 142, 181, 183, 199, 204, 207, 226, 239, 358, 362; *114, 132, 182, 208*
Dee, P. I. 212; *238*
Deenen, J. *322*
Dehlinger, U. 339; *338*
Delbrück, M. 215
Demaret, J. *396*
Demeur, M. *238, 322*
Dennison, D. 337
Deridder, G. *396*
Deucher, Adolphe xx
Dewar, J. 50, 51
Diesselhorst 116
Dilworth, C. C. *238*
Dirac, P. A. M. viii, xi, xvi, xvii, xix, xx, 133, 134, 138, 141, 147, 148, 149, 152, 157, 167, 176, 183, 184, 186, 189, 190, 191, 198, 199, 205, 207, 218, 219, 220, 223, 225, 226, 227, 229, 230, 233, 235, 240, 250, 251, 254, 255, 257, 259, 261, 269, 272, 273, 274, 275, 308, 313, 318, 321, 379; *132, 182, 208, 238, 266, 268*
Do, Tan-si *322*
Dobronravoff, N. 130
Donder, Th. de xix, 115, 133, 151, 152, 181, 183, 205, 207, 226, 239; *114, 132, 182, 208, 238*
Dorfman, J. 183, 194, 199, 203; *182*
Döring, W. 348, 349
Drude, Paul 64, 116, 117, 120, 129
Duane, W. 136, 138
Duhem, P. 42
Dukas, Helen xxvi
Dulong, P. L. 21, 52, 62, 64, 65, 67, 72, 88, 89
Dungen, F. H. van den xxviii, 339, 384
Dürr, H. P. 299, 307, 312, 321; *298*
Dymond, E. G. 145
Dyson, F. J. 269, 273, 274, 295, 304, 321, 370, 371; *268*

Eckart, C. 147
Ehlers, J. *396*

Ehrenfest, P. xvii, xxiii, 95, 96, 97, 105, 109, 110, 111, 112, 133, 135, 142, 146, 152, 156, 160, 164, 171, 172, 179, 230, *94, 132*
Ehrenhaft. F. 49, 50
Einstein, Albert vii, viii, xi, xiii, xiv, xvi, xvii, xviii, xx, xxi, xxii, xxiii, xxiv, xxv, xxvi, xxvii, xxxi, xxxii, 4, 5, 6, 9, 10, 13, 15, 18, 24, 28, 35, 36, 37, 39, 40, 42, 44, 45, 46, 48, 50, 51, 52, 53, 56, 57, 58, 59, 61, 62, 63, 64, 65, 66, 67, 68, 69, 70, 72, 75, 83, 85, 89, 91, 92, 95, 96, 99, 103, 104, 105, 108, 109, 111, 130, 133, 134, 135, 139, 140, 141, 142, 146, 149, 152, 153–79, 181, 183, 184, 185, 202, 203, 205, 207, 210, 213, 239, 250, 263, 284, 381, 383, 384, 385, 401; *12, 74, 132, 182*
Einstein, Elsa xxiv, xxv, xxvi, xxxi
Einstein, Mileva xxi
Elizabeth, Queen of the Belgians xxiv, xxv, xxvi, xxvii, xxxi, xxxii
Elliott, J. P. 323, 326, 333; *322*
Elliott, R. J. 370, 371
Ellis, C. D. 207, 210, 215, 221, 222; *208*
Elster, J. 244, 261
Emden, F. 391
Epstein, Paul 38, 101, 185
Errera, J. 183; *182, 208, 238*
Erwin, A. R. 320
Ettinghausen 129, 130
Evrard-Musette, M. *298*
Ewald, P. P. 82, 136, 137, 139
Exner, F. 44

Faraday, M. 50, 103, 119, 204, 390
Fawcett, E. 367, 369
Fedoroff, J. S. 85
Feenberg, F. 330, 337
Feldman, D. 295
Fitzgerald, G. F. xxi
Fermi, E. xix, xx, 115, 149, 183, 184, 190, 194, 196, 199, 207, 211, 215, 222, 225, 226, 229, 231, 232, 233, 235, 246, 249, 250, 259, 263, 277, 278, 280, 300, 308, 323, 326, 327, 347, 358, 359, 361, 362, 363, 367, 368, 369, 371, 372, 377; *182, 208*
Ferretti, B. 247, 299; *238, 298*
Feynman, R. P. xxix, xxxii, 230, 255, 261, 269, 270, 271, 272, 273, 274, 275, 276, 279, 287, 294, 295, 296, 305, 309, 318, 321, 327, 385; *268*
Field, G. 394
Fierz, M. 250, 261; *382*
Flamache, L. *238, 338*
Fletcher, H. 49, 51
Flexner, Abraham xxvi
Flores, J. *322*
Fock, V. 219, 235, 334, 357, 360, 361
Foldy, L. L. 337
Fournet, J. L. 345

Fowler, R. H. 133, 138, 151, 152, 181; *132*
Fowler, W. A. 389, 392, 393, 395; *388*
Franck, J. 96, 154
Frank, A. 193
Frank, F. C. 339, 348, 349, 350, 351, 353, 354; *338*
Franklin, B. 102
Franz 91, 106, 107, 115, 121, 123, 124, 365
Franzini, P. 331, 337
Frauenfelder, H. 323, 326, 327; *322*
Frautschi, S. 304, 305, 306, 321
Frazer, W. R. 296
French, J. B. 261, 323, 326, 337; *322*
Frenkel, J. 341, 348, 349
Fresnel, Augustin 181
Friedel, J. 357, 371, 372, 373; *356*
Friedmann, A. 383, 385, 395, 403, 404
Friedrich, K. O. 76, 81, 288, 296
Frisch, O. R. xxviii; *238*
Fröhlich, H. 357, 362, 376, 377, 378, 379; *356*
Froissart, M. 299; *298*
Fubini, S. 299, 321; *298*
Fulco, R. J. 296
Fuld-Rouserez, M. *322*
Fumi, F. 357, 372; *356*
Fünfer, E. 218
Furlan, G. 321

Gamow, G. xx, xxx, 207, 210, 211, 212, 213, 215, 220, 221, 222, 223, 225, 226, 242, 260, 278, 281, 296; *208*
Gans, R. 120, 122, 185
Gapon 223
Gardner, W. L. 245, 260
Gaspart, R. *338*
Géhéniau, J. xx, 299, 323, 389, 397; *238, 298, 322, 356, 382, 388, 396*
Geiger, H. xx, 50, 51, 81, 100, 103, 142, 226
Geitel, H. F. 244, 261
Gell-Mann, Murray xxix, 269, 270, 271, 273, 279, 288, 290, 293, 294, 295, 296, 299, 301, 303, 304, 307, 308, 309, 310, 311, 320, 321, 327, 328, 336; *268*
Gentile, G. 199
George, C. *298*
Gerlach, W. 156, 183, 185, 198, 205; *182*
Germer, L. H. 134, 142, 145
Giacconi, R. 397, 398; *396*
Gibbs, J. W. 30, 31, 38, 340, 348, 349
Gijsman, H. M. 379, 380
Gilbert 218
Gillet, C. *322*
Goche, O. *238, 338, 356*
Goeppert-Mayer, M. 241, 324, 332, 336
Gold, T. 384, 385; *382*
Goldberger, M. L. 269, 270, 271, 289, 290, 291, 292, 296, 304, 321; *268*

Goldhabe, M. 321
Goldschmidt, Robert B. xv, 4, 5, 6, 7, 8, 9, 13, 75; *12, 74*
Goldschmidt, V. M. 260
Goldschmidt, Y. *238*
Goldschmidt-Clermont, Y. 261
Goldstone, J. 311, 312, 319, 321
Gordan, P. A. 327
Gordon, W. 198
Gorter, C. J. 188, 269, 299, 323, 357, 383, 389, 397; *268, 356, 382, 388, 396*
Goudsmit, S. A. 202
Gouy, G. 44, 75; *74*
Graaff, R. J. van de 212, 245
Gray, J. A. 140, 142
Greenstein, J. L. 394, 395
Gribov, V. N. 305, 306, 321
Griswold, T. W. 370, 371
Gróh, J. 125, 126
Groodt, A. de xxv
Groot, S. R. de *396*
Grossmann, Marcel xxii
Groven, L. *238, 338*
Grüneisen, E. 67, 72, 75, 88, 89, 90, 91, 92, 121, 122, 362; *74*
Guggenheim, E. A. 340
Guillaume, C. E. 201
Guinier, A. 339, 345; *338*
Gurney 222, 226
Gürsey, F. 320
Guthrie 353
Guye, Charles Eugène xxi, xxxi, 133, 181, 183, 207; *132*

Haag, R. 288, 296, 299, 312; *298*
Haas, Arthur 33, 34, 38, 40
Haas, W. J. de xxiii, 95, 102, 108, 109, 183, 188, 194, 199, 202, 204, 205, 362, 367, 368, 369; *94, 182*
Hagedorn, R. 385
Hägg, G. 344
Hall, E. H. 115, 119, 122, 129, 130, 377, 379; *114*
Halpern, L. *396*
Hamilton, W. R. 16, 17, 19, 27, 38, 134, 157
Hamilton, J. 299; *298*
Harrison, B. K. 399
Hartree, D. R. and W. 219, 334, 357, 358, 359, 360, 361, 378
Hasenöhrl, F. 6, 9, 13, 37, 38, 61, 75, 80, 88; *12, 74*
Havighurst 136
Hawking, S. W. 401, 402
Haxel, O. 324, 336
Hayashi, C. 392, 393
Hecht, K. T. 323, 326, 333, 336; *322*
Heckmann, O. 384; *382*
Hedvall, J. A. 344

Heinrich, L. 260
Heise, J. *396*
Heisenberg, Werner VII, VIII, IX, XI, XVI, XVII,
    XVIII, XIX, XX, XXIV, XXIX, XXX, 110, 133, 134,
    135, 138, 144, 146, 147, 148, 149, 150, 151,
    152, 157, 158, 167, 179, 181, 183, 184, 191,
    194, 197, 198, 199, 203, 205, 207, 211, 213,
    215, 217, 222, 223, 224, 225, 226, 227, 228,
    229, 230, 231, 233, 234, 235, 236, 251, 257,
    258, 261, 272, 277, 283, 286, 290, 294, 295,
    296, 299, 303, 304, 312, 313, 316, 319, 321,
    323, 330, 389, 397; *132, 182, 208, 268, 298,*
    322, 388
Heitler, W. XIX, 135, 184, 197, 225, 239, 251,
    252, 253, 255, 259, 269, 271, 283, 284, 285,
    286, 296, 369, 371; *238, 268*
Helm 42
Helmholtz, H. von 48, 103
Hemptinne, M. de *238*
Henderson, C. 354
Henin, F. *298*
Henley, E. M. 337
Henriot, E. 115, 133, 183, 207, 239; *114, 132,*
    182, 208, 238, 338
Hensberg, H. *396*
Hensberge, G. *396*
Herapath, J. 42
L'Héritier, M. 244, 246, 261
Herman, R. *298, 396*
Herring, C. 371
Hertz, P. 96, 154
Hertzsprung, E. 387
Herzen, E. XVIII, 3, 8, 9, 133, 183, 207; *12, 74,*
    94, 114, 132, 182, 208
Hess, V. F. 244, 261
Heuvel, E. P. J. van den 398; *396*
Hevesy, G. de 115, 122, 125, 126, 127; *114*
Hewish, A. *396*
Higgs, P. W. 312, 321
Hilbert, David 38, 109, 148, 286, 287, 294, 315,
    383, 384
Hindenburg, P. von XXV
Hirn, G. A. 3
Hjalmar 138
Hofstadter, R. 274, 295, 299, 404; *298, 396*
Hollomon, J. H. *338*
Homes, G. A. *338*
Hooke, R. 346, 347
Hopf, L. XXI, 18
Horn, D. 321
Hossain, A. 337
Hostelet, G. 3; *12, 74*
Houston 362, 363
Houzeau, J. C. 3
Hove, L. van 269, 288, 296, 397; *238, 268*
Hoyle, F. 384, 385, 387, 389, 390, 391, 392,
    393, 395; *382, 388*

Huang, K. 371, 372
Hubble, E. P. 383, 385, 401, 403
Huisman, M. XXXI
Hulsizer, R. 261
Hulst, H. C. van de 384, 386, 387; *382*
Hultgren, A. 345
Humblet, J. *322*
Hume-Rothery, W. 371, 372
Hund, F. 186, 188, 193, 337
Huntington, H. B. 372, 373
Huygens, C. 92

Infeld, L. 261
Inglis, D. R. *322*
Inoue, T. 242, 260
Isacker, Van *238*
Ising 198
Iwanenko, D. 223, 232, 235

Jacobi 143, 147, 150
Jaeger 116
Jan, J. P. 379, 380
Jaumotte, André XII
Jeans, J. H. XIII, XIV, 6, 8, 13, 14, 17, 20, 21, 22,
    23, 24, 25, 26, 34, 38, 39, 75, 95; *12, 74*
Jensen, J. H. O. 324, 332, 336
Joffé, A. XXX, 115, 120, 122, 124, 125, 126, 127,
    130, 207, 239; *114, 208*
Joliot, F. XVIII, XX, 207, 210, 213, 215, 216, 217,
    223, 231, 239; *208*
Jones, H. 357, 372; *356*
Jordan, P. 134, 135, 147, 149, 157, 228, 229,
    235, 272
Josephson, B. 275
Judd, B. R. 323, 326, 329, 336; *322*

Kahn, F. D. 389, 390; *388*
Källén, G. 269, 271, 288, 294, 295, 299, 313,
    314; *268, 298*
Kamerlingh Onnes, H. XIII, XIV, XV, XXIV, 9,
    13, 14, 20, 48, 51, 53, 61, 75, 83, 88, 95, 105,
    106, 107, 108, 115, 118, 119, 127, 128, 192;
    12, 74, 94
Kampen, N. G. van 290
Kapitza, P. 183, 184, 194, 199, 203, 204, 205;
    182
Kármán, Th. von 70, 72, 89, 91, 92
Kaufmann, W. XXI, 100
Keesom, H. W. 47, 48, 51, 115, 118, 122, 128,
    129; *114*
Kelvin, Lord 48, 62, 65, 99, 103, 117
Kemmer, N. 232, 233, 236
Kerr, R. P. 401
Khalatnikov, I. M. *396*
Kibble, T. W. B. 312
Kiessling 344
Kipfer, P. *238, 356*

Kippenhahn, R. 398, 400
Kirchhoff, G. R. 15, 25, 29, 70, 381
Kirkendall 372
Kittel, C. 357, 369, 370, 371; *356*
Klein, M. J. 72, 144, 198, 217, 220, 221
Klein, O. 230, 235, 269, 278, 296, 384, 385,
  395; *238, 268, 382*
Klemens, P. G. 364
Kneser, H. 38
Knight, W. D. 369, 371, 372
Knipping, P. 76, 81
Knudsen, M. xv, 5, 6, 9, 11, 13, 15, 40, 41, 42,
  75, 95, 103, 115, 133, 183; *12, 74, 94, 132*
Koch, Cäsar xx, xxi, xxiv
Kochendörfer, A. 353, 354
Kolhörster, W. 244, 261
Korringa, J. 369, 371
Kossel, W. 348, 349
Köster, W. 339, 346; *338*
Kramer, P. 323, 326; *322*
Kramers, H. A. vii, xx, 133, 134, 135, 138, 139,
  140, 141, 142, 146, 152, 157, 179, 181, 183,
  188, 191, 198, 199, 204, 205, 207, 239, 256,
  257, 261, 273, 289, 291, 296, 304, 339; *132,
  182, 208, 238*
Kroll, N. M. 261, 321
Krönig, A. 42
Kronig, R. 138, 290, 291, 296
Kuenen 105
Kukaskin, B. V. *382*
Kurlbaum, F. 25, 32

Ladenburg, Rudolf xxi
Lamb, F. K. 399
Lamb, W. E. 256, 257, 261, 273, 274, 286, 294,
  295
Landau, L. D. 196, 199, 269, 270, 277, 278,
  280, 295, 296, 305, 374, 376, 399, 400
Landé, A. 147, 186, 202, 203, 223
Langevin, P. xiii, xiv, xvi, xvii, xviii, xix, xx,
  xxvi, xxx, 2, 3, 6, 9, 11, 13, 14, 24, 38, 53,
  59, 60, 61, 75, 80, 88, 95, 102, 105, 106, 108,
  112, 115, 119, 122, 124, 127, 128, 129, 133,
  152, 181, 183, 185, 191, 192, 193, 194, 196,
  200, 207, 213, 215; *12, 74, 94, 114, 132, 182,
  208*
Langmuir, I. 101, 133; *132*
Larmor, J. 5, 6, 8, 10, 28, 40, 95, 98, 99; *94*
Larsson, A. 138, 139
Lattes, C. M. G. 245, 260, 261
Laue, M. von xvi, 75, 76, 81, 82, 83, 84, 86, 88,
  136, 139; *74*
Laval, J. 339, 346, 347, 348; *338*
Lavanchy, C. xxi
Lawrence, E. O. 207, 210, 212, 215, 245; *208*
Leclercq-Willain, C. *322*
Ledoux, P. *382, 388, 396*

Leduc 130
Lee, T. D. 269, 277, 278, 280, 281, 295, 296,
  321, 328, 337
Lefébure, Charles xx, xxvi, xxvii, xxxii, 181
Lehmann, H. P. 48, 269, 271, 287, 288, 295,
  296
Leibniz, G. W. von 176
Lemaitre, G. 384, 385; *382, 388*
Lenard, A. 56
Leprince-Ringuet, L. 244, 246, 261, 264; *238*
Leroy, J. *396*
Leun, C. van der *322*
Leuven, P. van *322*
Lévy, M. 299, 321; *298*
Lewis, G. N. 139, 142
Lewis, H. 261
Libert-Heinnemann *322*
Liempt, van 342
Lindblad, B. 390, *388*
Linde 345
Lindemann, F. A. (later Viscount Cherwell)
  xxi, xxii, 13, 24, 52, 53, 58, 60, 66, 67, 68, 72,
  75, 83, 88, 91, 115, 118, 120, 121, 122, 124,
  127; *12, 74*
Liouville, J. 16, 19, 27, 38
Lipkin, H. J. 323, 326, 328; *322*
Livens, S. H. 120, 122
Logunov, A. 309, 321
Lomer, W. M. 353, 354; *338*
London, H. 135, 184, 225, 367, 368
Lorentz, H. A. xiii, xiv, xv, xvi, xvii, xxii,
  xxiii, xxiv, xxxi, 4, 5, 6, 8, 9, 10, 13, 14, 15,
  16, 17, 18, 19, 24, 26, 36, 37, 38, 39, 48, 53,
  59, 70, 75, 80, 83, 87, 88, 91, 92, 95, 96, 98,
  99, 102, 103, 105, 112, 115, 116, 117, 118,
  119, 120, 122, 124, 128, 130, 133, 134, 138,
  141, 145, 151, 152, 181, 183, 198, 251, 253,
  254, 258, 274, 283, 284, 286, 287, 289, 304,
  305, 307, 309, 312, 314, 364, 368; *12, 74, 94,
  114, 132*
Lovell, Sir A. C. Bernard 384, 386, 389; *382,
  388*
Low, F. E. 269, 296, 299; *298*
Löwdin, P. O. 357, 360, 361; *356*
Lummer, O. 24, 25, 32
Lurie, D. 321
Luttinger, P. 258, 261

MacDonald, D. K. C. 357, 362, 364; *356*
Mach, E. 42
Macq, P. *322*
Maglic, B. C. 320
Magnus, A. 66
Majorana 224, 225
Mallard 87
Mandelstam, S. 269, 270, 271, 292, 295, 299,
  304, 305, 321; *268, 298*

Mangin 48
Manneback, C. 183; *182*
Marbo, Camille xiv
March, A. 251
Marsden, E. 100, 101, 103, 104
Marshak, R. E. 242, 260, 278, 296, 299; *298*
Marton, L. *238*
Massey, H. S. W. 214
Mathews, P. M. *396*
Matthias, B. T. 357, 378; *356*
Matthiessen, A. 362, 364
Maxwell, J. C. xiii, 6, 13, 16, 20, 21, 22, 23, 24, 28, 29, 32, 33, 34, 39, 42, 43, 45, 51, 59, 60, 61, 70, 88, 97, 98, 99, 115, 117, 141, 184, 195, 254, 258
Mayer, Walther xxvi
Mayne, F. *298*
McClure, W. 337
McCrea, W. H. *382*
McManus, H. 255, 261
Mehra, J. vii, xii, 72, 384; *322, 396*
Meissner, W. 357; *356*
Meitner, Lise xx, 207, 213, 215, 217, 222, 226; *208, 238*
Mendeleyev, D. I. 77, 96, 100
Mendelssohn, K. A. 357, 365, 366; *356*
Meyer, E. 56, 60
Michel, L. 282, 299; *298*
Michelson, A. A. 43, 95, 139; *94*
Mieghem, J. van *388*
Mie, G. 90, 92
Millikan, R. A. 42, 49, 95, 96, 102, 104, 112, 119; *94*
Minkowski, R. 389, 392, 393, 394, 395; *388*
Mitte, H. 321
Miyazawa, H. 296
Molenaar 350
Møller, C. xx, xxvii, 235, 236, 258, 259, 269, 299, 323, 339, 357, 383, 389, 397; *238, 268, 298, 322, 338, 356, 382, 388, 396*
Mond, Ludwig 2
Moreau, J. J. 48
Morgan, W. W. 384, 387; *382*
Morley, E. W. 139
Moseley, H. G. J. 77, 82, 84, 100, 104
Moshinsky, M. 323, 326, 334, 336; *322*
Mossotti 87
Mott, N. F. 207, 215, 222, 269, 339, 353, 354, 357, 362, 371, 372, 384, 389; *208, 356*
Mottelson, B. R. 323, 325, 326, 335; *322*
Moulin, M. 48, 51
Mulliken, R. S. 361

Nagamiya, T. 375, 376
Nambu, Y. 269, 296, 321; *268*
Narlikar, J. V. 395
Nathan, O. xxxi

Neddermeyer, S. M. 217, 232, 235, 242, 260
Néel, L. 357, 374, 375, 376; *356*
Ne'eman, Y. 293, 301, 308, 320, 328, 336, 400; *396*
Nernst, Walther xiii, xiv, xv, xxi, xxii, xxv, 4, 5, 6, 7, 10, 13, 14, 19, 20, 32, 38, 40, 42, 51, 52, 53, 62, 65, 66, 67, 68, 70, 72, 75, 80, 83, 85, 88, 89, 91, 92, 118, 129, 130; *12, 74*
Nernst, Mme 8, 9
Neumann, J. von 147, 149, 315
Newton, I. 28, 79, 251, 381
Newman, E. T. 401
Nicholson, J. W. 120, 122
Nishijima, K. 301
Nishina, Y. 217, 220, 221, 230, 235
Norden, H. xxxi
Nordheim, L. 362
Nordsieck, A. 235, 255, 261
Novikov, I. D. 395, 397, 402
Nuttal, J. M. 226

Occhialini, G. P. S. 216, 218, 261; *238*
Oehme, R. 296
Ohm, G. S. 115, 118, 121, 126, 129
Okubo, S. 320
Okun, L. B. 299; *298*
Ölander, G. 344
Oliphant, M. 212
Olive, D. xviii, 321
Omnes, R. 299, 385; *298, 396*
Onsager, L. 357, 368, 369; *356*
Oort, J. H. 384, 386, 389, 390; *382, 388*
Opechowski, W. 256, 261
Oppenheimer, J. R. xxvii, 218, 239, 247, 257, 258, 259, 261, 269, 299, 339, 357, 383, 389, 402; *238, 266, 268, 382, 388*
D'Or, L. 3
O'Raiferteigh, L. 296
Ornstein, L. S. 89, 92
Orowan, E. 339, 354; *338*
Ostwald, W. xv, xx, 4, 42
Overhauser, A. W. 370, 371

Pacini, F. 397, 400; *396*
Pais, A. 261, 269, 271, 279, 280, 281, 282, 303, 321; *268*
Palacios 192, 194
Pancini. E. 242, 260
Pandharipande, V. R. 397, 400; *396*
Paschen, F. 24
Patterson, 138
Pauli, W. vii, viii, xi, xvi, xvii, xviii, xix, xx, xxix, xxx, 96, 110, 133, 135, 138, 142, 145, 146, 147, 152, 157, 181, 183, 184, 185, 186, 189, 191, 194, 195, 196, 197, 198, 199, 203, 207, 210, 218, 219, 220, 226, 227, 229, 230, 231, 235, 239, 250, 257, 259, 261, 272, 277,

287, 295, 296, 318, 319, 323, 325, 328, 332, 334, 339, 357, 360, 369, 383; *132, 182, 208, 238, 266, 356, 382*

Peierls, R. E. 207, 215, 220, 222, 226, 239, 251, 253, 255, 269, 295, 363, 364, 368, 369; *208, 238, 266, 268*

Pelseneer, J. xi

Peltier 53, 117

Penrose, R. 401, 402; *396*

Perrin, Francis xix, xx, 142, 207, 213, 215, 222, 223, 225, 269, 299, 323, 339, 357, 383, 389, 397; *208, 238, 268, 298, 322, 388, 396*

Perrin, Jean xiii, xiv, 6, 9, 10, 13, 15, 42, 43, 44, 45, 46, 47, 48, 49, 50, 51, 95, 102, 103; *12, 94*

Perry 202

Peterman, A. 295

Peters, B. 261

Pethick, C. J. 399; *396*

Petit 21, 52, 62, 64, 65, 67, 72, 88, 89

Phillip 217

Piccard, A. 115, 129, 133, 183, 194, 204, 205, 207; *114, 132, 182, 208*

Piccioni, O. 242, 260

Pines, D. 357, 358, 359, 360, 397, 399; *356, 396*

Pippard, A. P. 323, 357, 366, 368, 397; *356*

Planck, Max xiii, xiv, xvi, xxi, xxii, xxiv, xxv, xxx, 4, 5, 6, 8, 13, 14, 15, 17, 18, 19, 20, 23, 24, 25, 26, 27, 28, 33, 34, 36, 37, 38, 39, 40, 50, 53, 54, 55, 58, 59, 60, 61, 63, 64, 65, 68, 69, 70, 75, 79, 89, 96, 104, 110, 111, 133, 134, 135, 146, 147, 153, 154, 155, 157, 168, 205, 315, 385, 401; *12, 132*

Plesset 218

Podolsky, B. 173, 174, 179, 235

Pohl, R. 58, 60

Poincaré, Henri xiii, xiv, xxii, xxx, xxxii, 4, 9, 13, 18, 19, 24, 38, 39, 53, 56, 59, 61, 96, 98, 293; *12*

Poisson, S. D. 87, 147, 340

Pomeranchuk, Ia. 291, 296, 306, 320

Pontecorvo, B. 296

Pope, W. J. 3, 75, 85, 86, 88; *74*

Powell, C. F. 232, 239, 242, 243, 244, 246, 249, 259, 261; *238*

Prendergast, K. H. *388*

Present, R. D. 330

Price, W. C. xxxi

Prigogine, I. xi, xxviii, 314, 315; *238, 268, 298, 356, 388, 396*

Pringsheim, E. 24, 32, 58, 60

Proca, A. xx, 233, 236, 250, 261

Prout 103

Pryce, M. H. L. 232

Puppi, G. 278, 296

Quesne, C. 326, 334; *322*

Quincke 194

Racah, G. 329, 337

Radicati, L. A. 299, 320, 323, 325, 326, 329, 330, 331, 332, 337; *298, 322*

Ramsay, Sir W. xv

Rarita, W. 250, 261

Rasetti, F. xx

Rathenau, G. W. 339, 341; *338*

Ratnowsky, M. xxi

Ratnowsky, S. 89, 92

Ravensdale, T. xxxi

Rayet, M. *322*

Rayleigh, Lord 5, 6, 7, 8, 10, 13, 14, 15, 16, 19, 20, 24, 26, 34, 48, 51, 65, 355

Read, W. T. 340, 343, 350, 351, 353, 354

Rees, M. J. 397, 399; *396*

Regener, E. 49, 51

Regge, T. xxix, 299, 304, 305, 306, 307, 321; *298*

Reiche, F. 72

Reid, A. 145

Reid, R. xiv

Reidemeister, G. *322*

Reinganum, M. 34, 40

Retherford, R. C. 261, 295

Reuter, G. E. H. 367, 369

Rhodes 362

Richardson, Sir O. W. xix, 23, 95, 109, 115, 117, 118, 122, 123, 124, 127, 133, 142, 152, 181, 183, 191, 199, 202, 207, 239; *94, 114, 132, 182, 208, 238, 266*

Richarz, F. 88, 92

Righi, A. 130

Ripka, G. *322*

Ritz, W. 26, 40, 111, 146

Rivier, D. 296

Rochester, D. 244, 261

Röhm, F. 353, 354

Röntgen, W. K. 6, 9, 28, 35, 36, 54, 56, 58, 59, 75, 81, 126

Rose, M. E. 235

Rosen, N. 173, 174, 179, 251

Rosenblum, M. S. 184, 204, 207, 222, 261; *208*

Rosenfeld, Léon xi, xvii, xviii, xx, xxix, xxx, xxxi, xxxii, 207, 230, 235, 239, 240, 246, 247, 248, 249, 265, 269, 299, 323, 297; *208, 238, 268, 298, 382, 396*

Rosenhain, W. 115, 122, 124, 125; *114*

Rossi, B. xx, 218, 261, 389, 393; *388*

Rubens, H. xiv, xxi, xxii, 9, 13, 14, 25, 53, 75, 80; *12, 74*

Rudberg, E. 339, 344, 345; *338*

Ruderman, M. A. 370, 371; *396*

Ruelle, D. 299; *298*

Ruffini, R. 400, 402; *396*

Russell, H. N. 189, 387

Rutherford, E. (later Lord Rutherford) xiii, xiv, xv, xvi, xviii, xxiii, xxx, xxxi, 6, 8, 9,

13, 24, 42, 49, 50, 51, 53, 59, 75, 76, 78, 80,
81, 84, 95, 96, 98, 99, 100, 101, 102, 103, 104,
106, 107, 108, 111, 112, 115, 122, 124, 127,
154, 204, 207, 210, 211, 212, 215, 221, 222,
227; *12, 74, 94, 114, 208*
Ryter, C. *396*

Sachs, R. 296
Sacton, J. 328
Sakata, S. 242, 260, 299, 320, 321, 328, 337;
298
Sakita, B. 320
Sakurai, J. J. 279, 296, 316, 321
Salam, A. 269, 278, 280, 296; *268*
Salpeter, E. E. 309, 389, 392; *388*
Sandage, A. T. 323, 384, 389, 397; *382, 388*
Sargent, B. W. 277, 278, 296
Sato, H. *396*
Saunders 189
Schafroth, R. 377, 378
Schatzmann, E. 389; *382, 388, 396*
Scherrer, P. 122; *238*
Schidlof, A. 33, 34, 40
Schild, A. *396*
Schleicher, Kurt von xxv
Schlieder, S. 321
Schmid, C. 321
Schmidt, M. 389, 392, 393, 394, 395, 397, 403;
388, 396
Schomblond, C. *298*
Schönflies 85
Schottky 341
Schrieffer, J. R. 376, 378
Schrödinger, E. vii, xvi, 115, 122, 124, 125,
127, 133, 134, 135, 142, 143, 144, 146, 148,
149, 150, 151, 152, 156, 157, 207, 224, 225,
230, 251, 269, 272, 313, 383; *114, 132, 208
238, 266*
Schücking, E. 384
Schuster, P. 5, 6, 8, 54
Schwarzschild, K. 381, 395, 401
Schwinger, J. 229, 230, 249, 250, 258, 259, 261,
269, 272, 273, 274, 295, 299, 311, 314; *268,
298*
Sedding 46
Seeger, A. 357, 380; *356*
Seeley, Evelyn xxv, xxxi
Seelig, C. xiv, xxii, xxxi
Seeliger, H. 5, 6, 9
Segré 249
Seitz, F. 339, 350, 372, 373; *338*
Semet, Florimond 1
Sengier, J. *396*
Serber, R. xxviii, 239, 245, 246, 249, 250; *238*
Serpe, J. 256, 261
Serpukhov 306
Shaham, J. *396*

Shapley, H. *382*
Shockley, W. 339, 340, 343, 351, 352; *338*
Shull, C. G. 357, 373, 374; *356*
Siedentopf, H. F. W. 44
Siegbahn, M. 95, 96, 138, 139; *94*
Siemens 4
Simon 142
Slater, J. C. 139, 140, 141, 142, 179, 188, 196,
197, 361
Slichter, C. P. 370, 371
Smit, J. 357, 379; *356*
Smith, C. S. 339, 340, 343, 379; *338*
Smith, F. G. *396*
Smith, F. W. F. (Earl of Birkenhead) xxxi
Smoluchowski, M. von 42, 44, 47, 51
Snyder, H. S. 261, 402
Soddy, F. 101, 104
Sohncke, L. 85
Solomon, J. xix, xx
Solovine, Maurice xxvi, xxxii
Solvay, Alexandre 1
Solvay, Alfred 1, 2
Solvay, Armand xviii
Solvay, Ernest vii, viii, xi, xiii, xiv, xv, xviii,
xx, xxii, xxvii, xxxi, 1, 2, 3, 4, 5, 6, 7, 8, 9,
10, 11, 75, 183; *12, 94*
Solvay, Ernest-John xviii, xxviii
Solvay, Jacques xi, xii, xxviii
Solvay, Mme J. xii
Sommerfeld, Arnold xiii, xiv, xxi, xxiii, xxiv,
xxx, 13, 15, 36, 38, 39, 40, 53, 54, 55, 56, 57,
58, 59, 61, 70, 75, 76, 83, 84, 96, 101. 135,
145, 156, 183, 184, 185, 186, 187, 188, 189,
192, 194, 205, 289, 362, 367; *12, 74, 182*
Sondheimer, E. H. 362, 367, 369
Speiser, D. *298*
Speziali, P. xxxi
Spiegel, E. *388*
Spinoza 176
Spitzer, L. 389, 391, 392; *388*
Stahel, E. xx, 207; *208, 238*
Stark, J. 28, 35, 40, 55, 96, 101, 121, 122
Starr, C. 374, 376
Stas, J. S. 3
Staver, T. 359, 360
Stefan 24, 25, 26
Steinberger, J. 246
Stenström, K. W. 138
Stern, O. 156, 183, 198; *182*
Stewart, J. Q. 109
Stokes 42, 49, 54, 57, 59
Stoney, G. Johnstone 103
Stoops, R. 136
Stranski, I. N. 348, 349
Strömgren, B. 389; *388*
Stückelberg, E. C. G. 261, 272, 274, 296
Styvendael, M. van *238*

Sudarshan, E. C. G. 278, 296, 299, 316, 317, 318, 321; *298, 396*
Suess, H. E. 260, 324, 336
Swann, W. F. G. 120, 122
Swings, P. *388, 396*
Symanzik, K. 271, 287, 288, 295, 296

Takahashi, Y. 296
Talmi, I. *322*
Tamm, I. E. 232, 235, 269, 299, 319, 323, 389
Tassel, E. xxiii, xxviii, xxxi, 3
Tavkhelidze, A. 299, 307, 308, 309, 321; *298*
Taylor 341
Teller, Edward xxviii, 239, 240, 241, 242, 246, 259, 263, 278, 281, 296, 369, 371, 383; *238*
Thirring, W. 290, 296
Thomas, L. H. 137, 139, 198, 225, 371
Thomson, George P. 134, 145
Thomson, J. J. xvi, 6, 8, 26, 28, 33, 35, 39, 40, 42, 49, 59, 75, 76, 77, 78, 79, 80, 99, 100, 102, 103, 107, 108, 121, 122; *74*
Toll 290
Tomonaga, S. 227, 230, 258, 259, 261, 269, 272, 273, 274, 295, 299, 323, 389, 397; *268*
Tonnelat, M. A. 239, 251; *238*
Townes, C. H. 369, 371
Townsend, J. S. E. 48, 49, 58, 60, 78, 81
Trautman, A. *396*
Treiman, S. 296
Tuve, M. A. xx

Uhlenbeck, G. E. 202
Umezawa, H. 299; *298*

Veneroni, M. *322*
Verhaeghe, J. L. *322*
Verschaffelt, J. E. xix, xxviii, 95, 115, 122, 133, 181, 183, 203, 205, 207, 239; *74, 94, 114, 132, 182, 208, 238*
Villars, F. 287, 296
Vleck, J. H. van 183, 184, 186, 188, 189, 193, 194, 199, 357; *182, 356*
Voigt, W. 75, 76, 87, 88, 346, 348
Volger, J. 380
Volmer, M. 348, 349
Volta, A. 135, 158

Waals, jr. J. D. van der 5, 6, 9, 10, 17, 20, 43, 47, 50, 105, 380
Wakano 399
Walecka, J. D. 337
Waller, I. 138, 139
Walton, E. T. S. 207, 210, 212, 213; *208*
Warburg, Emil xiv, xv, xxii, 11, 13, 14, 25, 42, 75; *12, 74*
Ward, J. C. 273, 295
Waterman, A. T. 121, 122

Waterston, J. J. 42
Weaver, L. 337
Weber, F. H. 62, 72, 402
Webster 216
Wefelmeier, W. 337
Weigert, A. 398, 400
Weinberg, S. 299
Weiss, P. xxii, 15, 48, 50, 60, 61, 75, 80, 88, 95, 105, 106, 183, 184, 185, 188, 192, 193, 194, 199, 200, 201, 202, 204, 374; *74, 94, 182*
Weissberger, W. I. 282, 296, 320
Weisskopf, V. F. 230, 235, 259, 261, 273, 295
Weizsäcker, C. F. von xix, 260, 299, 324, 391; *298*
Wentzel, G. xix, 141, 235, 236, 254, 255, 261, 269, 295; *268*
Westgren 344
Weyl, H. 326
Wheeler, J. A. 236, 255, 261, 290, 324, 336, 384, 385, 397, 399, 400, 401; *382, 396*
Wick, G. C. 269, 280, 296, 319; *268*
Wiechert, E. 54
Wiedemann 91, 106, 107, 115, 121, 123, 124, 365
Wien, W. xiii, xxi, 5, 6, 9, 13, 15, 19, 23, 24, 25, 26, 38, 39, 55, 59, 61, 75, 83, 91, 118, 121, 122; *12, 74*
Wiener, N. 44, 147
Wightman, A. S. 271, 280, 287, 288, 295, 296, 299; *268, 298*
Wigner, E. P. xxix, 149, 188, 229, 235, 269, 280, 290, 296, 299, 323, 325, 326, 329, 330, 331, 332, 333, 336, 337, 358, 360, 397; *268, 298, 322*
Wildermuth, K. 337
Wilkinson, D. H. 323, 326, 327, 337; *322*
Williams, E. J. xx, 141, 345
Williams, J. *396*
Wilson, C. T. R. 49, 78, 101, 133, 141, 142, 181, 205, 211, 212, 244, 249, 261, 362, 368; *132*
Wilson, H. A. 49, 76, 120, 122, 240, 260
Wilson, J. 397
Wolfenstein, L. 303, 321
Wolfskehl, P. 109
Woltjer, H. R. 192
Woltjer, L. 389, 390, 391, 397, 403; *388, 396*
Wood, R. W. 75, 76, 80, 84, 92; *74*
Wouthuijsen, S. A. 299; *298*
Wright 57, 58, 60
Wu, C. S. 278, 296, 328, 337

Yamazaki, K. 321
Yang, C. N. 269, 278, 280, 281, 295, 296, 328, 337
Yoshida, K. 375, 376
Young, G. 326